Problem Solving in Chemical and Biochemical Engineering with POLYMATH,™ Excel, and MATLAB®

Second Edition

Problem Solving in Chemical and Biochemical Engineering with POLYMATH,™ Excel, and MATLAB®

Second Edition

Michael B. Cutlip

University of Connecticut

Mordechai Shacham

Ben-Gurion University of the Negev

PRENTICE
HALL

Upper Saddle River, NJ • Boston • Indianapolis • San Francisco
New York • Toronto • Montreal • London • Munich • Paris • Madrid
Cape Town • Sydney • Tokyo • Singapore • Mexico City

Many of the designations used by manufacturers and sellers to distinguish their products are claimed as trademarks. Where those designations appear in this book, and the publisher was aware of a trademark claim, the designations have been printed with initial capital letters or in all capitals.

The authors and publisher have taken care in the preparation of this book, but make no expressed or implied warranty of any kind and assume no responsibility for errors or omissions. No liability is assumed for incidental or consequential damages in connection with or arising out of the use of the information or programs contained herein.

The publisher offers excellent discounts on this book when ordered in quantity for bulk purchases or special sales, which may include electronic versions and/or custom covers and content particular to your business, training goals, marketing focus, and branding interests. For more information, please contact:

U.S. Corporate and Government Sales
(800) 382-3419
corpsales@pearsontechgroup.com

For sales outside the United States please contact:

International Sales
international@pearsoned.com

 This Book Is Safari Enabled

The Safari® Enabled icon on the cover of your favorite technology book means the book is available through Safari Bookshelf. When you buy this book, you get free access to the online edition for 45 days.

Safari Bookshelf is an electronic reference library that lets you easily search thousands of technical books, find code samples, download chapters, and access technical information whenever and wherever you need it.

To gain 45-day Safari Enabled access to this book:

- Go to http://www.prenhallprofessional.com/safarienabled
- Complete the brief registration form
- Enter the coupon code MARC-N9PQ-2Q7Y-5TQ2-KJFA

If you have difficulty registering on Safari Bookshelf or accessing the online edition, please e-mail customer-service@safaribooksonline.com.

Visit us on the Web: www.prenhallprofessional.com

Library of Congress Cataloging-in-Publication Data
Cutlip, Michael B.
  Problem solving in chemical and biochemical engineering with polymath,
Excel, and Matlab / Michael B. Cutlip, Mordechai Shacham. — 2nd ed.
      p. cm.
  Prev. ed.: Problem solving in chemical engineering with numerical methods.
Upper Saddle River, NJ : Prentice Hall PTR, 1999.
  Includes bibliographical references and index.
  ISBN 978-0-13-148204-3 (pbk. : alk. paper)
  1. Chemical engineering--Problems, exercises, etc. 2. Biochemical
engineering—Problems, exercises, etc. 3. Chemical engineering—Data
processing. 4. Biochemical engineering—Data processing. 5. Problem
solving. 6. Numerical analysis. 7. Microsoft Excel (Computer file) 8.
MATLAB I. Shacham, Mordechai. II. Cutlip, Michael B. Problem solving in
chemical engineering with numerical methods. III. Title.
  TP168.C88 2008
  660.01'51—dc22
                                        2007027494

ISBN-13: 978-0-13-148204-3
ISBN-10:   0-13-148204-1

Text printed in the United States on recycled paper at Courier in Stoughton, Massachusetts.
First printing, August 2007

*To our parents: Wilma Sampson Cutlip, Sidney B. Cutlip,
Lusztig Erzsèbet, and Schwarczkopf Zoltán*

Chapter 3 Regression and Correlation of Data   57

Chapter 4 Problem Solving with Excel   101

Chapter 8 Fluid Mechanics   283

Chapter 9 Heat Transfer   333

Chapter 10 Mass Transfer   383

Chapter 11 Chemical Reaction Engineering   445

Chapter 12   Phase Equilibria and Distillation   523

Chapter 13   Process Dynamics and Control   565

Chapter 14   Biochemical Engineering   617

Book Overview

This book provides extensive problem-solving instruction and suggestions, numerous examples, and many complete and partial solutions in the main subject areas of chemical and biochemical engineering and related disciplines. Problem solutions are clearly developed using fundamental principles to create mathematical models. An equation-oriented approach that enables computer-based problem solving on personal computers is utilized. Efficient and effective problem solving is introduced employing numerical methods for linear equations, nonlinear equations, ordinary and partial differential equations, linear and nonlinear regressions, and polynomial curve fitting. Basic to advanced problem solving is covered utilizing a novel integrated approach with three widely used mathematical software packages: POLYMATH, Excel, and MATLAB. Readers may choose to focus on one or more of these software packages or utilize another mathematical software package.

The book and a dedicated web site (**www.problemsolvingbook.com**) furnish all necessary problem information, software files, and additional enrichment materials. For advanced applications, unique software tools are provided for solving complex problems such as parameter estimation in dynamic systems and solution of constrained systems of algebraic equations.

Intended Audience

This book is intended for individuals who are interested in solving problems in chemical and biochemical engineering and in related fields by using mathematical software packages on personal computers. It can serve as a textbook for students in conjunction with college- and university-level courses, and it can be a companion reference book for individual students. For professionals, it can be an invaluable reference book that also allows extensive self-study in problem solving using the most widely used software packages.

Background

Prior to the introduction of the personal computers and mathematical software packages in the early 1980's, desktop calculations for engineering problem solving were mainly carried out with hand-held calculators. Sometimes mainframe computers were utilized, which required source code programming. Since then the emphasis has gradually moved to computer-based (or computer-enhanced) problem solving or CBPS on desktop or notebook computers. By the time the first edition of this book was published in 1999, it became evident that CBPS can be a very important, or possibly the most important, application of the computer in scientific and engineering education and in industrial practice.

The first edition of this book provided examples to the use of CBPS in core chemical engineering subject areas using the POLYMATH software package. Shortly after the publication of the first edition, we carried out several comparison studies in order to determine what types of software packages should be included in the "toolbox" of the engineering student and the practicing engineer that would enable the effective and efficient solution of practical problems. We arrived at the conclusion that three types of software are needed. There is a need for a numerical problem solver, such as POLYMATH, that accepts the model equations close to their mathematical forms and provides their numerical solution with very minimal user intervention. Additionally, there is also a need to be able to use spreadsheet software, such as Excel, because of its wide use in business and industry. Software like Excel is also used for the organization and presentation of information in tabular and graphical forms and for database management-related operations. Software packages that support programming, such as MATLAB, are needed to implement algorithms which are required in graduate research and advanced mathematics, programming, control, and numerical analysis courses.

It is increasingly important for today's engineering student and forward-looking engineering professionals to be proficient in the use of several software packages, and thus we greatly expanded the book so that it now includes solutions in Excel and MATLAB, in addition to POLYMATH. New problems have been introduced that demonstrate how the special capabilities of each of these packages can best be utilized for efficient and effective problem solving.

The POLYMATH Numerical Computation Package

The POLYMATH package provides convenient solutions to most numerical analysis problems, including the problems that are presented in this book. We authored and published the first PC version of POLYMATH in 1984, and it has been in use since then in over one hundred universities and selected industrial sites world wide. The version available at the time of the publication of the book, POLYMATH 6.1, was released in 2006. This package contains the following programs:

- Ordinary Differential Equations Solver
- Nonlinear Algebraic Equations Solver
- Linear Algebraic Equations Solver
- Polynomial, Multiple Linear, and Nonlinear Regression Program

The programs are extremely easy to use, and all options are menu driven. Equations are entered in standard form with user-defined notation. Results are presented in graphical or tabular form. A sophisticated calculator and a general unit conversion utility are available within POLYMATH.

The new and unique capability of the latest POLYMATH to automatically export any problem to Excel and MATLAB with a single keypress is extensively utilized within this book. Automatic export to Excel includes all intrinsic functions and logical variables. A POLYMATH ODE_Solver Add-In is included for solving ordinary differential equations in Excel. Upon export to MATLAB, the equations are ordered in the computational sequence, the intrinsic functions and logical statements are converted, and a MATLAB function is generated. Template files to run the functions are available in the HELP section of POLYMATH or from the book web site.

Current information on the latest POLYMATH software is available from

www.polymath-software.com

Many departments and some universities have obtained site licenses for POLYMATH. These licenses allow installation in all computer labs, and individual copies can be provided to all students, faculty and staff for use on personal computers. Detailed information is available from

academic@polymath-software.com

Use of This Book

This book is intended to serve as a companion text for the engineering student, the faculty instructor, or the practicing engineer. The instructions in the practical use of mathematical software package on representative problems from most chemical and biochemical engineering subject areas provide direct insight into problem setup and various practical aspects of numerical problem solving.

For the undergraduate student at the early stages of his/her studies, the book can serve as the textbook for learning to categorize the problems according to the numerical methods that should be used for efficient and effective solutions. It provides basic instruction in the use of three popular and widely used software packages: POLYMATH, Excel, and MATLAB. Emphasis is on setting up problems and effectively obtaining the necessary solutions.

In addition to providing general numerical solving capabilities, the text gives problems in most subject areas so that it can serve as a reference book in most courses, as it provides example problems that can be illustrative of problems that may be assigned in the various courses. The book also provides help with problem solving in advanced level for problems often encountered in undergraduate and graduate research such as nonlinear regression, parameter estimation in differential systems, solving two-point boundary value problems and partial differential equations, constrained equation solving, and optimization.

For the practicing engineer, the book serves as resource book in computer-based problem solving. It provides a solid foundation in problem solving and can develop basic and advanced skills in the utilization of spreadsheets. Practical problems illustrate various problem solving approaches that can be implemented for problem formulation, problem solving, analysis, presentation of results, and documentation. Of particular interest is the coverage of the correlation and regression of data with statistical analysis. All of the book's problems can be solved with the Excel spreadsheet software that is widely used in industry.

Engineering faculty can use the book to introduce numerical methods into an individual course, a sequence of courses, or an entire departmental curriculum. This book provides supplementary problems that can be assigned to students in order to introduce numerical problem solving which is avoided in most textbooks. Many of the problems can be easily extended to open-ended problem solving so that critical thinking skills can be developed. The numerical solutions can be used to answer many "what if" type questions so that students can be encouraged to think about the implications of the problem solutions. The book can also be used as a companion textbook for an introductory computer programming course or a comprehensive course in numerical analysis.

Book Organization

All the chapters of the book, except the introductory Chapter 1, are built around problems that serve to provide practical applications in a particular subject area. Most of the problems presented in the book have the same general format for the convenience of the reader. The concise problem topic is followed by a listing of the engineering concepts demonstrated by the problem. Then the numerical methods utilized in the solution are indicated just before the detailed problem statement. Each of the problems presents the detailed equations and parameter values that are necessary for solution, including the appropriate units in a variety of systems, with Systéme International d'Unités (SI) being the most commonly used. Because of the wide variety of problems posed in this book, the notation used has been standardized according to one of the major Prentice Hall textbooks in the various subject areas whenever possible. Physical properties are either given directly in the problem or in the appendices.

The book is divided into two parts. In the first part, which includes the first six chapters, subjects of general interest are presented, some on an introductory level and some on an advanced level. In Chapter 1, Introduction, the history of CBPS is briefly reviewed and guidelines are provided for categorizing problems according to the numerical techniques that should be used for their solution. Chapter 2, Basic Principles and Calculations, serves a dual purpose. The chapter introduces the reader to the subject material that is typically taught in a first chemical engineering course (in most universities called Material and Energy Balance, or Stoichiometry). Additionally, this chapter demonstrates the use of POLYMATH for solving simple problems belonging to the main categories discussed in the book, namely single nonlinear algebraic equations, systems of linear algebraic equations, linear and polynomial regression, and systems of ordinary differential equations (ODEs). In Chapter 3, Regression and Correlation of Data, the application of POLYMATH for analysis and regression of data using advanced statistical techniques is demonstrated. Chapter 4, Problem Solving with Excel, introduces the reader to the engineering and scientific problem solving capabilities of Excel using problems belonging to the same categories as in Chapter 2. The automatic export capabilities of POLYMATH to Excel are discussed. More advanced topics such as solution of systems of nonlinear algebraic equations (NLEs) and optimization with constraints (nonlinear programming) are also presented. In Chapter 5, Problem Solving with MATLAB, MATLAB is

used to solve the problems presented in Chapter 4. The capability of POLY-
MATH to automatically generate MATLAB m-files are presented and provided
templates for MATLAB problem solutions are demonstrated and utilized. In
Chapter 6, Advanced Techniques in Problem Solving, the problem solutions deal
with advanced topics such as two-point boundary value problems, systems of differential-algebraic equations, partial differential equations, and parameter estimation in systems of differential equations.

The second part of the book (Chapters 7 through 14) is organized according
to the particular subject areas such as Thermodynamics (Chapter 7), Fluid
Mechanics (Chapter 8), and so forth. The content of these chapters is presented
in the typical order of coverage in college or university-level courses.

New Content in the Second Edition

The contents of the book were almost doubled by adding six new chapters to the
eight chapters of the first edition. The introductory Chapter 1 was added in order
to help the reader in a very critical step of the problem solving—the characterization of the problem in terms of the solution method that has to be used.

After studying and verifying the importance of various software packages
in effective and efficient problem solving, the two chapters dealing with the use
of Excel and MATLAB were added. These chapters also introduce the new capability of the POLYMATH software to automatically convert a problem solution
into Excel worksheets and MATLAB m-files. This considerably shortens the
learning curve associated with the initial use of these packages.

Since the first edition was published, biochemical engineering has gained
importance and is now being taught in most colleges and universities. The new biochemical engineering chapter and selected problems in other chapters provide a
wide selection of problems in this important subject area. New chapters on "Phase
Equilibria and Distillation" (Chapter 12) and "Process Dynamics and Control"
(Chapter 13) have been added.

Companion Web Site

Readers of the book are encouraged to make full use of the companion web site that
will be maintained and extended by the book's authors. This web site enable downloads
of programs files which are used in the various book chapters for the three software
packages: POLYMATH, Excel, and MATLAB. Additional educational problems, learning
resources, corrections and updates to this book, and new materials are provided.

www.problemsolvingbook.com/

The web site also allows book owners to purchase and immediately download the latest POLYMATH software at significant discounts from the already
highly discounted POLYMATH Educational version software. This enables book
users to have the very latest software at very reasonable cost.

Instructors who are using the book have special access to all problems as
well as substantial educational and enrichment materials through the companion web site. This include suggestions as to the book use in individual courses,
sequences of courses, and throughout a departmental curriculum. Details about
this access are provided in Chapter 1 from the authors.

Recommendation for Book Use in Various Courses

There are many ways this book can be utilized in a variety of engineering and related courses. Some of the problem suggestions for courses are listed here.

1. Basic Principles and Calculations (or Material and Energy Balance or Stoichiometry): All of Chapter 2, Problems 4.1 and 5.1.
2. Thermodynamics: Problems 2.1 and 2.2, 2.5 through 2.13, 3.1 through 3.4, 3.8, 3.9, 3.14, 4.1, 4.4, 4.5, 5.1, 5.4, 5.5, 6.6, all of Chapter 7, and Reference No. 9 in Table 1.
3. Fluid Mechanics: All of Chapter 8, Problems 4.2, 5.2, 10.15, and Reference No. 2 in Table 1.
4. Heat Transfer: All of Chapter 9 and Problems 2.16, 3.5, 3.6, 3.7, 6.8, 10.12, 11.22, 11.23, 11.24, 11.25, 13.6 through 13.12, 13.14, and Reference No. 4 in Table 1.
5. Mass Transfer: All of Chapter 10 and Problems 6.5, 11.4, 11.26, 14.5, 14.16, and 14.17.
6. Chemical Reaction Engineering: All of Chapters 11 and 14, Problems 3.10 through 3.13, 4.3, 4.5, 5.3, 5.5, 6.1 through 6.6, 10.5 through 10.7, 10.11, 10.14, 10.15, 13.3, 13.12, 13.14, 13.15, and References No. 1 and 3 in Table 1.
7. Phase Equilibria and Distillation: All of Chapter 12, Problems 2.10, 2.11, 2.12, 3.8, 3.9, 3.14, 6.8, 7.8, through 7.12, and Reference No. 6 in Table 1.
8. Process Dynamics and Control: All of Chapter 13, Problems 2.14, 2.15, 2.16, 6.1, 6.2, 6.3, 6.8, 6.9, 8.14 through 8.17, 9.11 through 9.14, 10.3, 10.4, 10.13, 10.14, 11.5, 11.18, 11.20, 11.21, 11.22, 11.28, 12.10, 12.11, 14.1, 14.4, 14.6 through 14.10, 14.12, 14.13, 14.16, 14.17, and Reference No. 1 in Table 1.
9. Biochemical Engineering: All of Chapter 14 and Problems 2.3, 6.1, 6.9, 11.20, 11.27, 11.28, 12.11, 13.14, and 13.15.
10. Advanced Mathematics, Numerical Methods, and Systems of NLEs: Problems 4.5, 5.5, 6.4, 6.6, 6.7, 7.12, 7.13, 10.2, 13.2, and 13.3. Note that a MATLAB function for solving systems of constrained NLEs is provided.

    ODEs—Boundary Value Problems: Problems 6.5, 6.6, 8.1 through 8.4, 8.18, 9.2 through 9.7 10.1, 10.3, 10.5 through 10.10, 10.12, and 14.5.

    Differential Algebraic Equations: Problems 6.8 and 13.10, and Reference No. 6 in Table 1.

    Stiff Systems of ODEs: Problems 6.2, 6.3 and 6.4.

    Partial Differential Equations: Problems 6.9, 8.17, 8.18, 9.12, 9.13, 9.14, 10.13, 10.14, and 10.15.

    Nonlinear Regression: Problems 2.12, 2.13, 4.4, 5.4, 11.7, 14.7, and 14.8.
11. Applied Statistics: All of Chapter 3, Problem 6.10 and References Nos. 3 and 8 in Table 1.
12. Nonlinear Programming: Problems 4.5, 5.5, 6.9, 14.4, 14.6, 14.11 through 14.16.

13. Process Safety: References No. 7, 10, and 12 in Table 1.

14. Environmental Engineering: Reference No. 11 in Table 1.

15. Introduction to Computer Based Problem Solving: This course can compliment or replace the traditional programming course. This book can serve as the primary textbook for such a course. Content can include Chapters 1 and 2, Problems 3.3, 3.4, 3.5, 3.6, 4.1, 4.2, 4.4, 5.1, 5.2, and 5.4. Problems 2.14 and 8.8 can be used to introduce NLEs and ODEs at the introductory level as replacements for Problems 4.3 and 5.3.

Table 1 Additional Problem References

No.	Title and Reference
1	"Exothermic CSTR's—Just How Stable are the Multiple Steady States?" *Chem. Eng. Educ.*, *28*(1), 30-35 (1994).
2	"Numerical Experiments in Fluid Mechanics with a Tank and Draining Pipe," *Comput. Appl. Eng. Educ,* *2*(3), 175-183 (1994).
3	"Correlation and Over-correlation of Heterogeneous Reaction Rate Data," *Chem. Eng. Educ.*, *29*(1) 22-25, 45 (1995).
4	"The Wind-Chill Paradox: Four Problems in Heat Transfer," *Chem. Eng. Educ.*, *30*(4), 256-261 (1996).
5	"Replacing the Graph Paper by Interactive Software in Modeling and Analysis of Experimental Data," *Comput. Appl. Eng. Educ.,**4*(3), 241-251 (1996).
6	"What To Do If Relative Volatilities Cannot Be Assumed To Be Constant?—Differential-Algebraic Equation Systems in Undergraduate Education," *Chem. Eng. Educ.*, *31*(2), 86-93 (1997).
7	"Prediction and Prevention of Chemical Reaction Hazards—Learning by Simulation," *Chem. Eng. Educ.*, *35*(4), 268-273 (2001).
8	Letter to the Editor Concerning "An Undergraduate Course in Applied Probability and Statistics," *Chem. Eng. Educ.*, *36*(4), 263, 277 (2002).
9	"An Exercise for Practicing Programming in the ChE Curriculum—Calculation of Thermodynamic Properties Using the Redlich-Kwong Equation of State," *Chem. Eng. Educ.*, *27*(2), 148 (2003).
10	Letter to the Editor Concerning "Evaluations of Kinetic Parameters and Critical Runaway Conditions in the Reaction System of Hexamine-Nitric Acid to Produce RDX in a Non-Isothermal Batch Reactor," *Journal of Loss Prevention in the Process Industries*, *17*(6), 513-514 (2004).
11	"Applications of Mathematical Software Packages for Modeling and Simulations in Environmental Engineering Education," *Environment Modeling and Software*, *20*, 1307-1313 (2005).
12	"Combining HAZOP with Dynamic Simulation—Applications for Safety Education," *Journal of Loss Prevention in the Process Industries*, *19*, 754 (2006).

Chemical and Biochemical Engineering Departments

Academic departments are encouraged to consider adopting this book during the first introductory course in chemical and/or biochemical engineering and then utilizing the book as a supplement for many of the following courses in the curriculum. This allows an integrated approach to the use of numerical methods throughout the curriculum. This approach can be helpful in satisfying the ABET requirements for appropriate computer use in undergraduate studies.

A first course in numerical methods can also utilize many of the problems as relevant examples. In this application, the book will supplement a standard numerical methods textbook. Students will find the problems in this book to be more interesting than the strictly mathematical or simplified problems presented in many standard numerical analysis textbooks.

Acknowledgments

We would like to express our appreciation to our wives and families who have shared the burden of this effort which took longer than anticipated to complete. We particularly thank Professor H. Scott Fogler for his encouragement with this book effort and with the continuing development of the POLYMATH package. Many thanks are due to Professor Neima Brauner for her help in developing many of the problems.

Additionally, we appreciate the input and many suggestions of our students, who have been subjected to preliminary versions of the problems and have endured the various pre-release versions of the POLYMATH software over the years.

During the twenty-three years that POLYMATH has been in use, many of our colleagues provided advice and gave us help in revising and improving this software package. In particular, we would like to acknowledge the assistance of Professors N. Brauner, B. Carnahan, D. J. Cooper, H. S. Fogler, D. M. Himmelblau, D. S. Kompala, S. E. LeBlanc, E. M. Rosen, and J. D. Seader. H. S. Fogler has also provided some of the problems included in the book.

Continuing development of the POLYMATH program has been a continuing process. The initial programming and algorithm implementations were carried out and maintained for the first ten years by Orit Shacham. She spent many hours and most of her vacations fixing bugs and writing new code for still another version of POLYMATH. She always amazed us by the speed and precision with which she converted ideas into computer code.

For the last seven years, POLYMATH has been coauthored and programmed by Michael Elly. He has developed a very intuitive and user-friendly interface for interactive problem solving. He has also exhibited a unique capability for effectively implementing rather difficult algorithmic challenges. His creativity, organization, speed, and accuracy continue to impress us. We are happy to have him as a continuing member of our team.

The first draft of the first edition this book was typed (and retyped) by Michal Shacham. She took several months of vacation from her job to learn to use various word processors and graphical programs. The draft she typed became the basis for class testing and refinement of the book. Nancy Neborsky

Pickering learned the FrameMaker desktop publishing package and professionally entered the initial materials into the book format for the first edition.

The authoring of this book and the POLYMATH computer software are both very expensive endeavors in both resources and time. We are indebted to our universities, the University of Connecticut and the Ben-Gurion University of the Negev, for the continuous support we have received. The CACHE Corporation (Computer Aids for Chemical Engineering Education) has been very helpful in advancing and promoting the academic use of POLYMATH.

Michael B. Cutlip
Storrs, Connecticut

Mordechai Shacham
Beer-Sheva, Israel

Problem Solving with Mathematical Software Packages

1.1 EFFICIENT PROBLEM SOLVING—THE OBJECTIVE OF THIS BOOK

As an engineering student or professional, you are almost always involved in numerical problem solving on a personal computer. The objective of this book is to enable you to solve numerical problems that you may encounter in your student or professional career in a most effective and efficient manner. The tools that are typically used for engineering or technical problem solving are mathematical software packages that execute on personal computers on the desktop. In order to solve your problems most efficiently and accurately, you must be able to select the appropriate software package for the particular problem at hand. Then you must also be proficient in using your selected software tool.

In order to help you achieve these objectives, this book provides a wide variety of problems from different areas of chemical, biochemical, and related engineering and scientific disciplines. For some of these problems, the complete solution process is demonstrated. For some problems, partial solutions or hints for the solution are provided. Other problems are left as exercises for you to solve.

Most of the chapters of the book are organized by chemical and biochemical engineering subject areas. The various chapters contain between five and twenty-eight problems that represent many of the problem types that require a computer solution in a particular subject area. All problems presented in the book have the same general format for your convenience. The concise problem topic is first followed by a listing of the engineering or scientific concepts demonstrated by the problem. Then the numerical methods utilized in the solution are indicated just before the detailed problem statement is presented. Typically a particular problem presents all of the detailed equations that are necessary for solution, including the appropriate units in a variety of systems, with Système International d'Unités (SI) being the most commonly used. Physical properties are either given directly in the problem or in the appendices. Complete and partial solutions are provided to many of the problems. These solutions will help you learn to formulate and then to solve the unsolved problems in the book as well as the problems that you will face in your student and/or professional career.

. Three widely used mathematical software packages are used in this book for solving the various problems: POLYMATH,[*] Excel,[†] and MATLAB.[‡] Each of these packages has specific advantages that make it the most appropriate for solving a particular problem. In some cases, a combined use of several packages is most desirable. These mathematical software packages that solve the problems utilize what are called "numerical methods." This book presents the fundamental and practical approaches to setting up problems that can then be solved by mathematical software that utilizes numerical methods. It also gives much practical information for problem solving. The details of the numerical methods are beyond the scope of this book, and reference can be made to textbook by Constantinides and Mostoufi.[1] More advanced and extensive treatment of numerical methods can be found in the book of Press et al.[2]

The first step in solving a problem using mathematical software is to prepare a mathematical model of the problem. It is assumed that you have already learned (or you will learn) how to prepare the model of a problem from in particular subject area (such as thermodynamics, fluid mechanics, or biochemical engineering). A general approach advocated in this book is to start with a very simple model and then to make the model more complex as necessary to describe the problem. Engineering and scientific fundamentals are important in model building. The first step in the solution process is to characterize the problem according to the type of the mathematical model that is formulated: a system of algebraic equations or a system of ordinary differential equations, for example. When the problem is characterized in these terms, the software package that efficiently solves this type of problems can be utilized. Most of the later part of this chapter is devoted to learning how to characterize a problem in such mathematical terms.

In order to put the use of mathematical software packages for problem solving into proper perspective, it is important and interesting to review the history in which manual problem solving has been replaced by numerical problem solving.

1.2 FROM MANUAL PROBLEM SOLVING TO USE OF MATHEMATICAL SOFTWARE

The problem solving tools on the desktop that were used by engineers prior to the introduction of the handheld calculators (i.e., before 1970) are shown in Figure 1–1. Most calculations were carried out using the slide rule. This required carrying out each arithmetic operation separately and writing down the results of such operations. The highest precision of such calculations was to three decimal digits at most. If a calculational error was detected, then all the slide rule and arithmetic calculations had to be repeated from the point where the error occurred. The results of the calculations were typically typed, and hand-drawn

---

[*] POLYMATH is a product of Polymath Software (http://www.polymath-software.com).
[†] Excel is a trademark of Microsoft Corporation (http://www.microsoft.com).
[‡] MATLAB is trademark of The Math Works, Inc. (http://www.mathworks.com).

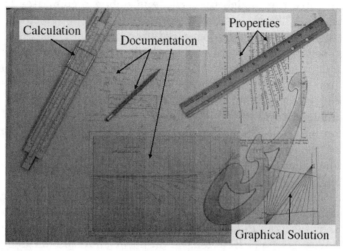

Figure 1–1 The Engineer's Problem Solving Tools Prior to 1970

graphs were often prepared. Temperature- and/or composition-dependent thermodynamic and physical properties that were needed for problem solving were represented by graphs and nomographs. The values were read from a straight line passed by a ruler between two points. The highest precision of the values obtained using this technique was only two decimal digits. All in all, "manual" problem solving was a tedious, time-consuming, and error-prone process.

During the slide rule era, several techniques were developed that enabled solving realistic problems using the tools that were available at that time. Analytical (closed form) solutions to the problems were preferred over numerical solutions. However, in most cases, it was difficult or even impossible to find analytical solutions. In such cases, considerable effort was invested to manipulate the model equations of the problem to bring them into a solvable form. Often model simplifications were employed by neglecting terms of the equations which were considered less important. "Short-cut" solution techniques for some types of problems were also developed where a complex problem was replaced by a simple one that could be solved. Graphical solution techniques, such as the McCabe-Thiele and Ponchon-Savarit methods for distillation column design, were widely used.

After digital computers became available in the early 1960's, it became apparent that computers could be used for solving complex engineering problems. One of the first textbooks that addressed the subject of numerical solution of problems in chemical engineering was that by Lapidus.[3] The textbook by Carnahan, Luther and Wilkes[4] on numerical methods and the textbook by Henley and Rosen[5] on material and energy balances contain many example problems for numerical solution and associated mainframe computer programs (written in the FORTRAN programming language). Solution of an engineering problem using digital computers in this era included the following stages: (1) derive the model equations for the problem at hand, (2) find the appropriate numerical method (algorithm) to solve the model, (3) write and debug a computer language pro-gram (typically FORTRAN) to solve the problem using the selected algorithm, (4) validate the results and prepare documentation.

Problem solving using numerical methods with the early digital computers was a very tedious and time-consuming process. It required expertise in numeri-cal methods and programming in order to carry out the 2nd and 3rd stages of the problem-solving process. Thus the computer use was justified for solving only large-scale problems from the 1960's through the mid 1980's.

Mathematical software packages started to appear in the 1980's after the introduction of the Apple and IBM personal computers. POLYMATH version 1.0, the software package which is extensively used in this book, was first published in 1984 for the IBM personal computer.

Introduction of mathematical software packages on mainframe and now personal computers has considerably changed the approach to problem solving. Figure 1–2 shows a flow diagram of the problem-solving process using such a package. The user is responsible for the preparation of the mathematical model (a complete set of equations) of the problem. In many cases the user will also need to provide data or correlations of physical properties of the compounds involved. The complete model and data set must be fed into the mathematical software package. It is also the user's responsibility to categorize the problem type. The problem category will determine the type of numerical algorithm to be used for the solution. This issue will be discussed in detail in the next section.

The mathematical software package will then solve the problem using the selected numerical technique. The results obtained together with the model defi-nition can serve as partial or complete documentation of the problem and its solution.

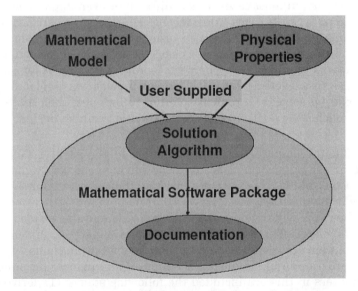

Figure 1–2 Problem Solving with Mathematical Software Packages

1.3 CATEGORIZING PROBLEMS ACCORDING TO THE SOLUTION TECHNIQUE USED

Mathematical software packages contain various tools for problem solving. In order to match the tool to the problem in hand, you should be able to categorize the problem according to the numerical method that should be used for its solution. The discussion in this section details the various categories for which representative examples are included in the book. Note that the study of the following categories (a) through (e) is highly recommended prior to using Chapters 7 through 14 of this book that are associated with particular subject areas. Categories (f) through (n) are advanced topics that should be reviewed prior to advanced problem solving.

(a) Consecutive Calculations

These calculations do not require the use of a special numerical technique. The model equations can be written one after another. On the left-hand side a variable name appears (the output variable), and the right-hand side contains a constant or an expression that may include constants and previously defined variables. Such equations are usually called "explicit" equations. A typical example for such a problem is the calculation of the pressure using the van der Waals equation of state.

$$R = 0.08206$$

$$T_c = 304.2$$

$$P_c = 72.9$$

$$T = 350$$

$$V = 0.6 \qquad\qquad\qquad (1\text{-}1)$$

$$a = (24/64)((R^2 T_c^2)/P_c)$$

$$b = (RT_c)/(8P_c)$$

$$P = (RT)/(V-b) - a/V^2$$

The various aspects associated with the solution of this type of problem are described in detail in Problems 4.1 and 5.1 (Molar Volume and Compressibility from Redlich-Kwong Equation). In those completely solved problems, the advantages of the different software packages (POLYMATH, Excel, and MATLAB) in the various stages of the solution process are also demonstrated. To gain the most benefit from using this book, you should proceed now to study Problems 4.1 and 5.1 and then return to this point.

(b) System of Linear Algebraic Equations

A system of linear algebraic equations can be represented by the equation:

$$\mathbf{A}\mathbf{x} = \mathbf{b} \tag{1-2}$$

where $\mathbf{A}$ is an $n \times n$ matrix of coefficients, $\mathbf{x}$ is an $n \times 1$ vector of unknowns and $\mathbf{b}$ an $n \times 1$ vector of constants. Note that the number of equations is equal to the number of the unknowns. A detailed description of the various aspects of the solution of systems of linear equations is provided in Problem 2.4 (Steady-State Material Balances on a Separation Train).

(c) One Nonlinear (Implicit) Algebraic Equation

A single nonlinear equation can be written in the form

$$f(x) = 0 \tag{1-3}$$

where f is a function and x is the unknown. Additional explicit equations, such as those shown in Section (a), may also be included. Solved problems associated with the solution of one nonlinear equation are presented in Problems 2.1 (Molar Volume and Compressibility Factor from Van Der Waals Equation), 2.9 (Gas Volume Calculations using Various Equations of State), 2.10 (Bubble Point Calculation for an Ideal Binary Mixture), and 2.13 (Adiabatic Flame Temperature in Combustion). These problems should be reviewed before proceeding further. The use of the various software packages for solving single nonlinear equations is demonstrated in solved Problems 4.2 and 5.2 (Calculation of the Flow Rate in a Pipeline). Please study those solved problems as well.

(d) Multiple Linear and Polynomial Regressions

Given a set of data of measured (or observed) values of a dependent variable: y_i versus n independent variables x_{1i}, x_{2i}, ... x_{ni}, multiple linear regression attempts to find the "best" values of the parameters $a_0, a_1, ...a_n$ for the equation

$$\hat{y}_i = a_0 + a_1 x_{1,i} + a_2 x_{2,i} + ... + a_n x_{n,i} \tag{1-4}$$

where $\hat{y}_i$ is the calculated value of the dependent variable at point i. The "best" parameters have values that minimize the squares of the errors

$$S = \sum_{i=1}^{N} (y_i - \hat{y}_i)^2 \tag{1-5}$$

where N is the number of available data points.

In polynomial regression, there is only one independent variable x, and Equation (1-4) becomes

$$\hat{y}_i = a_0 + a_1 x_i + a_2 x_i^2 + ... + a_n x_i^n \tag{1-6}$$

Multiple linear and polynomial regressions using POLYMATH are demonstrated in detail in solved Problems 3.3 (Correlation of Thermodynamic and Physical Properties of n-Propane) and 3.5 (Heat Transfer Correlations from Dimensional Analysis). The use of Excel and MATLAB for the same purpose is demonstrated respectively in Problems 4.4 and 5.4 (Correlation of the Physical Properties of Ethane). These examples should be studied before proceeding further.

(e) Systems of First-Order Ordinary Differential Equations (ODEs) – Initial Value Problems

A system of n simultaneous first-order ordinary differential equations can be written in the following (canonical) form

$$\frac{dy_1}{dx} = f_1(y_1, y_2, \ldots y_n, x)$$

$$\frac{dy_2}{dx} = f_2(y_1, y_2, \ldots y_n, x) \tag{1-7}$$

$$\vdots$$

$$\frac{dy_n}{dx} = f_n(y_1, y_2, \ldots y_n, x)$$

where x is the independent variable and y_1, y_2, ... y_n are dependent variables. To obtain a unique solution of n simultaneous first-order ODEs, it is necessary to specify n values of the dependent variables (or their derivatives) at specific values of the independent variable. If those values are specified at a common point, say x_0,

$$y_1(x_0) = y_{1,0}$$

$$y_2(x_0) = y_{2,0} \tag{1-8}$$

$$\vdots$$

$$y_n(x_0) = y_{n,0}$$

then the problem is categorized as an initial value problem.

The solution of systems of first-order ODE initial value problems is demonstrated in Problems 2.14 (Unsteady-state Mixing in a Tank) and 2.16 (Heat Exchange in a Series of Tanks) where POLYMATH is used to obtain the solution. The use of Excel and MATLAB for systems of first-order ODEs is demonstrated respectively in Problems 4.3 and 5.3 (Adiabatic Operation of a Tubular Reactor for Cracking of Acetone).

(f) System of Nonlinear Algebraic Equations (NLEs)

A system of nonlinear algebraic equations is defined by

$$f(x) = 0 \tag{1-9}$$

where f is an n vector of functions, and x is an n vector of unknowns. Note that the number of equations is equal to the number of the unknowns. Solved problems in the category of NLEs are Problems 8.11 (Flow Distribution in a Pipeline Network) and 6.6 (Expediting the Solution of Systems of Nonlinear Algebraic Equations). More advanced treatment of systems of nonlinear equations (obtained when solving a constrained minimization problem), is demonstrated along with the use of various software packages in Problems 4.5 and 5.5 (Complex Chemical Equilibrium by Gibbs Energy Minimization).

(g) Higher Order ODEs

Consider the n-th order ordinary differential equation

$$\frac{d^n z}{dx^n} = G\left(z, \frac{dz}{dx}, \frac{d^2 z}{dx^2}, \dots \frac{d^{n-1} z}{dx^{n-1}}, x\right)$$

(1-10)

This equation can be transformed by a series of substitution to a system of n first-order equations (Equation (1-7)). Such a transformation is demonstrated in Problems 6.5 (Shooting Method for Solving Two-Point Boundary Value Problems) and 8.18 (Boundary Layer Flow of a Newtonian Fluid on a Flat Plate).

(h) Systems of First-Order ODEs—Boundary Value Problems

ODEs with boundary conditions specified at two (or more) points of the independent variable are classified as boundary value problems. Examples of such problems and demonstration of the solution techniques used can be found in Problems 6.4 (Iterative Solution of ODE Boundary Value Problem) and 6.5 (Shooting Method for Solving Two-Point Boundary Value Problems).

(i) Stiff Systems of First-Order ODEs

Systems of ODEs where the dependent variables change on various time (independent variable) scales which differ by many orders of magnitude are called "Stiff" systems. The characterization of stiff systems and the special techniques that are used for solving such systems are described in detail in Problem 6.2 (Solution of Stiff Ordinary Differential Equations).

(j) Differential-Algebraic System of Equations (DAEs)

The system defined by the equations:

$$\frac{d\mathbf{y}}{dx} = \mathbf{f}(\mathbf{y}, \mathbf{z}, x)$$

$$\mathbf{g}(\mathbf{y}, \mathbf{z}) = 0$$

(1-11)

with the initial conditions $\mathbf{y}(\mathbf{x}_0) = \mathbf{y}_0$ is called a system of differential-algebraic equations. Demonstration of one particular technique for solving DAEs can be found in Problem 6.7 (Solving Differential Algebraic Equations – DAEs).

(k) Partial Differential Equations (PDEs)

Partial differential equations where there are several independent variables have a typical general form:

$$\frac{\partial T}{\partial t} = \alpha \left(\frac{\partial^2 T}{\partial x^2} + \frac{\partial^2 T}{\partial y^2} \right)$$

(1-12)

A problem involving PDEs requires specification of initial values and boundary conditions. The use of the "Method of Lines" for solving PDEs is demonstrated in Problems 6.8 (Method of Lines for Partial Differential Equations) and 9.14 (Unsteady-State Conduction in Two Dimensions).

(l) Nonlinear Regression

In nonlinear regression, a nonlinear function g

$$\hat{y}_i = g(a_0, a_1 \ldots a_n, x_{1,i}, x_{2,i}, \ldots x_{n,i})$$

(1-13)

is used to model the data by finding the values of the parameters a_0, $a_1 \ldots a_n$ that minimize the squares of the errors shown in Equation (1-5). Detailed description of the nonlinear regression problem and the method of solution using POLYMATH can be found in Problem 3.1 (Estimation of Antoine Equation Parameters using Nonlinear Regression). The use of Excel and MATLAB for nonlinear regression is demonstrated in respective Problems 4.4 and 5.4 (Correlation of the Physical Properties of Ethane).

(m) Parameter Estimation in Dynamic Systems

This problem is similar to the nonlinear regression problem except that there is no closed form expression for $\hat{y}_i$, but the squares of the errors function to be minimized must be calculated by solving the system of first order ODEs

$$\frac{d\hat{y}}{dt} = \mathbf{f}(a_0, a_1 \ldots a_n, x_1, x_2, \ldots x_n, t)$$

(1-14)

Problem 6.9 (Estimating Model Parameters Involving ODEs using Fermentation Data) describes in detail a parameter identifications problem and demonstrates its solution using POLYMATH and MATLAB.

(n) Nonlinear Programming (Optimization) with Equity Constraints

The nonlinear programming problem with equity constrains is defined by:

$$\begin{aligned} &Minimize \quad f(\mathbf{x}) \\ &Subject\ to \quad \mathbf{h}(\mathbf{x}) = \mathbf{0} \end{aligned}$$

(1-15)

where f is a function, $\mathbf{x}$ is an n-vector of variables and $\mathbf{h}$ is an m-vector ($m<n$) of functions.

Problems 4.5 and 5.5 (Complex Chemical Equilibrium by Gibbs Energy Minimization) demonstrate several techniques for solving nonlinear optimization problems with POLYMATH, Excel, and MATLAB.

1.4 EFFECTIVE USE OF THIS BOOK

Readers who wish to begin solving realistic problems with mathematical software packages are recommended to initially study the items listed in categories (a) through (e) in the previous section. In addition, different subject areas will require completion of some of the advanced topics listed in categories (f) through (n) of the previous section. Table 1–1 shows the advanced topics associated with the various subjects covered in Chapters 7 to 14 of this book. For example, you will be able to achieve effective solutions for the problems associated with Chapter 8 (Fluid Mechanics) if you initially study the categories (f) System of Nonlinear Algebraic Equations (NLEs), (g) Higher Order ODEs, (h) Systems of First-Order ODEs—Boundary Value Problems, and (k) Partial Differential Equations (PDEs). It is recommended that the advanced topics listed in Table 1–1 be studied before you begin to work on a chapter related to a particular subject area.

Table 1–1 Advanced Topic Prerequisites for Chapters 7 through 14

Chapter No.	Subject Area	Advanced Topics Required
7	Thermodynamics	(f)
8	Fluid Mechanics	(f), (g), (h), and (k)
9	Heat Transfer	(g), (h), and (k)
10	Mass Transfer	(g), (h), and (k)
11	Chemical Reaction Engineering	(f), (g), (h), (l), and (m)
12	Phase Equilibria and Distillation	(f) and (j)
13	Process Dynamics and Control	(f), (g), (h), and (i)
14	Biochemical Engineering	(f), (g), (h), (i), (l), and (n)

Several of the advanced topics are typically considered in the advanced subject areas of "Numerical Methods," "Advanced Mathematics," and "Optimization." Using the problems pertinent to a particular topic in these advanced subject areas can be very beneficial in the learning process. The problems associated with the various topics are shown in Table 1–2, where solved or partially solved problem numbers are shown in bold numerals

Some problems may require special solution techniques. Solved problems that demonstrate some special techniques are listed in Table 1-3. If a problem matches one or more of the categories in this table, then an examination of the similar solved problem can be very helpful.

Table 1–2 List of Problems Associated with Advanced Topics[a]

	Topic	Pertinent Problem Numbers*
(f)	System of Nonlinear Algebraic Equations (NLEs)	**4.5**, **5.5**, 6.6, 7.13, 7.14, 8.9, **8.10**, 8.12, 8.13, 10.2,**12.1**, 12.2, **12.3**, 12.4, 12.5, **12.8**, 12.9, 14.8, 14.11, 14.15
(g)	Higher Order ODEs	**6.5**. **8.16**, 8.18, **9.2**, **9.5**, 10.11, **13.1**, 13.2, **13.5**, **13.7**, **13.12**, **14.5**
(h)	Systems of First-Order ODEs—Boundary Value Problems	**6.4**,**6.5**,**8.1**,8.2,8.3,8.4,8.18, 9.1, 9.2, 9.6, 9.7, **10.1**, **10.3**, **10.5**, 10.6,10.7, **10.8**, 10.9, 10.10, 10.12, **14.5**
(i)	Stiff Systems of First-Order ODEs	**6.1**, 6.2
(j)	Differential-Algebraic System of Equations (DAEs)	**6.7**, **12.10**, 12.11
(k)	Partial Differential Equations (PDEs)	**6.8**, 8.17, 9.12, 9.13, **9.14**, **10.13**, **10.14**, **10.15**
(l)	Nonlinear Regression	**3.1**, **3.2**, **3.3**, 3.4, **4.4**, **5.4**, 11.7, **13.4**, 13.8, **14.2**, 14.7, 14.8, 14.13, 14.14, 14.15
(m)	Parameter Estimation in Dynamic Systems	**6.9**
(n)	Nonlinear Programming (Optimization)	**4.5**, **5.5**, **6.9**, 14.4, 14.6, 14.11, **14.16**

[a]Solved and partially solved problems are indicated in bold.

Table 1–3 List of Problems Associated with Special Problem Solving Techniques

Differential and Algebraic Equations	Pertinent Problem Number(s)
Plotting Solution Trajectory for an Algebraic Equation Using the ODE Solver	7.1, 7.5
Using the l'Hôpital's Rule for Undefined Functions at the Beginning or End Point of Integration Interval	7.11, 7.12
Using "If" Statement to Avoid Division by Zero	8.1
Switching Variables On and Off during Integration	8.16
Retaining a Value when a Condition Is Satisfied	8.16
Generation of Error Function	8.17
Functions Undefined at the Initial Point	9.2, 9.5
Using "If" Statement to Match Different Equations to the Same Variable	9.2, 10.4, 8.6
Ill-Conditioned Systems	6.3
Conversion of a System of Nonlinear Equations into a Single Equation	6.3
Selection of Initial Estimates for Nonlinear Equations	6.6
Modification of Strongly Nonlinear Equations for Easier Solution	6.6
Conversion of a Nonlinear Algebraic Equation to a Differential Equation	7.1, 7.5
Data Modeling and Analysis	
Using Residual Plot for Data Analysis	2.5, 3.1, 3.3, 3.5, 3.8, 3.14
Using Confidence Intervals for Checking Significance of Parameters	2.5, 3.1, 3.3, 3.8
Transformation of Nonlinear Models for Linear Representations	2.5, 3.3, 3.5, 3.8
Checking Linear Dependency among Independent Variables	3.11

1.5 SOFTWARE USAGE WITH THIS BOOK

The problems presented within this text can be solved with a variety of mathematical software packages. The problems along with the appendices provide all the necessary equations and parameters for obtaining problem solutions. However, POLYMATH is extensively utilized to demonstrate problem solutions throughout the book because it is extremely easy to use, and because the equations are entered in basically the same mathematical form as they are written.

Recent enhancements to POLYMATH have enabled the content of this book to be expanded to more directly support problem solving with Excel and MATLAB. This is due to the ability of POLYMATH to export a problem solution to a working spreadsheet in Excel or the necessary m-file of the problem for MATLAB.

The export process to Excel results in a complete working spreadsheet with the same notation and logic as utilized in the POLYMATH problem solution code. In addition, all the intrinsic functions (such as log, exp, sin, etc.) are automatically converted. POLYMATH also provides an Excel Add-In which allows the solution of systems of ordinary differential equations within Excel. This is separate software which operates in Excel and enables the same solution algorithms available in POLYMATH to be utilized within Excel. Complete details and an introduction to problem solving in Excel are given in Chapter 4.

For MATLAB, the m-file for a particular POLYMATH program solution is automatically generated with translation of the program logic and the needed intrinsic functions. Also, the generated statements in the m-file are automatically ordered as required by MATLAB. This use of MATLAB and the use of the generation of m-files from POLYMATH are discussed in Chapter 5.

Within this book, the problems are discussed with reference to important and necessary equations. The solutions are presented using the POLYMATH coding as this provides a clear representation of the problem formulation along with the mathematical equations and logic necessary for problem solution. The Excel and MATLAB solutions can easily be achieved through the use of the POLYMATH program via the automated export capability. All of the completely worked or partially worked problems in the book are available in files for all three packages – POLYMATH, Excel, and MATLAB. Access to these files is provided from the book web site as discussed in the next section. You can obtain an inexpensive educational version of the latest POLYMATH software from the book's web site, **www.problemsolvingbook.com**.

Thus you have a choice of the software with which you wish to generate problem solutions while using this book. POLYMATH is highly recommended as it is used throughout the book and is widely accepted as the most convenient mathematical software for a student or a novice to learn and use. You are encouraged to learn to solve problems with POLYMATH, Excel, and MATLAB as this book will highlight and utilize some of the particular capabilities that each package enables.

1.6 WEB-BASED RESOURCES FOR THIS BOOK

A special web site is dedicated to the continuing support of this book

www.problemsolvingbook.com

 and this is identified throughout the book with the icon on the left. This special web site provides readers with the following materials:

1. Solution files for the worked and partially worked problems for POLYMATH, Excel, and MATLAB
2. Access to the latest Educational Version of POLYMATH at a special reduced price that is available only to book owners for educational use
3. MATLAB templates for general problem solving
4. Additional problems as they become available
5. Corrections to the printed book
6. Special resources for students
7. Special resources for instructors who are using this book
   Note: Instructors should obtain more details from one of the authors via e-mail from their home institution to

michael.cutlip@uconn.edu

shacham@bgu.ac.il

GENERAL REFERENCES

1. Constantinides, A., and Mostoufi, N., *Numerical Methods for Chemical Engineers with MATLAB Applications*, 1st ed., Upper Saddle River, NJ: Prentice Hall, 1999.
2. Press, W. H., Teukolsky, S, A., Vetterling, W. T., and Flannery, B. P., *Numerical Recipes in C++: The Art of Scientific Computing*, 2nd ed., Cambridge, England: Cambridge University Press, 2002.
3. Lapidus, L., *Digital Computation for Chemical Engineers*, New York: McGraw-Hill, 1962.
4. Carnahan, B., Luther, H. A. and Wilkes, J. O., *Applied Numerical Methods*, New York: Wiley, 1969.
5. Henley, E. J., and E. M. Rosen, *Material and Energy Balance Computation*, New York: Wiley, 1969.

Basic Principles and Calculations

2.1 MOLAR VOLUME AND COMPRESSIBILITY FACTOR FROM VAN DER WAALS EQUATION

2.1.1 Concepts Demonstrated

Use of the van der Waals equation of state to calculate molar volume and compressibility factor for a gas.

2.1.2 Numerical Methods Utilized

Solution of a single nonlinear algebraic equation.

2.1.3 Problem Statement

The ideal gas law can represent the pressure-volume-temperature (PVT) relationship of gases only at low (near atmospheric) pressures. For higher pressures, more complex equations of state should be used. The calculation of the molar volume and the compressibility factor using complex equations of state typically requires a numerical solution when the pressure and temperature are specified.

The van der Waals equation of state is given by

$$\left(P + \frac{a}{V^2}\right)(V - b) = RT \tag{2-1}$$

where

$$a = \frac{27}{64}\left(\frac{R^2 T_c^2}{P_c}\right) \tag{2-2}$$

and

$$b = \frac{RT_c}{8P_c} \tag{2-3}$$

The variables are defined by

P = pressure in atm

V = molar volume in L/g-mol

T = temperature in K

R = gas constant (R = 0.08206 atm·L/g-mol·K)

T_c = critical temperature (405.5 K for ammonia)

P_c = critical pressure (111.3 atm for ammonia)

Reduced pressure is defined as

$$P_r = \frac{P}{P_c} \tag{2-4}$$

and the compressibility factor is given by

$$Z = \frac{PV}{RT} \tag{2-5}$$

(a)   Calculate the molar volume and compressibility factor for gaseous ammonia at a pressure P = 56 atm and a temperature T = 450 K using the van der Waals equation of state.

(b)   Repeat the calculations for the following reduced pressures: P_r = 1, 2, 4, 10, and 20.

(c)   How does the compressibility factor vary as a function of P_r?

2.1.4   Solution

Equation (2-1) cannot be rearranged into a form where V can be explicitly expressed as a function of T and P. However, it can easily be solved numerically using techniques for nonlinear equations. In order to solve Equation (2-1) using the POLYMATH *Simultaneous Algebraic Equation Solver*, it must be rewritten in the form

$$f(V) = \left(P + \frac{a}{V^2}\right)(V - b) - RT \tag{2-6}$$

where the solution is obtained when the function is close to zero, $f(V) \approx 0$. Additional explicit equations and data can be entered into the POLYMATH program in direct algebraic form. The POLYMATH program will reorder these equations as necessary in order to allow sequential calculation.

The POLYMATH equation set for this problem is given in Table 2–1.In order to solve a single nonlinear equation with POLYMATH, an interval for the expected solution variable, V in this case, must be entered into the program. This interval can usually be found by consideration of the physical nature of the problem.

Table 2–1 Equation Set in the POLYMATH Nonlinear Equation Solver
(File **P2-01A.POL**) (Line numbers added for clarity)

Line	Equation
1	f(V)=(P+a/(V^2))*(V-b)-R*T
2	P=56
3	R=0.08206
4	T=450
5	Tc=405.5
6	Pc=111.3
7	Pr=P/Pc
8	a=27*(R^2*Tc^2/Pc)/64
9	b=R*Tc/(8*Pc)
10	Z=P*V/(R*T)
11	V(min)=0.4
12	V(max)=1

(a) For part (a) of this problem, the volume calculated from the ideal gas law can be a basis for specifying the required solution interval. The POLYMATH *Calculator* is convenient to calculate the molar volume from $V = RT/P$ at the specified temperature and pressure as follows:

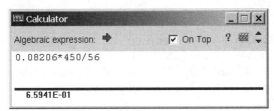

Thus the estimated molar volume using the ideal gas law is $V = 0.66$ L/g-mol. An interval for the expected solution for V can be entered as between 0.4 as the lower limit and 1.0 as the higher limit. The POLYMATH solution, which is given in Figure 2–1 for $T = 450$ K and $P = 56$ atm, yields $V = 0.5749$ L/g-mol, where the compressibility factor is $Z = 0.8718$.

www The POLYMATH problem solution file for part (a) is found in directory Chapter 2 with file named **P2-01A.POL**. This problem is also solved with Excel, Maple, MathCAD, MATLAB, Mathematica, and POLYMATH as Problem 1 in the Set of Ten Problems discussed on the book web site.

(b) Solution for the additional pressure values can be accomplished by changing the code for the equations in the POLYMATH program for P and P_r to

```
Pr=1
P=Pr*Pc
```

Additionally, the bounds on the molar volume V may need to be altered to obtain an interval where there is a solution. Subsequent program execution for the various P_r's is required.

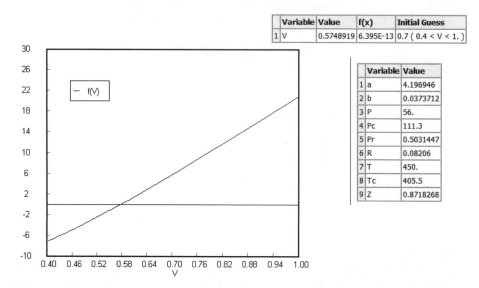

Variable	Value	f(x)	Initial Guess
1 V	0.5748919	6.395E-13	0.7 (0.4 < V < 1.)

	Variable	Value
1	a	4.196946
2	b	0.0373712
3	P	56.
4	Pc	111.3
5	Pr	0.5031447
6	R	0.08206
7	T	450.
8	Tc	405.5
9	Z	0.8718268

Figure 2–1 Plot of $f(V)$ versus V and Results Tables for van der Waals Equation
(File **P2-01A.POL**)

www   The POLYMATH problem solution file for part (b) is found in directory
Chapter 2 with file named **P2-01B.POL**. This problem is also solved with
Excel, Maple, MathCAD, MATLAB, Mathematica, and POLYMATH as
Problem 1 in the Set of Ten Problems discussed on the book web site.

 (c)   The calculated molar volumes and compressibility factors are summarized in Table 2–2. These calculated results indicate that there is a minimum in
the compressibility factor Z at approximately $P_r = 2$. The compressibility factor
then starts to increase and reaches $Z = 2.783$ for $P_r = 20$.

Table 2–2 Compressibility Factor for Gaseous Ammonia at 450 K

P(atm)	P_r	V	Z
56	0.503	.574892	0.871827
111.3	1.0	.233509	0.703808
222.6	2.0	.0772676	0.465777
445.2	4.0	.0606543	0.731261
1113.0	10.0	.0508753	1.53341
2226.0	20.0	.046175	2.78348

2.2 MOLAR VOLUME AND COMPRESSIBILITY FACTOR FROM REDLICH-KWONG EQUATION

2.2.1   Concepts Demonstrated

Use of the Redlich-Kwong equation of state to calculate molar volume and compressibility factor for a gas.

2.2.2   Numerical Methods Utilized

Solution of a single nonlinear algebraic equation.

2.2.3   Problem Statement

The Redlich-Kwong equation of state is given by

$$P = \frac{RT}{(V-b)} - \frac{a}{V(V+b)\sqrt{T}} \tag{2-7}$$

where

$$a = 0.42747\left(\frac{R^2 T_c^{5/2}}{P_c}\right) \tag{2-8}$$

$$b = 0.08664\left(\frac{RT_c}{P_c}\right) \tag{2-9}$$

The variables are defined by

P = pressure in atm

V = molar volume in L/g-mol

T = temperature in K

R = gas constant (R = 0.08206 atm·L/g-mol·K)

T_c = the critical temperature (405.5 K for ammonia)

P_c = the critical pressure (111.3 atm for ammonia)

Repeat Problem 2.1 using the Redlich-Kwong equation of state.

2.3 STOICHIOMETRIC CALCULATIONS FOR BIOLOGICAL REACTIONS

2.3.1   Concepts Demonstrated

Use of elemental balances to calculate the stoichiometric coefficients using respiratory quotient, RQ, in general biological reactions.

2.3.2   Numerical Methods Utilized

Solution of a system of linear equations.

2.3.3   Problem Statement

A simplified biological conversion reaction can be written for a carbohydrate reacting with oxygen and ammonia to form cellular material and only water and carbon dioxide products as

$$\mathrm{CH}_m\mathrm{O}_n + a\mathrm{O}_2 + b\mathrm{NH}_3 \rightarrow c\mathrm{CH}_\alpha\mathrm{O}_\beta\mathrm{N}_\delta + d\mathrm{H}_2\mathrm{O} + e\mathrm{CO}_2 \qquad \textbf{(2-10)}$$

Thus the reactant carbohydrate and the product of cellular material contain only one gram atom of carbon. When complete elemental analyses of the carbohydrate reactant and the cellular product are known, the elemental balances on Equation (2-10) can be written as

Carbon Balance: $1 = c + e$ **(2-11)**

Hydrogen Balance: $m + 3b = c\alpha + 2d$ **(2-12)**

Oxygen Balance: $n + 2a = c\beta + d + 2e$ **(2-13)**

Nitrogen Balance: $b = c\delta$ **(2-14)**

This is a system of four linear equations with five unknowns and may be completely defined adding an additional relationship between the unknowns.

The respiratory quotient, RQ, is defined as

$$\mathrm{RQ} = \frac{e}{a} \qquad \textbf{(2-15)}$$

and this equation can be added to the system defined by Equations (2-11) through (2-14).

(a)   Glucose substrate, $\mathrm{C}_6\mathrm{H}_{12}\mathrm{O}_6$, reacts with oxygen and ammonia to form a bacteria, $\mathrm{CH}_2\mathrm{O}_{0.27}\mathrm{N}_{0.25}$, water, and carbon dioxide with a respiratory quotient of 1.5. What are the stoichiometric coefficients for this reaction when it is written in the form of Equation (2-10)?

(b)   Repeat the calculations for part (a) with a respiratory quotient of 2.0.

(c)   Repeat the calculations of part (a) when benzoic acid substrate, $\mathrm{C}_6\mathrm{H}_5\mathrm{COOH}$, forms the same bacteria under anaerobic conditions where there is no gaseous oxygen present.

2.3.4 Solution (Partial)

(a) It is first necessary to express glucose in the form of Equation (2-10) as $C_1H_2O_1$. Equation (2-15) for the respiratory quotient can be written as

$$1.5a = e \qquad\qquad \textbf{(2-16)}$$

The problem then become a system of linear equations which can be rewritten in the form for the POLYMATH Linear Equation Solver where $m = 2$, $n = 1$, $\alpha = 2$, $\beta = 0.27$, and $\delta = 0.25$ as shown below.

$$
\begin{aligned}
c + e &= 1 \qquad\qquad \textbf{(2-17)}\\
3b - 2c - 2d &= -2\\
2a - 0.27c - d - 2e &= -1\\
b - 0.25c &= 0\\
1.5a - e &= 0
\end{aligned}
$$

The equation set must be entered into the matrix form as required by POLY-MATH as shown in Figure 2–2.

	a	b	c	d	e	beta
Number of linear equations	5					
Matrix of Coefficients and beta vector of constants						
1	0	0	1	0	1	1
2	0	3	-2	-2	0	-2
3	2	0	-0.27	-1	-2	-1
4	0	1	-0.25	0	0	0
5	1.5	0	0	0	-1	0

Figure 2–2 POLYMATH Matrix for Linear Equations of Problem 2.3(a) (File **P2-03A.POL**)

The POLYMATH Linear Equations Solver report summarizes the problem and the solution as shown in Figure 2–3.

POLYMATH Report
Linear Equations

Linear Equations Solution

	Variable	Value
1	a	0.2316476
2	b	0.1631321
3	c	0.6525285
4	d	0.5921697
5	e	0.3474715

The equations
[1] $c + e = 1$
[2] $3 \cdot b - 2 \cdot c - 2 \cdot d = -2$
[3] $2 \cdot a - 0.27 \cdot c - d - 2 \cdot e = -1$
[4] $b - 0.25 \cdot c = 0$
[5] $1.5 \cdot a - e = 0$

Coefficients matrix and beta vector

	a	b	c	d	e	beta
1	0	0	1.	0	1.	1.
2	0	3.	-2.	-2.	0	-2.
3	2.	0	-0.27	-1.	-2.	-1.
4	0	1.	-0.25	0	0	0
5	1.5	0	0	0	-1.	0

General
Number of equations: 5

Figure 2–3 POLYMATH Solution Report for Problem 2.3(a)

 The problem file is found in directory Chapter 2 with the file named **P2-03A.POL**.

2.4 STEADY-STATE MATERIAL BALANCES ON A SEPARATION TRAIN

2.4.1   Concepts Demonstrated

Material balances on a steady-state process with no recycle.

2.4.2   Numerical Methods Utilized

Solution of simultaneous linear equations.

2.4.3   Problem Statement

Paraxylene, styrene, toluene, and benzene are to be separated with the array of
distillation columns shown in Figure 2–4.

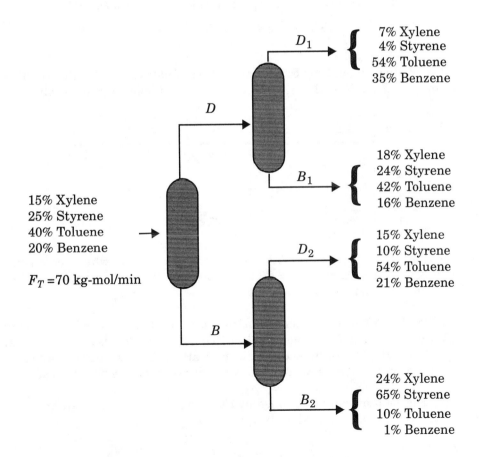

Figure 2–4   Separation Train

(a) Calculate the molar flow rates of D_1, D_2, B_1, and B_2.
(b) Reduce the original feed flow rate to the first column in turn for each one of the components by first 1%, then 2%, and calculate the corresponding flow rates of D_1, B_1, D_2, and B_2. Explain your results.
(c) Determine the molar flow rates and compositions of streams B and D for part (a).

2.4.4   Solution (Partial)

(a) Material balances on individual components yield

$$\text{Xylene: } 0.07D_1 + 0.18B_1 + 0.15D_2 + 0.24B_2 = 0.15 \times 70$$

$$\text{Styrene: } 0.04D_1 + 0.24B_1 + 0.10D_2 + 0.65B_2 = 0.25 \times 70$$

$$\text{Toluene: } 0.54D_1 + 0.42B_1 + 0.54D_2 + 0.10B_2 = 0.40 \times 70$$

$$\text{Benzene: } 0.35D_1 + 0.16B_1 + 0.21D_2 + 0.01B_2 = 0.20 \times 70$$

The coefficients and the constants in these equations can be directly introduced into the POLYMATH *Linear Equation Solver* in matrix form as follows:

Name	x1	x2	x3	x4		b
1	0.07	0.18	0.15	0.24		10.5
2	0.04	0.24	0.1	0.65		17.5
3	0.54	0.42	0.54	0.1		28
4	0.35	0.16	0.21	0.01		14

The solution is x1 = 26.25, x2 = 17.5, x3 = 8.75, and x4 = 17.5, which corresponds to the unknown flow rates of D_1 = 26.25 kg-mol/min, B_1 = 17.5 kg-mol/min, D_2 = 8.75 kg-mol/min, and B_2 = 17.5 kg-mol/min.

www

The POLYMATH problem solution file for part (a) is found in directory Chapter 2 with file named **P2-04A.POL**. This problem is also solved with Excel, Maple, MathCAD, MATLAB, Mathematica, and POLYMATH as Problem 2 in the Set of Ten Problems discussed on the book web site.

(b) The solution can be obtained by changing the vector of constants in the POLYMATH input as required in this problem.

2.5 FITTING POLYNOMIALS AND CORRELATION EQUATIONS TO VAPOR PRESSURE DATA

2.5.1 Concepts Demonstrated

Use of polynomials, the Clapeyron equation, and the Riedel equation to correlate vapor pressure versus temperature data.

2.5.2 Numerical Methods Utilized

Regression of polynomials of various degrees and linear regression of correlation equations with variable transformations.

2.5.3 Problem Statement

Table 2–3 presents data of vapor pressure versus temperature for benzene. Some design calculations require these data to be correlated accurately by algebraic equations.

Table 2–3 Vapor Pressure of Benzene (Ambrose[1])

Temperature, T (K)	Pressure, P (Pa)	Temperature, T (K)	Pressure, P (Pa)
290.08	8634.0	353.47	102040.0
302.39	15388.0	356.19	110850.0
311.19	22484.0	358.87	120140.0
318.69	30464.0	362.29	132780.0
325.1	38953.0	365.23	144530.0
330.54	47571.0	367.90	155800.0
334.89	55511.0	370.53	167600.0
338.94	63815.0	373.15	180060.0
342.95	72985.0	375.84	193530.0
346.24	81275.0	378.52	207940.0
349.91	91346.0	381.32	223440.0

Polynomial Regression Expression

A simple polynomial is often used as an empirical correlation equation. This can be written in general form as

$$P(x) = a_0 + a_1 x + a_2 x^2 + a_3 x^3 + \dots + a_n x^n \tag{2-18}$$

where $a_0 \dots a_n$ are parameters, also called coefficients, to be determined by regression and n is the degree of the polynomial. Typically the degree of the polynomial is selected that gives the best data correlation when using a least-squares objective function.

The Clapeyron equation is given by

$$\log(P) = A + \frac{B}{T}$$
(2-19)

where T is the absolute temperature in K and both A and B are the parameters of the equation that are typically determined by regression.

The Riedel equation (Perry et al.[2]) has the form

$$\log(P) = A + \frac{B}{T} + C\log(T) + DT^{\beta}$$
(2-20)

where T is the absolute temperature in K and A, B, C, and D are parameters determined by regression. β in the above equation is an integer exponent that is typically set to a value of 2.

(a)   Correlate the data with polynomials of different degrees by assuming that the absolute temperature in K is the independent variable and P in Pa is the dependent variable. Determine what degree of polynomial fits the data best.

(b)   Correlate the data using the Clapeyron equation.

(c)   Correlate the data using the Riedel equation.

(d)   Discuss which of the preceding correlations best represents the given data set.

2.5.4   Solution

(a) Data Correlation by a Polynomial   The POLYMATH *Polynomial, Multiple Linear, and Nonlinear Regression Program* can be used to solve this problem. First, the data must be entered and a name should be assigned to each variable (column). Let us denote the column of temperature in K as TK and the column of pressure as P. These columns can be used to obtain the polynomials that represent the data of P (dependent variable) versus data of TK (independent variable). The POLYMATH program simultaneously regresses the dependent variable using first- to fifth-degree polynomials of the form

$$P_{(calc)} = a_0 + a_1TK + a_2TK^2 + a_3TK^3 + a_4TK^4 + a_5TK^5$$
(2-21)

to the dependent variable data and presents the parameter values. The least-squares objective function that is minimized is given by

$$\sum_{i=1}^{N} (P_{(obs)} - P_{(calc)})^2$$
(2-22)

where N is the number of data points and (obs) and (calc) refer to observed and calculated values of the dependent variable P in this case.

Table 2–4 summarizes the results for this problem when the independent variable TK is fitted to dependent variable P.

Table 2–4 Coefficients and Variance of Different Degree Polynomials for Vapor Pressure Data

Degree	1	2	3	4	5
a_0	-7.436E+05	2.591E+06	-4.247E+06	1.406E+06	2.329E+06
a_1	2439.	-1.733E+04	4.40E+04	-2.376E+04	-3.76E+04
a_2		29.13	-153.5	150.2	232.9
a_3			0.1805	-0.4225	-0.6694
a_4				0.0004477	0.0008151
a_5					-2.182E-07
Var.	4.08E+08	9.595E+06	3.616E+04	1748.	1.184E+05

In addition to the coefficients, Table 2–4 presents the value of the variance (σ^2) for each polynomial. The variance is one of the indicators that can help to indicate what degree of the polynomial best represents the data. The variance is mathematically defined as

$$\sigma^2 = \sum_{i=1}^{N} \frac{(P_{(obs)} - P_{(calc)})^2}{\nu} \tag{2-23}$$

where ν represents the degrees of freedom determined by taking the number of data points and subtracting the number of model parameters. For a polynomial, $\nu = N - (n + 1)$, where N is the number of data points and n is the degree of the polynomial.

In this case the fourth-degree polynomial has the smallest variance, and this is one indication that this polynomial fits the data best. The POLYMATH program automatically highlights this polynomial by putting a frame around it. An additional indication for the goodness of the fit is the plot of the calculated curve (using the fourth-degree polynomial) together with the experimental data points, as shown in Figure 2–5, which indicates that there is close agreement between the experimental values (circles) and the calculated values (curve).

When dealing with models (equations) with many parameters (e.g., five parameters in the fourth-degree polynomial), it is important to consider the confidence intervals of the parameter values. While the statistical definition of a confidence interval is outside the scope of this introduction, it generally can be said that it represents the uncertainty associated with a particular parameter. (Problem 3.14 provides a detailed explanation of confidence intervals and their calculation.) The confidence intervals for the parameters of the fourth-degree polynomial can be obtained by requesting the *statistical analysis* option from the POLYMATH program. The parameter values together with the 95% confidence

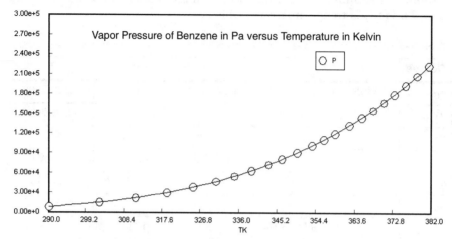

Figure 2–5 Observed Data Points and Calculated Curve for the Fourth-Degree
Polynomial

intervals are shown in Table 2–5.

Table 2–5 Parameter and Confidence Interval Values for the Fourth-Degree Polynomial

Parameter	Value	0.95 Confidence Interval
a_0	1.406E+06	6.342E+05
a_1	−2.376E+04	7594
a_2	150.2	34.00
a_3	−0.4225	0.06749
a_4	0.0004477	5.011E-05

The confidence interval indicates that there is uncertainty with regard to
parameter a_0, and that this value should actually be presented as
$a_0 = (1.406 \pm 0.634) \times 10^6$. The confidence interval is also a function of the precision of the data, the number of the data points, and the agreement between the
model (equation) and the data. A poor fit between the model and the data is often
indicated by confidence intervals that include the value '0' (zero) for one or more
parameters inside the interval. It can be seen in Table 2–5 that none of the
parameter confidence intervals include zero for the fourth-degree polynomial.
The statistical output from the POLYMATH program for the fifth-degree polynomial demonstrates that all parameter confidence intervals include zero, and this
indicates a less satisfactory representation of the data.

An additional important indicator for the fit between the experimental data and the model is the 'residual plot' In such a plot the error in the dependent variable, defined as error = $P_{(obs)} - P_{(calc)}$, is plotted versus $P_{(obs)}$. For a good fit, the error must be randomly distributed with zero mean. The residual plot for the fourth-degree polynomial is shown in Figure 2–6. This plot shows that the error

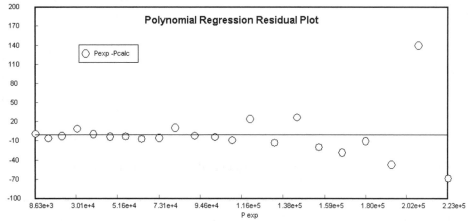

Figure 2–6 Residual Plot for Vapor Pressure Data Represented by a Fourth-Degree Polynomial

is not distributed randomly; rather it increases for higher pressure values. The maximal error is ~10 Pa in the low pressure region and increases to ~140 Pa in the high pressure region. This indicates that even the best polynomial does not represent the vapor pressure data well throughout the entire pressure range.

(b) Clapeyron Equation Data Correlation

Data correlation with the Clapeyron equation can utilize two additional transformed variables (columns) in POLYMATH, which are defined by the relationships $\log P = \log(P)$ and $Trec = 1/TK$. A request for linear regression when the first (and only) independent variable column is $Trec$ and the dependent variable column is $\log P$ yields the following plot and numerical results from POLYMATH, as shown in Figure 2–7. The confidence interval on the parameters A and B is small and the variance is small. However, a detailed examination of Figure 2–7 indicates that the data points should not be represented by a linear relationship. This observation is reinforced by the residual plot shown in Figure 2–8, where the experimental data set exhibits a curvature that is not predicted by the Clapeyron equation.

It should be noted that the variance calculated using dependent variable $\log(P)$ values (as in this case) cannot be compared with the variance calculated using dependent variable P (as calculated for the polynomials). Comparison of variances requires the same form of a variable to be utilized.

(c) Riedel Equation Data Correlation

The Riedel equation correlation requires two additional columns for transformed variables, $\log T = \log(TK)$ and $T2 = TK \times TK$. Multiple linear regression with $Trec$, $\log T$, and $T2$ as the independent variables and $\log P$ as the dependent vari-

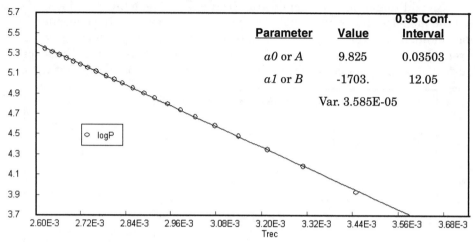

Figure 2–7 Observed Vapor Pressure Data and the Clapeyron Equation
Representation

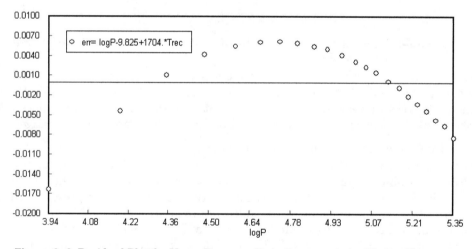

Figure 2–8 Residual Plot for Vapor Pressure Data Represented with the Clapeyron
Equation

able yields the plot and numerical results presented in Figure 2–9. When there are two or more independent variables, as in this case, POLYMATH places the individual data points on the x axis. As Figure 2–9 shows, there is fairly good agreement between the experimental and calculated values of $\log(P)$. The confidence intervals are much wider than for the polynomials or the Clapeyron equation. The error distribution of the residual plot given in Figure 2–10 is more random than for either the polynomial or the Clapeyron equation.

(d) Comparison of Data Correlations An overall comparison of the different models can be made with a variance based on P (instead of $\log P$). This is accomplished by calculating Equation (2-23) for each model. These variances are shown in Table 2–6.

Riedel equation
logP = a0 + a1*Trec + a2*logT + a3*T2

Parameter	Value	0.95 Conf. Interval
a0 or A	40.188	3.491
a1 or B	-2957	123.1
a2 or C	-10.72	1.274
a3 or D	4.03E-06	8.188E-07

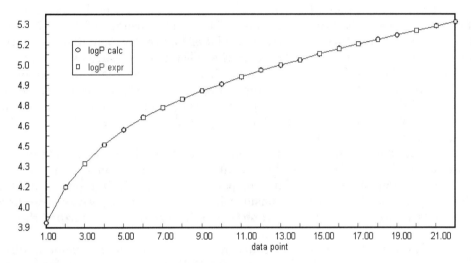

Figure 2–9 Observed Vapor Pressure Data and Values Calculated Using the Riedel Equation

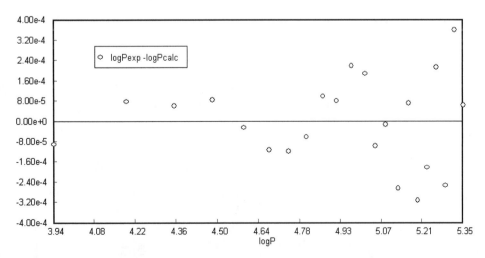

Figure 2–10 Residual Plot for Vapor Pressure Data Represented by Riedel Equation

Table 2–6 Variance Based on P for the Different Correlations

Equation	σ^2
Fourth-degree polynomial	1747.58
Clapeyron	2.34E+06
Riedel	1543.56

It may be concluded that the Clapeyron equation is clearly inappropriate for representing this data set because of the high variance of Table 2–6 and because of the curvature of the residual plot of Figure 2–8. The Riedel equation has the lowest variance when based on the calculated pressure as presented in Table 2–6. The fourth-degree polynomial has the intermediate variance of the three equations with the largest error at higher pressures, as seen in the residual plot of Figure 2–6. The Riedel equation has a residual plot that is more normally distributed, but it is presented in Figure 2–10 in a log(P) scale and thus larger errors will exist at larger pressures.

The Riedel equation has the lowest variance and is recommended for this data correlation. The Riedel model may also be more useful in situations where extrapolation must be made. While the polynomial is useful in representing this data set, it must be used with considerable care as *it is a purely empirical model that should never be used outside of the region of the input data*.

The POLYMATH problem solution file for this problem is found in directory Chapter 2 with file named **P2-05.POL**. A similar problem is also solved with Excel, Maple, MathCAD, MATLAB, Mathematica, and POLYMATH as Problem 3 in the Set of Ten Problems discussed on the book web site.

2.6 VAPOR PRESSURE CORRELATIONS FOR SULFUR COMPOUNDS IN PETROLEUM

2.6.1   Concepts Demonstrated

Use of polynomials, the Clapeyron equation, and the Riedel equation to correlate vapor pressure versus temperature data.

2.6.2   Numerical Methods Utilized

Regression of polynomials of various degrees and linear regression of correlation equations with variable transformations.

2.6.3   Problem Statement

Tables B–1 to B–4 (in Appendix B) provide data of vapor pressure (P in mm Hg) versus temperature (T in °C) for various sulfur compounds present in petroleum. Descriptions of the Clapeyron and Riedel equations are found in Problem 2.5.

(a) Use polynomials of different degrees to represent the vapor pressure data for one of the compounds in Tables B–1 to B–4. Consider $T\,(°K)$ as the independent variable and P as the dependent variable. Determine the degree and the parameters of the best-fitting polynomial for your selected compound.

(b) Correlate the data with the Clapeyron equation.

(c) Correlate the data with the Riedel equation.

The POLYMATH data files for Tables B–1 to B–4 are found in directory Tables with files named **B-01A.POL** to **B-04E.POL**.

2.7 MEAN HEAT CAPACITY OF *n*-PROPANE

2.7.1 Concepts Demonstrated

Calculation of mean heat capacity from heat capacity versus temperature data.

2.7.2 Numerical Methods Utilized

Regression of polynomials of various degrees to data and integration of fitted polynomials between definite limits.

2.7.3 Problem Statement

The mean heat capacity $(\overline{C}_p)$ between two temperatures T_{ref} and T can be calculated from

$$\overline{C}_p = \frac{\displaystyle\int_{T_{ref}}^{T} C_p\, dT}{T - T_{ref}} \qquad (2\text{-}24)$$

Use the data in Table 2–7 to complete the empty boxes in the column for the mean heat capacity of *n*-propane. Use 25 °C (298.15 K) as T_{ref}.

Table 2–7 Heat Capacity of Gaseous Propane (Thermodynamics Research Center,[3] with permission)

No.	Temperature K	Heat Capacity kJ/kg-mol · K	Mean Heat Capacity kJ/kg-mol · K
1	50	34.16	
2	100	41.30	
3	150	48.79	
4	200	56.07	
5	273.15	68.74	
6	300	73.93	
7	400	94.01	
8	500	112.59	
9	600	128.70	
10	700	142.67	
11	800	154.77	

No.	Temperature K	Heat Capacity kJ/kg-mol·K	Mean Heat Capacity kJ/kg-mol·K
12	900	163.35	
13	1000	174.60	
14	1100	182.67	
15	1200	189.74	
16	1300	195.85	
17	1400	201.21	
18	1500	205.89	

2.7.4   Solution (Suggestions)

Approach (1)   The preferred approach for solving this problem is to fit polynomials of various degrees to the C_p versus T data. The best-fitting polynomial is then selected as outlined in Problem 2.5. Once the parameters of the best polynomial have been obtained, the analytical expression for the integral of Equation (2-24) can easily be derived. The expression for the integral can be evaluated by breaking it into several terms so that each term can be entered into a separate column in the data table of the polynomial curve-fitting program. The sum of these separate columns divided by $(T - T_{ref})$ yields the mean heat capacity values. This approach uses the data table of the curve fitting program much like a spreadsheet.

Approach (2)   A cubic spline or polynomial can be employed to evaluate Equation (2-24) by using the option Analysis/Integration menu of the Data Table of the POLYMATH Regression program. This approach requires the calculation of each data point separately, so it is less convenient than the previous approach.

The POLYMATH problem data file is found in directory Chapter 2 with file named **P2-07.POL**.

2.8 VAPOR PRESSURE CORRELATION BY CLAPEYRON AND ANTOINE EQUATIONS

2.8.1   Concepts Demonstrated

Use of the Clapeyron and Antoine equations for vapor pressure correlation and estimation of latent heat of vaporization from the Clapeyron equation.

2.8.2   Numerical Methods Utilized

Linear regressions after proper transformations to a linear expression.

2.8.3   Problem Statement

The Clapeyron equation is commonly used to correlate vapor pressure (P_v) with absolute temperature (T) in °C where ΔH_v is the latent heat of vaporization.

$$\log P_v = -\frac{\Delta H_v}{RT} + B \qquad (2\text{-}25)$$

Another common vapor pressure correlation is the Antoine equation, which utilizes three parameters A, B, and C, with P_v typically in mm Hg and T in °C.

$$\log P_v = A + \frac{B}{T + C} \qquad (2\text{-}26)$$

A particular chemical is to be liquefied and stored in gas cylinders in an outside storage shed. The following data were obtained in the laboratory bomb calorimeter measurements. In this calorimeter, the liquid was slowly heated in a sealed container while the temperature and pressure of Table 2–8 were recorded.

(a) Determine the heat vaporization and the constant B using the Clapeyron equation.

(b) Assuming the yearly low and hot temperatures in the storage shed are 10°F and 120°F, calculate the expected vapor pressures at these temperature extremes.

(c) How do your answers to (b) change when you use the Antoine correlation given by Equation (2-26)?

(d) What do you think about storing the cylinders outside?

Table 2–8 Vapor Pressure Data

T (°C)	17	18	19	21	25	27	28
P (mm Hg)	13.6	22.21	35.54	85.98	413.23	832.62	1160.23

2.8.4   Solution (Partial)

Finding the heat of vaporization and B in Equation (2-25) requires fitting a straight line to the experimental data. This is accomplished by the regression of $\log(P_v)$ versus $1/T$, where T is the absolute temperature. This is explained in more detail in Problem 2.5. After the parameters of the regression have been determined, the values of ΔH_v and B can be calculated.

The Antoine equation must first be linearized. This is accomplished by multiplication of both sides of Equation (2-26) by $T + C$, yielding

$$(T + C)\log P_v = A(T + C) + B \qquad \qquad \textbf{(2-27)}$$

Equation (2-27) can be rearranged:

$$\log P_v = A + (AC + B)/T - C \log P_v/T \qquad \qquad \textbf{(2-28)}$$

Evaluation of the parameters of Equation (2-27) can be accomplished by defining one new dependent and two new independent variables (columns) given by

$$\log P = \log(P_v), \; \text{Trec} = 1/T \; \text{and} \; \text{logPonT} = \log(P_v)/T$$

Linear regression with Trec and logPonT as independent and logP as dependent variables will yield the desired parameters.

The linearization of the Antoine equation, in the form of Equation (2-28), is somewhat problematic, in a statistical sense, since the original dependent variable P_v appears in both sides of the equation. However, this linearization usually yields acceptable results. Nonlinear regression will be used in Problem 3.1 to calculate the Antoine equation parameters, and this is the preferred approach in a statistical sense.

Once the constants of the Clapeyron and Antoine equations have been found, the equations can be used to calculate the vapor pressure at different temperatures.

 The POLYMATH problem data file is found in directory Chapter 2 with file named **P2-08.POL**.

2.9 GAS VOLUME CALCULATIONS USING VARIOUS EQUATIONS OF STATE

2.9.1   Concepts Demonstrated

Gas volume calculations using the ideal gas, van der Waals, Soave-Redlich-Kwong, Peng-Robinson, and Beattie-Bridgeman equations of state.

2.9.2   Numerical Methods Utilized

Solution of a single nonlinear algebraic equation.

2.9.3   Problem Statement

It is proposed to use a steel tank to store carbon dioxide at 300 K. The tank is 2.5 m^3 in volume, and the maximum pressure it can safely withstand is 100 atm.

(a)   Determine the maximum number of moles of CO_2 that can be stored in the tank using the equations of state which are discussed in the text that follows.
(b)   Assuming that the Beattie-Bridgeman equation is the most accurate, what is the percent error in the calculated number of moles using the other correlations?
(c)   Repeat (b) for different values of T_r (T/T_C) and P_r (P/P_C). How do the accuracies of the different correlations change with T_r and P_r?

Ideal Gas

$$PV = RT \qquad \text{(2-29)}$$

where

P = pressure in atm

V = molar volume in L/g-mol

T = temperature in K

R = gas constant (R = 0.08206 L·atm/g-mol·K)

van der Waals equation

See Equations (2-1) through (2-3).

Soave-Redlich-Kwong equation   (see Himmelblau[5] or Felder[6])

$$P = \frac{RT}{V-b} - \left[\frac{\alpha a}{V(V+b)}\right] \qquad \text{(2-30)}$$

where

$$a = 0.42747 \left(\frac{R^2 T_C^2}{P_C} \right)$$

$$b = 0.08664 \left(\frac{RT_C}{P_C} \right)$$

$$\alpha = [1 + m(1 - \sqrt{T/T_C})]^2$$

$$m = 0.48508 + 1.55171\omega - 0.1561\omega^2$$

T_C = the critical temperature (304.2 K for CO_2)

P_C = the critical pressure (72.9 atm for CO_2)

ω = the acentric factor (0.225 for CO_2)

Peng-Robinson[7] equation

$$P = \frac{RT}{V-b} - \left[\frac{a(T)}{V(V+b) + b(V-b)} \right] \qquad \textbf{(2-31)}$$

where

$$b = 0.07780 \frac{RT_C}{P_C}$$

$$a(T) = 0.45724 \frac{R^2 T_C^2}{P_C} \alpha(T)$$

$$\alpha(T) = [1 + k(1 - (T/T_C)^{0.5})]^2$$

$$k = 0.37464 + 1.54226\omega - 0.26992\omega^2$$

Beattie-Bridgeman[4] equation

$$P = \frac{RT}{V} + \frac{\beta}{V^2} + \frac{\gamma}{V^3} + \frac{\delta}{V^4} \qquad \textbf{(2-32)}$$

where

$$\beta = RTB_0 - A_0 - \frac{Rc}{T^2}$$

$$\gamma = RTB_0 b + A_0 a - \frac{RcB_0}{T^2}$$

$$\delta = \frac{RB_0 bc}{T^2}$$

and A_0, a, B_0, b, and c are constants that depend on the particular gas.

For CO_2, A_0 = 5.0065; a = 0.07132; B_0 = 0.10476; b = 0.07235; and c = 66.0×10^4.

2.9.4   Solution (Partial)

One solution to this problem is to find the volume of 1 mole of CO_2 at the specified temperature and pressure for each equation of state and then calculate the moles in the 2.5 m^3 volume of the tank.

The first equation of state (ideal gas) can be solved directly. In order to be consistent with the rest of the equations, this one can be rewritten as an implicit expression

$$f(V) = PV - RT \qquad\qquad (2\text{-}33)$$

Equation (2-33) together with the specified numerical values of P, T, and R can be entered into the POLYMATH *Simultaneous Algebraic Equation Solver* as shown in Table 2–9. This set of equations yields the solution where V = 0.2462 L/g-mol, and the resultant number of moles in the vessel *nmoles* = 10.155 kg-mol.

Table 2–9 Equation Set in the POLYMATH Nonlinear Equation Solver (File **P2-09A.POL**)

Line	Equation
1	f(V)=P*V-R*T
2	P=100
3	R=0.08206
4	T=300
5	nmoles=2.5*1000/V
6	V(min)=0.01
7	V(max)=1

Table 2–10 Equation Set in the POLYMATH Nonlinear Equation Solver (File **P2-09B.POL**)

Line	Equation
1	f(V)=(P+a/(V*V))*(V-b)-R*T
2	P=100
3	R=0.08206
4	T=300
5	nmoles=2.5*1000/V
6	Tc=304.2
7	Pc=72.9
8	a=27*R^2*Tc^2/(Pc*64)
9	b=R*Tc/(8*Pc)
10	V(min)=0.01
11	V(max)=1

The van der Waals equation can be solved similarly as follows with the code shown in Table 2–10. The solution obtained using the van der Waals equation is V = 0.0796 L/g-mol and the number of moles = 31.418 kg-mol. Calculations involving the additional equations of state can be carried out in a similar manner.

The POLYMATH problem solution files for part (a) are found in directory Chapter 2 with files named **P2-09A1.POL** and **P2-09A2.POL**.

2.10 BUBBLE POINT CALCULATION FOR AN IDEAL BINARY MIXTURE

2.10.1 Concepts Demonstrated

Calculation of bubble point in an ideal binary mixture.

2.10.2 Numerical Methods Utilized

Solution of a single nonlinear algebraic equation.

2.10.3 Problem Statement

(a) Calculate the bubble point temperature and equilibrium composition associated with a liquid mixture of 10 mol% n-pentane and 90 mol% n-hexane at 1 atm.

(b) Repeat the calculations for liquid mixtures containing 0 mol% up to 100 mol% of n-pentane.

(c) Plot the bubble point temperature and mol% of n-pentane in the vapor phase as a function of the mol% in the liquid phase.

The vapor pressure of n-pentane, P_A^*, in mm Hg can be calculated from the Antoine equation:

$$\log P_A^* = 6.85221 - \frac{1064.63}{T + 232.0} \tag{2-34}$$

where T is the temperature in °C.

The vapor pressure of n-hexane, P_B^*, can be calculated from the Antoine equation:

$$\log P_B^* = 6.87776 - \frac{1171.53}{224.366 + T} \tag{2-35}$$

2.10.4 Solution

At the bubble point, the sum of the partial vapor pressures of the components must equal the total pressure, which in this case is 1 atm or 760 mm of Hg. Denoting x_A as the mole fraction of n-pentane in the liquid mixture and x_B as the mole fraction of n-hexane, the nonlinear equation to be solved for the bubble point temperature is given by

$$f(T_{bp}) = x_A P_A^* + x_B P_B^* - 760 \tag{2-36}$$

At the solution, $f(T_{bp})$ should become very small $[f(T_{bp}) \approx 0]$.

The vapor phase mole fraction of n-pentane, y_A, and the mole fraction of n-hexane, y_B, can be calculated from Raoult's law given by the equations

$$y_A = x_A P_A^* / 760 \qquad \textbf{(2-37)}$$

and

$$y_B = x_B P_B^* / 760 \qquad \textbf{(2-38)}$$

The solution of this problem involves finding the root of a single nonlinear equation given by Equation (2-36), where P_A^* and P_B^* are calculated from rearranged Equations (2-34) and (2-35). The complete set of equations can be entered into the POLYMATH *Simultaneous Algebraic Equation Solver* as given in Table 2–11.

Table 2–11 Equation Set in the POLYMATH Nonlinear Equation Solver
(File **P4-01A.POL**)

Line	Equation
1	f(Tbp)=xA*PA+xB*PB-760
2	xA=0.1
3	PA=10^(6.85221-1064.63/(Tbp+232))
4	PB=10^(6.87776-1171.53/(224.366+Tbp))
5	xB=1-xA
6	yA=xA*PA/760
7	yB=xB*PB/760
8	Tbp(min)=30
9	Tbp(max)=69

A single nonlinear equation requires limits on the unknown between which the solution is to be found. Such limits can usually be found based on the physical nature of the problem. In this case, for example, the normal boiling points of n-pentane (36.07 °C) and n-hexane (68.7 °C) can be the basis for the lower and upper limits. After entering these values, the results are obtained as summarized in Table 2–12 and Figure 2–11.

Table 2–12 Tabulated Results for the Bubble Point Calculation

Solution		
Variable	Value	f(v)
T_{bp}	63.6645	2.075E-09
P_A	1784.05	
P_B	646.217	
x_A	0.1	
x_B	0.9	
y_A	0.234743	
y_B	0.765257	

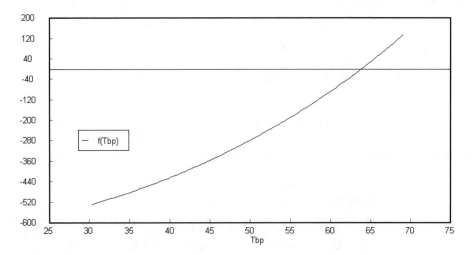

Figure 2–11 Graph of Solution for Bubble Point Temperature

The results indicate that the bubble point temperature is 63.66 °C and at this temperature the vapor is composed of 23.48 mol% of n-pentane and 76.52 mol% of n-hexane.

The calculations can be repeated for different mol%'s of n-pentane by changing x_A in the set of equations and resolving the problem with POLYMATH.

It is important to note that the bubble point can also be considered as the temperature at which the individual mole fractions in the gas phase sum to 1.0 for the given liquid phase composition. Thus this problem can alternately be solved by solving the nonlinear equation given next as an alternate for Equation (2-36).

$$f(T_{bp}) = y_A + y_B - 1 \tag{2-39}$$

The POLYMATH problem solution file for part (a) is found in directory Chapter 2 with file named **P2-10A.POL**.

2.11 DEW POINT CALCULATION FOR AN IDEAL BINARY MIXTURE

2.11.1   Concepts Demonstrated

Calculation of dew point for an ideal binary mixture.

2.11.2   Numerical Methods Utilized

Solution of a single nonlinear algebraic equation.

2.11.3   Problem Statement

(a)   Calculate the dew point temperature and the equilibrium liquid composition of a gas mixture containing 10 mol% n-pentane, 10 mol% n-hex-
ane, and the balance nitrogen (noncondensable) at 1 atm.

(b)   Repeat the calculation for smaller amounts of nitrogen, as indicated in
Table 2–13.

Table 2–13 Dew Point Calculation for Binary Mixture

y_A	y_B	T_{dp}	x_A	x_B
0.1	0.1			
0.2	0.2			
0.3	0.3			
0.4	0.4			
0.5	0.5			

2.11.4   Solution (Partial)

At the dew point, the mole fractions of the components in the liquid phase, x_A for
n-pentane and x_B for n-hexane, sum to 1.0. This can be expressed by

$$f(T_{dp}) = x_A + x_B - 1 \qquad \text{(2-40)}$$

The liquid composition at this temperature can be calculated using Raoult's law:

$$x_A = 760(y_A/P_A^*)$$
$$x_B = 760(y_B/P_B^*) \qquad \text{(2-41)}$$

where y_A and y_B refer to the gas phase mole fractions. The vapor pressures of
n-pentane and n-hexane, P_A^* and P_B^*, can be calculated from Equations (2-34)
and (2-35) as discussed in Problem 2.10.

2.12 BUBBLE POINT AND DEW POINT FOR AN IDEAL MULTICOMPONENT MIXTURE

2.12.1 Concepts Demonstrated

Vapor liquid equilibrium calculations for a multicomponent mixture.

2.12.2 Numerical Methods Utilized

Solution of a single nonlinear algebraic equation.

2.12.3 Problem Statement

A multicomponent mixture, with composition shown in Table 2–14, is being transported in a closed container under high pressure. It has been suggested that transporting the mixture in refrigerated tanks under atmospheric pressure would be less expensive.

(a) Find out whether this suggestion is practical by calculating the bubble and dew point temperatures of the solution at atmospheric pressure.
(b) Repeat these calculations for different pressures.
(c) What is the effect of the pressure on the difference between the bubble and dew point temperatures?

The vapor pressure of the different components can be calculated from the Antoine equation given by Equation (2-26). The Antoine equation constants are given in Table 2–14.

Table 2–14 Liquid Composition and Antoine Constants for the Multicomponent Mixture

	Mole Fraction	A	B	C
Methane	0.1	6.61184	-389.93	266.0
Ethane	0.2	6.80266	-656.4	256.0
Propane	0.3	6.82973	-813.2	248.0
n-Butane	0.2	6.83029	-945.9	240.0
n-Pentane	0.2	6.85221	-1064.63	232.0

2.12.4 Solution

See Problems 2.10 and 2.11 for the methods of solution.

2.13 ADIABATIC FLAME TEMPERATURE IN COMBUSTION

2.13.1 Concepts Demonstrated

Material and energy balances on an adiabatic system and calculation of adiabatic flame temperature.

2.13.2 Numerical Methods Utilized

Solution of a single nonlinear algebraic equation and use of logical variable during solution.

2.13.3 Problem Statement

When natural gas is burned with air, the maximum temperature that can be reached (theoretically) is the adiabatic flame temperature (AFT). This temperature depends mainly on the composition of the natural gas and the amount of air used in the burner. Natural gas consists mainly of methane, ethane, and nitrogen. The composition is different for natural gas found in various locations.

Determine the AFT for the following conditions, and plot the AFT as a function of mol% CH_4 and the stoichiometric molar air to fuel ratio. The composition of natural gas is given in Table 2–15. The air-to-fuel ratios vary between 0.5 to 2.0. It can be assumed that the air and natural gas enter the burner at room temperature and atmospheric pressure. What composition and air-to-fuel ratio leads to the highest AFT?

Table 2–15 Composition of Natural Gas

Compound	mol%
CH_4	65 – 95
C_2H_6	3 – 33
N_2	2

The molar heat capacity of the reactants and the combustion products can be calculated from the equation

$$C_p^* = \alpha + \beta T + \gamma T^2 \tag{2-42}$$

where T is in K and C_p^* is in cal/g-mol·K. The constants of this equation for the different components are shown in Table 2–16 as given by Smith and Van Ness.[8] The heat of combustion is -212798 cal/g-mol for CH_4 and -372820 cal/g-mol for C_2H_6, as reported by Henley.[4] Assume that both the air and the natural gas enter at the temperature of 298 K and that the N_2 content of the natural gas is always 2.0 mol%. Air is 21 mol% O_2.

Table 2–16 Molar Heat Capacity of Gases (Smith and Van Ness[8] with permission.)

	α	$\beta \times 10^3$	$\gamma \times 10^6$
CH_4	3.381	18.044	-4.30
C_2H_6	2.247	38.201	-11.049
CO_2	6.214	10.396	-3.545
H_2O	7.256	2.298	0.283
O_2	6.148	3.102	-0.923
N_2	6.524	1.25	-0.001

2.13.4 Solution

The stoichiometric equations are

$$CH_4 + 2O_2 \rightarrow CO_2 + 2H_2O$$

$$C_2H_6 + 7/2O_2 \rightarrow 2CO_2 + 3H_2O$$

The actual to theoretical molar air-to-fuel ratio can be denoted by x with the inlet mole fractions of CH_4 and C_2H_6 denoted by y and z, respectively. For 1 mol of natural gas, there would be 0.02 mol N_2, y mol CH_4, and z mol C_2H_6. Therefore, the total moles of air required to react completely with the 1 mol of natural gas would be given by $(2y + [7/2]z)/0.21$.

Material balances for the different compounds using a 1 mol natural gas basis are shown in Table 2–17 for both fuel-rich ($x < 1$) and fuel-lean ($x > 1$) situations.

Table 2–17 Material Balance on the Reacting Species

	Moles in the product ($x < 1$)		Moles in the product ($x > 1$)	
	Expression	For $y = 0.75$	Expression	For $y = 0.75$
CH_4	$y(1-x)$	$0.75(1-x)$	0	0
C_2H_6	$z(1-x)$	$0.23(1-x)$	0	0
CO_2	$(y+2z)x$	$1.21x$	$y+2z$	1.21
H_2O	$(2y+3z)x$	$2.19x$	$2y+3z$	2.19
O_2	0	0	$\left(2y+\dfrac{7}{2}z\right)(x-1)$	$2.305(x-1)$
N_2	$0.02+3.76x\left(2y+\dfrac{7}{2}z\right)$	$0.02+8.67x$	$0.02+3.76x\left(2y+\dfrac{7}{2}z\right)$	$0.02+8.67x$

Since both the gas and the air enter at 298 K, this temperature can be used as a reference for enthalpy calculations. The enthalpy change for the product gases from T = 298 K up to the adiabatic flame temperature T_f can be calculated from the following expression:

$$\Delta H_P = \sum_{i=1}^{6} \alpha_i n_i (T_f - 298) + \frac{1}{2} \sum_{i=1}^{6} \beta_i n_i (T_f^2 - 298^2) + \frac{1}{3} \sum_{i=1}^{6} \gamma_i n_i (T_f^3 - 298^3) \quad \textbf{(2-43)}$$

where ΔH_P is the enthalpy change per mole of natural gas fed, and n_i is the number of moles of the different compounds, as shown in Table 2–17.

For x < 1 the general energy balance can be written as

$$f(T_f) = -212798xy - 372820xz + \Delta H_P = 0 \quad \textbf{(2-44)}$$

For $x > 1$ the same equation can be used with the value $x = 1$.

Equations (2-43) and (2-44), together with the data from Tables 2–16 and 2–17, can be entered into the POLYMATH *Simultaneous Algebraic Equation Solver.*

The coding for the example case where y = 0.75 and x = 0.5 is shown in Table 2–18. Note that POLYMATH uses "if ... then ... else ..." statements to provide the logic for the correct value of the molar air-to-fuel ratio.

Table 2–18 Equation Set in the POLYMATH Nonlinear Equation Solver
(File **P2-13.POL**)

Line	Equation
1	f(T)=212798*y*x+372820*z*x+H0-Hf
2	y=0.75
3	x=0.5
4	z=1-y-0.02
5	CH4=if(x<1)then(y*(1-x))else(0)
6	C2H6=if(x<1)then(z*(1-x))else(0)
7	CO2=if(x<1)then((y+2*z)*x)else(y+2*z)
8	H2O=if(x<1)then((2*y+3*z)*x)else(2*y+3*z)
9	N2=0.02+3.76*(2*y+7*z/2)*x
10	alp=3.381*CH4+2.247*C2H6+6.214*CO2+7.256*H2O+6.524*N2
11	bet=18.044*CH4+38.201*C2H6+10.396*CO2+2.298*H2O+1.25*N2
12	gam=-4.3*CH4-11.049*C2H6-3.545*CO2+0.283*H2O-0.001*N2
13	H0 = alp*298+bet*0.001*298*298/2+gam*10^(-6)*298^3/3
14	Hf=alp*T+bet*0.001*T^2/2+gam*1E-6*T^3/3
15	T(min)=1000
16	T(max)=3000

For this case, the flame temperature is calculated to be T = 2198.0 K. Additional adiabatic flame temperature calculations can be made for specified inlet natural gas compositions and air-to-fuel ratios in a similar manner.

 The POLYMATH problem solution file for the example case is found in directory Chapter 2 with file named **P2-13.POL**.

2.14 UNSTEADY-STATE MIXING IN A TANK

2.14.1 Concepts Demonstrated

Unsteady-state material balances.

2.14.2 Numerical Methods Utilized

Solution of simultaneous ordinary differential equations.

2.14.3 Problem Statement

A large tank is used for removing a small amount of settling solid particles (impurities) from brine in a steady-state process. Normally, a single input stream of brine (20% salt by weight) is pumped into the tank at the rate of 10 kg/min and a single output stream is pumped from the tank at the same flow rate. Normal operation keeps the level constant with the total mass in the tank at 1000 kg which is well below the maximum tank capacity.

At a particular time ($t = 0$) an operator accidentally opens a valve, which causes pure water to flow continuously into the tank at the rate of 10 kg/min (in addition to the brine feed), and the level in the tank begins to rise.

Determine the amount of both water and salt in the tank as a function of time during the first hour after the pure water valve has been opened. Assume that the outlet flow rate from the tank does not change and the contents of the tank are well mixed at all times.

2.14.4 Solution

A mass balance on the total mass in the tank yields

Accumulation = Input – Output

$$\frac{dM}{dt} = 10 + 10 - 10 = 10 \qquad \textbf{(2-45)}$$

where M is the mass in kg.

A mass balance on the salt in the tank yields

$$\frac{dS}{dt} = 10(0.2) - 10\left(\frac{S}{M}\right) = 2 - 10\left(\frac{S}{M}\right) \qquad \textbf{(2-46)}$$

where S is the weight of salt in the tank in kg. Note that S/M represents the mass fraction of salt that is leaving the tank at any time t. This is also the mass fraction of salt within the tank since the tank is well mixed. Thus both M and S are functions of time for this problem. At $t = 0$, the initial conditions are that $M = 1000$ kg and $S = 200$ since the brine contains 20% salt by mass.

Entering Equations (2-45) and (2-46) together with the initial values into the POLYMATH *Simultaneous Differential Equation Solver* yields the following equation presented in Table 2–19.

Table 2–19 POLYMATH Equation Set (File **P2-14.POL**)

Line	Equation
1	d(M)/d(t)=10
2	d(S)/d(t)=2-10*S/M
3	SaltPC=100*S/M
4	t(0)=0
5	M(0)=1000
6	S(0)=200
7	t(f)=60

After the problem is solved, the POLYMATH screen display presented in Figure 2–12 summarizes the initial, maximal, minimal, and final values of all the problem variables. This numerical solution indicates that total amount of brine in the tank has increased from the initial 1000 kg to 1600 kg after one hour (60% increase). The amount of salt increased much more moderately to 222.5 kg from the initial value of 200 kg.

The different variables can be plotted as a function of time or other variables. For example, the amount of salt, S, as a function of time, t, is shown in Figure 2–13. The increase in the weight of salt in the tank is interesting and not expected. This can be explained by considering that the added water serves to dilute the salt solution in the tank, and the constant outflow from the tank carries out a smaller amount of salt. Since the input of salt to the tank is constant, the amount of salt in the tank increases. This is perhaps made clearer by calculating the percentage of salt in the tank during this same time period. Note that this is also the percentage of salt in the outlet stream from the tank. Let's define an algebraic equation for this percentage of salt.

$$\text{SaltPC} = 100\frac{S}{M} \tag{2-47}$$

This algebraic equation can be added to the problem, and the resulting graph of the SaltPC variable is given in Figure 2–14. This shows that the mass percentage of salt is reduced from 20% initially to 13.9% after 20 minutes. Thus, the addition of the pure water input dilutes the brine in the tank and reduces the percent of salt in the output stream. Since the output flow rate was unchanged,

Calculated values of DEQ variables

	Variable	Initial value	Minimal value	Maximal value	Final value
1	t	0	0	60.	60.
2	M	1000.	1000.	1600.	1600.
3	S	200.	200.	222.5	222.5

Figure 2–12 Intermediate Results Summary from POLYMATH

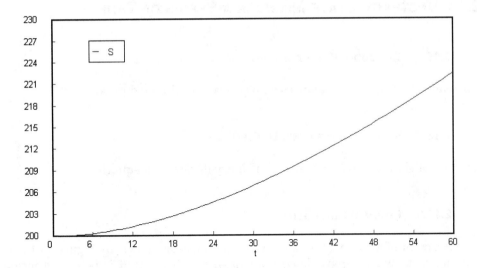

Figure 2–13 Amount of Salt in Tank as a Function of Time

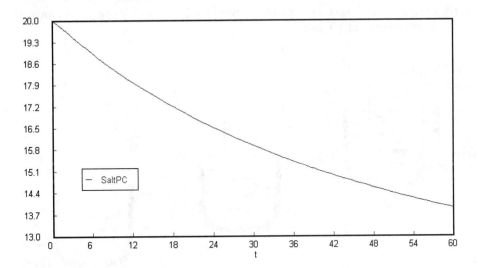

Figure 2–14 Percentage of Salt in the Tank as a Function of Time

the mass of salt in the tank increased.

 The POLYMATH problem solution file is found in directory Chapter 2 with file named **P2-14.POL**.

2.15 UNSTEADY-STATE MIXING IN A SERIES OF TANKS

2.15.1   Concepts Demonstrated

Unsteady-state material balances on a series of well-mixed tanks.

2.15.2   Numerical Methods Utilized

Solution of simultaneous first-order ordinary differential equations.

2.15.3   Problem Statement

A series of three well-mixed settling tanks (as shown in Figure 2–15) is used for settling solid particles (impurities) from brine that is being fed into a process. Under normal steady-state operation, brine (containing 20% salt by mass) is entering and exiting each tank at the flow rate of 10 kg/min. The three tanks contain 1000 kg brine each, with 20% salt by mass. At a particular time ($t = 0$), an additional valve is opened through which pure water flows into the first tank at the rate of 10 kg/min.

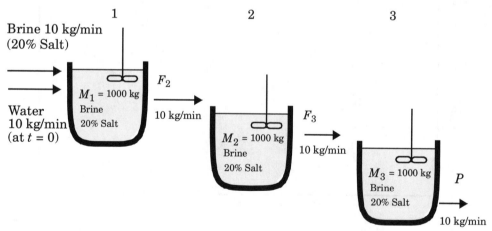

Figure 2–15 A Series of Settling Tanks under Normal Steady-State Operation

(a) Assuming that the rest of the flow rates remain the same and that the contents of the tanks are well mixed, determine and plot the amount and mass percent of the salt in the three tanks during the first hour after the opening of the pure water valve.

(b) What will be the mass percent of the salt in the outlet streams of each of the three tanks after the one-hour period?

2.16 HEAT EXCHANGE IN A SERIES OF TANKS

2.16.1   Concepts Demonstrated

Unsteady-state energy balances and dynamic response of well-mixed heated tanks in series.

2.16.2   Numerical Methods Utilized

Solution of simultaneous first order ordinary differential equations.

2.16.3   Problem Statement

Three tanks in sequence are used to preheat a multicomponent oil solution before it is fed to a distillation column for separation. Each tank is initially filled with 1000 kg of oil at 20°C. Saturated steam at a temperature of 250°C condenses within coils immersed in each tank. The oil is fed into the first tank at the rate of 100 kg/min and overflows into the second and the third tanks at the same flow rate. The temperature of the oil fed to the first tank is 20°C. The tanks are well mixed so that the temperature inside the tanks is uniform, and the outlet stream temperature is the temperature within the tank. The heat capacity, C_p, of the oil is 2.0 kJ/kg·°C. For a particular tank, the rate at which heat is transferred to the oil from the steam coil is given by the expression

$$Q = UA(T_{steam} - T) \tag{2-48}$$

where UA is the product of the heat transfer coefficient and the area of the coil.

UA = 10 kJ/min·°C for each tank

T = temperature of the oil in the tank in °C

Q = rate of heat transferred in kJ/min

(a)   Determine the steady-state temperatures in all three tanks. What time interval will be required for T_3 to reach 99% of this steady-state value during startup?

(b)   After operation at steady state, the oil feed is stopped for three hours. What are the highest temperatures that the oil in each tank will reach during this period?

(c)   After three hours the oil feed is restored. How long will it take to achieve 101% of steady state value for T_3? Will all steady-state temperatures be the same as before in part (a)?

2.16.4   Solution

The sequence of heating tanks is depicted in Figure 2–16.

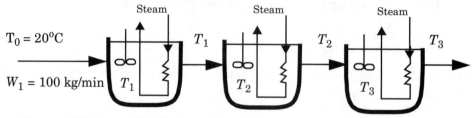

Figure 2–16 Series of Tanks for Oil Heating

Energy balances should be made on each of the individual tanks. In these balances, the mass flow rate to each tank will remain at the same fixed value. Thus $W = W_1 = W_2 = W_3$. The mass in each tank will be assumed constant as the tank volume and oil density are assumed to be constant. Thus $M = M_1 = M_2 = M_3$. For the first tank, the energy balance can be expressed by

$$\text{Accumulation} = \text{Input} - \text{Output}$$

$$MC_p\frac{dT_1}{dt} = WC_pT_0 + UA(T_{\text{steam}} - T_1) - WC_pT_1 \tag{2-49}$$

Note that the unsteady-state mass balance is not needed for tank 1 or any other tanks since the mass in each tank does not change with time. The preceding differential equation can be rearranged and explicitly solved for the derivative, which is the usual format for numerical solution.

$$\frac{dT_1}{dt} = [WC_p(T_0 - T_1) + UA(T_{\text{steam}} - T_1)]/(MC_p) \tag{2-50}$$

Similarly, for the second tank,

$$\frac{dT_2}{dt} = [WC_p(T_1 - T_2) + UA(T_{\text{steam}} - T_2)]/(MC_p) \tag{2-51}$$

For the third tank,

$$\frac{dT_3}{dt} = [WC_p(T_2 - T_3) + UA(T_{\text{steam}} - T_3)]/(MC_p) \tag{2-52}$$

Equations (2-50) to (2-52), together with the numerical data and initial values given in the problem statement, can be entered into the POLYMATH *Simultaneous Differential Equation Solver*.

(a) The initial start-up will be from a temperature of 20°C in all three tanks; thus this is the appropriate initial condition for each tank temperature. The final value or steady-state value can be determined by solving the equations to steady state by giving a large time interval for the numerical solution. Alternately, one could set the time derivatives to zero and solve the resulting algebraic equations. In this case, it is easiest just to solve numerically the differential equations to a large value of t, where steady state is achieved. The POLYMATH coding for this aspect of this problem is given in Table 2–20.

Table 2–20 POLYMATH Equation Set (File **P2-16A.POL**)

Line	Equation
1	d(T1)/d(t)=(W*Cp*(T0-T1)+UA*(Tsteam-T1))/(M*Cp)
2	d(T2)/d(t)=(W*Cp*(T1-T2)+UA*(Tsteam-T2))/(M*Cp)
3	d(T3)/d(t)=(W*Cp*(T2-T3)+UA*(Tsteam-T3))/(M*Cp)
4	W=100
5	Cp=2.0
6	T0=20
7	UA=10.
8	Tsteam=250
9	M=1000
10	t(0)=0
11	T1(0)=20
12	T2(0)=20
13	T3(0)=20
14	t(f)=200

The time to reach steady state is usually considered to be the time to reach 101% of the final steady-state value for the variable that is decreasing and responds most slowly. For this problem, T_3 increases the most slowly, and the steady-state value is found to be 51.317 °C. In POLYMATH, this can be easily done by displaying the output in tabular form for T_1, T_2, and T_3 so that the approach to steady state can accurately be observed. Thus part (a) can be completed by determining the time when T_3 reaches 0.99(51.317) or 50.804. Again, the tabular form of the output is useful in determining this time.

(b) The steady-state temperatures from part (a) can be used as initial conditions to this problem. The flow rate, W, should be set to zero and the problem solved numerically to a time of three hours entered as 180 minutes.

(c) The return to steady state can be simulated by changing the flow rate, W, to its original value and continuing the numerical integration to large values of time, t, to achieve steady state again. The time for T_3 to be reduced to 101% of the steady-state value can be considered as the time to achieve steady state. This can be determined from the numerical solution to steady state.[6]

The POLYMATH problem solution file for part (a) is found in directory Chapter 2 with file named **P2-16A.POL**. This problem is also solved with Excel, Maple, MathCAD, MATLAB, Mathematica, and POLYMATH as Problem 6 in the Set of Ten Problems discussed on the book web site.

REFERENCES

1. Ambrose, D. "Reference Values of the Vapor Pressure of Benzene and Hexafluorobenzene," *J. Chem. Thermo.*, *13*, 1161, (1981).
2. Perry, R. H., Green, D. W., and Malorey, J. D., eds. *Perry's Chemical Engineers Handbook*, 6th ed., New York: McGraw-Hill, 1984.
3. Thermodynamics Research Center API44 Hydrocarbon Project, *Selected Values of Properties of Hydrocarbon and Related Compounds*, Texas A&M University, College Station, TX, 1978. (available from NIST, Gaithersburg, MD)
4. Henley, E. J., and Rosen, E. M. *Material and Energy Balance Computation.* New York: Wiley, 1969.
5. Himmelblau, D. M. *Basic Principles and Calculations in Chemical Engineering*, 6th ed., Englewood Cliffs, NJ: Prentice Hall, 1996.
6. Felder, R. M., and Rousseau, R. W. *Elementary Principles of Chemical Processes*, 2nd ed., New York: Wiley, 1986.
7. Peng, D. Y., and Robinson, D. B. *Ind. Eng. Chem. Fundam.*, *15*, 59 (1976).
8. Smith, J. M., and Van Ness, H. C. *Introduction to Chemical Engineering Thermodynamics*, 2nd ed., New York: McGraw-Hill, 1959.

Regression and Correlation of Data

3.1 ESTIMATION OF ANTOINE EQUATION PARAMETERS USING NONLINEAR REGRESSION

3.1.1 Concepts Demonstrated

Direct use of the Antoine equation to correlate vapor pressure versus temperature data.

3.1.2 Numerical Methods Utilized

Nonlinear regression of a general algebraic expression with determination of the overall variance and confidence intervals of individual parameters.

3.1.3 Excel Options and Functions Demonstrated

The Antoine equation is a widely used vapor pressure correlation that utilizes three parameters, A, B, and C. It is often expressed by

$$P_v = 10^{\left(A + \frac{B}{T+C}\right)} \tag{3-1}$$

where P_v is the vapor pressure in mm Hg and T is the temperature in °C.

Vapor pressure data for propane (Pa) versus temperature (K) is found in Table B–5 in Appendix B. Convert this data set to vapor pressure in mm Hg and temperature in °C. Then

(a) Determine the parameters of the Antoine equation and the corresponding 95% confidence intervals for the parameters from the given data set by using nonlinear regression on Equation (3-1).
(b) Calculate the overall variance for the Antoine equation.
(c) Prepare a residual plot for the Antoine equation.
(d) Assess the precision of the data and the appropriateness of the Antoine equation for correlation of the data set.

3.1.4   Problem Definition

The form of the Antoine equation to be considered in this problem is to be regressed in its nonlinear form. An alternative treatment is to use multiple linear regression on the logarithmic form of this equation, but this transformation is not fully suitable, as discussed in Problem 2.8.

Nonlinear Regression

General nonlinear regression can be used to determine the parameters of an explicit algebraic equation, or model equation, as defined by Equation (3-2):

$$y = f(x_1, x_2, ..., x_n; a_1, ..., a_m) \tag{3-2}$$

where the single dependent variable is y, the n independent variables are $x_1, x_2, ..., x_n$, and the m parameters are $a_1, a_2, ..., a_m$. It is usually assumed that the experimental errors in the preceding equation are normally distributed with constant variance. Nonlinear regression algorithms determine the parameters for a particular model by minimizing the least-squares objective function (LS) given by

$$LS = \sum_{i=1}^{N} \left(y_{i(\text{obs})} - y_{i(\text{calc})} \right)^2 \tag{3-3}$$

where N is the number of data points and (obs) and (calc) refer to observed and calculated values of the dependent variable.

The overall estimate of the model variance, σ^2, is calculated from

$$\sigma^2 = \frac{\sum_{i=1}^{N} \left(y_{i(\text{obs})} - y_{i(\text{calc})} \right)^2}{\nu} = \frac{LS}{\nu} \tag{3-4}$$

where ν is the degrees of freedom, which is equal to the number of data points less the number of model parameters, $(N - m)$. Thus an algorithm that minimizes the least-squares objective function also minimizes the variance. Often the variance is used to compare the goodness of fit for various models.

A widely used graphical presentation that indicates any systematic difficulties with a particular model is called the residuals plot. This is simply the error, ε_i, in the model plotted versus the observed value of the independent variable, $y_{i(\text{obs})}$, with the error given by

$$\varepsilon_i = \left(y_{i(\text{obs})} - y_{i(\text{calc})} \right) \tag{3-5}$$

The POLYMATH *Regression and Data Analysis Program* can be used to solve this problem directly, as the needed options are readily available.

(a) Nonlinear Regression of the Antoine Equation The first step is to enter the data by assigning a name to each variable (column). The column for the vapor pressure in Pa will be designated as P and the column for the temperature in K will be denoted as TK. Since the Antoine equation is to correlate the vapor pressure in mm Hg and the temperature in °C, then a new column denoted as Pv and another new column denoted as TC can be created that calculate the pressure and temperature in units of mm Hg and °C by

$$Pv = .0075002{*}P \qquad\qquad (3\text{-}6)$$

$$TC \ = \ TK - 273.15 \qquad\qquad (3\text{-}7)$$

Once the needed data columns are available within the POLYMATH program, then the solution option for nonlinear regression can be selected.

The general form of the nonlinear equation using the available variables should be entered as

$$Pv \ = \ 10\char`\^(A + B/(TC + C)) \qquad\qquad (3\text{-}8)$$

The initial estimates for the parameters of this equation should be based on good approximations. In this case, the data of Table 2–14 suggest that the values of A = 6, B = -1000, and C = 200 are appropriate. Figure 3–1 shows the results. The 95% confidence intervals as calculated by POLYMATH are given in Table 3–1. (See Problem 3.14 for more on confidence intervals.)

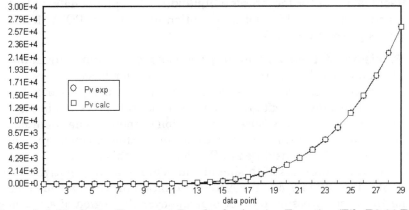

Figure 3–1 Nonlinear Regression Results for Antoine Equation (File **P3-01.POL**)

Table 3–1 95% Confidence Intervals for Parameters of the Antoine Equation for Propane
(File **P3-01.POL**)

Parameter	Initial Guess	Value	95% Confidence
A	6.	7.264138	0.0595187
B	-1000.	-1046.159	39.29494
C	200	281.5827	6.152435

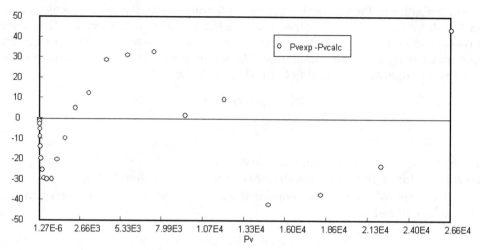

Figure 3–2 Residual Plot of the Antoine Equation for Propane (File **P3-01.POL**)

(b) Overall Variance for the Antoine Equation The overall model variance is 507.9 as calculated by POLYMATH. The variance can also be calculated from Equation (3-4).

(c) Residuals Plot for the Antoine Equation   This plot is defined by Equation (3-5), and creation of this plot is an option available in POLYMATH. The residual plot is given in Figure 3–2.

(d) Precision of Data and Appropriateness of Antoine Equation   The residual plot indicates that there is a slight tendency for larger errors for large values of the vapor pressure, P_v. Certainly the small vapor pressure data has smaller errors. This may indicate that the experimental errors are not independent of the measured vapor pressure. A regression of the Antoine equation using $\log P$ as the dependent variable might be useful to consider as an additional nonlinear regression. (This is considered in Problem 3.2 of this chapter.)

The confidence intervals of this nonlinear regression are relatively narrow, which indicates that this form of the Antoine equation provides an adequate correlation for the vapor pressure data of propane over the region of experimental data. Note that POLYMATH allows the calculated model output to be placed in the data table, allowing a direct numerical comparison of P_{obs} and P_{calc}.

The problem solution file is found in directory Chapter 3 with file name **P3-01.POL**. A similar problem is also solved with Excel, Maple, MathCAD, MATLAB, Mathematica, and POLYMATH as Problem 3 in the Set of Ten Problems discussed on the book web site.

3.2 ANTOINE EQUATION PARAMETERS FOR VARIOUS HYDROCARBONS

3.2.1   Concepts Demonstrated

Direct use of the Antoine equation to correlate vapor pressure versus temperature data.

3.2.2   Numerical Methods Utilized

Nonlinear regression of a general algebraic expression with determination of the overall variance and confidence intervals of individual parameters.

3.2.3   Excel Options and Functions Demonstrated

Table B–6 (Appendix B) presents vapor pressure data for a variety of hydrocarbons as function of temperature. A particular design procedure requires an equation that correlates these data.

(a) Select one of the hydrocarbons in Table B–6. Find the Antoine equation parameters, the corresponding 95% confidence intervals of the parameters, and the overall variance. Prepare the residual plot. The regression should be carried out as discussed in Problem 3.1.

(b) Repeat part (a) with nonlinear regression with the dependent variable as log P.

(c) Repeat part (a) using linear regression as described in Problem 2.6.

(d) Assess the appropriateness of the resulting Antoine equations from (a), (b), and (c) for correlation of the data set. Which equation should be selected?

The problem data files for Table B–6 are found in directory Tables with file names **B-07A.POL** through **B-07F.POL**.

3.3 CORRELATION OF THERMODYNAMIC AND PHYSICAL PROPERTIES OF *n*-PROPANE

3.3.1   Concepts Demonstrated

Correlations for heat capacity, thermal conductivity, viscosity, and latent heat of vaporization.

3.3.2   Numerical Methods Utilized

Linear and nonlinear regression of data with linearization and transformation functions.

3.3.3   Excel Options and Functions Demonstrated

Tables B–7 through B–10 present values for different properties of propane (heat capacity, thermal conductivity for gas, viscosity, and latent heat of vaporization for liquid) as a function of temperature.

> Determine appropriate correlations for the properties of propane listed in Tables B–7 through B–10 using suggested expressions given in the chemical engineering literature.

3.3.4   Problem Definition

A variety of regressions will be used in this problem solution. All of these are available in the POLYMATH *Regression and Data Analysis Program*. Previous problems that illustrate the regressions that will be used in this solution include Problem 2.5 for polynomial fitting, Problems 2.5 and 2.8 for linear regression, and Problem 3.1 for nonlinear regression. It is important to select among available correlations by using key statistical indicators, such as variance, 95% confidence intervals, and residual plots. This solution indicates how this selection can be accomplished.

Theoretical considerations as well as experience have shown which type of correlations are the best to represent temperature dependence of different physical and thermodynamic properties. This information is available in the literature and will be used to develop the various correlations in this problem solution.

(a) Heat Capacity for a Gas   Heat capacity for propane gas is given in Table B–7 for temperatures between 50 K and 1500 K. According to Perry et al.,[1] the heat capacities of gases are most commonly represented as a simple polynomial:

$$C_p = a_0 + a_1 T + a_2 T^2 + a_3 T^3 + \ldots \tag{3-9}$$

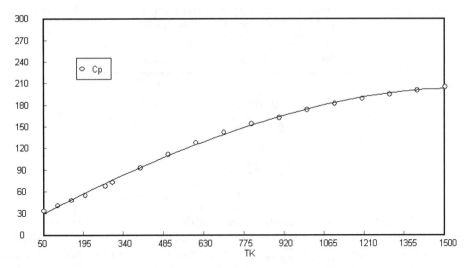

Figure 3–3 Third Degree Polynomial Representation for Heat Capacity of *n*-Propane
(File **P3-03A.POL**)

Entering the data from Table B–7 into the POLYMATH *Regression and Data Analysis Program* and selecting the option to "fit a polynomial" yields the result shown in Figure 3–3 and summarized in Table 3–2.

Table 3–2 Polynomial Coefficients and Variance for *n*-Propane (File **P3-03A.POL**)

Degree	2	3	4	5
a_0	17.74	20.52	26.76	31.05
a_1	0.2178	0.1965	0.1230	0.05151
a_2	−6.164e−05	−2.685e−05	0.000189	0.0005088
a_3		−1.513e−08	−2.357e−07	−7.900e−07
a_4			7.229e−11	4.7874e−10
a_5				−1.06e−13
variance	6.816	6.007	1.867	0.6622

All of the polynomials with the exception of the first degree visually represent the data in a similar way to Figure 3–3. However, the results in Table 3–2 show that the variance of the fifth-degree polynomial is significantly lower than that of the other polynomials, indicating that this correlation represents the data best. Verification that all the parameters of the fifth-degree polynomial are significantly different from zero can be accomplished with the 95% confidence levels for the parameters available from POLYMATH and summarized in Table 3–3. Since all the confidence intervals are smaller in absolute value than the respective parameter values, this indicates that all of the terms of the polynomial are useful in the data correlation.

Table 3–3 Parameter Values and 95% Confidence Intervals for the Fifth-Degree Polynomial Representing n-Propane Heat Capacity Data

Parameter	Value	0.95 Confidence Interval
a_0	31.04	2.749
a_1	0.05151	0.03535
a_2	0.0005088	0.0001431
a_3	$-7.900e{-}07$	$2.378e{-}07$
a_4	$4.787e{-}10$	$1.714e{-}10$
a_5	$-1.06e{-}13$	$4.45e{-}14$

In addition to the examination of the confidence intervals, it is very useful to examine the residuals plot which is given in Figure 3–4 for the fifth-degree polynomial. The error, $C_{p(\text{obs})} - C_{p(\text{calc})}$, is found from the plot to be randomly distributed, indicating that there are no definite trends in the error, which further supports the polynomial correlation of the measured heat capacity data.

It is interesting to compare the residual plot of Figure 3–4 with the residual plot of the third-degree polynomial given in Figure 3–5. In this latter correlation there is a clear oscillatory pattern with much larger errors, which indicates that more parameters are probably needed to represent the data satisfactorily.

(b) Thermal Conductivity Thermal conductivity for gaseous propane is given in Table B–8 for the temperature range 231 K to 600 K. Perry et al.[1] note that over small temperature ranges the thermal conductivity of low-pressure gases can be fairly well correlated by a linear equation, which is also a first-degree polynomial.

The plots of all the polynomials with second and higher degree are very

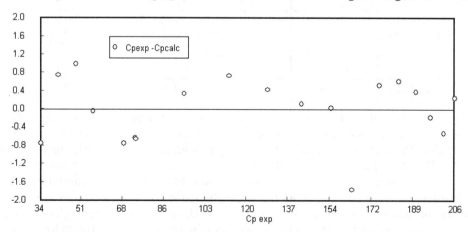

Figure 3–4 Residual Plot for Heat Capacity Represented by Fifth-Degree Polynomial (File **P3-03A.POL**)

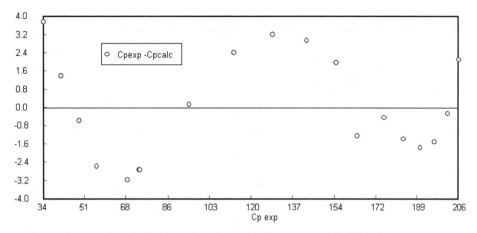

Figure 3–5 Residual Plot for Heat Capacity Represented by Third-Degree
Polynomial (File **P3-03A.POL**)

similar in their visual representation of the thermal conductivity data. A critical
analysis requires the results tabulated in Table 3–4, which show the coefficient
values, 95% confidence intervals, and variances for the polynomials of first to
third degree.

Table 3–4 Polynomial Representation of Gaseous Thermal Conductivity Data (File **P3-03B.POL**)

	First Degree		Second Degree		Third Degree	
	Value	95% Confidence Interval	Value	95% Confidence Interval	Value	95% Confidence Interval
a_0	−0.02317	0.002485	−0.0050072	0.001492	0.006234	0.0009192
a_1	0.0001382	5.831E-06	4.260E-05	7.63E-06	−4.753E-05	7.234E-06
a_2			1.16E-07	9.194E-09	3.444E-07	1.814E-08
a_3					−1.841E-10	1.459E-11
σ^2	2.017E-06		5.003E-08		1.161E-09	

This table indicates that there is a substantial decrease in the value of the
variances between the first- and second-degree polynomials. There is also a variance decrease between the second- and third-degree polynomials. None of the
95% confidence intervals of the coefficients include zero. Thus these results
based on variance and confidence intervals indicate that the third-degree polynomial is preferred for correlating these data. Figures 3–6 and 3–7 show the residual plots for the first-degree and the third-degree polynomials, respectively. For
the first-degree polynomial there is a clear trend in the residuals, while for the
third-degree polynomial the error is randomly distributed. Thus the overall conclusion based upon both the numerical and graphical statistical results is that

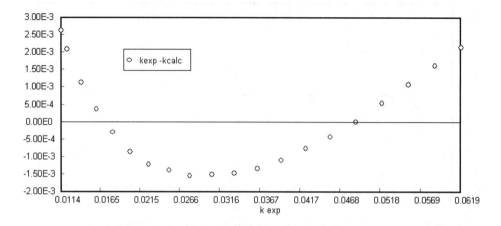

Figure 3–6 Residual Plot for the Thermal Conductivity Represented by a First-degree Polynomial (File **P3-03B.POL**)

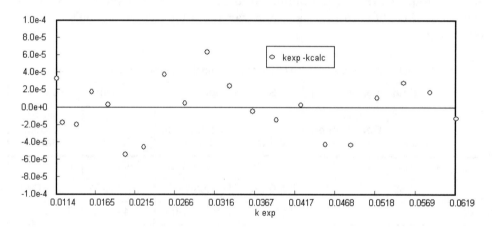

Figure 3–7 Residual Plot for Thermal Conductivity Represented by a Third-Degree Polynomial (File **P3-03B.POL**)

the third-degree polynomial should be used to represent this thermal conductivity data.

For a wide range of temperature, Perry et al.[1] recommend the correlation of thermal conductivity, k, with $(T)^n$, where T is the absolute temperature and $n \approx 1.8$. This correlation can be directly evaluated with nonlinear regression using the form of Equation (3-10), where both c and n are parameters.

$$k = cT^n \qquad\qquad \textbf{(3-10)}$$

The resulting nonlinear correlation result from POLYMATH is given in Figure 3–8.

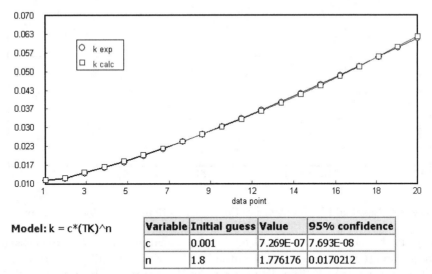

Model: k = c*(TK)^n

Variable	Initial guess	Value	95% confidence
c	0.001	7.269E-07	7.693E-08
n	1.8	1.776176	0.0170212

Figure 3–8 Nonlinear Correlation for Thermal Conductivity of *n*-Propane (File **P3-03B.POL**)

It is interesting that the exponent, n, is indeed approximately 1.8, as suggested by Perry et al.[1] The entire correlation, including 95% confidence intervals, is found to be

$$k = (7.269 \pm 0.769) \times 10^{-7} \times T^{(1.776 \pm 0.017)} \tag{3-11}$$

Figure 3–9 shows the residual plot for Equation (3-11) where the residuals do not appear to be randomly distributed. The data exhibit a certain curvature that cannot be represented by an equation containing only two parameters, such as Equation (3-11).

The third-degree polynomial and the nonlinear equation for representing this thermal conductivity correlation may be compared by considering the overall variance of each correlation. The variance of the third-degree polynomial is 1.161E-09 from Table 3–4 as compared to the variance of the nonlinear equation which is 7.185E-08. The latter variance is calculated by POLYMATH or from Equation (3-4) using the sum of squares from Figure 3–8 and dividing by 8 for the degrees of freedom (10 data points – 2 parameters). Thus, both the variance and the residual plot comparisons indicate that the third-degree polynomial provides a better correlation for this thermal conductivity data set.

(c) Liquid Viscosity The recommended correlation for viscosity of liquids by Perry et al.[1] is similar to the Antoine equation for vapor pressure:

$$\log(\mu) = A + B/(T + C) \tag{3-12}$$

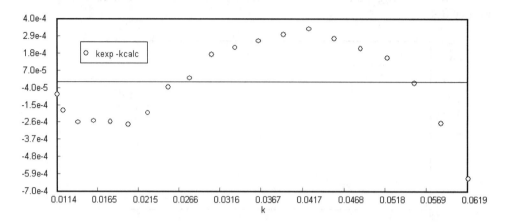

Figure 3–9 Residual Plot for Thermal Conductivity Represented by Nonlinear
Equation (File **P3-03B.POL**)

where μ is the viscosity and A, B and C are parameters. If T is expressed in
Kelvin, parameter C can be approximated by $C = 17.71 - 0.19T_b$, where T_b is the
normal boiling point in Kelvin. For n-propane the normal boiling point is 231 K;
thus the approximate value of C is –26.18.

The nonlinear regression option of POLYMATH can be used directly on
Equation (3-12) for calculating the parameters from the data of Table B–9. Alternatively, Equation (3-12) can be linearized as it was done for vapor pressure in
Problem 2.8 to determine A, B, and C. Since the nonlinear regression requires
good initial estimates of the parameters and estimates are not available for A
and B, a multiple linear regression that requires no initial estimates will be used
to obtain these estimates. A subsequent nonlinear regression will then be performed with these as initial estimates.

Thus, Equation (3-12) can be linearized by using the value of C to be –26.18.
Thus the POLYMATH program needs the variable transformations given by
logmu = log(mu) and invTKplusC = 1/(TK - 26.18). The linearized expression with
these new variables becomes

$$logmu = A + B*invTKplusC \qquad (3\text{-}13)$$

A linear regression with the only independent variable column as invTKplusC
and the dependent variable column as logmu gives A = –4.510 and B = 159.9.
These values and C = –26.18 provide the needed initial estimates for the nonlinear regression of Equation (3-12).

The nonlinear regression results are presented in Figure 3–10, where the
apparent representation by the nonlinear model appears adequate. However, the
residual plot of Figure 3–11 indicates a clear oscillatory trend in the error distribution. This suggests that a three-parameter equation is probably not adequate
for representing viscosity data over such a wide range of temperature.

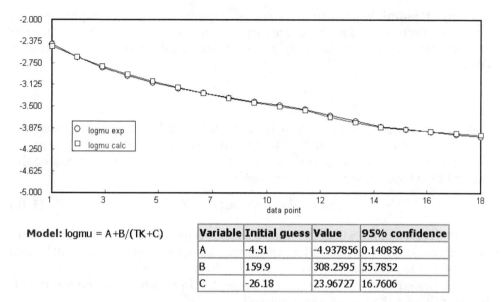

Model: logmu = A +B/(TK+C)

Variable	Initial guess	Value	95% confidence
A	-4.51	-4.937856	0.140836
B	159.9	308.2595	55.7852
C	-26.18	23.96727	16.7606

Figure 3–10 Observed and Calculated Viscosity Data. Antoine Equation Parameters Calculated by Nonlinear Regression.

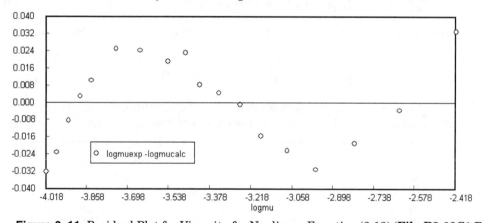

Figure 3–11 Residual Plot for Viscosity for Nonlinear Equation (3-12) (**File P3-03C1.P**

A four-parameter equation used in Reid et al.[2] provides another possible correlation equation for viscosity of liquids:

$$\log(\mu) = A + B/T + C\log T + DT^2 \tag{3-14}$$

The four parameters of Equation (3-14) and the 95% confidence intervals can be determined with multiple linear regression to be $A = -12.678 \pm 1.46$, $B = 369.3 \pm 29.0$, $C = 3.327 \pm 0.594$, and $D = (-9.098 \pm 1.261) \times 10^{-6}$. This correlation gives a normally distributed residual plot indicating a more satisfactory correlation of liquid velocity than Equation (3-12).

(d) Heat of Vaporization Heat of vaporization data for propane are shown in Table B–10 for the temperature range of 85.47 K to 360 K. Heat of vaporization can be correlated by an equation based on the Watson relation (Perry et al.[1]):

$$\Delta H = A(T_C - T)^n \qquad (3\text{-}15)$$

Watson's recommended value for n is 0.38, but n can be found by regression of the experimental data. The critical temperature of propane is 369.83 K.

The equation can be directly used in nonlinear regression, or it can be linearized by taking the log of each side yielding

$$\log(\Delta H) = \log A + n \log(T_C - T) \qquad (3\text{-}16)$$

Both regressions will be conducted so the original data from Table B–10 are entered as TK and deltaH. The transformations logdeltaH = log(deltaH), and logTCmTK = log(369.83 - TK) are used for the linear regression.

The linear regression results are presented in Figure 3–12 and the resulting equation for the heat of vaporization is

$$\Delta H = 2.926 \times 10^6 (369.83 - T)^{0.3765} \qquad (3\text{-}17)$$

The nonlinear regression using the linear regression results as initial estimates is as follows:

$$\Delta H = 2.969 \times 10^6 (369.83 - T)^{0.3736} \qquad (3\text{-}18)$$

Note that both equations are very similar. The slight differences in the parameter values arise because the least-squares objective function in the linear regres-

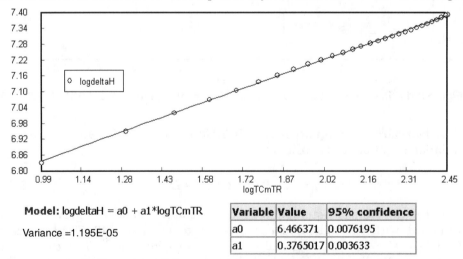

Model: logdeltaH = a0 + a1*logTCmTR		Variable	Value	95% confidence
Variance =1.195E-05		a0	6.466371	0.0076195
		a1	0.3765017	0.003633

Figure 3–12 Heat of Vaporization Data Correlation by Linear Regression (File **P3-03D.POL**)

sion is calculated from $\log(\Delta H)$, whereas the least-squares objective function in the nonlinear regression is calculated from ΔH.

The residual plot from the nonlinear regression presented in Figure 3–13 shows a cyclic trend, indicating that the data are not represented well correlated by this equation. A similar trend is found for the linear regression as well.

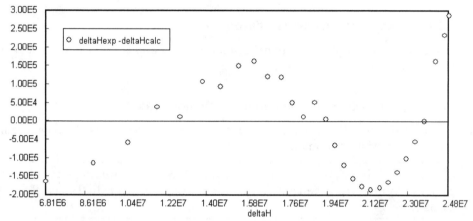

Figure 3–13 Heat of Vaporization Data Correlation by Nonlinear Regression (File **P3-03D.POL**)

A fourth-degree polynomial represents the data more adequately as is summarized in Figure 3-14.

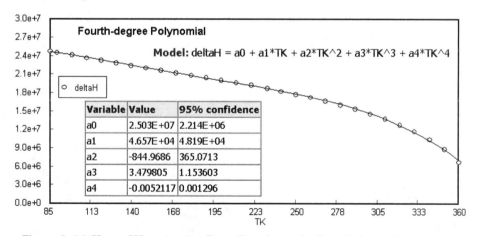

Figure 3–14 Heat of Vaporization Data Correlation by Fourth-degree Polynomial (File **P3-03D.POL**)

The problem solution files are found in directory Chapter 3 with file names **P3-03A.POL**, **P3-03B.POL**, **P3-03C1.POL**, **P3-03C2.POL** and **P3-03D.POL**. The data files for Table B–7 through Table B–10 are found in directory Tables with file names **B-08.POL** through **B-11.POL**.

3.4 TEMPERATURE DEPENDENCY OF SELECTED PROPERTIES

3.4.1   Concepts Demonstrated

Presentation and correlation of thermodynamic and physical property data.

3.4.2   Numerical Methods Utilized

Fitting different types of curves to experimental data and transformation of variables to obtain linear expressions.

3.4.3   Excel Options and Functions Demonstrated

Tables B–11 through B–14 present values of different properties of various compounds as a function of temperature.

> Select one of the compounds and determine the best correlations to represent the temperature dependency of the properties given in Tables B–11 through B–14.

 The problem data files for Table B–11 through Table B–14 are found in directory Tables with file names **B-12A.POL—B-12D.POL, B-13A.POL— B-13G.POL, B-14A.POL—B-14F.POL**, and **B-15A.POL—B-15F.POL**.

www

3.5 HEAT TRANSFER CORRELATIONS FROM DIMENSIONAL ANALYSIS

3.5.1 Concepts Demonstrated

Correlation of heat transfer data using dimensionless groups.

3.5.2 Numerical Methods Utilized

Linear and nonlinear regression of data with linearization and use of transformation functions.

3.5.3 Excel Options and Functions Demonstrated

An important tool in the correlation of engineering data is the use of dimensional analysis. This treatment leads to the determination of the independent dimensionless numbers, which may be important for a particular problem. Linear and nonlinear regression can be very useful in determining the correlations of dimensionless numbers with experimental data.

A treatment of heat transfer within a pipe has been considered by Geankoplis[3] using the Buckingham method, and the result is that the Nusselt number is expected to be a function of the Reynolds and Prandtl numbers.

$$Nu = f(Re, Pr) \quad \text{or} \quad \frac{hD}{k} = f\left(\frac{Dv\rho}{\mu}, \frac{C_p\mu}{k}\right) \tag{3-19}$$

A typical correlation function suggested by Equation (3-19) is

$$Nu = aRe^b Pr^c \tag{3-20}$$

where a, b, and c are parameters that can be determined from experimental data.

A widely used correlation for heat transfer during turbulent flow in pipes is the Sieder-Tate[4] equation:

$$Nu = 0.023 Re^{0.8} Pr^{1/3} (\mu/\mu_w)^{0.14} \tag{3-21}$$

in which a dimensionless viscosity ratio has been added. This ratio (μ/μ_w) is the viscosity at the mean fluid temperature to that at the wall temperature.

Table B–15 gives some of the data reported by Williams and Katz[5] for heat transfer external to 3/4-inch outside diameter tubes where the Re, Pr, (μ/μ_w) and Nu dimensionless numbers have been measured.

> (a) Use multiple linear regression to determine the parameter values of
>   the functional forms of Equations (3-20) and (3-21) that represent the
>   data of Table B–15.
> (b) Repeat part (a) using nonlinear regression.
> (c) Which functional form and parameter values should be recommended
>   as a correlation for this data set? Justify your selection.

3.5.4   Problem Definition

For convenience, let the functional forms of Equations (3-20) and (3-21) be written as

$$Nu = a_i Re^{b_i} Pr^{c_i} \tag{3-22}$$

$$Nu = d_i Re^{e_i} Pr^{f_i} Mu^{g_i} \tag{3-23}$$

where $i = 1$ indicates parameter values from linear regression and $i = 2$ indicates
parameter values from nonlinear regression. *Mu* represents the viscosity ratio.
The POLYMATH *Polynomial, Multiple Linear and Nonlinear Regression Program* can be used to carry out the linear and nonlinear regressions.

(a) A linear regression of either Equation (3-22) or (3-23) requires a transformation into a linear form. This is easily accomplished by taking the ln (natural logarithm) of each side of each equation. The resulting transformed equations
are

$$\ln Nu = \ln a_1 + b_1 \ln Re + c_1 \ln Pr \tag{3-24}$$

$$\ln Nu = \ln d_1 + e_1 \ln Re + f_1 \ln Pr + g_1 \ln Mu \tag{3-25}$$

The data of Table B–15 can be entered directly into POLYMATH under columns defined as Re, Pr, Mu, and Nu. Additional columns can be created that provide the needed ln's for the linear regression by using lnRe = ln(Re), lnPr = ln(Pr),
lnMu = ln(Mu) and lnNu = ln(Nu). The results of the linear regressions are summarized in Table 3–5. Note that the reported values for a_1 and d_1 have been calculated from the regression results that give the ln values of these parameters. The
calculated values from the current linear regression can be automatically
entered into the data sheet by pressing "s" from the Display Option menu, which
gives the regression results.

The linear regression results in Table 3–5 for Equations (3-24) and (3-25)
are consistent when comparing parameters a with d, b with e, and c with f. The
95% confidence intervals are all relatively small except for a, d and g, which have
large intervals that include zero.

Table 3–5 Summary of Parameter Values from Regressions

Parameter	Linear Regression Equations (3-24) and (3-25)		Nonlinear Regression Equations (3-22) and (3-23)	
	Value	95% Confidence Interval	Value	95% Confidence Interval
a	0.6623	3.878	0.1655	0.009780
b	0.5395	0.1160	0.6636	0.005149
c	0.2454	0.1139	0.3414	0.01418
d	0.5347	5.634	0.1491	0.001117
e	0.5588	0.1507	0.6733	0.0007076
f	0.2524	0.1230	0.3286	0.003538
g	−0.06772	0.3164	−0.1778	0.03055

(b) Direct nonlinear regressions of both Equations (3-22) and (3-23) can be completed where the converged values from the linear regressions of part (a) can be used as initial parameter estimates. The results are also summarized in Table 3–5. Again the results are consistent when comparing parameters a with d, b with e, and c with f. The 95% confidence intervals are all relatively small except for g which has a large interval that includes zero.

(c) There are a number of considerations when selecting the most appropriate correlation.

Confidence Intervals A major indicator is the 95% confidence interval of each parameter. When the confidence interval is very large relative to the parameter, this suggests that the parameter may not be important in the correlation and perhaps should be set to zero. This seems to be the case in *both* the linear and nonlinear regressions carried out in parts (a) and (b) for the parameter g, which is the exponent of Mu. This is an indication that Mu should not be included in the correlation. An alternate explanation is that Mu may be dependent upon other variables in the regression. An examination of this will be considered later in this section.

Residual Plots It is always very useful to examine residual plots of regressions to determine if there are any obvious trends, as the errors should be random. A typical residual plot is shown in Figure 3–15 for the nonlinear regression of Equation (3-22). This residual plot demonstrates that the error is randomly distributed. Since there are no unusual error patterns in any of the linear and nonlinear regressions considered here, the residual plots are not helpful in selecting a correlation in this case.

Comparison of Variances A comparison of variances for these correlations deserves special attention because the dependent variable in the linear regression is $\ln(Nu)$ while in the nonlinear regression it is Nu. It is necessary to

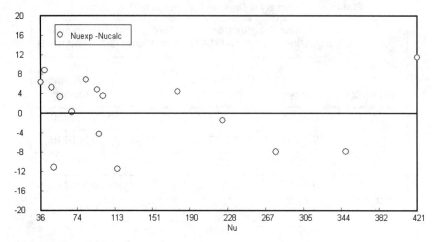

Model: Nu = d2*Re^e2*Pr^f2*Mu^g2

Variable	Initial guess	Value	95% confidence
d2	0.53	0.149055	0.0011173
e2	0.56	0.6733455	0.0007076
f2	0.25	0.3286108	0.0035381
g2	-0.067	-0.177765	0.0305463

Figure 3–15 Residual Plot for Heat Transfer Data Using Nonlinear Regression with Equation (3-22)

use a variance based on the same variable for comparisons. The variance based on the Nusselt number for this problem is defined as

$$\sigma^2 = \frac{1}{\nu} \sum_{i=1}^{N} \left(Nu_{(\text{obs})} - Nu_{(\text{calc})}\right)^2 \tag{3-26}$$

where ν is the degrees of freedom, which is equal to the number of data points less the number of model parameters, $(N - m)$. A relative error variance for this problem can also be defined as

$$\sigma_r^2 = \frac{1}{\nu} \sum_{i=1}^{N} \left(\frac{Nu_{(\text{obs})} - Nu_{(\text{calc})}}{Nu_{(\text{obs})}}\right)^2 \tag{3-27}$$

These variances can be calculated within POLYMATH after the various regressions have been completed by defining columns to evaluate the terms in the summation functions. There is a convenient option within POLYMATH that can sum individual columns. The resulting calculations are summarized in Table 3–6.

An examination of the *variance column* in Table 3–6 indicates that both correlations obtained with nonlinear regression are superior to both correlations

Table 3–6 Calculated Variances for Heat Transfer Correlations

Regression Equation	Variance σ^2	Relative Variance σ_r^2
Linear Regression Equation (3-24)	265.6	0.01028
Linear Regression Equation (3-25)	234.7	0.01135
Nonlinear Regression Equation (3-22)	68.28	0.01684
Nonlinear Regression Equation (3-23)	66.68	0.01477

obtained with linear regression. This is because higher Nu values have a greater influence on the regression. Consideration of the *relative variance column* suggests that both linear regressions result in the lowest relative variance. This is because the $\ln(Nu)$ is used as the dependent variable in the regression, which lessens the effect of data points with larger Nu values. This indicates the general conclusion that *logarithmic transformations are useful if relative errors are to be minimized.*

Thus the selection of the regression equation depends on the experimental errors as to whether they are relative or proportional to the measured Nu values. Since this information is not known about this data set, the selection cannot be made on the variance or relative variance calculations.

Possible Interdependency of Variables The linear and nonlinear regressions assume that the independent variables do not depend on each other. Possible dependency between assumed "independent" variables can be examined by plotting one variable with another. In this case, the large confidence interval on Mu, which includes zero, is an indication that Mu may be related to other variables. A regression of Mu versus $\ln(Re)$ shows definite dependence, as shown in Figure 3–16. Since the viscosity ratio apparently was not changed independently of Re during the experiments, its effect on Nu cannot be isolated. The conclusion with regard to the power of the viscosity ratio in Equation (3-23) is that the data of Katz and Williams are insufficient for determining an exponent for this ratio.

Final Correlation Since the nonlinear regression gives the Nu directly and the variance as calculated by Equation (3-26) is usually employed in regression, the recommended correlation for the data set is given by

$$Nu = (0.1655 \pm 0.00978)Re^{(0.6636 \pm 0.005149)}Pr^{(0.3414 \pm 0.01418)} \tag{3-28}$$

More experiments should be performed to investigate further the Mu ratio and to determine the best variance to use in the regression of this data set.

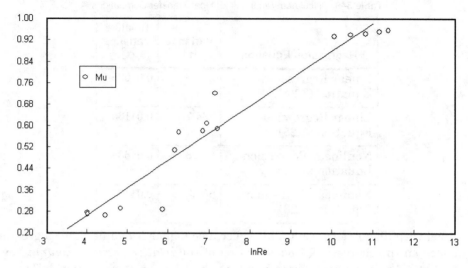

Model: Mu = a0 + a1*lnRe

Variable	Value	95% confidence
a0	-0.1419525	0.116736
a1	0.1024107	0.0150536

Figure 3–16 Plot of the Viscosity Ratio (μ/μ_w) versus ln of Reynolds Number
ln(Re)

The problem solution file is found in directory Chapter 3 with file names
P3-05AB1.POL and **P3-05AB2.POL**.

3.6 HEAT TRANSFER CORRELATION OF LIQUIDS IN TUBES

3.6.1 Concepts Demonstrated

Correlation of heat transfer data using dimensionless groups.

3.6.2 Numerical Methods Utilized

Linear and nonlinear regression of data with linearization and use of transformation functions.

3.6.3 Excel Options and Functions Demonstrated

Tables B–16 through B–18 give data reported by Sieder and Tate[4] for heat transfer in a concentric tube heat exchanger with a 0.75-inch No. 16 BWG inside tube and a 1.25-inch outside iron pipe. The measurements were made for three different types of oil. The dimensionless groups identified in the table as Nu, Re, Pr are the Nusselt, Reynolds and Prandtl numbers, respectively. The column labeled (μ/μ_w) is the ratio of viscosity at the mean fluid temperature to that at the wall temperature.

(a) Determine the most appropriate correlation for the heat transfer data in Tables B–16 through B–18. Evaluate expressions that have the general form of the Sieder-Tate equation[4] written in the form

$$Nu = a_0 Re^{a_1} Pr^{a_2} (\mu/\mu_w)^{a_3}$$

where a_0, a_1, a_2, and a_3 are parameters to be determined from the total data set.

(b) Compare your results with the equation suggested by Sieder and Tate, which is given by

$$Nu = 0.023 Re^{0.8} Pr^{1/3} (\mu/\mu_w)^{0.14}$$

3.6.4 Problem Definition (Comment)

Challenging problems involving nonlinear regression may require increasing the maximum number of iteration for a particular algorithm or restarting the algorithm from the current (unconverged) parameter values.

The problem data files are found in directory Tables and designated **B–16.POL** through **B–18.POL**.

3.7 HEAT TRANSFER IN FLUIDIZED BED REACTOR

3.7.1   Concepts Demonstrated

Correlation of heat transfer data using dimensionless groups.

3.7.2   Numerical Methods Utilized

Linear and nonlinear regression of data with linearization.

3.7.3   Excel Options and Functions Demonstrated

Dow and Jacob[6] proposed the following dimensionless equation as a suitable representation for experimental data dealing with heat transfer between a vertical tube and a fluidized air solid mixture:

$$Nu = a_1 \left(\frac{D_t}{L}\right)^{a_2} \left(\frac{D_t}{D_p}\right)^{a_3} \left(\frac{1-\varepsilon}{\varepsilon} \frac{\rho_s C_s}{\rho_g C_g}\right)^{a_4} \left(\frac{D_t G}{\mu_g}\right)^{a_5} \quad \text{(3-29)}$$

where

$$Nu = \frac{h_m D_t}{k_g} = \text{Nusselt number}$$

$$h_m = \text{heat transfer coefficient}$$

$$D_t = \text{tube diameter}$$

$$D_p = \text{solid-particle diameter}$$

$$L = \text{heated fluidized bed length}$$

$$\varepsilon = \text{void fraction of fluid bed}$$

$$G = \text{gas mass velocity}$$

$$k_g, \rho_g, C_g, \mu_g = \text{properties of gas phase}$$

$$C_s, \rho_s = \text{properties of solid phase}$$

(a)   Use the data in Table B–19 to calculate the parameter values a_1, a_2, ... a_5 for Equation (3-29) using linear regression on $\log(Nu)$ and nonlinear regression on Nu.

(b)   Which regression is the most useful in correlation of the data? Justify your selection with confidence intervals, variances, and residual plots.

(c)   Explain the differences between the two types of regression utilized in this problem.

The problem data file is found in directory Tables and designated B–19.

3.8 CORRELATION OF BINARY ACTIVITY COEFFICIENTS USING MARGULES EQUATIONS

3.8.1   Concepts Demonstrated

Estimation of parameters in the Margules equations for the correlation of binary activity coefficients.

3.8.2   Numerical Methods Utilized

Linear and nonlinear regression, transformation of data for regression; calculation and comparison of confidence intervals, residual plots, and sum of squares.

3.8.3   Excel Options and Functions Demonstrated

The Margules equations for correlation of binary activity coefficients are

$$\gamma_1 = \exp[x_2^2(2B - A) + 2x_2^3(A - B)] \tag{3-30}$$

$$\gamma_2 = \exp[x_1^2(2A - B) + 2x_1^3(B - A)] \tag{3-31}$$

where x_1 and x_2 are mole fractions of components 1 and 2, respectively, and γ_1 and γ_2 are activity coefficients. Parameters A and B are constant for a particular binary mixture.

Equations (3-30) and (3-31) can be combined to give the excess Gibbs energy expression:

$$g = G_E/RT = x_1\ln\gamma_1 + x_2\ln\gamma_2 = x_1x_2(Ax_2 + Bx_1) \tag{3-32}$$

Activity coefficients at various mole fractions are available for the benzene and n-heptane binary system in Table 3–7, from which g in Equation (3-32) can

Table 3–7 Activity Coefficients for the System Benzene(1) and n-Heptane(2)[a]

No.	x_1	γ_1	γ_2
1	0.0464	1.2968	0.9985
2	0.0861	1.2798	0.9998
3	0.2004	1.2358	1.0068
4	0.2792	1.1988	1.0159
5	0.3842	1.1598	1.0359
6	0.4857	1.1196	1.0676
7	0.5824	1.0838	1.1096
8	0.6904	1.0538	1.1664

Table 3–7 (Continued) Activity Coefficients for the System Benzene(1) and *n*-Heptane(2)[a]

No.	x_1	γ_1	γ_2
9	0.7842	1.0311	1.2401
10	0.8972	1.0078	1.4038

[a]Brown, *Australian J. Sci. Res.*, A5, 530 (1952).

be calculated. A multiple linear regression without the free parameter can be used to estimate the parameter values of *A* and *B*. Another method is to sum Equations (3-30) and (3-31) and use nonlinear regression on this sum to determine *A* and *B*.

(a)   Use multiple linear regression on Equation (3-32) with the data of Table 3–7 to determine *A* and *B* in the Margules equations for the benzene and *n*-heptane binary system.

(b)   Estimate *A* and *B* by employing nonlinear regression on a single equation that is the sum of Equations (3-30) and (3-31).

(c)   Compare the results of the regressions in (a) and (b) using parameter confidence intervals, residual plots, and sums of squares of errors (least-squares summations calculated with both activity coefficients).

3.8.4   Problem Definition

(a)   Linear Regression of Excess Gibbs Energy Equation   The data can be entered into the POLYMATH *Polynomial, Multiple Linear and Nonlinear Regression Program* with column headings designated as x1, gamma1 and gamma2. A column to calculate the mole fraction x2 must be defined through the transformation function x2 = 1 - x1. Next the Gibbs energy variable can be calculated by a column defined as g = x1*ln(gamma1) + x2*ln(gamma2). Equation (3-32) can be rearranged to a linear form as

$$g = Ax_1x_2^2 + Bx_1^2x_2 = a_1X_1 + a_2X_2 \tag{3-33}$$

Thus the final two transformation columns needed for multiple linear regression can be defined as X1 = x1*x2^2 and X2 = x1^2*x2.

All that remains is to request the "Linear regression without the free parameter" solution option in the POLYMATH program. Note that this regression option sets the a_0 parameter in the multiple linear relationship to zero, as desired in Equation (3-33). The multiple linear regression with X1 as the first independent column name, X2 as the second independent column name, and *g* as the dependent column name yields the results given in Figure 3–17. The confidence intervals on the parameters *A* and *B* are given in Table 3–8, and the residual plot is reproduced in Figure 3–18.

Table 3–8 Multiple Linear Regression Results for Margules Equation Parameters

Parameter	Value	95% Confidence Interval	Lower Limit	Upper Limit
1 or A	0.2511	0.008277	0.2428	0.2594
2 or B	0.4609	0.008484	0.4524	0.4694

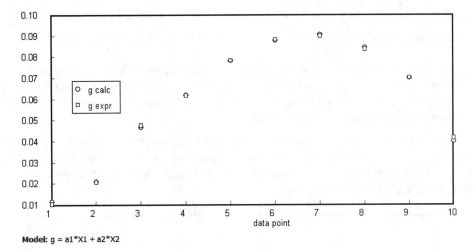

Model: g = a1*X1 + a2*X2

Figure 3–17 Multiple Linear Regression for Parameters of Margules Equations

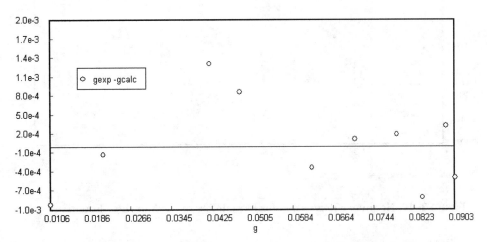

Figure 3–18 Residual Plot for Multiple Linear Regression for Margules Equations

The POLYMATH problem solution file for part (a) is found in directory Chapter 3 with file named **P3-08A.POL**.

(b) Nonlinear Regression for Sum of Equations (3-30) and (3-31)

Let's create a column for the summation equation, $(\gamma_1 + \gamma_2)$, and call it gsum, as it will provide the function values during the nonlinear regression. The function for regression is given as follows in mathematical form and equivalent POLYMATH coding.

$$\text{gsum} = \exp[x_2^2(2B - A) + 2x_2^3(A - B)] + \exp[x_1^2(2A - B) + 2x_1^3(B - A)]$$

gsum = exp(x2^2*(2*B-A)+2*x2^3*(A-B))+exp(x1^2*(2*A-B)+2*x1^3*(B-A))

(3-34)

Note that this nonlinear objective function has approximately equal contributions from both γ_1 and γ_2 since these activity coefficients are close to unity throughout the data set, and thus the nonlinear regression will weigh both activity expressions approximately equally.

A nonlinear regression on the expression for gsum with the initial parameter estimates for A and B from Table 3–8 converges to the results shown in Figure 3–19. The residual plot is reproduced in Figure 3–20, and the confidence intervals are given in Table 3–9.

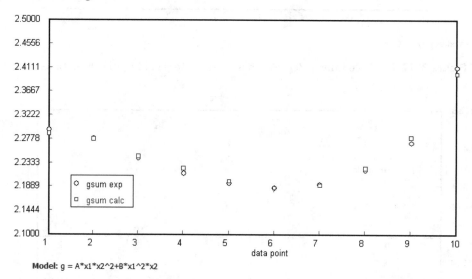

Model: g = A*x1*x2^2+B*x1^2*x2

Figure 3–19 Nonlinear Regression for Parameters of the Margules Equations

Table 3–9 Nonlinear Regression Results for Margules Equation Parameters

Parameter	Value	95% Confidence Interval
A	0.26077	0.012054
B	0.45157	0.015643

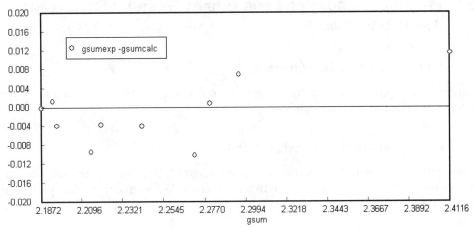

Figure 3–20 Residual Plot for Nonlinear Regression for Margules Equations

www

The POLYMATH problem solution file for part (b) is found in directory Chapter 3 with file name **P3-08B.POL**.

(c) Compare the Results of the Regressions in (a) and (b)

The basic results from the two regressions are quite similar as both correlations reproduce the input data fairly well, have narrow confidence intervals on the various parameters, and show fairly random distribution on the residual plots. A close examination, however, indicates that the multiple linear treatment has both a slightly better fit and lower confidence intervals.

Since the two regression expressions are quite different, the variance or sum of squares as calculated cannot be directly compared. A final comparison will involve an additional evaluation of the *sum of squares* of the errors for both activity coefficients as determined from the same form given by

$$SS = \sum_{i=1}^{N} [(\gamma_{1i(\text{obs})} - \gamma_{1i(\text{calc})})^2 + (\gamma_{2i(\text{obs})} - \gamma_{2i(\text{calc})})^2] \qquad (3\text{-}35)$$

where (obs) refers to the observed data values and (calc) refers to the calculated values from the regression. An evaluation of Equation (3-35) can be set up in POLYMATH by defining additional columns to calculate both gam1calc and gam2calc using the corresponding values of A and B from each regression. A final column definition providing the terms for each data point can be summed to yield SS (for multiple linear regression) = 1.222×10^{-3}, and SS (for nonlinear regression) = 7.993×10^{-4}

Thus both regressions provide highly accurate correlations for activity coefficients. While the nonlinear regression provides somewhat better values for the activity coefficients based on an evaluation of the sum of squares, the multiple linear regression parameter estimates have somewhat smaller confidence intervals. Residual plots for both show no definite trends.

3.9 MARGULES EQUATIONS FOR BINARY SYSTEMS CONTAINING TRICHLOROETHANE

3.9.1   Concepts Demonstrated

Estimation of parameters in the Margules equations for the correlation of binary activity coefficients.

3.9.2   Numerical Methods Utilized

Linear and nonlinear regression, transformation of data for regression, and calculations with comparisons involving confidence intervals, residual plots and sum of squares.

3.9.3   Excel Options and Functions Demonstrated

The binary activity coefficient data for binary systems containing 1,1,1-trichloroethane are summarized in Tables B–20 through B–22.

(a)   Use multiple linear regression of Equation (3-32) to determine A and B in the Margules equations for one of the binary systems containing 1,1,1-trichloroethane.

(b)   Estimate A and B by employing nonlinear regression on a single equation that is the sum of Equations (3-30) and (3-31).

(c)   Compare the results of the regressions in (a) and (b) using parameter confidence intervals, residual plots, and sums of squares of errors (least-squares summations calculated with both activity coefficients).

The data files for Tables B–20 through B–22 are found in directory Tables with file names **B-21.POL** through **B-23.POL**.

3.10 RATE DATA ANALYSIS FOR A CATALYTIC REFORMING REACTION

3.10.1 Concepts Demonstrated

Evaluation of catalytic reaction rate expressions for experimental data.

3.10.2 Numerical Methods Utilized

Linear and nonlinear regression, transformation of data for regression, and calculations with comparisons involving confidence intervals, residual plots and sum of squares.

3.10.3 Excel Options and Functions Demonstrated

Quanch and Rouleau[7] investigated different models for the reversible catalytic reforming reaction

$$CH_4 + 2H_2O \leftrightarrow CO_2 + 4H_2$$

The experimental results of reaction rate as function of partial pressure of the products at 350°C are given in Table 3–10.

Table 3–10 Reaction Rate Data for Catalytic Reforming Reaction

Partial Pressure (atm)				Reaction Rate of CO_2
CH_4	H_2O	CO_2	H_2	(g-mol/hr · gm) $\times 10^3$
0.06298	0.23818	0.00420	0.01669	0.13717
0.03748	0.26315	0.00467	0.01686	0.15584
0.05178	0.29557	0.00542	0.02079	0.20028
0.04978	0.23239	0.00177	0.07865	0.05700
0.04809	0.29491	0.00655	0.02464	0.20150
0.03849	0.24171	0.00184	0.06873	0.07887
0.03886	0.26048	0.00381	0.01480	0.14983
0.05230	0.26286	0.05719	0.01635	0.15988
0.05185	0.33529	0.00718	0.02820	0.26194
0.06432	0.24787	0.00509	0.02055	0.14426
0.09609	0.28457	0.00652	0.02627	0.20195

One of several models for this catalytic reaction, in which methane is adsorbed on the catalyst surface, is given by

$$r_{CO_2} = \frac{k_s K_{CH_4}\left(P_{CH_4}P_{H_2O}^2 - \dfrac{P_{CO_2}P_{H_2}^4}{K_P}\right)}{1 + K_{CH_4}P_{CH_4}} \tag{3-36}$$

where the overall equilibrium constant is known from thermodynamic calculation to be $K_P = 5.051 \times 10^{-5}$ atm^2. Thus there are only two parameters which need to be evaluated for this catalytic rate expression.

Another simpler model is simply for a reversible reaction in which there is no component adsorption on the catalyst. This rate expression is given by

$$r_{CO_2} = k_1\left(P_{CH_4}P_{H_2O}^2 - \frac{P_{CO_2}P_{H_2}^4}{K_P}\right) \tag{3-37}$$

where there is only one parameter, k_1.

Note that the rate, r_{CO_2}, is the positive net generation of CO_2, and that the rate of CH_4 would be the negative of this same rate.

> (a)   Find the values of parameters k_s and K_{CH_4} using nonlinear regression on the data from Table 3–10.
> (b)   Determine the value of parameter k_1.
> (c)   Which of these two rate equations best represents the given data set? Justify your selection.

The problem data file is found in directory Chapter 3 with file name **P3-10.POL**.

3.11 REGRESSION OF RATE DATA–CHECKING DEPENDENCY AMONG VARIABLES

3.11.1   Concepts Demonstrated

Correlation of reaction rate data with various reaction rate models.

3.11.2   Numerical Methods Utilized

Multiple linear regression with determination of parameter confidence intervals, residual plots, and identification of linear dependency among regression variables.

3.11.3   Excel Options and Functions Demonstrated

Table 3–11 presents rate data for the reaction $A \leftrightarrow R$, as reported by Bacon and Downie.[8] They suggested fitting the rate data with two reaction rate models. An irreversible model has the form of a first-order reaction

$$r_R = k_0 C_A \tag{3-38}$$

and a reversible model has the form of reversible first-order reactions

$$r_R = k_1 C_A - k_2 C_R \tag{3-39}$$

where r_R is the rate of generation of component R (gm-mol/dm$^3 \cdot$ s); C_A and C_R are the respective concentrations of components A and R (gm-mol/dm^3); and k_0, k_1, and k_2 are reaction rate coefficients (s^{-1}).

(a) Calculate the parameters of both reaction rate expressions using the data in Table 3–11.
(b) Compare the two models and determine which one better correlates the rate data.
(c) Determine if the two variables, C_A and C_R, are correlated.
(d) Discuss the practical significance of any correlation among the regression variables.

Table 3–11   Reaction Rate Data from Bacon and Downie[8] (with permission)

Run No.	r_R	C_A	C_R
	gm-mol/dm$^3 \cdot$ s $\times 10^8$	gm-mol/dm$^3 \times 10^4$	gm-mol/dm$^3 \times 10^4$
1	1.25	2.00	7.98
2	2.50	4.00	5.95
3	4.05	6.00	4.00

Table 3–11 (Continued) Reaction Rate Data from Bacon and Downie[8] (with permission)

Run No.	r_R	C_A	C_R
	gm-mol/dm³· s × 10⁸	gm-mol/dm³ × 10⁴	gm-mol/dm³ × 10⁴
4	0.75	1.50	8.49
5	2.80	4.00	5.99
6	3.57	5.50	4.50
7	2.86	4.50	5.47
8	3.44	5.00	4.98
9	2.44	4.00	5.99

3.11.4   Problem Definition

(a)   Regression of Rate Expressions   Both rate expressions are in a standard form for linear regression, and POLYMATH *Regression and Data Analysis Program* has an option for regression without the free parameter. *POLYMATH hint*: For convenience, the data values can be entered from the table into columns without the power of 10, and subsequent transformation can be used to scale all the data to their proper values.

(b)   Comparison of Rate Models   The linear regressions of both expressions are summarized in Table 3–12, where the two-parameter model is shown to have a lower variance than the one-parameter model. The residual plots for both equations appear to be randomly distributed. However, the confidence interval on k_2 is large and includes zero and negative values, so the two-parameter model is of questionable value relative to the one-parameter model.

Table 3–12 Multiple Linear Regression Results

Parameter	Value	0.95 Conf. Interval	Variance
k_0	6.551×10^{-5}	2.73×10^{-6}	2.34×10^{-18}
k_1	6.867×10^{-5}	4.51×10^{-6}	1.728×10^{-18}
k_2	-2.630×10^{-6}	3.18×10^{-6}	

(c)   Checking for Correlation of Variables   A simple test for correlation among problem variables is to carry out linear regression to determine if one variable is linearly related to another. In this case, a regression of C_R versus C_A provides the correlation shown in Figure 3–21. This plot clearly shows that the two experimental variables, C_R and C_A, are linearly related.

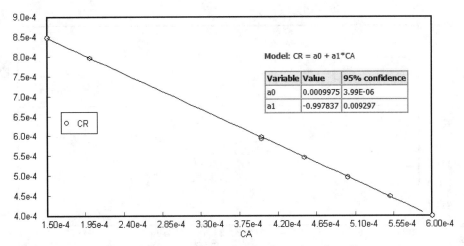

Figure 3–21 Regression of C_R versus C_A for the Reaction $A \leftrightarrow R$

(d) Significance of any Correlation among the Regression Variables

The discovery of this linear relationship means that the previous regression of Equation (3-39) is not valid because it assumes independence of all variables. In a new regression, the known relationship between C_R and C_A can be simplified to Equation (3-40) and then utilized in a new regression.

$$C_R = 0.001 - C_A \tag{3-40}$$

Introducing Equation (3-40) into Equation (3-39) gives

$$r_R = k_1 C_A - k_2(0.001 - C_A) = (-0.001)k_2 + (k_1 + k_2)C_A \tag{3-41}$$

which is a linear relationship. Denoting $a_0 = -0.001 k_2$ and $a_1 = k_1 + k_2$ allows a linear regression to be done, which gives the results in Table 3–13 for the expression

$$r_R = a_0 + a_1 C_A \tag{3-42}$$

The parameter values in the table indicate that a_0 is not significantly different from zero (the confidence interval is larger than the parameter value itself) so it can be removed from the correlation, yielding the irreversible model of Equation (3-38). Table 3–12 indicates that $k_0 = 6.551 \times 10^{-5} \pm 2.74 \times 10^{-6}$. The

Table 3–13 Regression Results for Equation (3-42)

Parameter	Value	0.95 Conf. Interval
a_0	-2.626×10^{-09}	3.167×10^{-09}
a_1	7.13×10^{-05}	7.38×10^{-06}

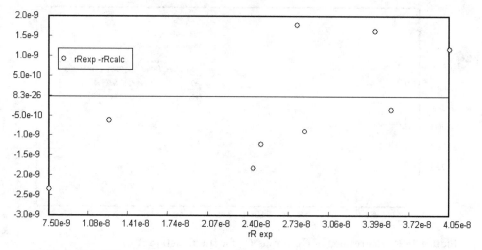

Model: rR = a0 + a1*CA

Variable	Value	95% confidence
a0	-2.626E-09	3.167E-09
a1	7.13E-05	7.38E-06

Figure 3–22 Residual Plot for Reaction Rate Data Represented by Equation (3-38)

residual plot for this irreversible model showing random distribution of the errors is presented in Figure 3–22. All of this seems to support the Bacon and Downie[8] conclusion that the irreversible model is to be preferred.

Actually, no conclusion can be derived from the data and the regression results because of the linear dependency between C_R and C_A. Because of this dependency, no independent information on k_2 is available, and the calculation of k_2 from the relationship $a_0 = -0.001 k_2$ cannot provide a reasonable estimate for k_2. This is supported by the large confidence limits on a_0 as summarized in Table 3–13.

This problem brings out the most important consideration in the design of experiments which is always to change the variables independent of one another. Otherwise, some important information will be lost due to linear or other dependencies between variables. In this example, the concentrations always sum to the same total, which will happen for binary gas mixtures at a given temperature and pressure. Certainly the use of a diluent would have been very appropriate during the experimental measurement of the reaction rates in this case.

 The problem data file is found in directory Chapter 3 with file name **P3-11.POL**.

www

3.12 REGRESSION OF HETEROGENEOUS CATALYTIC RATE DATA

3.12.1 Concepts Demonstrated

Correlation of heterogeneous catalytic reaction rate data with a rate expression.

3.12.2 Numerical Methods Utilized

Multiple linear and nonlinear regression with linearization of expressions and transformation of variables, and identification of possible dependency among regression variables.

3.12.3 Excel Options and Functions Demonstrated

Table 3–14 presents reaction rate data for the heterogeneous catalytic reaction given by $A \rightarrow B$.

Table 3–14 Reaction Rate Data for Heterogeneous Catalytic Reaction

Number	P_A (atm)	P_B(atm)	$r \times 10^3$	Number	P_A (atm)	P_B(atm)	$r \times 10^3$
1	1	0	5.1	5	0.6	0.4	6
2	0.9	0.1	5.4	6	0.5	0.5	6.15
3	0.8	0.2	5.55	7	0.4	0.6	6.3
4	0.7	0.3	5.85	8	0.3	0.7	6.45

The following equation has been suggested to correlate the data:

$$r = \frac{k_1 P_A}{(1 + K_A P_A + K_B P_B)^2} \tag{3-43}$$

where k_1, K_A, and K_B are coefficients to be determined by regression. Note that Equation (3-43) can be linearized by rearranging, inverting, and taking the square root.

$$\left(\frac{P_A}{r}\right)^{1/2} = \frac{1}{\sqrt{k_1}} + \frac{K_A}{\sqrt{k_1}}P_A + \frac{K_B}{\sqrt{k_1}}P_B \tag{3-44}$$

(a) Determine how many parameters of Equations (3-43) and (3-44) should be estimated by regressing the data in Table 3–14.

(b) Calculate the parameters using linear and nonlinear regression and compare the results obtained by these two methods.

The problem data file for Table 3–14 is found in directory Chapter 3 with file name **P3-12.POL**.

3.13 VARIATION OF REACTION RATE CONSTANT WITH TEMPERATURE

3.13.1   Concepts Demonstrated

Correlation of the change in reaction rate constant with temperature using the Arrhenius equation.

3.13.2   Numerical Methods Utilized

Multiple linear and nonlinear regression with comparison of regression results using variances, confidence intervals, and residual plots.

3.13.3   Excel Options and Functions Demonstrated

The catalytic hydrogenation of ethylene over copper magnesia catalyst has been studied in a continuous flow tubular reactor by Wynkoop and Wilhelm,[9] whose experiments were carried out at various temperatures. Some of the data are tabulated in Table B–23 as the reaction rate constant k with units of (g-mol/cm^3·s·atm) versus the temperature T in (°C).

The change of k as a function of temperature can be expressed by the Arrhenius equation

$$k = A \exp[-E/(RT)] \tag{3-45}$$

where typically T is the absolute temperature, R is the gas constant (1.987 cal/g-mol·K), E is the activation energy (typically with units of cal/g-mol), and A is the frequency factor with units of the rate constant.

A convenient alternative form of the Arrhenius expression for the rate constant is

$$k = k_0 \exp[E/R(1/T_0 - 1/T)] \tag{3-46}$$

where T_0 is some arbitrary absolute temperature where $k = k_0$. Note that $k_0 = A \exp(-E/(RT_0))$.

(a)   Use both linear and nonlinear regression to find the Arrhenius parameters in Equations (3-45) and (3-46). Set $T_0 = 298$ K in Equation (3-46).

(b)   Which equation and regression gives the most accurate correlation of the data? Explain your choice.

(c)   Compare your most accurate correlation results with the values reported by Wynkoop and Wilhelm[9] as $A = 5960$ g mol/cm^3·s·atm and $E = 13320$ cal/g-mol.

The POLYMATH data files for Table B–23 is found in directory Tables with file name **B-24.POL**.

3.14 CALCULATION OF ANTOINE EQUATION PARAMETERS USING LINEAR REGRESSION

3.14.1 Concepts Demonstrated

Direct use of the Antoine equation to correlate vapor pressure versus temperature data.

3.14.2 Numerical Methods Utilized

Multiple linear regression with determination of the overall variance and confidence intervals of individual parameters.

3.14.3 Excel Options and Functions Demonstrated

Multiple Linear Regression   Multiple linear regression can be defined as fitting a linear function of the form given in Equation (3-47)

$$y_{(\text{calc})} = a_0 + a_1 x_1 + a_2 x_2 + \dots a_n x_n \tag{3-47}$$

to N observed data points, where $x_1, x_2 \dots x_n$ are n independent variables, $a_0, a_1 \dots a_n$ are $n + 1$ parameters, $y_{(\text{calc})}$ is the estimated value of the dependent variable, and $y_{(\text{obs})}$ is the observed value of the dependent variable. The parameters of Equation (3-47) can be calculated by solving the following system of linear equations:

$$\mathbf{X}^{\mathrm{T}}\mathbf{X}\mathbf{A} = \mathbf{X}^{\mathrm{T}}\mathbf{Y} \tag{3-48}$$

where $\mathbf{X}$ is the matrix of the observed value of the independent variables, $\mathbf{A}$ is the vector of the parameters, and $\mathbf{Y}$ is the vector of observed values of the dependent variable. Thus

$$\mathbf{X} = \begin{bmatrix} 1 & x_{1,1} & x_{2,1} & \cdots & x_{n,1} \\ 1 & x_{1,2} & x_{2,2} & \cdots & x_{n,2} \\ \cdot & \cdot & & \cdot & \cdot \\ \cdot & \cdot & & \cdot & \cdot \\ \cdot & \cdot & & \cdot & \cdot \\ 1 & x_{1,N} & x_{2,N} & \cdots & x_{n,N} \end{bmatrix} \quad \mathbf{A} = \begin{bmatrix} a_0 \\ a_1 \\ \cdot \\ \cdot \\ \cdot \\ a_n \end{bmatrix} \quad \text{and} \quad \mathbf{Y} = \begin{bmatrix} y_{1(\text{obs})} \\ y_{2(\text{obs})} \\ \cdot \\ \cdot \\ \cdot \\ y_{N(\text{obs})} \end{bmatrix} \tag{3-49}$$

Note that the first index in the elements of the matrix $\mathbf{X}$ is the variable number and the second index is the observed data point number. The total number of observed data points is N.

The variance σ^2 can be calculated from Equation (3-4), where $y_{i(\text{obs})}$ is the observed, $y_{i(\text{calc})}$ is the estimated value of the dependent variable, and ν is the degrees of freedom given by $[N - (n + 1)]$.

The exact parameter values of Equation (3-47) denoted by $\beta_0, \beta_1 \ldots \beta_n$ should be located inside the interval

$$a_{i-1} - \sqrt{\alpha_{ii}}\sigma t_v \le \beta_{i-1} < a_{i-1} + \sqrt{\alpha_{ii}}\sigma t_v \qquad \text{(3-50)}$$

where α_{ii} are the diagonal elements of the $(\mathbf{X}^T\mathbf{X})^{-1}$ matrix. t_v is the statistical t distribution value corresponding to the degrees of freedom given by v at the desired percent confidence level and σ is the standard deviation (square root of the variance). The confidence interval is given by the term $\sqrt{\alpha_{ii}}\sigma t_v$.

The error, which is the difference between the observed and calculated values of the dependent variable, can be calculated from Equation (3-5).

Calculate the Antoine equation parameters of Equation (3-1) and the various statistical indicators for the propane vapor pressure data of Table B–5. Report the parameters for the vapor pressure in psia and the temperature in °F. (This problem is similar to Problem 3.1, but the parameters have different units.) The fundamental calculations for linear regression are to be carried out during the solution. The following sequence is to be used:

(a) Transform the data so that the Antoine equation parameters can be calculated using multiple linear regression.
(b) Find the matrices $\mathbf{X}^T\mathbf{X}$ and $\mathbf{X}^T\mathbf{Y}$.
(c) Solve system of equations to obtain the vector $\mathbf{A}$.
(d) Calculate the variance, the diagonal elements of $(\mathbf{X}^T\mathbf{Y})^{-1}$ and the 95% confidence intervals of the parameters (use the t distribution values provided in Table A-4).
(e) Prepare a residual plot (plot of ε_i versus $y_{i(\text{obs})}$).
(f) Assess the precision of the data and the appropriateness of the Antoine equation for correlation of the data.

3.14.4 Problem Definition

Most of the steps of the solution can be carried out using the POLYMATH *Regression and Data Analysis Program*. To start the solution, the data from Table B–5 must be entered into the program.

(a) The Antoine equation can be linearized as shown in Problem 2.8. An alternative linear form of the equation is

$$T \log P_v = (AC + B) + AT - C \log P_v \qquad \text{(3-51)}$$

where log represents the logarithm to the base of 10. This form of the equation will be used in this example. The original data should be entered into POLYMATH and transformed into the variables (columns) shown in Table 3–15 as specified by $y = T \log P_v$, $x1 = T$ and $x2 = \log P_v$.

Table 3–15 Transformed Variables for the Antoine Equation Regression

y	x1	x2
−60.7227	−70	0.867467
−59.26	−60	0.987666
−55.0185	−50	1.10037
−48.3806	−40	1.20952
−39.2249	−30	1.3075
−28.0967	−20	1.40483
−14.9693	−10	1.49693
0	0	1.58206
16.6276	10	1.66276
34.8859	20	1.74429
54.6454	30	1.82151
75.6838	40	1.89209
98.1421	50	1.96284
121.787	60	2.02979
146.54	70	2.09342
172.378	80	2.15473
199.336	90	2.21484
227.184	100	2.27184
256.122	110	2.32838
285.625	120	2.38021

(b) For the case of multiple linear regression with two independent variables x1 and x2 and one dependent variable y, the matrix $\mathbf{X}^T\mathbf{X}$ and the vector $\mathbf{X}^T\mathbf{Y}$ can be written

$$\mathbf{X}^T\mathbf{X} = \begin{bmatrix} N & \sum x_{1,i} & \sum x_{2,i} \\ \sum x_{1,i} & \sum x_{1,i}^2 & \sum x_{1,i}x_{2,i} \\ \sum x_{2,i} & \sum x_{1,i}x_{2,i} & \sum x_{2,i}^2 \end{bmatrix} \quad \mathbf{X}^T\mathbf{Y} = \begin{bmatrix} \sum y_i \\ \sum x_{1,i}y_{i(\text{obs})} \\ \sum x_{2,i}y_{i(\text{obs})} \end{bmatrix} \quad \textbf{(3-52)}$$

The sums can be calculated by summing the numbers in the respective columns. Five new columns should be defined in order to obtain the sums of $x_{1,i}$, $x_{1,i}^2$, $x_{2,i}$, $x_{2,i}^2$, $x_{1,i}y_{i(obs)}$, and $x_{2,i}y_{i(obs)}$. The terms, as they have been entered into the POLYMATH *Linear Equation Solver Program*, are as follows:

$$\begin{bmatrix} 20 & 500 & 34.5131 \\ 500 & 79000 & 1383.28 \\ 34.5131 & 1383.28 & 63.6854 \end{bmatrix} \begin{bmatrix} a_0 \\ a_1 \\ a_2 \end{bmatrix} = \begin{bmatrix} 1383.28 \\ 159281 \\ 3339.67 \end{bmatrix} \tag{3-53}$$

Note that in order to obtain accurate results, the numbers should be entered with at least six significant decimal digits.

(c)   The solution to the general Equation (3-48) as given by Equation (3-53) is obtained using POLYMATH *Linear Equations Solver Program*. The results are

$$a_0 = 677.892 \qquad a_1 = 5.22878 \qquad a_2 = -428.502 \tag{3-54}$$

The inverse $\mathbf{X}^T\mathbf{X}$ matrix can be calculated using the POLYMATH *Linear Equation Solver Program* with the same $\mathbf{X}^T\mathbf{X}$ matrix but changing the vector of constants to $(1, 0, 0)$, $(0, 1, 0)$, and $(0, 0, 1)$ in turn. Thus one diagonal element of the inverse matrix is obtained for each solution. The diagonal elements obtained are $\alpha_{11} = 43.0499$, $\alpha_{22} = 0.00113993$, and $\alpha_{33} = 18.3651$.

(d)   The variance, σ^2, can be calculated by defining a new column given by

$$var = (y - (677.892 + 5.22878 * x1 - 428.502 * x2)) \wedge 2 \tag{3-55}$$

Summation of this column and division of the sum by $\nu = 20 - 3 = 17$ yields $\sigma^2 = 0.349557$. The respective t distribution value, from Table A-4, is $t = 2.1098$ (17 degrees of freedom, 95% confidence interval). Thus the respective confidence intervals as given by $\sqrt{\alpha_{ii}}\sigma t$ in Equation (3-50) are

for $a_0 = (43.0499 * 0.349557)^{1/2} 2.1098 = 8.184$;

for $a_1 = (0.00113993 * 0.349557)^{1/2} 2.1098 = 0.04211$

and for $a_2 = (18.3651 * 0.349557)^{1/2} 2.1098 = 5.3456$

(e)   Each ε_i can be calculated by using Equation (3-55) without the power of 2. A residual plot can be obtained during the linear regression of ε versus y, as shown in Figure 3–23.

(f)   The residual plot shows that the errors are larger for small or negative values of y that correspond to low vapor pressure. To obtain a more accurate correlation, more precise measurements should be carried out in the low-pressure (low-temperature) range.

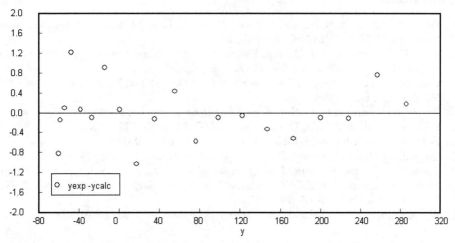

Figure 3–23 Residual Plot for Vapor Pressure Data Represented by Linearized
Antoine Equation

The even distribution of the errors around zero and the narrow confidence intervals on the parameter indicate that the Antoine equation adequately correlates the vapor pressure data for propane in the region where measurements were made.

 The problem data files are found in directory Chapter 3 with file names **P3-14A.POL** and **P3-14B.POL**.

REFERENCES

1. Perry, R. H., Green, D. W., and Malorey, J. D., eds. *Perry's Chemical Engineers Handbook*, 6th ed, New York: McGraw-Hill, 1984.
2. Reid, R. C., Prausnitz, J. M., and Poling, B. F. *The Properties of Gases and Liquids*, 4th ed. New York: McGraw-Hill, 1987.
3. Geankoplis, C. J. *Transport Processes and Unit Operations*, 3rd ed. Englewood Cliffs, NJ: Prentice Hall, 1993.
4. Sieder, E. N., and Tate, G. E. *Ind. and Eng. Chem.*, *28*, 1429 (1936).
5. Williams, R. B., and Katz, D. L. *Trans. ASME*, *74*, 1307–1320 (1952).
6. Dow, W. M., and Jacob, M. *Chem. Eng. Progr.*, *47*, 637 (1951).
7. Quanch, Q. P., and Rouleau, J. *Appl. Chem. Biotechnol.*, *25*, 445 (1975).
8. Bacon, D. W., and Downie, J. *Evaluation of Rate Data–III*, in Crynes, B. L., and Fogler, H. S., eds., *AICHEMI Modular Instruction: Series E, Kinetics,* Vol 2. New York: AICHE, 1981, pp. 65–74.
9. Wynkoop, R., and Wilhelm, R. H. *Chem. Eng. Progr.*, *46*, 300 (1950).

Problem Solving with Excel

4.1 MOLAR VOLUME AND COMPRESSIBILITY FROM REDLICH-KWONG EQUATION

4.1.1 Concepts Demonstrated

Analytical solution of the cubic Redlich-Kwong equation for compressibility factor and calculation of the molar volume at various reduced temperature and pressure values.

4.1.2 Numerical Methods Utilized

Solution of a set of explicit equations.

4.1.3 Excel Options and Functions Demonstrated

Explicit solution involving definition of constants and arithmetic formulas, arithmetic functions, creating series, absolute and relative addressing, if statements and logical functions, two-input data tables and XY (scatter) plots.

4.1.4 Problem Definition

The R-K equation is usually written (Shacham et al.)[1]

$$P = \frac{RT}{V-b} - \frac{a}{V(V+b)\sqrt{T}} \tag{4-1}$$

where

$$a = 0.42747\left(\frac{R^2 T_c^{5/2}}{P_c}\right) \tag{4-2}$$

$$b = 0.08664\left(\frac{RT_c}{P_c}\right) \tag{4-3}$$

and

P = pressure in atm
V = molar volume in liters/g-mol
T = temperature in K
R = gas constant (R = 0.08206 (atm·liter/g-mol·K))
T_c = critical temperature in K
P_c = critical pressure in atm

The compressibility factor is given by

$$z = \frac{PV}{RT} \qquad \text{(4-4)}$$

Equation (4-1) can be written, after considerable algebra, in terms of the compressibility factor as a cubic equation (see Seader and Henley)[2]

$$f(z) = z^3 - z^2 - qz - r = 0 \qquad \text{(4-5)}$$

where

$$r = A^2 B \qquad \text{(4-6)}$$

$$q = B^2 + B - A^2 \qquad \text{(4-7)}$$

$$A^2 = 0.42747 \left(\frac{P_R}{T_R^{5/2}} \right) \qquad \text{(4-8)}$$

$$B = 0.08664 \left(\frac{P_R}{T_R} \right) \qquad \text{(4-9)}$$

in which P_r is the reduced pressure (P/P_c) and T_r is the reduced temperature (T/T_c).

Equation (4-5) can be solved analytically for three roots. Some of these roots are complex. Considering only the real roots, the sequence of calculations involves the steps

$$C = \left(\frac{f}{3} \right)^3 + \left(\frac{g}{2} \right)^2 \qquad \text{(4-10)}$$

where

$$f = \frac{-3q - 1}{3} \qquad \text{(4-11)}$$

$$g = \frac{-27r - 9q - 2}{27} \qquad \text{(4-12)}$$

If $C > 0$ there is one real solution for z given by

$$z = D + E + 1/3 \qquad \text{(4-13)}$$

where

$$D = (-g/2 + \sqrt{C})^{1/3} \qquad (4\text{-}14)$$

$$E = (-g/2 - \sqrt{C})^{1/3} \qquad (4\text{-}15)$$

If C < 0, there are three real solutions

$$z_k = 2\sqrt{\frac{-f}{3}}\cos\left[\left(\frac{\phi}{3}\right) + \frac{2\pi(k-1)}{3}\right] + \frac{1}{3} \quad k = 1, 2, 3 \qquad (4\text{-}16)$$

where

$$\phi = a\cos\sqrt{\frac{g^2/4}{(-f^3)/27}} \qquad (4\text{-}17)$$

In the supercritical region when $T_r \geq 10$, two of these solutions are negative, so the maximal z_k is selected as the true compressibility factor.

(a) Use POLYMATH to calculate the volume of steam (critical temperature is $T_c = 647.4$ K and critical pressure is $P_c = 218.3$ atm) at $T_r = 1.0$ and $P_r = 1.2$. Compare your result with the value obtained from a physical property data base ($V = 0.052456$ L/g-mol). Also complete the calculation for $T_r = 3.0$ and $P_r = 10$ ($V = 0.0837$ L/g-mol). Carry out both calculations only if the parameter $C > 0$.

(b) Calculate the compressibility factor and the molar volume of steam using Excel for the reduced temperatures and reduced pressures listed in Table 4–1. Prepare a table and a plot of the compressibility factor versus P_r and T_r as well as a table and a plot of the molar volume versus pressure and T_r. The pressure and the volume should be in a logarithmic scale in the second plot.

Table 4–1 Reduced Pressures and Temperatures for Calculation

P_r	P_r	P_r	P_r	P_r	T_r
0.1	2	4	6	8	1
0.2	2.2	4.2	6.2	8.2	1.2
0.4	2.4	4.4	6.4	8.4	1.5
0.6	2.6	4.6	6.6	8.6	2.0
0.8	2.8	4.8	6.8	8.8	3.0
1	3	5	7	9	
1.2	3.2	5.2	7.2	9.2	
1.4	3.4	5.4	7.4	9.4	
1.6	3.6	5.6	7.6	9.6	
1.8	3.8	5.8	7.8	9.8	
				10	

4.1.5   Solution

(a) The set of explicit equations that is entered into the POLYMATH Nonlinear Equations Solver program for solution is shown in Table 4–2.

Table 4–2 Equation Set in the POLYMATH Nonlinear Equation Solver (File **P2-01A.POL**)

Line	Equation
1	R = 0.08206 # Gas constant (L-atm/g-mol-K)
2	Tc = 647.4 # Critical temperature (K)
3	Pc = 218.3 # Critical pressure (atm)
4	a = 0.42747 * R ^ 2 * Tc ^ (5 / 2) / Pc # Eq.(4-2), RK equation constant
5	b = 0.08664 * R * Tc / Pc # Eq.(4-3),RK equation constant
6	Pr = 1.2 # Reduced pressure (dimensionless)
7	Tr = 1 # Reduced temperature (dimensionless)
8	r = Asqr * B # Eq.(4-6)
9	q = B ^ 2 + B - Asqr # Eq.(4-7)
10	Asqr = 0.42747 * Pr / (Tr ^ 2.5) # Eq.(4-8)
11	B = 0.08664 * Pr / Tr # Eq.(4-9)
12	C = (f/3) ^ 3 + (g / 2) ^ 2 # Eq.(4-10)
13	f = (-3 * q - 1) / 3 # Eq.(4-11)
14	g = (-27 * r - 9 * q - 2) / 27 # Eq.(4-12)
15	z = If (C > 0) Then (D + E + 1 / 3) Else (0) # Eq.(4-13), Compressibility factor
16	D = If (C > 0) Then ((-g / 2 + sqrt(C)) ^ (1 / 3)) Else (0) # Eq.(4-14)
17	E1 = If (C > 0) Then (-g / 2 - sqrt(C)) Else (0) # Eq.(4-15)
18	E = If (C > 0) Then ((sign(E1) * (abs(E1)) ^ (1 / 3))) Else (0) # Eq.(4-15)
19	P = Pr * Pc # Pressure (atm)
20	T = Tr * Tc # Temperature (K)
21	V = z * R * T / P # Molar volume (L/g-mol)

Note that the row numbers have been added only to help with the explanation; they are not part of the POLYMATH input. Some explanation is included in the POLYMATH input in form of optional comments (text that starts with the "#" sign and ends with the end of the line). In this particular problem the calculations can be carried out sequentially; thus all the equations are entered as explicit equations of the form: x = an expression, where x is a variable name. A variable name must start with an English letter and may contain English letters, numbers, and the underscore sign "_". Note that no special characters, subscripts or superscripts, Greek letters, parentheses, and arithmetic operators (such as +, /, etc.) are allowed.

In expressions, the multiplications sign "*" must be explicitly typed everywhere it is needed. For division, the / (backslash) operator is used; for exponentiation, the "^" operator is used; and for calculating square root, the "sqrt" function is used. POLYMATH supports only the use of the round parentheses "( )". It is important to use enough pairs of parentheses, especially when division is involved, to obtain the correct sequence of calculations.

The equations can be entered into POLYMATH in any order as POLYMATH reorders the equations so that variables are calculated, appearing on the left-hand side of the equal sign, before they appear in an expression on the right-hand side of the equal sign. In the set of equations given in Table 4–2, for example, POLYMATH will first calculate f in line 13 and then g in line 14 before calculating C in line 12.

The calculation of the compressibility factor for the case where C > 0 is carried out by the equations in lines 15-18 in Table 4–2. Calculations with variables E1 and E in lines 17 and 18 deal with possible negative cube roots. Note that the POLYMATH if statement ensures that the variables are calculated if C > 0, otherwise zero value is substituted for them. The syntax of the if statement is:

x = if (condition) then (expression 1) else (expression 2)

The condition may include the following operators: and, or (Boolean operators), >, <, >=, <=, == (equals). The expressions may be any formula, including another "if" statement. Note that Equations (4-16) and (4-17) for calculating the compressibility factor if C < 0 are not included in the equation set of Table 4–2.

The POLYMATH solution obtained for $T_r = 1.0$ and $P_r = 1.2$ is $V = 0.052298$ L/g-mol; this is about 0.2% different from the value from the physical property data base. The compressibility factor at this point is: $z = 0.25788$. At $T_r = 3.0$ and $P_r = 10$ the POLYMATH solutions obtained are $V = 0.083655$ L/g-mol (the same as the given value) and $z = 1.14586$.

(b) The POLYMATH equation set can be exported to Excel by opening an Excel workbook, entering the POLYMATH equation editor window, and pressing the Excel icon or F4 key. The Excel Worksheet that is automatically generated is partially shown in Figure 4–1. Column A indicates the type of the equations (all "Explicit" in this particular case). Column B shows the variable names, as defined in the POLYMATH file while column C gives their numerical values. Column C actually contains the Excel formulas for calculating the variable values,

	A	B	C	D	E	F
1	**POLYMATH NLE Migration Document**					
2		Variable	Value		*Polymath Equation*	Comments
3	Explicit Eqs	R	0.08206		R=0.08206	Gas constant (L-atm/g-
4		Tc	647.4		Tc=647.4	Critical temperature (K)
5		Pc	218.3		Pc=218.3	Critical pressure (atm)
6		a	140.619862		a=0.42747 * R ^ 2 * Tc ^ (5 / 2) / Pc	Eq. (4-2), RK equation c
7		b	0.02108477		b=0.08664 * R * Tc / Pc	Eq. (4-3),RK equation c
8		Pr	1.2		Pr=0.1	Reduced pressure (dim
9		Tr	1		Tr=1	Reduced temperature (c
10		r	0.05333184		r=Asqr * B	Eq. (4-6)
11		q	-0.3981867		q=B ^ 2 + B - Asqr	Eq. (4-7)
12		Asqr	0.512964		Asqr=0.42747 * Pr / (Tr ^ 2.5)	Eq. (4-8)
13		B	0.103968		B=0.08664 * Pr / Tr	Eq. (4-9)
14		C	1.7186E-05		C=(f/3) ^ 3 + (g / 2) ^ 2	Eq. (4-10)
15		f	0.06485332		f=(-3 * q - 1) / 3	Eq. (4-11)
16		g	0.00532297		g=(-27 * r - 9 * q - 2) / 27	Eq. (4-12)
17		z	0.25788001		z=If (C > 0) Then (D + E + 1 / 3) Else (0)	Eq. (4-13), Compressib.
18		D	0.11406621		D=If (C > 0) Then ((-g / 2 + sqrt(C)) ^ (1 / 3)) Else (0)	Eq. (4-14)
19		E1	-0.0068071		E1=If (C > 0) Then (-g / 2 - sqrt(C)) Else (0)	Eq. (4-15)
20		E	-0.1895195		E=If (C > 0) Then ((sign(E1) * (abs(E1)) ^ (1 / 3))) Else (0)	Eq. (4-15)
21		P	261.96		P=Pr * Pc	Pressure (atm)
22		T	647.4		T=Tr * Tc	Temperature (K)
23		V	0.05229822		V=z * R * T / P	Molar volume (L/g-mol)

Figure 4–1 POLYMATH Equation Set Exported to Excel (File **P4-01B1.XLS**)

	A	B	C	D	E	F
2		Variable	Value		*Polymath Equation*	*Comments*
3	Explicit Eqs	R	=0.08206		*R=0.08206*	*Gas constant (L-a*
4		Tc	=647.4		*Tc=647.4*	*Critical temperatu*
5		Pc	=218.3		*Pc=218.3*	*Critical pressure (*
6		a	=(((0.42747 * (C3 ^ 2)) * (C4 ^ (5 / 2))) / C5)		*a=0.42747 * R ^ 2 * Tc ^ (5 / 2) / Pc*	*Eq. (4-2), RK equa*
7		b	=(((0.08664 * C3) * C4) / C5)		*b=0.08664 * R * Tc / Pc*	*Eq. (4-3), RK equa*

Figure 4–2 Some of the Excel Formulas of the Exported Problem (File **P4-01B1.XLS**)

but these formulas can be seen only when pointing on a particular cell or when selecting the "View Formulas" option from the Excel "Tools/Options/View" drop-down menu.

Columns "E" and "F" present the POLYMATH equations and comments (not completely shown) for documentation purposes. It is important to remember that only the Excel formulas, stored in column "C," are used for calculations.

Some of the Excel formulas generated are shown in Figure 4–2. Several points are worth noting regarding these formulas: 1) Only the right-hand side of the equations is included in the Excel formula. The value obtained is assigned to the particular cell where the formula resides (it is not assigned to a particular variable). 2) When the formula contains an expression, it must start with the equal (=) sign. If it contains only a numerical constant (like the value 0.08206), the omission of the equal sign is permitted. 3) The Excel formulas are very similar to the POLYMATH equations except that the variable names are replaced by the addresses of the cells where the particular variables are being calculated. 4) The Excel "If" statement is different from the POLYMATH "If" statement. The calculation of the compressibility factor given by z in cell C17 is carried out, for example, by the formula

=IF((C14 > 0),((C18 + C20) + (1 / 3)),0)

The molar volume and compressibility factor obtained by the Excel formulas for $T_r = 1.0$ and $P_r = 1.2$ are the same as obtained by POLYMATH (see Table 4–2); thus the correctness of the formulas has been verified. Now the calculations can be carried out for all the T_r and P_r values shown in Table 4–1. This is accomplished by the "two-variable data table" tool of Excel.

First the framework of the Excel Table is prepared as shown in Figure 4–3 by entering the desired P_r values listed into separate rows in column G (only

	G	H	I	J	K	L
1			Compressibility Factor (z)			
2		Tr=1	Tr=1.2	Tr=1.5	Tr=2.0	Tr=3.0
3		1	1.2	1.5	2	3
4	0.1					
5	0.2					
6	0.4					
7	0.6					
8	0.8					
0	1					

Figure 4–3 Preparation of a "Two-Variable Data Table" for Calculating Compressibility Factor Values (File **P4-01B1.XLS**)

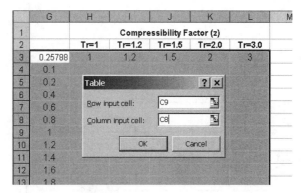

Figure 4–4 Selection of Row and Column Input Cells for the Excel Data Table
(File **P4-01B1.XLS**)

part of the values are shown) and the T_r values are entered into separate columns in the 3rd row. The address of the calculated value of the compressibility factor (C17, see Figure 4–1) could be entered in the upper corner on the left side of the table (cell H3). Since the compressibility factor should calculated only if the variable C > 0, the cell content should be modified to display only meaningful values. This is achieved with an "If" statement in cell G3.

=IF(C17>0,C17,"Irrelevant")

Note that the headings entered in the row 2 are not essential parts of the table, but they are used for "Legend" in the graph to be prepared.

After entering the P_r and T_r values and the address of the target result, the entire area of the table is selected and the Excel Table option from the Data menu is chosen, as shown in Figure 4–4. The address of the parameter T_r (C9, see Figure 4–1) is specified as the Row Input Cell, since the T_r values are entered in a row, and the address of the parameter P_r (C8) is specified as the Column Input Cell.

After clicking on the OK button, the Excel Table is filled with the compressibility factors corresponding to the desired reduced temperatures and reduced pressures. Partial results of the calculations are shown in Figure 4–5.

	G	H	I	J	K	L
1		**Compressibility Factor (z)**				
2		Tr=1	Tr=1.2	Tr=1.5	Tr=2.0	Tr=3.0
3	0.25788	1	1.2	1.5	2	3
4	0.1	0.965162	0.979972	0.990293	0.996817	1.000162
5	0.2	0.928637	0.959637	0.980652	0.993718	1.000356
6	0.4	0.849068	0.918005	0.961605	0.987783	1.000842
7	0.6	0.756568	0.875036	0.942949	0.982211	1.001457
8	0.8	0.638741	0.830724	0.924788	0.97702	1.002201
9	1	0.346664	0.785203	0.907245	0.972226	1.003072

Figure 4–5 Partial Results for Compressibility Factor Calculation for various
P_r and T_r (File **P4-01B1.XLS**)

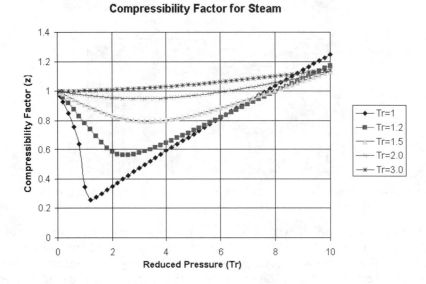

Figure 4–6 Compressibility Factor of Steam versus P_r and T_r. (File **P4-01B1.XLS**)

The generated table in the Excel worksheet can be used for preparing the plot (of type: XY, scatter) of the compressibility factor z (on the Y axis) versus P_r (on the X axis) and T_r (parameter). Figure 4–6 shows the resulting Excel plot.

The molar volume at various P_r and T_r values can be calculated by generating a two-input data table similar to the one shown in Figure 4–5. In this case, the address of the calculated value of the molar volume (C23, see Figure 4–1) is entered in the upper corner on the left side of the table. After the table is generated a new column containing the pressure values is added to the left of the column which contains the P_r values, as shown in Figure 4–7. This table can be used for preparing the plot (of type: XY, scatter) of the molar volume (on the Y axis) versus P (on the X axis) and T_r (parameter). After the plot is prepared, the "Format axis" options for both the X and Y axes have to be used to change the scales to logarithmic. The resultant plot is shown in Figure 4–8.

H3		f_x =IF(C14>0,C23,"Irrelevant")				
	G	H	I	J	K	L	M
2	P	Pr	Tr=1	Tr=1.2	Tr=1.5	Tr=2.0	Tr=3.0
3		0.052298	1	1.2	1.5	2	3
4	21.83	0.1	2.348825	2.861839	3.614977	4.851721	7.302004
5	43.66	0.2	1.129969	1.401228	1.789891	2.418319	3.651712
6	87.32	0.4	0.516574	0.670219	0.877563	1.201937	1.826743
7	130.98	0.6	0.306865	0.425899	0.573692	0.796772	1.218577
8	174.64	0.8	0.194306	0.303248	0.421982	0.594421	0.914611
9	218.3	1	0.084364	0.229305	0.331182	0.473203	0.732325

Figure 4–7 Two-Input Table for Molar Volume (File **P4-01B2.XLS**)

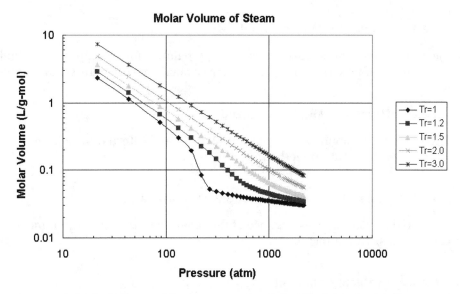

Figure 4–8 Molar Volume of Steam versus P and T_r (File **P4-01B2.XLS**)

The problem solution files are found in directory Chapter 4 and designated **P4-01A.POL**, **P4-01B1.XLS**, and **P4-01B2.XLS.**

4.2 CALCULATION OF THE FLOW RATE IN A PIPELINE

4.2.1   Concepts Demonstrated

Application of the general mechanical energy balance for incompressible fluids, and calculation of flow rate in a pipeline for various pipe diameters and lengths.

4.2.2   Numerical Methods Utilized

Solution of a single nonlinear algebraic equation and alternative solution using the successive substitution method.

4.2.3   Excel Options and Functions Demonstrated

Absolute and relative addressing, use of the "goal seek" tool, programming of the successive substitution technique.

4.2.4   Problem Definition

Figure 4–9 shows a pipeline that delivers water at a constant temperature $T = 60°F$ from point 1 where the pressure is $p_1 = 150$ psig and the elevation is $z_1 = 0$ ft to point 2 where the pressure is atmospheric and the elevation is $z_2 = 300$ ft.

The density and viscosity of the water can be calculated from the following equations.

$$\rho = 62.122 + 0.0122T - 1.54\times10^{-4}T^2 + 2.65\times10^{-7}T^3 - 2.24\times10^{-10}T^4 \qquad \textbf{(4-18)}$$

$$\ln \mu = -11.0318 + \frac{1057.51}{T + 214.624} \qquad \textbf{(4-19)}$$

where T is in °F, ρ is in lb_m/ft^3, and μ is in $lb_m/ft \cdot s$.

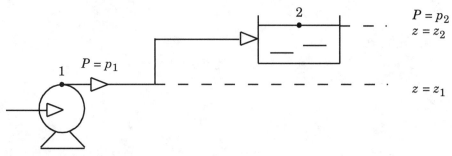

Figure 4–9 Pipeline at Steady State

(a)   Calculate the flow rate q (in gal/min) for a pipeline with effective length of L = 1000 ft and made of nominal 8-inch diameter schedule 40 commercial steel pipe. (Solution: v = 11.61 ft/s, gpm = 1811 gal/min)
(b)   Calculate the flow velocities in ft/s and flow rates in gal/min for pipelines at 60°F with effective lengths of L = 500, 1000, ... 10,000 ft and made of nominal 4-, 5-, 6- and 8-inch schedule 40 commercial steel pipe. Use the successive substitution method for solving the equations for the various cases and present the results in tabular form. Prepare plots of flow velocity v versus D and L, and flow rate q versus D and L.
(c)   Repeat part (a) at temperatures T = 40, 60, and 100°F and display the results in a table showing temperature, density, viscosity, and flow rate.

4.2.5   Equations and Numerical Data

The general mechanical energy balance on an incompressible liquid applied to this case yields

$$-\frac{1}{2}v^2 + g\Delta z + \frac{g_c \Delta P}{\rho} + 2\frac{f_F L v^2}{D} = 0 \qquad \text{(4-20)}$$

where v is the flow velocity in ft/s, g is the acceleration of gravity given by g = 32.174 ft/s^2, $\Delta z = z_2 - z_1$ is the difference in elevation (ft), g_c is a conversion factor (in English units g_c = 32.174 ft·lb$_m$/lb$_f$·s^2), $\Delta P = P_2 - P_1$ is the difference in pressure lb$_m$/ft^2), f_F is the Fanning friction factor, L is the length of the pipe (ft) and D is the inside diameter of the pipe (ft). The use of the successive substitution method requires Equation (4-20) to be solved for v as

$$v = \sqrt{\left(g\Delta z + \frac{g_c \Delta P}{\rho}\right) \bigg/ \left(0.5 - 2\frac{f_F L}{D}\right)} \qquad \text{(4-21)}$$

The equation for calculation of the Fanning friction factor depends on the Reynold's number, Re = $v\rho D/\mu$, where μ is the viscosity in lb$_m$/ft·s. For laminar flow (Re < 2100), the Fanning friction factor can be calculated from the equation

$$f_F = \frac{16}{\text{Re}} \qquad \text{(4-22)}$$

For turbulent flow (Re > 2100) the Shacham[3] equation can be used

$$f_F = 1/16\left\{\log\left[\frac{\varepsilon/D}{3.7} - \frac{5.02}{Re}\log\left(\frac{\varepsilon/D}{3.7} + \frac{14.5}{Re}\right)\right]\right\}^2 \qquad \text{(4-23)}$$

where ε/D is the surface roughness of the pipe (ε = 0.00015 ft for commercial steel pipes).

The flow velocity in the pipeline can be converted to flow rate by multiplying it by the cross section are of the pipe, the density of water (7.481 gal/ft^3), and factor (60 s/min). Thus q has units of (gal/min). The inside diameters (D) of nominal 4-, 5-, 6-, and 8-inch schedule 40 commercial steel pipes are provided in Appendix Table D–5.

4.2.6   Solution

(a) The problem is set up first for solving for one length and one diameter value with POLYMATH. The POLYMATH Nonlinear Algebraic Equation Solver is used for this purpose. It should be emphasized that Equation (4-21) (or Equation (4-20)) cannot be solved explicitly for the velocity in the turbulent region as in that region the friction factor is a complex function of the Reynolds number (and the velocity, see Equation (4-23)). Thus Equation (4-21) should be input as an "implicit" (nonlinear) equation. The implicit equations are entered in the form: f(x) = an expression, where x is the variable name, and f(x) is an expression that should have the value of zero at the solution. Bounds for the unknown x should be provided. Minimal and maximal values between which the function is continuous and one or more roots are probably located should be provided. For the velocity calculation, the following equation and bounds are used:

```
f(v) = v - sqrt((32.174 * deltaz + deltaP * 144 * 32.174 / rho) / (0.5 - 2 * fF * L / D))
       # Flow velocity (ft/s)
v(min) = 1
v(max) = 20
```

Note that the program looks for a solution where f(v) = 0; thus, there is no need to write this out explicitly. The complete set of equations is shown in Table 4–3.

Table 4–3 Equation Input to the POLYMATH Nonlinear Equation Solver Program (File **P4-02A.POL**)

Line	Equation
1	f(v) = v - sqrt((32.174 * deltaz + deltaP * 144 * 32.174 / rho) / (0.5 - 2 * fF * L / D)) # Flow velocity (ft/s)
2	fF = If (Re < 2100) Then (16 / Re) Else (1 / (16 * (log(eoD / 3.7 - 5.02 * log(eoD / 3.7 + 14.5 / Re) / Re)) ^ 2)) # Fanning friction factor (dimensionless)
3	eoD = epsilon / D # Pipe roughness to diameter ratio (dimensionless)
4	Re = D * v * rho / vis # Reynolds number (dimesionless)
5	deltaz = 300 # Elevation difference (ft)
6	deltaP = -150 # Pressure difference (psi)
7	T = 60 # Temperature (deg F)
8	L = 1000 # Effective length of pipe (ft)
9	D = 7.981 / 12 # Inside diameter of pipe (ft)
10	pi = 3.1416 # The constant pi
11	epsilon = 0.00015 # Surface rougness of the pipe (ft)
12	rho = 62.122 + T * (0.0122 + T * (-1.54e-4 + T * (2.65e-7 - T * 2.24e-10))) # Fluid density (lb/cu. ft.)
13	vis = exp(-11.0318 + 1057.51 / (T + 214.624)) # Fluid viscosity (lbm/ft-s)
14	q = v * pi * D ^ 2 / 4 * 7.481 * 60 # Flow rate (gal/min)
15	v(min) = 1
16	v(max) = 20

The solution obtained by POLYMATH is the same as specified in the problem statement ($v = 11.61$ ft/s, $q = 1811$ gal/min).

(b) The POLYMATH equation set can be exported to Excel by opening an Excel Workbook, reactivating the POLYMATH equation editor window and the pressing the Excel icon or the F4 function key. The Excel worksheet generated is summarized in Table 4–4.

Column A indicates the type of the equations in the problem. In this case there are explicit equations in rows 3 to 16. In cell C16 an initial estimate for the implicit variable (unknown) is specified. Cell C17 specifies the implicit equation whose value should approach zero at the solution.

Table 4–4 POLYMATH Equation Set Exported to Excel

	A	B	C	D	E
1	POLYMATH NLE Migration Document				
2		Variable	Value		Polymath Equation
3	Explicit Eqs	fF	0.00387711		fF=If (Re < 2100) Then (16 / Re) Else (1 / (16 * (log(eoD / 3.7 - 5.02 * log(eoD / 3.7 + 14.5 / Re) / Re)) ^ 2))
4		eoD	0.000225536		eoD=epsilon / D
5		Re	572291.1788		Re=D * v * rho / vis
6		deltaz	300		deltaz=300
7		deltaP	-150		deltaP=-150
8		T	60		T=60
9		L	1000		L=1000
10		D	0.665083333		D=7.981 / 12
11		pi	3.1416		pi=3.1416
12		epsilon	0.00015		epsilon=0.00015
13		rho	62.35393696		rho=62.122 + T * (0.0122 + T * (-1.54e-4 + T * (2.65e-7 - T * 2.24e-10)))
14		vis	0.000760873		vis=exp(-11.0318 + 1057.51 / (T + 214.624))
15		q	1637.35643		q=v * pi * D ^ 2 / 4*7.481 * 60
16	Implicit Vars	v	10.5		v(0)=10.5
17	Implicit Eqs	f(v)	-1.067599475		f(v)=v - sqrt((32.174 * deltaz + deltaP * 144 * 32.174 / rho) / (0.5 - 2 * fF * L / D))

The implicit equation for velocity v can be solved with Excel by first selecting the "Goal Seek" utility from the "Tools" dropdown menu. In the "Goal Seek" communication window, the target cell (C17 in this case), its desired value (zero) and the variable to be changed (in cell C16) have to be specified as shown in Figure 4–10. After pressing OK, the value v = 11.61 is obtained with function value f(v) = –1.02684E-05, thus the solution is the same as obtained by POLYMATH.

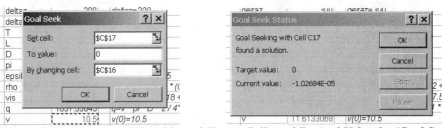

Figure 4–10 Selection of Variable and Target Cells and Desired Value for "Goal Seek"

The solutions of the set of equations for a large number of pipe lengths and diameter values is most efficiently accomplished with the Excel "Two input Data Table" capability. "Goal Seek" cannot be effectively applied to create such a data table. The use of an iterative method such as the "successive substitution" method is recommended.

The iteration function of the successive substitution method for calculation of the flow velocity is given by

$$v_{i+1} = F(v_i) \quad i = 0, 1,... \tag{4-24}$$

where i is the iteration number, F is the function in the right side of Equation (4-21) and v_0 is the initial estimate for the flow velocity. An error estimate at iteration i is provided by

$$\varepsilon_i = |v_i - v_{i+1}| \tag{4-25}$$

The solution is acceptable when the error is small enough, typically $\varepsilon_i < 1 \times 10^{-5}$.

The successive substitution calculations can be organized for row by row iterations in another location on the spreadsheet. This requires that some of the rows (and formulas) of Table 4–4 be changed. The expressions which are functions of the unknown velocity v (fF, Re, and q) should be grouped separately from the constants and placed in rows 13 to 15. This can be accomplished by cutting and pasting the entire row in the Excel code as needed. The rows that contain expressions that are independent of velocity v should be placed in rows 3 through 12. The variable addresses for cells containing these variables should be replaced by absolute addresses in the formulas in cells C13-C16 (See Table 4–5 where the "$" in $C$9, for example, indicates absolute cell address.). The expression for $f(v)$ in cell C17 must be replace by an expression to calculate v_{i+1} (Equation (4-21)). An additional formula for calculating ε_i must be added in cell C18.

The modified cell formulas are shown in Table 4–5. Introducing $v_0 = 10.5$ in cell C16 yields $v_1 = 11.57$ in cell C17. Thus the error for the first successive substitution is $\varepsilon_0 = 1.068$ as calculated in cell C18 and shown in Table 4–6.

Table 4–5 Cell Contents after Modifications

	A	B	C
1	POLYMATH NLE Migration Document		
2		Variable	
3		eoD	=(C10 / C8)
4		deltaz	=300
5		deltaP	=-150
6		T	=60
7		L	=1000
8		D	=0.66508
9		pi	=3.1416
10		epsilon	=0.00015
11		rho	=(62.122 + (C6 * (0.0122 + (C6 * (-0.000154 + (C6 * (0.000000265 - (C6 * 0.000000000224)))))))

	A	B	C
12		vis	=EXP((-11.0318 + (1057.51 / (C6 + 214.624))))
13		q	=((((C16 * $C$9) * ($C$8 ^ 2)) / 4) * 7.481) * 60)
14		fF	=IF((C15 < 2100),(16 / C15),(1 / (16 * (LOG10((($C$3 / 3.7) - ((5.02 * LOG10((($C$3 / 3.7) + (14.5 / C15)))) / C15))) ^ 2))))
15		Re	=((($C$8 * C16) * $C$11) / $C$12)
16		v(i)	10.5
17	Iteration	v(i+1)	=SQRT((((32.174 * $C$4) + ((($C$5 * 144) * 32.174) / $C$11) / (0.5 - (((2 * C14) * $C$7) / $C$8))))
18		err	=ABS(C17-C16)

Table 4–6 Cell Calculations after Modifications

	A	B	C
1	POLYMATH NLE Migration Document		
2		Variable	
3		eoD	0.000225537
4		deltaz	300
5		deltaP	-150
6		T	60
7		L	1000
8		D	0.66508
9		pi	3.1416
10		epsilon	0.00015
11		rho	62.35393696
12		vis	0.000760873
13		q	1637.340021
14		fF	0.003877114
15		Re	572288.3105
16		v(i)	10.5
17	Iteration	v(i+1)	11.56756289
18		err	1.067562894

It is convenient to calculate the iterations for the successive substitution method in another part of the spreadsheet by placing all the variables related to a single iteration in one row (instead of one column). This can be accomplished by copying the range B13:C18 to range starting in cell H2 using the "Paste Special" and then the "Transpose" options, found under the Excel "Edit" dropdown menu. The result gives the relevant heading in row 2 and the calculations for the first iteration in row 3 as shown in the top part of Figure 4–11.

The cell range H2:M2 is a transposed copy of the cell range C13:C18 of Table 4–6. The iteration number heading has been manually added in cell G2 as "It.No." and the value of "0" is placed in G3. The first iteration is identified as "1" in column G4. The cell range H3:M3 is then copied and pasted into the same column location in the 4th row. The **relative cell address** of v_{i+1} from the third row is manually substituted into the v_i column in the 4th row; thus the **formula** appearing in cell K4 is "=L3" and the calculation is shown in the bottom part of Figure 4–11.

	H	I	J	K	L	M
1						
2	fF	Re	q	v(i)	v(li+1)	err
3	0.003877	572291.2	1637.356	10.5	11.5676	1.067599

	G	H	I	J	K	L	M
1							
2	It. No.	fF	Re	q	v(i)	v(li+1)	err
3	0	0.003877	572291.2	1637.356	10.5	11.5676	1.067599
4	1	0.003849	630479.5	1803.837	11.5676	11.61158	0.04398

Figure 4–11 Creation of Iteration Table for Successive Substitution within Excel

The result of this first iteration shown in Figure 4–11 indicates that the convergence rate of the successive substitution method is very fast as the initial error of $\varepsilon_0 = 1.068$ is reduced to $\varepsilon_1 = 0.044$. Additional iterations can be carried out by copying the cell range H4:M4 and pasting this range into as many rows as the number of iterations desired. For the particular pipe diameter and length values used, the estimated error is below 10^{-7} after four iterations. However, to be on the "safe side," the number of iterations can be increased to ten. The results of 10 iterations are given in Figure 4–12.

The calculations for all the pipe diameter and length values specified in the problem statement can be automatically carried out within Excel by using the "Two Input Data Table." First, the framework of the Excel Table is prepared in another area of the spreadsheet by entering the column headings horizontally starting in G17 and entering the pipe diameters in feet in the cells immediately below. Then the length values (from 500 to 10,000 in increments of 500) into separate rows starting in cell G19. (See Figure 4–13 where only part of the values are shown.) Cell G18 must contain the flow velocity needed for the "Two Input Data Table." The absolute **addresses** of the calculated value of the converged flow velocity ($L$13) and the associate error estimate ($M$13) must be used. Since the velocity value is acceptable as a solution only if the estimated error is $< 10^{-5}$, the following "If " statement is introduced in cell G18:

=IF($M$13<0.00001,$L$13,"No Convergence")

Note that the headings entered in the 17th row are not essential parts of the

	G	H	I	J	K	L	M
1							
2	It. No.	fF	Re	q	v(i)	v(li+1)	err
3	0	0.003877	572291.2	1637.356	10.5	11.5676	1.067599
4	1	0.003849	630479.5	1803.837	11.5676	11.61158	0.04398
5	2	0.003848	632876.6	1810.695	11.61158	11.61325	0.001672
6	3	0.003848	632967.8	1810.955	11.61325	11.61331	6.34E-05
7	4	0.003848	632971.2	1810.965	11.61331	11.61332	2.4E-06
8	5	0.003848	632971.3	1810.966	11.61332	11.61332	9.1E-08
9	6	0.003848	632971.3	1810.966	11.61332	11.61332	3.45E-09
10	7	0.003848	632971.3	1810.966	11.61332	11.61332	1.31E-10
11	8	0.003848	632971.3	1810.966	11.61332	11.61332	4.95E-12
12	9	0.003848	632971.3	1810.966	11.61332	11.61332	1.88E-13
13	10	0.003848	632971.3	1810.966	11.61332	11.61332	7.11E-15

Figure 4–12 Results of 10 Iterations within the Excel Spreadsheet

	G18	▼		f_x =IF($M$13<0.00001,$L$13,"No Convergence")		
	G	H	I	J	K	L	M
17	L (ft)	D=4"	D=5"	D=6"	D=8"		
18	11.61332	0.336	0.421	0.505	0.665		
19	500						
20	1000						
21	1500						
22	2000						

Figure 4–13 Setting Up a Two Input Data Table in Excel

table but are used for "Legend" in the graph to be prepared. The "Data Table" is created by first **highlighting the region** of the table but not including the heading cells, G18:K38. Then click on "Table" under the Data pull-down menu. In the communication box, the cell C8 (diameter in ft) is specified as "Row input cell" and cell C7 (pipe length in ft) is specified as column input cell. The resulting solutions with the specified error tolerance for the indicated range of pipe diameter and length values are partially shown in Figure 4–14.

The "XY Scatter" plot of the flow velocity versus pipe length and diameter shown in Figure 4–15 can be created using the "Chart" options from the "Insert" dropdown menu. The plot of flow rate q versus D and L can be created in a similar manner to the flow velocity plot. This is most easily accomplished by copying the previously created two-dimensional table to another location such as shown in the left side of Figure 4–16. The Excel "Data Table" formula cell (M18 in this

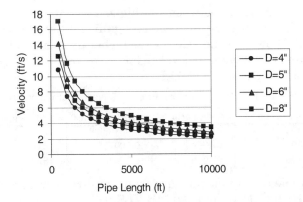

Figure 4–14 Creating the Two Input Data Table for Flow Velocity v in Excel (File **P4-02B.XLS**)

Figure 4–15 Flow Velocity Plot versus Pipe Length and Diameter (File **P4-02B.XLS**)

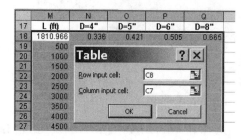

	M	N	O	P	Q
17	L (ft)	D=4"	D=5"	D=6"	D=8"
18	1810.966	0.336	0.421	0.505	0.665
19	500	427.5047	780.5269	1274.226	2656.357
20	1000	294.4635	536.5943	873.8108	1810.966
21	1500	236.9825	431.8006	702.8761	1454.833
22	2000	203.1197	370.175	602.5637	1246.775
23	2500	180.1895	328.4767	534.7535	1106.435
24	3000	163.3622	297.8873	485.0335	1003.659
25	3500	150.3465	274.2303	446.592	924.2548
26	4000	139.8974	255.2392	415.7371	860.5516
27	4500	131.2732	239.5645	390.2723	807.9925

Figure 4–16 Creating the Two Input Data Table for Flow Rate q in Excel (File **P4-02B.XLS**)

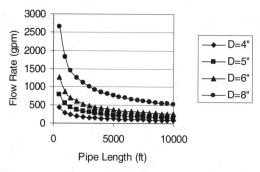

Figure 4–17 Flow Rate versus Pipe Length and Diameter (File **P4-02B.XLS**)

case) must be modified by entering the absolute address of the flow rate q as $J$13.

=IF($M$13<0.00001,$J$13,"No Convergence")

The "Data Table" generation is then repeated and the result is shown on the right side of Figure 4–16. The corresponding plot of the flow rate versus pipe length and diameter is shown in Figure 4–17.

(c) The values of the density, viscosity, and flow rate at various temperatures using the property equations can be carried out using either POLYMATH or EXCEL, and the results are summarized in Table 4–7. The density changes very little (less than 1.2%) between the highest and lowest temperature, while the change of the viscosity is much more significant (over 50%). At the highest temperature, the flow rate and velocity are higher by about 7.5% than at the lowest temperature. The Reynolds number increases by over 125% when the temperature increases from 40°F to 100°F.

Table 4–7 Quantities Calculated from Property Equations at Various Temperatures

Temperature (°F)	Density (lb/cu. ft.)	Viscosity (lb_m/ft·s)	Flow Rate (gal/min)
40	62.380	0.001030	1784
60	62.354	0.000761	1811
100	62.045	0.000466	1876

The problem solution files are found in directory Chapter 4 and designated **P4-02A.POL** and **P4-02B.XLS**.

4.3 ADIABATIC OPERATION OF A TUBULAR REACTOR FOR CRACKING OF ACETONE

4.3.1   Concepts Demonstrated

Calculation of the conversion and temperature profiles in an adiabatic tubular reactor.

4.3.2   Numerical Methods Utilized

Solution of simultaneous ordinary differential equations.

4.3.3   Excel Options and Functions Demonstrated

Use of POLYMATH and the POLYMATH ODE Solver Add-In for Excel to solve differential equations.

4.3.4   Problem Definition

The irreversible, vapor-phase cracking of acetone (A) to ketene (B) and methane (C) that is given by the reaction

$$CH_3COCH_3 \rightarrow CH_2CO + CH_4$$

is carried out adiabatically in a tubular reactor. The reaction is first order with respect to acetone and the specific reaction rate can be expressed by

$$\ln k = 34.34 - \frac{34222}{T} \tag{4-26}$$

where k is in s^{-1} and T is in K. The acetone feed flow rate to the reactor is 8000 kg/hr, the inlet temperature is $T = 1150$ K and the reactor operates at the constant pressure of $P = 162$ kPa (1.6 atm). The volume of the reactor is 4 m^3.

4.3.5   Equations and Numerical Data

The material balance equations for the plug-flow reactor are given by

$$\frac{dF_A}{dV} = r_A \tag{4-27}$$

$$\frac{dF_B}{dV} = -r_A \tag{4-28}$$

$$\frac{dF_C}{dt} = -r_A \tag{4-29}$$

where F_A, F_B, and F_C are flow rates of acetone, ketene, and methane in g-mol/s, respectively and r_A is the reaction rate of A in g-mol/m^3·s. The reaction is first order with respect to acetone, thus

$$r_A = -kC_A \tag{4-30}$$

where C_A is the concentration of acetone in g-mol/m^3. For a gas phase reactor, using the appropriate units of the gas constant, the concentration of the acetone in g-mil/m^3 is obtained by

$$C_A = \frac{1000 y_A P}{8.31 T} \tag{4-31}$$

The mole fractions of the various components are given by

$$y_A = \frac{F_A}{F_A + F_B + F_C} \qquad y_B = \frac{F_B}{F_A + F_B + F_C} \qquad y_C = \frac{F_C}{F_A + F_B + F_C} \tag{4-32}$$

The conversion of acetone can be calculated from

$$x_A = \frac{F_{A0} - F_A}{F_{A0}} \tag{4-33}$$

An enthalpy (energy) balance on a differential volume of the reactor yields

$$\frac{dT}{dV} = \frac{-r_A(-\Delta H)}{F_A C_{pA} + F_B C_{pB} + F_C C_{pC}} \tag{4-34}$$

where ΔH is the heat of the reaction at temperature T (in J/g-mol) and C_{pA}, C_{pB}, and C_{pC} are the molar heat capacities of acetone, ketene and methane (in J/g-mol·K). Fogler[4] provides the following equations for calculating the heat of reaction and the molar heat capacities.

$$\Delta H = 80770 + 6.8(T - 298) - 0.00575(T^2 - 298^2) - 1.27 \times 10^{-6}(T^3 - 298^3) \tag{4-35}$$

$$C_{pA} = 26.6 + 0.183 T - 45.86 \times 10^{-6} T^2 \tag{4-36}$$

$$C_{pB} = 20.04 + 0.0945 T - 30.95 \times 10^{-6} T^2 \tag{4-37}$$

$$C_{pC} = 13.39 + 0.077 T - 18.71 \times 10^{-6} T^2 \tag{4-38}$$

(a) Calculate the flow-rates (in g-mol/s) and the mole fractions of acetone, ketene and methane along the reactor. Use POLYMATH to calculate and plot the conversion and reactor temperature (in K) versus volume.

(b) The conversion in the reactor in part (a) is very low in adiabatic operation because the reactor content cools down very quickly. It is suggested that feeding nitrogen along with the acetone might be beneficial in maintaining a higher temperature. Modify the POLYMATH equation set to enable adding nitrogen to the feed, transfer the equations to Excel and compare the final conversions and temperatures for the cases where 28.3, 18.3, 8.3, 3.3 and 0.0 g-mol/s nitrogen is fed into the reactor (the total molar feed rate is 38.3 g-mol/s in all the cases).

4.3.6   Solution (Partial)

The POLYMATH ordinary differential equations solver is used for solving this problem. Equations (4-26) to (4-38) and other needed equations can be entered into POLYMATH without any significant changes. Note that these equations can be entered in any order as they will be ordered during the problem solution. The feed flow rate to the reactor F_{A0} has to be specified in units of g-mol/s. The molecular weight of acetone (58 g/g-mol) is used for this conversion. The complete POLYMATH problem is summarized in Table 4–8.

Table 4–8 Equation Input to the POLYMATH Ordinary Differential Equation Solver (File **P4-03A.POL**)

Line	Equation
1	d(FA)/d(V) = rA # Differential mass balance on acetone
2	d(FB)/d(V) = -rA # Differential mass balance on ketene
3	d(FC)/d(V) = -rA # Differential mass balance on methane
4	d(T)/d(V) = (-deltaH) * (-rA) / (FA * CpA + FB * CpB + FC * CpC) # Differential enthalpy balance
5	XA = (FA0 - FA) / FA0 # Conversion of acetone
6	rA = -k * CA # Reaction rate in mol/m3-s
7	P = 162 # Pressure kPa
8	FA0 = 38.3 # Feed rate of acetone in mol/s
9	CA = yA * P * 1000 / (8.31 * T) # Concentration of acetone in mol/m3
10	yA = FA / (FA + FB + FC) # Mole fraction of acetone
11	yB = FB / (FA + FB + FC) # Mole fraction of ketene
12	yC = FC / (FA + FB + FC) # Mole fraction of methane
13	k = 8.2E14 * exp(-34222 / T) # Reaction rate constant in s-1
14	deltaH = 80770 + 6.8 * (T - 298) - .00575 * (T ^ 2 - 298 ^ 2) - 1.27e-6 * (T ^ 3 - 298 ^ 3)
15	CpA = 26.6 + .183 * T - 45.86e-6 * T ^ 2
16	CpB = 20.04 + 0.0945 * T - 30.95e-6 * T ^ 2
17	CpC = 13.39 + 0.077 * T - 18.71e-6 * T ^ 2
18	FB(0) = 0 # Feed rate of ketene in mol/s
19	FA(0) = 38.3 # Feed rate of acetone in mol/s
20	FC(0) = 0 # Feed rate of methane in mol/s
21	T(0) = 1035 # Inlet reactor temperature in K
22	V(0) = 0 # Reactor volume in m3
23	V(f) = 4

The POLYMATH solution that is summarized in Table 4–9 indicates that the inlet temperature of 1035 K is reduced to 907.54 K within the reactor as the reaction is endothermic. Consequently the specific reaction rate, k, and the reaction rate with respect to acetone, $-r_A$, are reduced by more that two orders of magnitude. This results in a low conversion of the acetone, only 15.7%.

Table 4–9 POLYMATH Results for Problem 4.3 (a)

	Variable	Initial value	Minimal value	Maximal value	Final value
1	CA	18.83535	12.68959	18.83535	12.68959
2	CpA	166.8786	154.9084	166.8786	154.9084
3	CpB	84.69309	80.3113	84.69309	80.3113
4	CpC	73.04238	67.86058	73.04238	67.86058
5	deltaH	7.876E+04	7.876E+04	7.977E+04	7.977E+04
6	FA	38.3	28.44647	38.3	28.44647
7	FA0	38.3	38.3	38.3	38.3
8	FB	0	0	9.853527	9.853527
9	FC	0	0	9.853527	9.853527
10	k	3.580818	0.0344545	3.580818	0.0344545
11	P	162.	162.	162.	162.
12	rA	-67.44594	-67.44594	-0.4372133	-0.4372133
13	T	1035.	907.5422	1035.	907.5422
14	V	0	0	4.	4.
15	xA	0	0	0.2572723	0.2572723
16	yA	1.	0.5907454	1.	0.5907454
17	yB	0	0	0.2046273	0.2046273
18	yC	0	0	0.2046273	0.2046273

(b) The addition of the inert gas nitrogen to the reactor feed requires the addition of an equation for heat capacity of nitrogen and modification to the energy balance.

$$C_{pN} = 6.25 + 0.00878T - 2.1 \times 10^{-8} T^2 \tag{4-39}$$

$$\frac{dT}{dV} = \frac{-r_A(-\Delta H)}{F_A C_{pA} + F_B C_{pB} + F_C C_{pC} + F_N C_{pN}} \tag{4-40}$$

It is also necessary to add an equation that allows the molar flow rate of nitrogen, F_N, to be calculated when the feed rate of acetone, F_{A0}, is specified.

$$F_N = 38.3 - F_{A0} \tag{4-41}$$

Additionally, the equations for the mole fractions need to be modified to include the molar flow rate of nitrogen. Also, the initial condition on the differential equation for F_{A0} must be specified with the current initial condition. The modified POLYMATH program for $F_{A0} = 10$ kg-mol/s shown in Figure 4–18 can be automatically exported to Excel by either pressing the F4 key or clicking the mouse on the Excel icon from the Differential Equation Solver window.

```
Differential Equations: 4   Auxiliary Equations: 15   ✔ Ready for solution

d(FA)/d(V) = rA # Differential mass balance on acetone
d(FB)/d(V) = -rA # Differential mass balance on ketene
d(FC)/d(V) = -rA # Differential mass balance on methane
d(T)/d(V) = (-deltaH) * (-rA) / (FA * CpA + FB * CpB + FC * CpC + FN * CpN) # Differential enthalpy balance
XA = (FA0-FA)/FA0 # Conversion of acetone
rA = -k * CA # Reaction rate in kg-mole/m3-s
FA0 = 10 # Feed rate of acetone in kg-mol/s
FN = 38.3 - FA0 # Feed rate of nitrogen in kg-mol/s
P = 162 # Pressure kPa
CA = yA * P * 1000 / (8.31 * T) # Concentration of acetone in k-mol/m3
yA = FA / (FA + FB + FC + FN) # Mole fraction of acetone
yB = FB / (FA + FB + FC + FN) # Mole fraction of ketene
yC = FC / (FA + FB + FC + FN) # Mole fraction of methane
k = 8.2E14 * exp(-34222 / T) # Reaction rate constant in s-1
deltaH = 80770 + 6.8 * (T - 298) - .00575 * (T ^ 2 - 298 ^ 2) - 1.27e-6 * (T ^ 3 - 298 ^ 3)# Heat of reaction in J/mol-K
CpA = 26.6 + .183 * T - 45.86e-6 * T ^ 2 # Heat capacity of acetone in J/mol-K
CpB = 20.04 + 0.0945 * T - 30.95e-6 * T ^ 2 # Heat capacity of ketene in J/mol-K
CpC = 13.39 + 0.077 * T - 18.71e-6 * T ^ 2 # Heat capacity of methane in J/mol-K
CpN = 6.25 + 8.78e-3 * T - 2.1e-8 * T ^ 2 # Heat capacity of nitrogen in J/mol-K
FB(0) = 0 # Feed rate of ketene in kg-mol/s
FA(0) = 10 # Feed rate of acetone in kg-mol/s
FC(0) = 0 # Feed rate of methane in kg-mol/s
T(0) = 1035 # Inlet reactor temperature in K
V(0) = 0 # Reactor volume in m3
V(f) = 4
```

Figure 4–18 The Revised POLYMATH Program Ready for Export to Excel
(File **P4-03A.POL**)

The revised Excel worksheet as automatically generated from the POLY-MATH program is shown in Figure 4–19.

	A	B	C	D	E	F
1	**POLYMATH DEQ Migration Document**					
2		**Variable**	**Value**		**Polymath Equation**	**Comments**
3	Explicit Eqs	XA	0		XA=(FA0-FA)/FA0	Conversion of acetone
4		rA	-17.60990721		rA=-k * CA	Reaction rate in kg-mole/h
5		FA0	10		FA0=10	Feed rate of acetone in kg
6		FN	28.3		FN=38.3 - FA0	Feed rate of nitrogen in kg
7		P	162		P=162	Pressure kPa
8		CA	4.917845344		CA=yA * P * 1000 / (8.31 * T)	Concentration of acetone
9		yA	0.261096606		yA=FA / (FA + FB + FC + FN)	Mole fraction of acetone
10		yB	0		yB=FB / (FA + FB + FC + FN)	Mole fraction of ketene
11		yC	0		yC=FC / (FA + FB + FC + FN)	Mole fraction of methane
12		k	3.580817609		k=8.2E14 * exp(-34222 / T)	Reaction rate constant in
13		deltaH	78758.21631		deltaH=80770 + 6.8 * (T - 298) - .00575 * (T ^ 2 - 298	Heat of reaction in J/mol-K
14		CpA	166.8786215		CpA=26.6 + .183 * T - 45.86e-6 * T ^ 2	Heat capacity of acetone
15		CpB	84.69308625		CpB=20.04 + 0.0945 * T - 30.95e-6 * T ^ 2	Heat capacity of ketene in
16		CpC	73.04238025		CpC=13.39 + 0.077 * T - 18.71e-6 * T ^ 2	Heat capacity of methane
17		CpN	15.31480428		CpN=6.25 + 8.78e-3 * T - 2.1e-8 * T ^ 2	Heat capacity of nitrogen
18	Integration Vars	FA	10		FA(0)=10	Differential mass balance
19		FB	0		FB(0)=0	Differential mass balance
20		FC	0		FC(0)=0	Differential mass balance
21		T	1035		T(0)=1035	Differential enthalpy balan
22	ODE Eqs	d(FA)/d(V)	-17.60990721		d(FA)/d(V) = rA	
23		d(FB)/d(V)	17.60990721		d(FB)/d(V) = -rA	
24		d(FC)/d(V)	17.60990721		d(FC)/d(V) = -rA	
25		d(T)/d(V)	-659.7507676		d(T)/d(V) = (-deltaH) * (-rA) / (FA * CpA + FB * CpB + FC * CpC + FN * CpN)	
26	Indep Var	V	0		V(0)=0 ; V(f)=4	

Figure 4–19 Generated Excel Problem as Exported from POLYMATH
(File **P4-03B.XLS**)

The Excel version of this problem separates the set of equations and data into four categories (see column A in Figure 4–19). Rows 3 to 17 contain explicit algebraic equations and constants. The initial values for the variables that are defined by differential equations are included in rows 18 to 21. The differential equations are defined in rows 22 to 25, and the initial value for the independent variable is specified in row 26.

The names of the variables are shown in column B, and the Excel formulas of the equations are included in column C. Column E presents the equations as they were entered into POLYMATH. The POLYMATH comments are also copied into column F. It should be emphasized that only the formulas in column C are used for calculations.

The POLYMATH ODE_Solver Add-In is used for solving the equations. It can be found in the "Tools" dropdown menu. Before using the ODE_Solver, it should be verified that the list of Add-Ins shows the "Ode_Solver" as as active and the "Solver Add-In" is not marked (non-active). This eliminates possible interference beween the two Add-Ins in some versions of Excel.

Selection of the POLYMATH ODE from the "Tools" menu brings up the communication box shown in Figure 4–20. Pressing the "Reload" button will automatically enter the problem into the ODE_Solver communication box. Otherwise, the ranges of the cells of the initial values and the differential equations must be entered as well as the address of the cell that contains the initial value of the independent variable and the final value (numerical) of the independent variable. Checking of the "Show Report" will place the solution output in a new worksheet. The "Intermediate Cells to Store" will output the vector of specified cells during the numerical integration.

When "Solve" is clicked, the POLYMATH ODE_Solver will start changing the independent variable value until it reaches its final value. During this process the values of the problem variables will be calculated and updated. At the

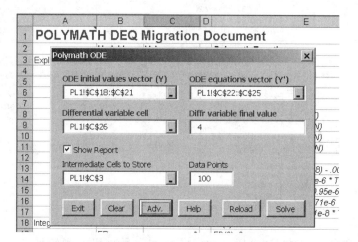

Figure 4–20 POLYMATH ODE_Solver Add-In Communication Box

18	Integration Vars	FA	6.864475993	FA(0)=10	Differential mass balance
19		FB	3.135524007	FB(0)=0	Differential mass balance
20		FC	3.135524007	FC(0)=0	Differential mass balance
21		T	911.8567009	T(0)=1035	Differential enthalpy balance
22	ODE Eqs	d(FA)/d(V)	-0.145866418	d(FA)/d(V) = rA	
23		d(FB)/d(V)	0.145866418	d(FB)/d(V) = -rA	
24		d(FC)/d(V)	0.145866418	d(FC)/d(V) = -rA	
25		d(T)/d(V)	-6.011513678	d(T)/d(V) = (-deltaH) * (-rA) / (FA * CpA + FB * CpB + FC * CpC + FN * CpN)	
26	Indep Var	V	4	V(0)=0 ; V(f)=4	

Figure 4–21 Final Values of Some of the Variables (Feed Flow Rate of Nitrogen is 28.3 g-mol/s)

end of the integration, the final values of the problem variables will be displayed. Some of those values for this problem are shown in Figure 4–21.

The "Show Report" option in the ODE solver communication box (see Figure 4–20) automatically creates a new worksheet that includes the table of initial, minimal, maximal, and final values of the integration variables. Additionally, a table of the values of the problem variables versus the independent variable is generated for the integration range. The number of data points displayed in the table of the detailed results is the number shown in the "Data Points" field of the communication box. If there is a need to include additional variables in the report, their cell range must be specified in the "Intermediate Cells to Store" field. In this case the value of x_A is of interest, so cell C3 was added to the list of variables to be stored.

The resulting "Report," automatically created on a new worksheet and partially shown in Figure 4–22, provides the initial, maximal, minimal, and final values of V (cell C26 from the problem worksheet), F_A, F_B, F_C, T (cells C18, C19, C20, and C21), and x_A (cell C3). Note that the names of the variables are not normally displayed as they are not essential components of the problem definition in Excel, but they have been added to the spreadsheet for clarity.

	A	B	C	D	E	F	G
1	**POLYMATH Report DEQ**						
2	Ordinary Differential Equations (RKF45).						
3							
4	**Calculated values of DEQ variables**						
5			Variable	Initial	Minimal	Maximal	Final
6		1	C26 or V	0	0	4	4
7		2	C18 or FA	10	6.864476	10	6.864476
8		3	C19 or FB	0	0	3.135524	3.135524
9		4	C20 or FC	0	0	3.135524	3.135524
10		5	C21 or T	1035	911.8567	1035	911.8567
11		6	C3 or xA	0	0	0.313552	0.313552

Figure 4–22 Partial View of DEQ Report Worksheet for $F_{A0} = 10$

A comparison of the results when pure acetone is fed to the reactor (Table 4–9) with the case when 10 g-mol/s acetone and 28.3 mol/s of nitrogen are fed into the reactor (Figure 4–22) shows that the addition of the nitrogen increases the conversion from $x_A = 0.257$ to $x_A = 0.314$. However, this increase comes at the

27	Intermediate data points					
28	t	FA	FB	FC	T	xA
29	1	0	10	0	0	1035	0
30	2	0.083061	9.186783	0.813217	0.813217	1004.17	0.077327
31	3	0.134176	8.934929	1.065071	1.065071	994.4717	0.102389
32	4	0.181147	8.76353	1.23647	1.23647	987.8297	0.119473
33	5	0.209688	8.677067	1.322933	1.322933	984.4662	0.128097
34	6	0.241688	8.591675	1.408325	1.408325	981.1358	0.136688
35	7	0.305688	8.447931	1.552069	1.552069	975.51	0.151889
36	8	0.337688	8.386281	1.613719	1.613719	973.0897	0.158358
37	9	0.369688	8.329929	1.670071	1.670071	970.8734	0.164247
38	10	0.401688	8.278062	1.721938	1.721938	968.8301	0.16965
39	11	0.465688	8.185344	1.814656	1.814656	965.1694	0.179268
40	12	0.497688	8.143562	1.856438	1.856438	963.5164	0.183588

Figure 4–23 Partial View of DEQ Report Worksheet Showing Intermediate Data
Points for FA0 = 10 (File **P4-03B.XLS**)

expense of almost fourfold reduction of the flow rates of the reactant and the
products.

The tabular results of the Excel "Report" are shown in Figure 4–23 where
variables have been entered to replace cell addresses in line 28 for clarity. This
Excel table can be used to prepare temperature and conversion profile plots for
the reactor (see Figures 4–24 and 4–25). It can be seen that even with the addition of the nitrogen, the main part of the reaction is carried out in the first quarter (1 m^3 volume) of the reactor where the conversion reaches $x_A = 0.23$. In the
additional 75% of the volume the conversion only increases to $x_A = 0.314$.

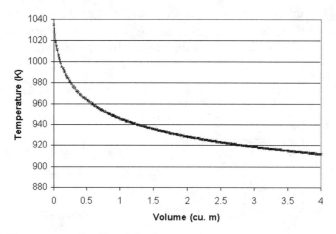

Figure 4–24 Temperature Profile in the Reactor for $F_N = 28.3$ g-mol/s
(File **P4-03B.XLS**)

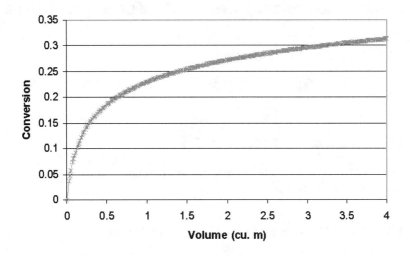

Figure 4–25 Conversion Profile in the Reactor for F_N = 28.3 g-mol/s
              (File **P4-03B.XLS**)

The problem solution files are found in directory Chapter 4 and designated **P4-03A.POL**, **P4-03B.POL**, and **P4-03B.XLS**.

4.4   CORRELATION OF THE PHYSICAL PROPERTIES OF ETHANE

4.4.1   Concepts Demonstrated

Correlations for heat capacity, vapor pressure, and liquid viscosity for an ideal gas.

4.4.2   Numerical Methods Utilized

Polynomial, multiple linear, and nonlinear regression of data with linearization and transformation functions.

4.4.3   Excel Options and Functions Demonstrated

Use of the Excel LINEST function for multiple linear and polynomial regression. Use of the Excel Add-In "Solver" for nonlinear regression.

4.4.4   Problem Definition

Determine appropriate correlations for heat capacity, vapor pressure, and liquid viscosity of ethane. The data files are given and also the data are available in Appendix F. Compare those correlations with the expressions suggested by the Design Institute for Physical Properties, DIPPR.[5]

(a)   Compare third-degree and fifth-degree polynomials for the correlation of the heat capacity data (Table F–1 of Appendix F) using both POLYMATH and Excel by examining the respective variances, confidence intervals, and residual plots.

(b)   Use Excel to compare the fifth-degree polynomial for the correlation of the heat capacity data (Table F–2 of Appendix F) with the two DIPPR recommended correlations for the appropriate temperature intervals.

(c)   Utilize multiple linear regression in Excel to fit the Wagner equation to the vapor pressure of ethane data found in Table F–3 of Appendix F. Comment on the applicability of the Wagner equation for correlating these data. Compare the correlation obtained by the Wagner equation with that of the Riedel equation recommended by DIPPR.

(d)   Use nonlinear regression to fit the Antoine equation to the liquid viscosity data of ethane data found in Table F–4 of Appendix F. Initial estimates of the nonlinear regression parameters should be obtained by linear regression. Verify nonlinear regression results in both POLYMATH and Excel. Compare the correlation obtained by the Antoine equation with that of the Riedel equation recommended by DIPPR.

4.4.5 Solution

This problem can be approached by first setting up the problem in POLYMATH and achieving a solution. Then the problem is exported to Excel from the POLYMATH program, and the same calculations in Excel are verified between the two software packages. Further use of Excel is emphasized in the detailed problem solution and the generation of the tabular and graphical results.

(a) The temperature dependency of the heat capacities of gases is commonly represented by simple polynomials of the form

$$C_p = a_0 + a_1 T + a_2 T^2 + a_3 T^3 + \ldots \qquad (4\text{-}42)$$

where C_p is the heat capacity in J/kg-mol·K, T is the temperature in K, and a_0, a_1,... are the coefficients (parameters) of the correlation determined by regression of experimental data. The degree of the polynomial which best represents the experimental data can be determined based on the variance, the correlation coefficient (R^2), the confidence intervals of the parameters, and the residual plot. The heat capacity data for ethane gas are given in Appendix F, Tables F–1 and F–2. There are 19 data points in Table F–1 but they encompass a wider temperature range (1450 K) than the 41 data points in Table F–2 that have a much smaller range of temperature range (400 K).

The data of Table F–1 can be fitted to a third-degree polynomial of the form given by Equation (4-42) by first using the POLYMATH Regression Program. The results of the polynomial obtained with POLYMATH are summarized in Figure 4–26, and the POLYMATH graphical result is given in Figure 4–27. The high value of the correlation coefficient ($R^2 = 0.9971$) as well as the plot of the calcu-

Model: Cp = a0 + a1*T_K + a2*T_K^2 + a3*T_K^3

Variable	Value	95% confidence
a0	2.505E+04	4141.482
a1	91.41495	24.82297
a2	0.0365871	0.0378178
a3	-2.989E-05	1.621E-05

General
Degree of polynomial = 3
Regression including a free parameter
Number of observations = 19

Statistics

R^2	0.9975999
R^2adj	0.9971199
Rmsd	446.1547
Variance	4.791E+06

Figure 4–26 Third-Degree Polynomial Coefficient and Statistics for the Heat Capacity Data of Appendix Table F–1 (File **P4-04A.POL**)

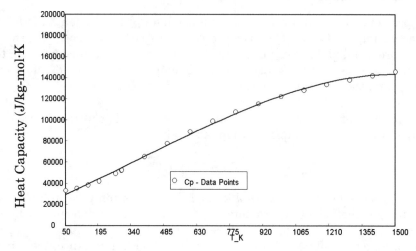

Figure 4–27 Third-Degree Polynomial Representation for Heat Capacity of Ethane (File **P4-04A.POL**)

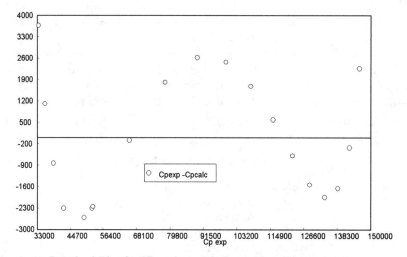

Figure 4–28 Residual Plot for Heat Capacity Represented by Third-Degree Polynomial for Data Set A (File **P4-04A.POL**)

lated and experimental values seems to indicate that the representation of the data by the third-degree polynomial is quite satisfactory. However, the residual plot of Figure 4–28 shows a clear cyclic pattern, and the error in representation of some of the points is >5%, which is well above the common experimental error in heat capacity data. In the case of a_2, the confidence interval is slightly larger in absolute value than the parameter itself. Thus the third-degree polynomial representation is unsatisfactory, and better representation should be sought.

	A	B	C	D	E	F	G	H
1	**POLYMATH Polynomial Regression Migration Document**							
2								
3	T_K	T_K^2	T_K^3	Cp	Cp calc	Cp residual	Cp residual ^2	
4	50	2500	125000	33390	29705.72299	-3684.277013	13573897.11	
5	100	10000	1000000	35650	34524.72357	-1125.276432	1266247.049	
6	150	22500	3375000	38660	39481.83074	821.8307351	675405.7571	
7	200	40000	8000000	42260	44554.63003	2294.630034	5265326.995	
8	273.16	74616.3856	20382211.89	49540	52139.00692	2599.006924	6754836.99	
9	298.15	88893.4225	26503573.92	52470	54762.87925	2292.879251	5257295.26	
10	300	90000	27000000	52720	54957.64721	2237.647211	5007065.039	

Figure 4–29 Columns Generated in the Excel Worksheet when a Third-Degree Polynomial Regression is Exported form POLYMATH to Excel (File **P4-**

	I	J	K	L	M	N
1						
2	Polynomial Regression of degree 3. Including a free parameter.					
3		a3	a2	a1	a0	
4	Coefficients	-3E-05	0.036587	91.41495	25047.24	
5	Std.dev.s	7.61E-06	0.017747	11.64851	1943.445	
6	R2, SE (y)	0.9976	2188.737	#N/A	#N/A	
7	95% conf. int.	1.62E-05	0.037818	24.82297	4141.482	
8	Variance	4790568				
9	Sum of Squares	71858513				
10	Model	Cp = a3 * T_K^3 + a2 * T_K^2 + a1 * T_K + a0				

Figure 4–30 Third-Degree Polynomial Coefficients and Statistics for the Heat Capacity Data of Table F–1 (File **P4-04A1.XLS**)

The calculations for the third-degree polynomial can easily be carried out within Excel. This is accomplished from POLYMATH by clicking on the Excel icon from POLYMATH Data Table after the problem is selected for the variable and the desired polynomial degree. Note that an Excel spreadsheet must be open on your computer in order for the "Export to Excel" to take place. The columns generated in the Excel worksheet, after exporting the problem from POLYMATH, are partially shown in Figure 4–29. The temperature and heat capacity data are found in columns A and D respectively, and the formulas for calculating various powers of T are placed in columns B and C. The Excel result is summarized in , which corresponds very closely to the POLYMATH solution.

In a similar manner, the problem for the fifth-degree polynomial can be set up in POLYMATH and exported to Excel. The resulting worksheet is partially presented in Figure 4–31 where the data columns are shown. The temperature and heat capacity data are found in columns A and F respectively, and the formulas for calculating various powers of T are placed in columns B through E.

Consider now the underlying calculations in the Excel worksheet that are shown in Figure 4–32. The first three rows of this table (cell range L4:Q6) are obtained from Excel's LINEST function. Thus the formula in that range of cells is given by

$\{$=LINEST(F4:F22,A4:E22,TRUE,TRUE)$\}$

where (F4:F22) is the range where the dependent variable, C_p, is stored, the second range (A4:E22) is the range where the independent variables (temperature

	A	B	C	D	E	F	G	H	I	J
1	**POLYMATH Polynomial Regression Migration Document**									
2										
3	T_K	T_K^2	T_K^3	T_K^4	T_K^5	Cp	Cp calc	Cp residual	Cp residual ^2	
4	50	2500	125000	6250000	312500000	33390	33762.18787	372.1878732	138523.8129	
5	100	10000	1000000	100000000	10000000000	35650	35206.23623	-443.7637733	196926.2865	
6	150	22500	3375000	506250000	75937500000	38660	38190.67548	-469.324517	220265.5022	
7	200	40000	8000000	1600000000	3.2E+11	42260	42355.99691	95.99690613	9215.405987	
8	273.16	74616.3856	20382211.89	5567605000	1.52085E+12	49540	49938.08766	398.0876623	158473.7869	

Figure 4–31 Fifth-Degree Polynomial Excel Worksheet for the Heat Capacity Data of Appendix Table F–1 (File **P4-04A2.XLS**)

	K	L	M	N	O	P	Q	R
1								
2	Polynomial Regression of degree 5. Including a free parameter.							
3		a5	a4	a3	a2	a1	a0	
4	**Coefficients**	-7.8E-11	3.68E-07	-0.00065	0.481171	-32.5818	34267.42	
5	**Std.dev.s**	8.82E-12	3.4E-08	4.71E-05	0.028361	7.005462	544.6506	
6	**R2, SE (y)**	0.999947	348.3187	#N/A	#N/A	#N/A	#N/A	
7	**95% conf. int.**	1.9E-11	7.34E-08	0.000102	0.061259	15.1318	1176.445	
8	**Variance**	121325.9						
9	**Sum of Squares**	1577237						
10	**Model**	Cp = a5 * T_K^5 + a4 * T_K^4 + a3 * T_K^3 + a2 * T_K^2 + a1 * T_K + a0						

Figure 4–32 Fifth-Degree Polynomial Coefficients and Statistics for the Heat Capacity Data of Table F–1 (File **P4-04A2.XLS**)

and its various powers) are stored. The first logical variable indicates if there is a free parameter (TRUE) in the expression, and the second logical variable indicates whether correlation statistics should be shown (TRUE) in addition to the parameter values.

The regression model parameters are shown in the 4th row of Figure 4–32. The respective parameter standard deviations σ_j, as provided by the LINEST function, are shown in row 5. The respective 95% confidence intervals are calculated in row 7 by multiplying the σ_j by the statistical t distribution value consistent with the number of degrees of freedom (the appropriate t value is inserted by the POLYMATH export utility). The confidence interval of the parameter a_0 is calculated, for example, using the formula

=2.017*Q5

The linear correlation coefficient ($R2 = 0.999947$) in cell L6 and the standard error on the dependent variable in cell M6 are also calculated by the LINEST function. The Variance is calculated in cell L8 (=(M6)^2), and the Sum of Squares of the Residuals in cell L9 (=SUM(I4:I44)) is calculated from the generated Excel table.

When changes are introduced in the data, the Excel results table (Figure 4–32) will be updated correctly unless there is a change in the number of data points. If the number of data points is reduced or increased, the data range for the LINST function must be changed, and a different t value (reflecting the change in the degrees of freedom) must be introduced.

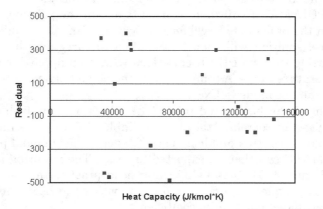

Figure 4–33 Residual Plot Created in Excel for Heat Capacity Represented by Fifth-Degree Polynomial for the Data Set A

The parameter values for the polynomial shown in Figure 4–32 are used to calculate the "Cp calc" values of Figure 4–31. For example, the formula to calculate "Cp calc" for T = 50 K is

=$L$4*A4^5+$M$4*A4^4+$N$4*A4^3+$O$4*A4^2+$P$4*A4^1+$Q$4

Note that these formulas are automatically generated by the POLYMATH software when the export to Excel is requested. The respective residuals, (Cpcalc-Cp), are calculated and placed in column H.

The residual plot, that can be created within Excel, is presented in Figure 4–33. The correlation coefficient is R^2 = 0.9999, and the variance has been significantly reduced. All of the confidence intervals are smaller in absolute value than the associated parameter values. The residual plot of Figure 4–33 indicates a random residual distribution with maximum error ~1%, which is very similar to the magnitude of the experimental error for this type of data. Thus it can be concluded that the fifth-degree polynomial adequately represents the heat capacity data of Table F–1 in Appendix F.

(b) DIPPR[5] recommends an equation for heat capacity of ethane for the temperature range from 200 K through 1500 K

$$C_p = A + B\left[\frac{C/T}{\sinh(C/T)}\right] + D\left[\frac{E/T}{\cosh(E/T)}\right] \tag{4-43}$$

with parameters A = 4.0326E+04, B = 1.3422E+05, C = 1.6555E+03, D = 7.3223E+04, and E = 7.5287E+02. For the more limited temperature range from 50 K through 200 K, DIPPR recommends using a second-degree polynomial

$$C_p = a_0 + a_1 T + a_2 T^2 \tag{4-44}$$

with the parameter values a_0=3.1742E+04, a_1= 2.6567E+01, and a_2 = 1.2927E-01.

A comparison of the heat capacity data correlations first requires the determination of the fifth-order polynomial for the ethane data of Table F–2 in Appendix F. POLYMATH will then be used to obtain the polynomial and subsequently export the problem to Excel for verification of the polynomial representation. The Excel solution will then be modified to carry out the heat capacity calculations using the two DIPPR equations with each applied over the recommended temperature range. A comparison of the polynomial with the DIPPR correlations will then be made in Excel.

The problem can be entered into POLYMATH and the fifth-order polynomial can be used to correlate the data of Table F–2 in the same manner as described in part (a) of this problem. The fifth-degree polynomial problem specified in POLYMATH can then be exported to Excel. The resulting Excel solution is shown in Figure 4–34. It is helpful and good practice to also carry out the POLYMATH polynomial regression in order to verify the Excel solution by comparing the   calculated polynomial coefficients.

The heat capacity values recommended by DIPPR (Equations (4-43) and (4-44)) and the corresponding residual calculations can easily be compared by inserting two new columns in the worksheet immediately to the right of the "Cp residual^2" column I in the Excel worksheet (see Figure 4–36). The five coefficients of Equation (4-43) are entered in the range of cells G48:K48 and the three coefficients of Equation (4-44) are stored in the range of cells G49:I49 as shown in Figure 4–35.

The calculated heat capacity values from the DIPPR equations can be entered in Column J with title "CpD calc" and the residuals are entered in column K with title "CpD residual." The formula for calculating CpD for the first 11 data points ($T \le 200K$) is given by the Excel equivalent to Equation (4-44).

$$=\$G\$49+\$H\$49*A4+\$I\$49*A4^2$$

Note that this formula refers to $T = 100$ K in Figure 4–36.

	K	L	M	N	O	P	Q	R
1								
2	Polynomial Regression of degree 5. Including a free parameter.							
3		a5	a4	a3	a2	a1	a0	
4	Coefficients	5.07E-09	-8.2E-06	0.004616	-0.99332	147.0626	27071.2	
5	Std.dev.s	1.46E-10	2.19E-07	0.000126	0.034084	4.342579	206.0041	
6	R2, SE (y)	0.999999	12.71245	#N/A	#N/A	#N/A	#N/A	
7	95% conf. int.	2.94E-10	4.43E-07	0.000254	0.068917	8.780695	416.5403	
8	Variance	161.6065						
9	Sum of Squares	5856.227						
10	Model	Cp = a5 * T_K^5 + a4 * T_K^4 + a3 * T_K^3 + a2 * T_K^2 + a1 * T_K + a0						

Figure 4–34 Fifth-Degree Polynomial Coefficients and Statistics from Excel for the Heat Capacity Data of Table F–2 of Appendix F

	A or a0	B or a1	C or a2	D	E
47	A or a0	B or a1	C or a2	D	E
48	4.0326E+04	1.3422E+05	1.6555E+03	7.3223E+04	7.5287E+02
49	3.1742E+04	2.6567E+01	1.2927E-01		

Figure 4–35 Coefficients of the DIPPR Equations (File **P4-04.XLS (Cp_Table B)**)

	A	B	C	D	E	F	G	H	I	J	K
1	**POLYMATH Polynomial Regression Migration Document**										
2											
3	T_K	T_K^2	T_K^3	T_K^4	T_K^5	Cp	Cp calc	Cp residual	Cp residual ^2	CpD calc	CpD resid
4	100	10000	1000000	1E+08	1E+10	35698	35695	-3.0314659	9.18978525	3.57E+04	-6.6
5	110	12100	1331000	1.46E+08	1.61E+10	36249	36260	10.8427755	117.565781	3.62E+04	-20.463
6	120	14400	1728000	2.07E+08	2.49E+10	36817	36826	8.65949312	74.98682118	3.68E+04	-25.472
7	130	16900	2197000	2.86E+08	3.71E+10	37401	37402	0.63596182	0.404447442	3.74E+04	-20.627
8	140	19600	2744000	3.84E+08	5.38E+10	38003	37996	-7.2374236	52.38030027	3.80E+04	-7.928
9	150	22500	3375000	5.06E+08	7.59E+10	38628	38615	-13.136289	172.5620765	3.86E+04	7.625
10	160	25600	4096000	6.55E+08	1.05E+11	39279	39265	-14.341419	205.6762906	3.93E+04	23.032
11	170	28900	4913000	8.35E+08	1.42E+11	39961	39950	-11.1779	124.945454	4.00E+04	33.293
12	180	32400	5832000	1.05E+09	1.89E+11	40680	40674	-5.9542598	35.4532099	4.07E+04	32.408
13	190	36100	6859000	1.3E+09	2.48E+11	41439	41440	1.0983953	1.206472241	4.15E+04	17.377
14	200	40000	8000000	1.6E+09	3.2E+11	42243	42250	6.88723704	47.43403401	4.22E+04	-16.8
15	210	44100	9261000	1.94E+09	4.08E+11	43092	43105	12.5185765	156.7147568	43221.87	129.86679
16	220	48400	1.1E+07	2.34E+09	5.15E+11	43989	44004	15.3587237	235.8903934	43981.94	-7.057476
17	230	52900	1.2E+07	2.8E+09	6.44E+11	44934	44949	15.0948477	227.8544275	44831.64	-102.3636
18	240	57600	1.4E+07	3.32E+09	7.96E+11	45924	45938	13.7958363	190.3251001	45764.11	-159.8942
19	250	62500	1.6E+07	3.91E+09	9.77E+11	46959	46969	9.97315603	99.46384127	46771.34	-187.6573
20	260	67600	1.8E+07	4.57E+09	1.19E+12	48036	48041	4.64171187	21.54548908	47844.67	-191.335

Figure 4–36 Addition of DIPPR Equation Calculations to Excel Spreadsheet (File **P4-04.XLS** (Cp_Table B))

The remaining data points use the Excel equivalent to Equation (4-43) as it is applied to temperatures greater than 200 K. This is shown below for cell H19 in Figure 4–36.

```
=$G$48+$H$48*(($I$48/A19)/SINH($I$48/A19))^2+$J$48*(($K$48/
A19)/COSH($K$48/A19))^2
```

The residuals for the DIPPR equations are calculated in Column K by entering the formula for the difference between the DIPPR result in Column J and the measured C_p in Column F.

The residuals of the heat capacity values calculated by fifth order polynomial in Column H and the DIPPR equations in Column K can be plotted in Excel as shown in Figure 4–37. The maximal error in polynomial representation is < 0.1% and the maximal error in the DIPPR correlation is about 0.5%. Note that

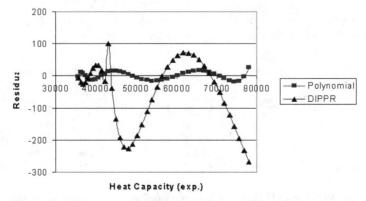

Figure 4–37 Residual Comparison of Heat Capacity Representation by a Fifth-Degree Polynomial and the DIPPR Equations for Table F–2 (File **P4-04.XLS**)

the larger error for DIPPR is expected as the DIPPR correlation of Equation (4-43) is for a much larger range of temperature. The residuals of both correlations show cyclic trends, and these trends can probably be attributed to prior smoothening of the experimental data.

(c) The Wagner equation is considered by many as the most appropriate model to represent the vapor pressure data over the full range between the triple point and critical point. The most widely used form of the Wagner equation is

$$\ln P_R = \frac{a\tau + b\tau^{1.5} + c\tau^3 + d\tau^6}{T_R} \tag{4-45}$$

where $T_R = T/T_C$ is the reduced temperature, $P_R = P/P_C$ is the reduced pressure, and $\tau = 1 - T_R$. For ethane, $T_C = 305.32$ K, $P_C = 4.8720E+06$ Pa and the triple point temperature is 90.352 K. Thus the data in Table F–3 of Appendix F cover almost the full range between the triple point and the critical point, and the Wagner equation is appropriate for correlation of these data.

The use of Excel for solving this problem is preceded by the use of POLYMATH to enter the data into the POLYMATH Data Table. The ability to easily transform data is utilized in POLYMATH to define additional columns in the Data Table as transformation functions defined by

```
TR = T / 305.32
lnPR = ln(P/4872000)
t = (1-TR)/TR
t15 = (1-TR)^1.5/TR
t3 = (1-TR)^3/TR
t6 = (1-TR)^6/TR
```

The resulting POLYMATH Data Table is partially shown in Figure 4–38.

These data transformations allow Multiple Linear Regression to fit the data to the Wagner equation with lnPr as the dependent variable and the independent variables t, t15, t3, and t6. Note that in this Multiple Linear Regression there should be no free parameter; thus, the POLYMATH Data Table option "through origin" should be marked. This problem is exported to Excel after it is setup in the POLYMATH Regression Data Table.

	R001 : C004	lnPr		X ✓	= ln(P_Pa / 4872000)			
	T_K	P_Pa	Tr	lnPr	t	t15	t3	t6
01	92	1.7	0.3013232	-14.86839	2.318696	1.938126	1.13187	0.3860338
02	94	2.8	0.3078737	-14.3694	2.248085	1.870275	1.07692	0.3570586
03	96	4.6	0.3144242	-13.87296	2.180417	1.805374	1.024827	0.3302303
04	98	7.2	0.3209747	-13.42493	2.11551	1.743244	0.9754096	0.305383
05	100	11	0.3275252	-13.00112	2.0532	1.683718	0.9285029	0.2823653

Figure 4–38 POLYMATH Data Table with Original and Transformed Data Columns (File **P4-04C.POL**)

	F	G	H	I	J	K	L	M	N
1	**Migration Document**								
2					Multiple Linear Regression. No free parameter.				
3	lnPr calc	lnPr residual	lnPr residual ^2			a4	a3	a2	a1
4	-14.8538	0.014542267	0.000211478		**Coefficients**	-1.2599194	-1.67115626	1.28954942	-6.4584809
5	-14.357	0.012431298	0.000154537		**Std.dev.s**	0.14859015	0.135235971	0.1086674	0.04802602
6	-13.8828	-0.009812224	9.62797E-05		**R2, SE (y)**	0.99999741	0.009668808	#N/A	#N/A
7	-13.4298	-0.004871516	2.37317E-05		**95% conf. int.**	0.29123669	0.265062504	0.21298811	0.094131
8	-12.9967	0.004373583	1.91282E-05		**Variance**	9.3486E-05			
9	-12.5824	-0.016578932	0.000274861		**Sum of Squares**	0.00962904			
10	-12.1856	-0.005438393	2.95761E-05		**Model**	lnPr = a4 * t + a3 * t15 + a2 * t3 + a1 * t6			

Figure 4–39 Wagner Equation Model Results for the Ethane Vapor Pressure (File **P4-04.XLS** (Vp_Regress))

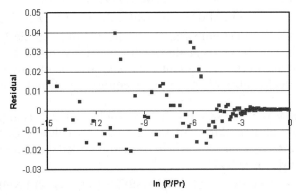

Figure 4–40 Residual Plot in Excel for Ethane Vapor Pressure Data Represented by the Wagner Equation (File **P4-04.XLS** (Vp_Regress))

The Excel results after export from POLYMATH for fitting the Wagner equation to the vapor pressure data are partially presented in Figure 4–39, and the residuals are plotted in Figure 4–40. The correlation coefficient is R^2 = 0.99999, and all the confidence intervals are smaller in absolute value than the associated parameter values. The residual plot shows random residual distribution, and the maximum error is <1%, which is very similar to the magnitude of the experimental error for this type of data. Thus it can be concluded that the Wagner equation adequately correlates the vapor pressure of ethane over the experimental temperature range.

The Riedel equation recommended by DIPPR for vapor pressure data of ethane is given by

$$\ln P = A - \frac{B}{T} + C \ln T + D T^E \qquad \text{(4-46)}$$

with the parameters A = 51.857, B = –2600, C = –5.13, D = 1.49E-05 and E = 2.

The comparison between the Wagner equation and the Riedel equation can be carried out by creating a new Excel worksheet that utilizes the Wagner equation variables and results given in the POLYMATH to Excel worksheet shown in Figure 4–39. Some of the information entered in this prepared worksheet is shown in Figure 4–41 (only four rows of data, out of the 107 data points in this case, are shown). The measured temperature and vapor pressure data are inserted in columns A and B.

	A	B	C	D	E	F	G	H	I	J	K
			A	B	C	D	E				
1											
2			5.1857E+01	-2.60E+03	-5.13E+00	1.49E-05	2.00E+00				
3	Temp (K)	Vapor Pressure (Pa)	lnPr	lnPr calc	lnPr residual	lnPr Calc DIPPR	lnPr Res DIPPR	Calc VP Wagner	VP Res Wagner	Calc VP DIPPR	VP Res DIPPR
4	92	1.7	-14.86839	-14.8538477	0.014542267	-14.8516	0.01677	1.7249	0.0249	1.72875	0.0287
5	94	2.8	-14.3694	-14.3569687	0.012431298	-14.3554	0.01403	2.83501	0.03501	2.83956	0.0396
6	96	4.6	-13.87296	-13.8827722	-0.00981222	-13.8817	-0.00876	4.55508	-0.0449	4.5599	-0.0401
7	98	7.2	-13.42493	-13.4298015	-0.00487152	-13.4292	-0.0043	7.16504	-0.035	7.16916	-0.0308

Figure 4–41 Worksheet for Comparison of Vapor Pressure Correlation by Wagner and DIPPR Equations (File **P4-04.XLS** (Vp_Compare))

The data of "lnPr" and "lnPr calc" (columns C and D in Figure 4–41) are copied from the POLYMATH migration worksheet that is partially shown in Figure 4–39. Note that in order to paste the "lnPr calc" values, the "Paste Special Values" should be used otherwise error messages will be obtained (and the data columns and the coefficients of the Wagner equation will not be copied into the new worksheet).

In the 2nd row, the numerical values of the Riedel equation parameters are entered with their names shown in the 1st row. In column E, the "lnPr Calc DIPPR" is calculated using the DIPPR recommended equation by manually entering the formula for cell D4.

=($C$2+$D$2/A4+$E$2*LN(A4)+$F$2*(A4)^$G$2)-LN(4872000)

Then this formula is copied to all the cells below for the entire data set.

The residual plot of the "lnPr Res DIPPR" in this case is very similar to the residual plot obtained for the Wagner equation (Figure 4–40). The comparison between the two equations is more meaningful if it is carried out with the help of the residual plots based on the pressure (instead of $\ln(P_R)$). The preparation of such a plot is left as an exercise for the reader.

(d) A recommended correlation for viscosity of liquids by Perry[6] is similar to the Antoine equation for vapor pressure and given by

$$\ln\mu = A + \frac{B}{T+C} \tag{4-47}$$

where μ is the viscosity and the parameters are A, B, and C. If T is expressed in degrees K, then parameter C can be approximated by $C = 17.71 - 0.19T_b$, where T_b is the normal boiling point in K. For ethane, the normal boiling point is 184.55 K, and thus the approximate value of C is -17.35.

Equation (4-47) is nonlinear and can be fitted to the experimental viscosity data of Table F–4 in Appendix F using general nonlinear regression. However, good initial estimates are necessary for the nonlinear regression. These can be obtained by linearizing Equation (4-47) using the approximate value of C for ethane to obtain

$$\ln \mu = A + \frac{B}{T - 17.35} = a_0 + a_1 \left(\frac{1}{T - 17.35} \right) = a_0 + a_1 X_1 \text{ or } Y = a_0 + a_1 X_1 \quad \textbf{(4-48)}$$

Thus, the linear form can be used in the POLYMATH Data Table containing the viscosity and temperature data by creating additional columns to calculate the transformed variables $Y = \ln \mu$ and $X_1 = 1/(T-17.35)$. A portion of the POLYMATH Data Table which utilizes these transformed variables and is set up for the linear regression of Equation (4-48) is shown in Figure 4–42. The results of the POLYMATH Linear Regression are shown in Figure 4–43. These results provide the initial estimates of $A = -11.1$, $B = 364.6$ and $C = -17.35$ for the nonlinear regression of Equation (4-47).

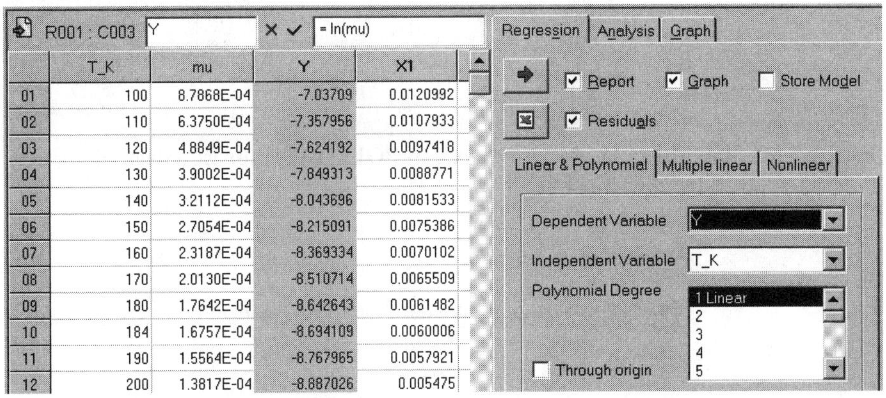

Figure 4–42 Setup of POLYMATH Linear Regression for Equation (4-48) (File **P4-04D1.POL**)

Model: Y = a0 + a1*X1

Variable	Value	95% confidence
a0	-11.09516	0.2912278
a1	364.5876	44.46516

Figure 4–43 Linear Regression Results from POLYMATH for Equation (4-48) (File **P4-04D1.POL**)

The nonlinear regression can be set up in POLYMATH and then exported to Excel. The setup of the POLYMATH Nonlinear Regression is shown in Figure 4–44 which gives the results that are summarized in Figure 4–45 where some 73 iterations were required.

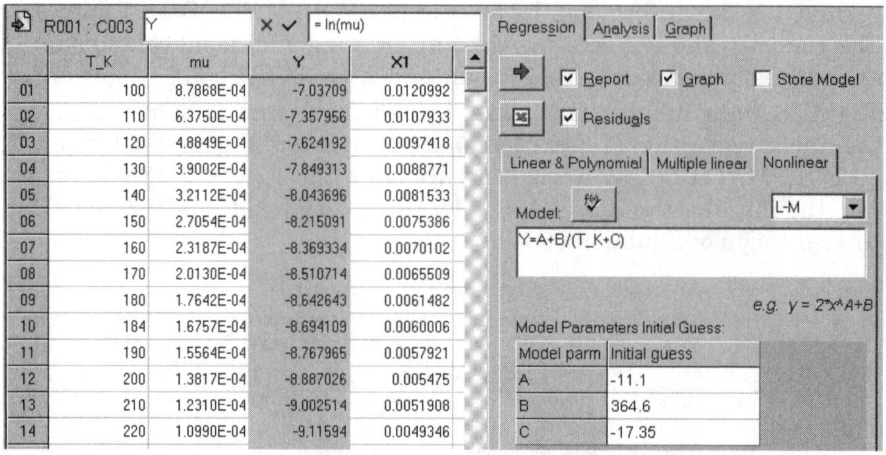

R001 : C003	Y		= ln(mu)				
	T_K	mu	Y	X1			
01	100	8.7868E-04	-7.03709	0.0120992			
02	110	6.3750E-04	-7.357956	0.0107933			
03	120	4.8849E-04	-7.624192	0.0097418			
04	130	3.9002E-04	-7.849313	0.0088771			
05	140	3.2112E-04	-8.043696	0.0081533			
06	150	2.7054E-04	-8.215091	0.0075386			
07	160	2.3187E-04	-8.369334	0.0070102			
08	170	2.0130E-04	-8.510714	0.0065509			
09	180	1.7642E-04	-8.642643	0.0061482			
10	184	1.6757E-04	-8.694109	0.0060006			
11	190	1.5564E-04	-8.767965	0.0057921			
12	200	1.3817E-04	-8.887026	0.005475			
13	210	1.2310E-04	-9.002514	0.0051908			
14	220	1.0990E-04	-9.11594	0.0049346			

Regression | Analysis | Graph

☑ Report   ☑ Graph   ☐ Store Model

☑ Residuals

Linear & Polynomial | Multiple linear | Nonlinear

Model: L-M ▼

Y=A+B/(T_K+C)

e.g. $y = 2*x^A+B$

Model Parameters Initial Guess:

Model parm	Initial guess
A	-11.1
B	364.6
C	-17.35

Figure 4–44 Nonlinear Regression Setup in POLYMATH for Equation (4-48) (File **P4-04D1.POL**)

Model: Y = A +B/(T_K+C)

Variable	Initial guess	Value	95% confidence
A	-11.1	-18.60756	5.789612
B	364.6	6568.593	7779.302
C	-17.3	476.3679	395.2097

Nonlinear regression settings
Max # iterations = 128

Precision

R^2	0.9886379
R^2adj	0.9875016
Rmsd	0.0199363
Variance	0.0105127

General

Sample size	23
Model vars	3
Indep vars	1
Iterations	73

Figure 4–45 Nonlinear Regression Result in POLYMATH for Equation (4-48) (File **P4-04D1.POL**)

The export of the POLYMATH setup for Nonlinear Regression to Excel by pressing the Excel icon gives the initial worksheet that is partially shown in Figure 4–46. Note that this problem in Excel must be solved by using the Excel Add-

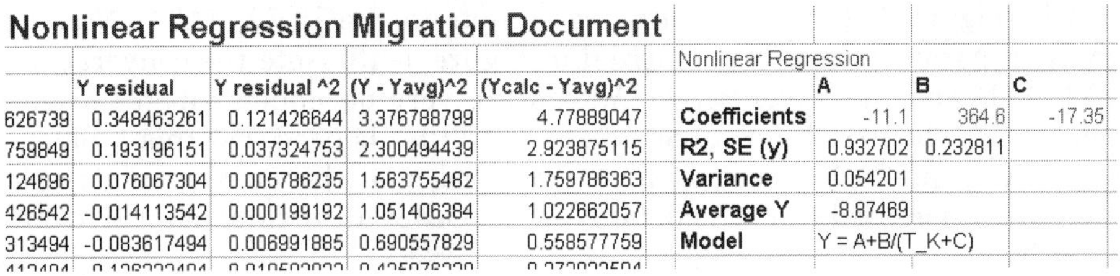

Nonlinear Regression Migration Document

	Y residual	Y residual ^2	(Y - Yavg)^2	(Ycalc - Yavg)^2	Nonlinear Regression	A	B	C
626739	0.348463261	0.121426644	3.376788799	4.77889047	Coefficients	-11.1	364.6	-17.35
759849	0.193196151	0.037324753	2.300494439	2.923875115	R2, SE (y)	0.932702	0.232811	
124696	0.076067304	0.005786235	1.563755482	1.759786363	Variance	0.054201		
426542	-0.014113542	0.000199192	1.051406384	1.022662057	Average Y	-8.87469		
313494	-0.083617494	0.006991885	0.690557829	0.558577759	Model	Y = A+B/(T_K+C)		
412494	0.126222494	0.010502022	0.425078220	0.272822584				

Figure 4–46 Nonlinear Regression Exported to Excel-Initial Worksheet

In called "Solver." This Add-In should be available from the drop-down menu in Excel under "Tools" and then "Add-Ins..."

The objective function for the nonlinear regression problem within Excel is the sum of squares of the Y residuals that is found in the cell at the base of the "Y residual ^2" column.

When Solver is called from the "Tools" menu in Excel to perform the nonlinear regression, an interface appears in which the "Solver Parameters" must be entered. Solver requires that the Target Cell be set as the sum of squares of the Y residuals which should be minimized. Also the Coefficients cells for A, B, and C must be identified in the "By Changing Cells" entry box. This is shown in Figure 4–47. In the "Equal To:" field of the Solver it is important to move the marking to Min (from the default Max marking). After a mouse click on the "Solve" button,

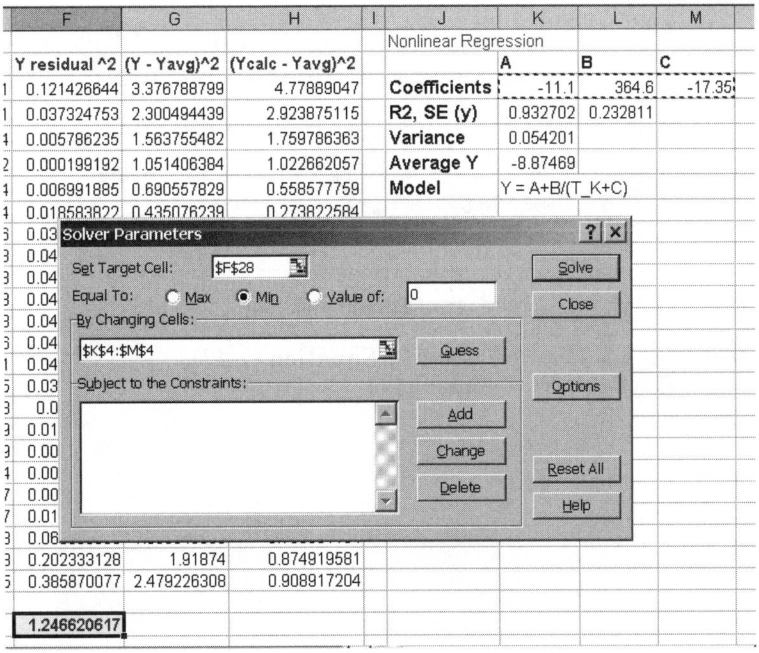

Figure 4–47 Use of the Excel Solver Add-In for Nonlinear Regression

the coefficients are changed to the converged values. In this Solver solution shown in Figure 4–48, the results are similar to the POLYMATH Nonlinear Regression parameters as summarized in Figure 4–45. Note the convergence of POLYMATH and the Excel Solver Add-In are very dependent upon the initial estimates and the particular numerical method that is used. For this problem in POLYMATH, the L-M algorithm has been used, and the number of iterations needed to be increased from the default value. Other algorithms may give different results.

The residual plot from Excel reproduced in Figure 4–49 has a cyclic pattern and considerable errors. This indicates that this model for correlation of ethane viscosity is not very satisfactory. Many more models do exist which could be fitted to these ethane data.

J	K	L	M
Nonlinear Regression			
	A	B	C
Coefficients	-18.5391	6476.254	471.6516
R2, SE (y)	0.988663	0.095612	
Variance	0.009142		
Average Y	-8.87016		
Model	Y = A+B/(T_K+C)		

Figure 4–48 Solver Results for the Excel Nonlinear Regression of Equation (4-47) (File **P4-04.XLS** [Antoine (2)])

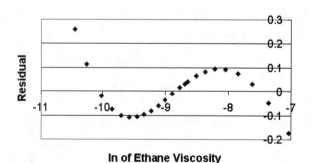

Figure 4–49 Residual Plot from Excel for Equation (4-47) (File **P4-04.XLS** [Antoine (2)])

The comparison with the Riedel equation using the parameters recommended by DIPPR follows the same procedure that was followed in connection with the vapor pressure data correlation and discussed in the solution to the previous part (c).

The Antoine and Riedel equation representations of the liquid viscosity are compared in Figure 4–50. The residuals of the Riedel equation seem to follow a cyclic pattern as do the residuals of the Antoine equation, but the errors are considerably smaller.

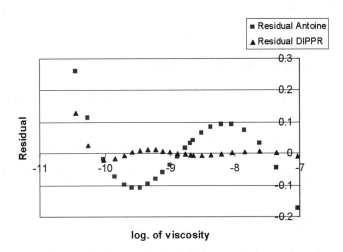

Figure 4–50 Comparison of Viscosity Represented by the Antoine Equation (4-47) and the Riedel Equation with the DIPPR Recommended Constants (File **P4-04.XLS** [Antoine (2)])

The problem solution files are found in directory Chapter 4 and designated **P4-04A.POL, P4-04B.POL, P4-04C.POL, P4-04D1.POL, P4-04D2.POL,** and **P4-04.XLS**.

4.5 COMPLEX CHEMICAL EQUILIBRIUM BY GIBBS ENERGY MINIMIZATION

4.5.1   Concepts Demonstrated

Formulation of a chemical equilibrium problem as a Gibbs energy minimization problem with atom balance constraints. Use of Lagrange multipliers to introduce the constraints into the objective function. Conversion of the minimization problem into a system of nonlinear algebraic equations.

4.5.2   Numerical Methods Utilized

Solution of a system of nonlinear algebraic equations with constraints.

4.5.3   Excel Options and Functions Demonstrated

Use of the Excel Add-In Solver for constrained minimization.

4.5.4   Problem Definition

Ethane reacts with steam to form hydrogen over a cracking catalyst at a temperature of $T = 1000$ K and pressure of $P = 1$ atm. The feed contains 4 moles of H_2O per mole of CH_4. Balzisher et al.[7] suggest that only the compounds shown in Table 4–10 are present in the equilibrium mixture (assuming that no carbon is deposited). The Gibbs energies of formation of the various compounds at the temperature of the reaction (1000K) are also given in Table 4–10. The equilibrium composition of the effluent mixture is to be calculated using these data.

Table 4–10 Compounds Present in Effluent of Steam Cracking Reactor[7]

No.	Component	Gibbs Energy kcal/gm-mol	Feed gm-mol	Effluent Initial Estimate
1	CH_4	4.61		0.001
2	C_2H_4	28.249		0.001
3	C_2H_2	40.604		0.001
4	CO_2	-94.61		0.993
5	CO	-47.942		1
6	O2	0		0.0001[a]
7	H_2	0		5.992
8	H_2O	-46.03	4	1
9	C_2H_6	26.13	1	0.001

[a]This initial estimate is more realistic and useful than the original published estimate of 0.007.

(a) Formulate the problem as a constrained minimization problem. Introduce the constraints into the objective function using Lagrange multipliers and differentiate this function to obtain a system of nonlinear algebraic equations.

(b) Use the POLYMATH "Constrained" solution algorithm to find the solution to this system of nonlinear equations. Start the iterations from the initial estimates shown in Table 4–10.

(c) Use Excel's "Solver" to solve the problem as a constrained minimization problem without the use of Lagrange multipliers and without differentiation of the objective functions. Compare the results with those obtained in (b).

4.5.5   Solution

The objective function to be minimized is the total Gibbs energy given by

$$\min_{n_i} \frac{G}{RT} = \sum_{i=1}^{c} n_i \left(\frac{G_i^0}{RT} + \ln \frac{n_i}{\sum n_i} \right) \tag{4-49}$$

where n_i is the number of moles of component i, c is the total number of compounds, R is the gas constant, and G_i^0 is the Gibbs energy of pure component i at temperature T. The minimization of Equation (4-49) must be carried out subject to atomic balance constraints

Oxygen Balance $g_1 = 2n_4 + n_5 + 2n_6 + n_7 - 4 = 0$ (4-50)

Hydrogen Balance $g_2 = 4n_1 + 4n_2 + 2n_3 + 2n_7 + 2n_8 + 6n_9 - 14 = 0$ (4-51)

Carbon Balance $g_3 = n_1 + 2n_2 + 2n_3 + n_4 + n_5 + 2n_9 - 2 = 0$ (4-52)

The identification of the various components is given in Table 4–10.

These three constraints can be introduced into the objective functions using Lagrange multipliers: λ_1, λ_2, and λ_3. The extended objective function is

$$\min_{n_i, \lambda_j} F = \sum_{i=1}^{c} n_i \left(\frac{G_i^0}{RT} + \ln \frac{n_i}{\sum n_i} \right) + \sum_{j=1}^{3} \lambda_j g_j \tag{4-53}$$

The condition for minimum of this function at a particular point is that all the partial derivatives of F with respect to n_i and λ_j vanish at this point. The partial derivative of F with respect to n_1, for example, is

$$\frac{\partial F}{\partial n_1} = \frac{G_1^0}{RT} + \ln \frac{n_1}{\sum n_i} + 4\lambda_2 + \lambda_3 = 0 \tag{4-54}$$

The other partial derivatives with respect to n_i can be obtained similarly. If it is expected that the amount of a particular compound at equilibrium is very close to zero, it is preferable to rewrite the equation in a form that does not require calculation of the logarithm of a very small number. Rearranging Equation (4-54), for example, yields

$$n_1 - \sum n_i \exp\left(\frac{G_1^0}{RT} + 4\lambda_2 + \lambda_3\right) = 0 \tag{4-55}$$

The partial derivatives of F with respect to λ_1, λ_2, and λ_3 are g_1, g_2, and g_3 respectively.

(b) The complete set of nonlinear equations, as entered into the POLY-MATH Nonlinear Algebraic Equation Solver, is shown in Table 4–11. There are 12 implicit equations associated with the 12 unknowns. In the POLYMATH input, the amount (moles) of a compound (n_i) is represented by the formula of the compound, for clarity. The equations associated with O_2 and C_2H_2 are written in the form of Equation (4-55) as preliminary tests have shown difficulty in convergence of the solution algorithm when the equations that contain logarithms of the amount of those compounds are used.

Table 4–11 Equation Input to the POLYMATH Nonlinear Equation Solver Program (File **P4-05B1.POL**)

No.	Equation
1	R = 1.9872
2	sum = H2 + O2 + H2O + CO + CO2 + CH4 + C2H6 + C2H4 + C2H2
3	f(lamda1) = 2 * CO2 + CO + 2 * O2 + H2O - 4 # Oxygen balance
4	f(lamda2) = 4 * CH4 + 4 * C2H4 + 2 * C2H2 + 2 * H2 + 2 * H2O + 6 * C2H6 - 14 # Hydrogen balance
5	f(lamda3) = CH4 + 2 * C2H4 + 2 * C2H2 + CO2 + CO + 2 * C2H6 - 2 # Carbon balance
6	f(H2) = ln(H2 / sum) + 2 * lamda2
7	f(H2O) = -46.03 / R + ln(H2O / sum) + lamda1 + 2 * lamda2
8	f(CO) = -47.942 / R + ln(CO / sum) + lamda1 + lamda3
9	f(CO2) = -94.61 / R + ln(CO2 / sum) + 2 * lamda1 + lamda3
10	f(CH4) = 4.61 / R + ln(CH4 / sum) + 4 * lamda2 + lamda3
11	f(C2H6) = 26.13 / R + ln(C2H6 / sum) + 6 * lamda2 + 2 * lamda3
12	f(C2H4) = 28.249 / R + ln(C2H4 / sum) + 4 * lamda2 + 2 * lamda3
13	f(C2H2) = C2H2 - exp(-(40.604 / R + 2 * lamda2 + 2 * lamda3)) * sum
14	f(O2) = O2 - exp(-2 * lamda1) * sum
15	H2(0) = 5.992
16	O2(0) = 0.0001 > 0
17	H2O(0) = 1
18	CO(0) = 1
19	CH4(0) = 0.001 > 0
20	C2H4(0) = 0.001 > 0
21	C2H2(0) = 0.001 > 0
22	CO2(0) = 0.993
23	C2H6(0) = 0.001 > 0
24	lamda1(0) = 10
25	lamda2(0) = 10
26	lamda3(0) = 10

The initial estimates suggested by Balzisher et al.[7] (shown in Table 4–10) are entered in lines 15 through 23 of the input to POLYMATH. Note that the initial estimate O2(0)=0.0001>0, for example, indicates that this variable is constrained to be always positive ("absolutely positive") during the problem solution. For this problem, only the "constrained" POLYMATH solution algorithm converges to the solution from the given initial estimates, while the other algorithms stop with the error message that the calculation of the logarithm of a negative number is attempted. The constrained algorithm also allows definition of some or all of the unknowns to be positive at the solution ("physically positive"). For O_2 in this problem, the corresponding "physically positive" format as specified in the initial estimate would be O2(0)=0.0001>=0. This format was not used in this problem.

The variables for which the initial estimates are specified in lines 16, 19, 20, 21, and 23 in Table 4–10 are marked as "absolutely positive." The rest of the variables do not approach zero so as to obtain negative values during the solution process (lines 15, 17, 18, and 19), or they are allowed to have both negative and positive values (lines 24, 25, and 26).

The POLYMATH solution for the equilibrium composition of the effluent mixture is shown in Table 4–12. These results indicate that the effluent contains

Table 4–12 POLYMATH Results for Equilibrium Composition of Effluent Stream (File **P4-05B1.POL**)

	Variable	Value	f(x)	Initial Guess
1	C2H2	3.157E-10	7.238E-25	0.001
2	C2H4	9.541E-08	-2.58E-13	0.001
3	C2H6	1.671E-07	-1.688E-13	0.001
4	CH4	0.0665638	0	0.001
5	CO	1.388517	2.442E-15	1.
6	CO2	0.5449182	-1.11E-15	0.993
7	H2	5.345225	1.11E-16	5.992
8	H2O	1.521646	-1.665E-15	1.
9	lamda1	24.41966	0	10.
10	lamda2	0.2530591	0	10.
11	lamda3	1.559832	0	10.
12	O2	5.459E-21	-5.687E-27	0.0001
13	R	1.9872		
14	sum	8.866871		

significant amounts of H_2 (5.345 moles per mole of C_2H_6 feed), H_2O, CO, CO_2 and CH_4 and contains only trace amounts of C_2H_6, C_2H_4, C_2H_2 and O_2. All the function values are smaller by several orders of magnitude than the respective variable values, indicating that the solution is correct. The same values were obtained also by Balzisher et al.,[7] who used a dedicated FORTRAN program to solve the same problem.

The value of the objective function at this point may be added to the POLYMATH problem by including Equation (4-49). The additional equations comprising Equation (4-49) are summarized in Table 4–13. The objective function value at the solution is –104.34, and this can be used to compare the POLYMATH solution with the Excel solution in part (c).

Table 4–13 Equations for Calculation of the Gibbs Energy Function (File **P4-05B2.POL**)

Line	Equation
1	G_O2 = O2 * ln(abs(O2 / sum))
2	G_H2 = H2 * ln(H2 / sum)
3	G_H2O = H2O * (-46.03 / R + ln(H2O / sum))
4	G_CO = CO * (-47.942 / R + ln(CO / sum))
5	G_CO2 = CO2 * (-94.61 / R + ln(CO2 / sum))
6	G_CH4 = CH4 * (4.61 / R + ln(abs(CH4 / sum)))
7	G_C2H6 = C2H6 * (26.13 / R + ln(abs(C2H6 / sum)))
8	G_C2H4 = C2H4 * (28.249 / R + ln(abs(C2H4 / sum)))
9	G_C2H2 = C2H2 * (40.604 / R + ln(abs(C2H2 / sum)))
10	ObjFun = G_H2 + G_H2O + G_CO + G_O2 + G_CO2 + G_CH4 + G_C2H6 + G_C2H4 + G_C2H2

(c) The use of the Excel Add-In "Solver"[*] to solve this problem requires only the Gibbs Energy objective function of Equation (4-49) as presented in Table 4–13 and the atomic material balance constraints given by Equations (4-50), (4-51), and (4-52). It is convenient to enter these needed equations into Excel via POLYMATH. The resulting equations from POLYMATH as exported to Excel are given in Table 4–14. Note that the Gibbs Energy objective function is in line 4, and the atomic constraints are in lines 5, 6, and 7. The equations in line 4 and lines 9 through 18 are components of the objective function (Equation (4-49)). Lines 19 through 27 provide the initial estimates for the nine problem variables.

Table 4–14 Gibbs Energy Minimization with Atom Balance Constraints as Exported from POLYMATH to Excel (Files **P4-05C.POL** and **P4-05C.XLS**)

	A	B	C	D	E
1	POLYMATH NLE Migration Document				
2		Variable	Value		Polymath Equation
3	Explicit Eqs	R	1.9872		R=1.9872
4		sum	8.996		sum=H2 + H2O + CO + O2 + CO2 + CH4 + C2H6 + C2H4 + C2H2
5		OxBal	-4.441E-16		OxBal=2 * CO2 + CO + 2 * O2 + H2O - 4
6		HydBal	0		HydBal=4 * CH4 + 4 * C2H4 + 2 * C2H2 + 2 * H2 + 2 * H2O + 6 * C2H6 - 14
7		CarBal	0		CarBal=CH4 + 2 * C2H4 + 2 * C2H2 + CO2 + CO + 2 * C2H6 - 2
8		eps	1E-21		eps=0.1e-20
9		G_O2	-0.0501104		G_O2=O2 * ln(abs((O2 + eps) / sum))

[*] The Excel Add-In "Solver" may require special installation from Microsoft Excel or Microsoft Office. If this Add-In is not available from the drop-down Tools/Add-Ins menu in Excel, please consult the Microsoft instructions to install this software.

	A	B	C	D	E
10		G_H2	-2.4348779		G_H2=H2 * ln(H2 / sum)
11		G_H2O	-25.360025		G_H2O=H2O * (-46.03 / R + ln(H2O / sum))
12		G_CO	-26.322183		G_CO=CO * (-47.942 / R + ln(CO / sum))
13		G_CO2	-49.464812		G_CO2=CO2 * (-94.61 / R + ln(CO2 / sum))
14		G_CH4	-0.0067847		G_CH4=CH4 * (4.61 / R + ln(CH4 / sum))
15		G_C2H6	0.00404462		G_C2H6=C2H6 * (26.13 / R + ln(abs((C2H6 + eps) / sum))
16		G_C2H4	0.00511094		G_C2H4=C2H4 * (28.249 / R + ln(abs((C2H4 + eps) / sum))
17		G_C2H2	0.01132823		G_C2H2=C2H2 * (40.604 / R + ln(abs((C2H2 + eps) / sum))
18		ObjFun	-103.61831		ObjFun=G_H2 + G_H2O + G_CO + G_O2 + G_CO2 + G_CH4 + G_C2H6 + G_C2H4 + G_C2H2
19		O2	0.007		O2=0.007
20		H2	5.992		H2=5.992
21		H2O	1		H2O=1
22		CO	1		CO=1
23		CH4	0.001		CH4=0.001
24		C2H4	0.001		C2H4=0.001
25		C2H2	0.001		C2H2=0.001
26		CO2	0.993		CO2=0.993
27		C2H6	0.001		C2H6=0.001

It should be noted that the Solver Add-In cannot find the minimum when Equation (4-49) is entered in its original form. Excel execution stops with error messages indicating that there is an attempt to calculate logarithm of a negative number. In order to prevent calculation of logarithm of a negative number or logarithm of zero, the expressions for calculating the partial Gibbs energy of some of the compounds have been changed according to

$$
\frac{\bar{r}_i}{T} = n_i \left\{ \frac{G_i^0}{RT} + \ln \left[\text{abs} \left(\frac{n_i + \varepsilon}{\sum n_i} \right) \right] \right\}
\tag{4-56}
$$

where ε is a very small number ($\varepsilon = 1 \times 10^{-21}$). This equation is used only for the compounds which present in trace amounts in the effluent: O_2, C_2H_2, C_2H_4, and C_2H_6.

The use of the Excel Solver Add-In for the solution of this minimization problem is best explained with reference to Figures 4–51, 4–52, and 4–53. After the "Solver" interface is requested within the Excel "Tools" drop-down menu as shown in Figure 4–51, the "Target Cell" to be minimized should be entered as cell C18 which contains the formula for the objective function (see Table 4–14). Next, the "Equal to:" option should be set to "Min". The cells to be changed are the cells which contain the numbers of moles of the various compounds which is indicated in Excel by $C$19:$C$27. The three constraint equations from the atomic material balance constraints given by Equations (4-50), (4-51), and (4-52) are entered by reference to the Excel cell containing each equation as shown in Figure 4–51. These balances should all be equal to zero at the solution. Addition of the first constraint is shown in Figure 4–52.

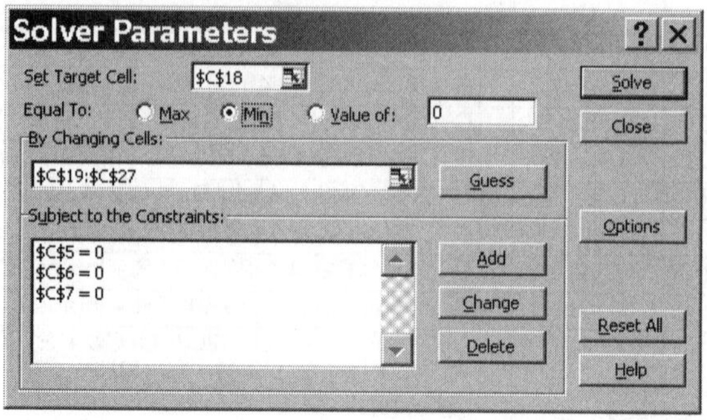

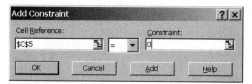

Figure 4–51 Solver Parameters for the Gibbs Energy Minimization Problem
(File **P4-05C.XLS**)

Figure 4–52 Addition of First Constraint for Solver Minimization
(File **P4-05C.XLS**)

The constraints regarding the positive value of the compounds can be specified in the "Options" communication box shown in Figure 4–53. Two default options must be changed with the "Options" button to obtain a solution. First, "Assume Non-Negative" must be selected, otherwise the "Solver" converges to a solution where some of the mole numbers are negative. Second, "Use Automatic Scaling" must also be selected in order to obtain a feasible solution. The Excel Solver solution is partially shown in Figure 4–54.

The results obtained by the POLYMATH constrained Nonlinear Equation

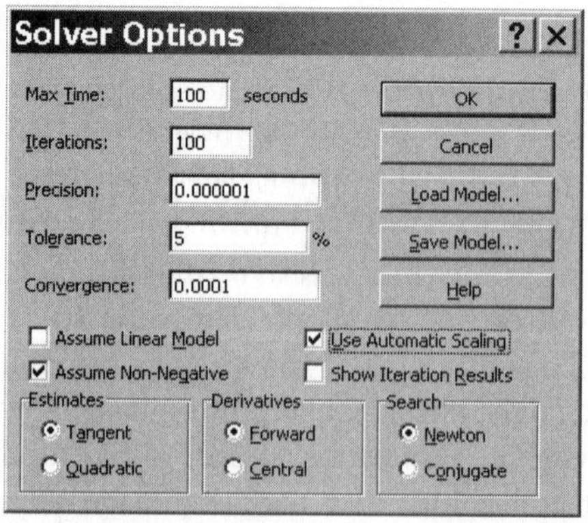

Figure 4–53 Solver Options for the Gibbs Energy Minimization Problem

18	ObjFun	-104.2757	ObjFun=G_H2 + G_H2O + G_CO + G_O2 + G_CO2 + G_CH4 + G_C2	
19	O2	0	O2=1e-20
20	H2	5.5304163	H2=5.992
21	H2O	1.466646	H2O=1
22	CO	1.4636555	CO=1
23	CH4	0.001396	CH4=0.001
24	C2H4	3.117E-06	C2H4=0.001
25	C2H2	0	C2H2=0.001
26	CO2	0.5348493	CO2=0.993
27	C2H6	4.651E-05	C2H6=0.001

Figure 4–54 Partial Excel Spreadsheet Showing Solver Minimization Solution

Solver, the Excel Solver constrained minimization algorithm, and the values reported by Balzisher et al.[7] are summarized in Table 4–15. Examination of the various results indicates that the minimum Gibbs energy obtained by POLYMATH is slightly lower than the value obtained by the Solver. The composition values obtained by POLYMATH are almost identical to those reported by Balzisher et al.[7] There are considerable differences between those values from POLYMATH and those obtained by the Excel Solver Add-In. There may be other optimization routines that can be used with Excel to achieve a closer agreement in the solution.

Table 4–15 Effluent Composition and Minimum of Gibbs Energy by Various Solution Techniques

No.	Component	Initial Estimate	POLYMATH	Excel Solver	Balzisher et. al[7]
1	CH4	0.001	0.066564	0.00149444	0.066456
2	C2H4	0.001	9.54E-08	1.0112E-06	9.41E-8
3	C2H2	0.001	3.16E-10	2.9847E-07	3.15E-10
4	CO2	0.993	0.544918	0.53441967	0.544917
5	CO	1	1.388517	1.46396259	1.3886
6	O2	0.007	5.46E-21	0	3.704E-21
7	H2	5.992	5.345225	5.52962977	5.3455
8	H2O	1	1.521646	1.46719804	1.5215
9	C2H6	0.001	1.67E-07	6.0332E-05	1.572E-7
	Gibbs Energy	-103.61831	-104.34	-104.27612	

The problem solution files are found in directory Chapter 4 and designated **P4-05B1.POL**, **P4-05B2.POL**, **P4-05C.POL**, and **P4-05C.XLS**.

www

REFERENCES

1. Shacham, M., Brauner, N., and Cutlip, M. B. "An Exercise for Practicing Programming in the ChE Curriculum–Calculation of Thermodynamic Properties Using the Redlich-Kwong Equation of State," Chem. Eng. Educ., 37(2), 148 (2003).
2. Seader, J. D. and Henley, E. J., *Separation Process Principles*, 2nd ed., New York, NY: McGraw-Hill, 2005.
3. Shacham, M., *Ind. Eng. Chem. Fund.*, *19*, 228-229 (1980).
4. Fogler, H. S., *Elements of Chemical Reaction Engineering*, 3rd ed. Upper Saddle N.J.: Prentice Hall, p. 523, 1999.
5. DIPPR, http://dippr.byu.edu/samplesite/
6. Perry, R. H., Green, D. W., and Malorey, J. D., eds. *Perry's Chemical Engineers Handbook*, 6th ed, New York: McGraw-Hill, 1984.
7. Balzisher, R. E., Samuels, M. R., and Eliassen, J. D., *Chemical Engineering Thermodynamics*, Englewood Cliffs, N.J.: Prentice Hall, 1972.

Problem Solving with MATLAB

5.1 MOLAR VOLUME AND COMPRESSIBILITY FROM REDLICH-KWONG EQUATION

5.1.1 Concepts Demonstrated

Analytical solution of the cubic Redlich-Kwong equation for compressibility factor and calculation of the molar volume at selected reduced temperatures and reduced pressures.

5.1.2 Numerical Methods Utilized

Solution of a set of explicit equations.

5.1.3 MATLAB Options and Functions Demonstrated

Interactive execution of m-files; definitions of scalars and arrays; clearing variables; arithmetic operations; math and logical functions; command window control functions: *clc*, *format* and *disp*; graphing functions: *plot*, *loglog*, *title*, *xlabel* and *ylabel*; control flow functions: *while*, *for*, *if*, *else*, and *end*; comments.

5.1.4 Problem Definition

Work Problem 4.1 using MATLAB.

(a) Use POLYMATH to generate the MATLAB formatted equation set for this problem. Convert the equation set into a MATLAB function which obtains T_r, P_r, T_c, and P_c as input parameters and returns to the calling point the values of the compressibility factor z and the molar volume V. Check the function by calculating the compressibility factor and the volume of steam (critical temperature, $T_c = 647.4$ K and critical pressure, $P_c = 218.3$ atm) at $T_r = 1.0$ and $P_r = 1.2$. Compare with the values obtained by POLYMATH ($z = 0.258$ and $V = 0.0523$ L/g-mol). Repeat the calculation for $T_r = 3.0$ and $P_r = 10$ ($z = 1.146$ and $V = 0.0837$ L/g-mol). Carry out the calculations only if the parameter C > 0.

(b) Calculate the compressibility factor and the molar volume of steam using MATLAB for the reduced temperatures and reduced pressures listed in Table 5–1. Prepare a table and a plot of the compressibility factor versus P_r and T_r as well as a table and a plot of the molar volume versus pressure and T_r. The pressure and the volume should be in a logarithmic scale in the second plot.

Table 5–1 Reduced Pressures and Temperature for Calculation

Pr	Pr	Pr	Pr	Pr	Tr
0.1	2	4	6	8	1
0.2	2.2	4.2	6.2	8.2	1.2
0.4	2.4	4.4	6.4	8.4	1.5
0.6	2.6	4.6	6.6	8.6	2.0
0.8	2.8	4.8	6.8	8.8	3.0
1	3	5	7	9	
1.2	3.2	5.2	7.2	9.2	
1.4	3.4	5.4	7.4	9.4	
1.6	3.6	5.6	7.6	9.6	
1.8	3.8	5.8	7.8	9.8	
				10	

5.1.5 Solution

(a) The equation set for this problem, as entered into the POLYMATH Nonlinear Equations Solver program, is shown in Table 5–2. Note that the row numbers have been added only to help with the explanations; they are not part of the POLYMATH input. Additional details are given in Problem 4.1 in Chapter 4.

Table 5–2 POLYMATH Equation Set for Problem 5.1 (File **P5-1A.POL**)

Line	Equation
1	R = 0.08206 # Gas constant (L-atm/g-mol-K)
2	Tc = 647.4 # Critical temperature (K)

Line	Equation
3	Pc = 218.3 # Critical pressure (atm)
4	a = 0.42747 * R ^ 2 * Tc ^ (5 / 2) / Pc # Eq. (4-2), RK equation constant
5	b = 0.08664 * R * Tc / Pc # Eq. (4-3),RK equation constant
6	Pr = 0.1 # Reduced pressure (dimensionless)
7	Tr = 1 # Reduced temperature (dimensionless)
8	r = Asqr * B # Eq. (4-6)
9	q = B ^ 2 + B - Asqr # Eq. (4-7)
10	Asqr = 0.42747 * Pr / (Tr ^ 2.5) # Eq. (4-8)
11	B = 0.08664 * Pr / Tr # Eq. (4-9)
12	C = (f/3) ^ 3 + (g / 2) ^ 2 # Eq. (4-10)
13	f = (-3 * q - 1) / 3 # Eq. (4-11)
14	g = (-27 * r - 9 * q - 2) / 27 # Eq. (4-12)
15	z = If (C > 0) Then (D + E + 1 / 3) Else (0) # Eq. (4-13), Compressibility factor (dimensionless)
16	D = If (C > 0) Then ((-g / 2 + sqrt(C)) ^ (1 / 3)) Else (0) # Eq. (4-14)
17	E1 = If (C > 0) Then (-g / 2 - sqrt(C)) Else (0) # Eq. (4-15)
18	E = If (C > 0) Then ((sign(E1) * (abs(E1)) ^ (1 / 3))) Else (0) # Eq. (4-15)
19	P = Pr * Pc # Pressure (atm)
20	T = Tr * Tc # Temperature (K)
21	V = z * R * T / P # Molar volume (L/g-mol)

POLYMATH has the convenient user option that enables output of the equation set for the current problem as ordered equations that are correctly formatted for MATLAB. All intrinsic and logical functions are automatically translated. This option must be enabled within POLYMATH by selecting the "Show MATLAB formatted problem in report" in the POLYMATH Settings menu as shown in Figure 5–1. This option then provides a MATLAB listing at the end of the POLYMATH Report that can be generated along with the problem solution as is partially shown in Figure 5–2.

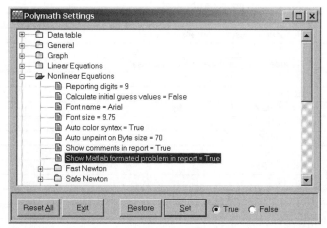

Figure 5–1 Setting the "Show MATLAB Formatted Problem Option" in POLYMATH

Matlab formatted problem

%Gas constant (L-atm/g-mol-K)
R = 0.08206;
%Critical temperature (K)
Tc = 647.4;
%Critical pressure (atm)
Pc = 218.3;
%Eq. (4-2), RK equation constant
a = 0.42747 * R ^ 2 * Tc ^ (5 / 2) / Pc;
%Eq. (4-3),RK equation constant
b = 0.08664 * R * Tc / Pc;
%Reduced pressure (dimensionless)
Pr = 0.1;

Figure 5–2 Partial View of MATLAB Formatted Output Contained in POLYMATH Report File

The "MATLAB formatted problem" from the POLYMATH Report can then be copied directly into the MATLAB editor. The equation set for this problem in the MATLAB editor is shown in Table 5–3. Note that POLYMATH orders the equations sequentially as required by MATLAB and converts any needed intrinsic functions and logical expressions.

Table 5–3 Ordered Equation Set Converted to MATLAB Format and Syntax by POLYMATH

Line	Command, % Comment
1	R = 0.08206;
2	%Critical temperature (K)
3	Tc = 647.4;
4	%Critical pressure (atm)
5	Pc = 218.3;
6	%Eq. (4-2), RK equation constant
7	a = 0.42747 * R ^ 2 * Tc ^ (5 / 2) / Pc;
8	%Eq. (4-3),RK equation constant
9	b = 0.08664 * R * Tc / Pc;
10	%Reduced pressure (dimensionless)
11	Pr = 0.1;
12	%Reduced temperature (dimensionless)
13	Tr = 1;
14	%Eq. (4-8)
15	Asqr = 0.42747 * Pr / (Tr ^ 2.5);
16	%Eq. (4-9)
17	B = 0.08664 * Pr / Tr;
18	%Eq. (4-6)
19	r = Asqr * B;
20	%Eq. (4-7)
21	q = B ^ 2 + B - Asqr;
22	%Eq. (4-12)
23	g = (-27 * r - (9 * q) - 2) / 27;
24	%Eq. (4-11)
25	f = (-3 * q - 1) / 3;

Line	Command, % Comment
26	%Eq. (4-10)
27	C = (f / 3) ^ 3 + (g / 2) ^ 2;
28	%Eq. (4-15)
29	if (C > 0)
30	E1 = 0 - (g / 2) - sqrt(C);
31	else
32	E1 = 0;
33	end
34	%Eq. (4-14)
35	if (C > 0)
36	D = (0 - (g / 2) + sqrt(C)) ^ (1 / 3);
37	else
38	D = 0;
39	end
40	%Eq. (4-15)
41	if (C > 0)
42	E = sign(E1) * abs(E1) ^ (1 / 3);
43	else
44	E = 0;
45	end
46	%Eq. (4-13), Compressibility factor (dimensionless)
47	if (C > 0)
48	z = D + E + 1 / 3;
49	else
50	z = 0;
51	end
52	%Pressure (atm)
53	P = Pr * Pc;
54	%Temperature (K)
55	T = Tr * Tc;
56	%Molar volume (L/g-mol)
57	V = z * R * T / P;

The POLYMATH conversion to MATLAB also places the comments in separate lines; however, these comments can be manually placed in the same line with the respective commands to obtain a more concise representation in Table 5–4. Every POLYMATH "if" statement is converted into a separate MATLAB "if" statement. Since all "if" statements for this problem all have the same condition, then these statements can be manually combined into one statement. The modified MATLAB code is shown in Table 5–4.

Table 5–4 Modified MATLAB file (File Prob_5_1a.m)

Line	Command, % Comment
1	R = 0.08206;%Gas constant (L-atm/g-mol-K)
2	Tc = 647.4;%Critical temperature (K)
3	Pc = 218.3;%Critical pressure (atm)

Table 5–4 (Continued) Modified MATLAB file (File **Prob_5_1a.m**)

Line	Command, % Comment
4	a = 0.42747 * R ^ 2 * Tc ^ (5 / 2) / Pc;%Eq. (4-2), RK equation constant
5	b = 0.08664 * R * Tc / Pc;%Eq. (4-3),RK equation constant
6	Pr = 0.1;%Reduced pressure (dimensionless)
7	Tr = 1;%Reduced temperature (dimensionless)
8	Asqr = 0.42747 * Pr / (Tr ^ 2.5);%Eq. (4-8)
9	B = 0.08664 * Pr / Tr;%Eq. (4-9)
10	r = Asqr * B;%Eq. (4-6)
11	q = B ^ 2 + B - Asqr;%Eq. (4-7)
12	g = (-27 * r - (9 * q) - 2) / 27;%Eq. (4-12)
13	f = (-3 * q - 1) / 3;%Eq. (4-11)
14	C = (f / 3) ^ 3 + (g / 2) ^ 2;%Eq. (4-10)
15	if (C > 0)
16	E1 = 0 - (g / 2) - sqrt(C);%Eq. (4-15)
17	D = (0 - (g / 2) + sqrt(C)) ^ (1 / 3);%Eq. (4-14)
18	E = sign(E1) * abs(E1) ^ (1 / 3);%Eq. (4-15)
19	z = D + E + 1 / 3;%Eq. (4-13), Compressibility factor (dimensionless)
20	else
21	z = 0;
22	end
23	P = Pr * Pc;%Pressure (atm)
24	T = Tr * Tc;%Temperature (K)
25	V = z * R * T / P;%Molar volume (L/g-mol)

The MATLAB code generated by POLYMATH and subsequently modified to be more concise can be used to carry out the calculation of the compressibility factor and molar volume for one set of reduced temperature and reduced pressure values at a time. Partial MATLAB results from the MATHLAB Workspace are shown in Figure 5–3.

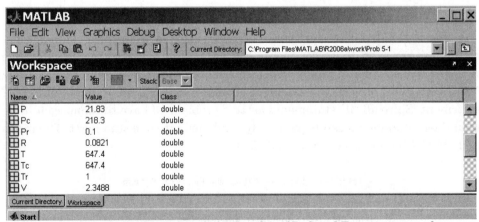

Figure 5–3 Partial MATLAB Result for One Set of Reduced Temperature and Pressure (File **Prob_5_1a.m**)

The repeated calculations with different T_r and P_r values requires the use of a *function* in the m-file that contains the equations. This is accomplished by adding a *function* statement as the first statement in the file

$$function \ [z, V]=RKfun(Tr,Pr,Tc,Pc)$$

where *function* is a MATLAB keyword which indicates that this is a function file. Tr, Pr, Tc, and Pc are the input arguments containing variable names whose numerical values are passed from the calling program into the function. z and V are output arguments whose numerical values are passed from the function to the calling program, and RKfun is a name selected by the user for the function. The input arguments are separated by commas and put inside round parentheses; the output arguments are put inside square brackets. Note that the function must be saved in a file which has the same name, **RKfun.m**. All the variables that are defined inside the function, except the input and output arguments, are local variables; thus, they are not recognized outside of the function.

The variables Tr, Pr, Tc, and Pc are passed to the function as input arguments, and their definition lines must therefore be deleted when creating the function. This completes the modifications required to obtain a MATLAB function from the MATLAB formatted output provided by POLYMATH. The final form of the function is shown in Table 5–5.

Table 5–5 MATLAB Function for Calculating Compressibility Factor and Molar Volume Using the Redlich-Kwong Equation of State (File **RKfun.m**)

Line	Command, % Comment
1	function [z, V]=RKfun(Tr,Pr,Tc,Pc)
2	R = 0.08206;%Gas constant (L-atm/g-mol-K)
3	a = 0.42747 * R ^ 2 * Tc ^ (5 / 2) / Pc;%Eq. (4-2), RK equation constant
4	b = 0.08664 * R * Tc / Pc;%Eq. (4-3),RK equation constant
5	Asqr = 0.42747 * Pr / (Tr ^ 2.5);%Eq. (4-8)
6	B = 0.08664 * Pr / Tr;%Eq. (4-9)
7	r = Asqr * B;%Eq. (4-6)
8	q = B ^ 2 + B - Asqr;%Eq. (4-7)
9	g = (-27 * r - (9 * q) - 2) / 27;%Eq. (4-12)
10	f = (-3 * q - 1) / 3;%Eq. (4-11)
11	C = (f / 3) ^ 3 + (g / 2) ^ 2;%Eq. (4-10)
12	if (C > 0)
13	E1 = 0 - (g / 2) - sqrt(C);%Eq. (4-15)
14	D = (0 - (g / 2) + sqrt(C)) ^ (1 / 3);%Eq. (4-14)
15	E = sign(E1) * abs(E1) ^ (1 / 3);%Eq. (4-15)
16	z = D + E + 1 / 3;%Eq. (4-13), Compressibility factor (dimensionless)
17	else
18	z = 0;
19	end
20	P = Pr * Pc;%Pressure (atm)
21	T = Tr * Tc;%Temperature (K)
22	V = z * R * T / P;%Molar volume (L/g-mol)

This function provides meaningful results only if $C > 0$. In such a case, the variables E1, D, E, and z are calculated (see lines 13 through 16 in Table 5–5). Otherwise E1, D, and E are not calculated, and z obtains zero value.

The *RKfun* function may be checked by introducing the values of T_r, P_r, T_c, and P_c provided in the problem statement. This can be done in the MATLAB "Command" window, as shown in Figure 5–4. Into the "Current directory" field the name of the directory where *RKfun* resides is entered. *RKfun* is called with the input parameter values provided in the problem statement. The input and the returned output parameter values are shown in the "Workspace" window. It can be seen that the resultant compressibility factor and molar volume values are the same as obtained previously by POLYMATH.

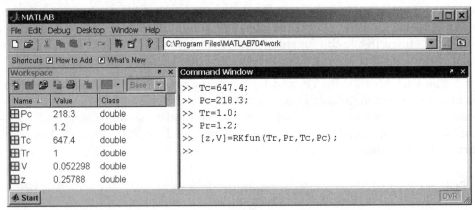

Figure 5–4 Checking the RKfun Function in the Command Window (File **RKfun.m**)

(b) A separate script file (a MATLAB m-file) must be prepared in order to carry out the calculations repeatedly with different P_r and T_r values, present the results in a tabular form, and plot the results. Part of this file is shown in Table 5–6.

Table 5–6 MATLAB Script for Calculating z and V and Presenting the Results in Tabular and Graphical Forms (File **Prob_5_1b.m**)

Line	Command, % Comment
1	%filename Prob_5_1b
2	clear, clc, format compact, format short g
3	Tc = 647.4; %Critical temperature (K)
4	Pc = 218.3; %Critical pressure (atm)
5	Tr_set=[1 1.2 1.5 2 3];
6	Pr_set(1) = 0.1;
7	Pr_set(2) = 0.2;
8	i = 2;
9	while Pr_set(i)<=10
10	i=i+1;
11	Pr_set(i)=Pr_set(i-1)+0.2;
12	end
13	n_Tr = size(Tr_set,2);

Line	Command, % Comment
14	n_Pr = size(Pr_set,2);
15	for i=1:n_Tr
16	Tr=Tr_set(i);
17	for j=1:n_Pr
18	Pr=Pr_set(j);
19	[z(j,i), V(j,i)] = RKfun(Tr,Pr,Tc,Pc) ;
20	if z(j,i)==0
21	disp([' No solution obtained for Tr = ' num2str(Tr) ' and Pr = ' num2str(Pr)]);
22	end
23	end
24	end
25	disp(' Compressibility Factor Versus Tr and Pr');
26	disp(' Tabular Results');
27	disp(' ');
28	disp(' Pr\Tr 1.0 1.2 1.5 2 3 ');
29	Res=[Pr_set' z];
30	disp(Res);
31	plot(Pr_set,z(:,1),'-',Pr_set,z(:,2),'+',Pr_set,z(:,3),'*',Pr_set,z(:,4),'x',Pr_set,z(:,5),'o');
32	legend('Tr=1','Tr=1.2','Tr=1.5','Tr=2','Tr=3');
33	title(' Compressibility Factor Versus Tr and Pr')
34	xlabel('Reduced Pressure (Pr)');
35	ylabel('Compressibility Factor (z)');
36	pause

The first commands in the script file **Prob_5_1b.m** (see Line 2 in Table 5–6) are command window and workspace control functions. Those are nonspecific to the problem at hand and are in the category of "good programming practice."

clear – clears the workspace from any leftover variables or functions from previous runs

clc – clears the command window

format compact – suppresses extra line feeds

format short g – prints numbers with the best of fixed or floating point format with 5 digits

Note that two commands in the same line are separated by a comma or a semi-colon. Additional information about any of the commands can be obtained by typing *help command* (*help format*, for example) in the command window.

In lines 3 and 4, the critical temperature and pressure for water are specified. In line 5, a row vector *Tr_set* that contains 5 reduced temperature values is specified. Square brackets are used to specify vectors and matrices, where the various numerical entries are separated by at least one space. A semi-colon is placed at the end of the command in order to eliminate the printing of the vector in the command window.

In lines 6 through 12, a row vector, *Pr_set*, is generated that contains all the reduced pressure values specified in the problem statement. The first two elements in this vector are specified in lines 6 and 7. The remaining elements are represented by a linear series with increment of 0.2. Those elements are generated by means of a *while* statement. The purpose of this statement is to repeat commands an indefinite number of times.

The general form of a *while* statement is

while expression, statements, end

and the *while* command is executed as long as the *expression* is *true*. In this particular case it is executed as long as the last *Pr* value is <= 10. The statements inside the *while* loop increase the counter *i* by one and place an additional *Pr* value as the last element in the vector *Pr_set*. The number of elements in the *TR_set* and *Pr_set* vectors are determined by means of the *size* function in lines 13 and 14, respectively.

The compressibility factor and molar volume values are calculated inside two nested *for* loops in lines 15 through 24. The general form of a *for* statement is

for variable = expr, statement, ..., statement end

The values of the expression are stored one at a time in the variable, and then the following statements, up to the *end*, are executed. In the particular case shown in Table 5–6, the variable *i* obtains the values 1, 2, 3 ... up to n_Tr, and the current value of *Tr* is set as the *i*-th element of the vector *Tr_set*. For every value of *i*, the variable *j* obtains the values 1, 2, 3 ... up to *n_Pr*, and the current value of *Pr* is set as the *j*-th element of the vector *Pr_set*. The function *RKfun* is called for all the *i* and *j* values, and the resultant *z* and *V* values are stored in two matrices where every row is associated with one Pr value and every column is associated with a *Tr* value (see line 19 in Table 5–6).

In lines 20 through 22, a warning is issued in the case where the compressibility factor is not calculated by *RKfun* (if C<=0). The warning is displayed by means of a *disp* statement. The *disp* command displays matrices. In the particular case shown in line 21, a row vector containing a text string comprising of plain text and numbers (Tr and Pr) converted to text is displayed.

In lines 25 through 28, the title and the column headings of the table are displayed. In line 29, the transposed *Pr_set* vector is attached to the *z* matrix as an additional column and the combined matrix (*Res*) is displayed in line 30.

Partial results from this program are summarized in Table 5–7. The results are exactly the same as obtained by Excel (see Figure 4–5 in Problem 4.1).

Table 5–7 Partial Results for Compressibility Factor Calculation for Various P_r and T_r Values

Compressibility Factor versus Tr and Pr				
Tabular Results				
Pr\Tr	1.0	1.2	1.5	2	3
0.1	0.96516	0.97997	0.99029	0.99682	1.0002
0.2	0.92864	0.95964	0.98065	0.99372	1.0004
0.4	0.84907	0.91801	0.96161	0.98778	1.0008
0.6	0.75657	0.87504	0.94295	0.98221	1.0015
0.8	0.63874	0.83072	0.92479	0.97702	1.0022

The commands used for plotting the compressibility factor versus the reduced temperature and the reduced pressure are shown in lines 31 through 35 of Table 5–6. Note the reduced pressure is put on the horizontal axis (x–axis), the compressibility factor on the vertical axis (y–axis), and the reduced temperature is a parameter.

The command *plot*(x,y,s) plots vector y versus vector x where the string *s* is used to define the line type, symbol, and color. The (x, y, s) triples can be repeated to put several different curves on the same plot. In line 31 of the MATLAB code, separate curves of *z* versus *Pr* are selected for plotting at the different Tr values. In the first triplet, the *Pr_set* vector is put on the x–axis, the first column of the *z* –vector (corresponding to *Tr* = 1) is put on the *y*-axis and the string '-' indicates that solid line should be used to plot this curve. There are five such triplets for the five *Tr* values. The commands *legend, title, xlabel,* and *ylabel* are self explanatory. The plot obtained using this set of commands is shown in Figure 5–5.

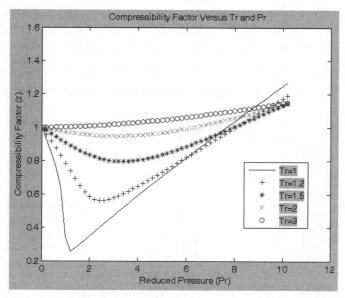

Figure 5–5 Compressibility Factor of Steam versus Pr and Tr

The pause command in line 36 causes the program to wait for a key-press before continuing to display another table and plot.

Commands for the final plot are shown in , which is a continuation of the MATLAB script file **Prob_5_1b.m**.

Table 5–8 Continuation of MATLAB *script* File (File **Prob_5_1b.m**)

Line	Command, %Comment
37	P_set=Pr_set.*Pc;
38	disp(' Molar Volume Versus Tr and P');
39	disp(' Tabular Results');
40	disp(' ');

Table 5–8 (Continued) Continuation of MATLAB *script* File (File **Prob_5_1b.m**)

Line	Command, %Comment
41	disp(' P\Tr 1.0 1.2 1.5 2 3 ');
42	Res=[P_set' V];
43	disp(Res);
44	loglog(P_set,V(:,1),'-',P_set,V(:,2),'+',P_set,V(:,3),'*',P_set,V(:,4),'x',P_set,V(:,5),'o');
45	legend('Tr=1','Tr=1.2','Tr=1.5','Tr=2','Tr=3');
46	title(' Molar Volume Versus Tr and P')
47	xlabel('Pressure (atm)');
48	ylabel('Molar Volume (L/g-mol)');

Preparation for a table and a plot of the molar volume versus pressure, and Tr first requires creation of a vector that contains the pressure values in atm. This can be accomplished by a single equation shown in line 37.

P_set = Pr_set.*Pc;

The dot before the multiplication sign indicates that every element of the Pr_set vector must be separately multiplied by the scalar *Pc*. Plotting of the molar volume in a logarithmic scale utilizes the *loglog* function as shown in line 44.

loglog(P_set,V(:,1),'-',P_set,V(:,2),'+',P_set,V(:,3),'*',P_set,V(:,4),'x', P_set,V(:,5),'o');

Figure 5–6 shows the resulting plot of the molar volume versus the pressure and Tr.

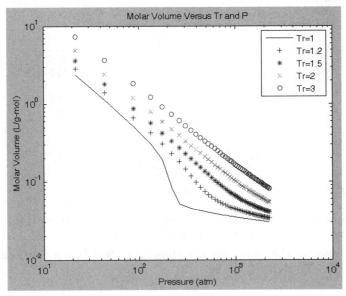

Figure 5–6 Molar Volume of Steam versus P and T_r.

The problem solution files are found in directory Chapter 5 and designated **P5-1A.POL**, **Prob_5_1a.m**, **RKfun.m**, and **Prob_5_1b.m**.

5.2 CALCULATION OF THE FLOW RATE IN A PIPELINE

5.2.1 Concepts Demonstrated

Application of the general mechanical energy balance for incompressible fluids, and calculation of flow rate in a pipeline for various pipe diameters and lengths.

5.2.2 Numerical Methods Utilized

Solution of a single nonlinear algebraic equation.

5.2.3 MATLAB Options and Functions Demonstrated

Use of the MATLAB *fzero* function for solving a nonlinear equation, logarithmic functions.

5.2.4 Problem Definition

Rework Problem 4.2 using MATLAB.

(a) Prepare a MATLAB program using the MATLAB output generated by POLYMATH for this problem creation. Incorporate the MATLAB *fzero* function to calculate the flow velocity. Determine the velocity for a pipeline with effective length of $L = 1000$ ft and made of nominal 8-inch diameter schedule 40 commercial steel pipe. (Solution: $v = 11.61$ ft/s)

(b) Modify the MATLAB program from part (a) to become a function that will take the flow velocity (v), pipe diameter (D), the pipe length (L), and the ambient temperature (T) as input parameters and will return to the calling point the value of $f(v)$ (as defined in Section 4.2.6a).

(c) Calculate the flow velocity and flow rate for pipelines with effective length of $L = 500, 1000, \ldots 10,000$ ft and made of nominal 4-, 5-, 6- and 8-inch schedule 40 commercial steel pipe. Use the *fzero* function for solving the equations for the various cases and present the results in tabular form. Prepare plots of flow velocity versus D and L and flow rate versus D and L.

5.2.5 Solution

(a) The set of POLYMATH equations for this problem are shown in Table 4–3 of Problem 4.2. The option within POLYMATH to "Show MATLAB formatted problem in report" must be enabled in the POLYMATH Settings menu under Nonlinear Equations in order to obtain the MATLAB equation set. This option is accessible by opening the icon marked as: "Setup preferences and numerical parameters" as discussed in Problem 5.1. The MATLAB formatted equation set is then displayed in the POLYMATH report page at the end of the solution results.

The resultant MATLAB code generated by POLYMATH should be copied into the MATLAB editor, and this is shown in Table 5–9 after the comments have been placed at the end of each line of code. This code includes a specified pipe length and a pipe diameter. The **OneNLEtemplate.m** file shown in Table 5–10 (and available also from POLYMATH's Help section) can be used to determine the flow velocity for the specified parameter values with the MATLAB *fzero* function for solving simultaneous nonlinear equations.

Table 5–9 MATLAB Code Generated by POLYMATH for Problem 5.2 (File **P5-2a1.m**)

Line	Command, %Comment
1	xguess = 10.5 ;
2	function fv = NLEfun(v);
3	T = 60; %Temperature (deg F)
4	epsilon = 0.00015; %Surface rougness of the pipe (ft)
5	rho = 62.122 + T * (0.0122 + T * (-1.54e-4 + T * (2.65e-7 - T * 2.24e-10)));%Fluid density (lb/cu. ft.)
6	deltaz = 300; %Elevation difference (ft)
7	deltaP = -150; %Pressure difference (psi)
8	vis = exp(-11.0318 + 1057.51 / (T + 214.624)); %Fluid viscosity (lbm/ft-s)
9	L = 1000; %Effective length of pipe (ft)
10	D = 7.981 / 12; %Inside diameter of pipe (ft)
11	pi = 3.1416; %The constant pi
12	eoD = epsilon / D; %Pipe roughness to diameter ratio (dimensionless)
13	Re = D * v * rho / vis; %Reynolds number (dimensionless)
14	if (Re < 2100) %Fanning friction factor (dimensionless)
15	fF = (16 / Re) ;
16	else
17	fF = (1 / (16 * (log10(eoD / 3.7 - 5.02 * log10(eoD / 3.7 + 14.5 / Re) / Re)) ^ 2));
18	end
19	q = v * pi * D ^ 2 / 4; %Flow rate (sq. ft./s)
20	gpm = q * 7.481 * 60; %Flow rate (gpm)
21	fv = v - sqrt((32.174 * deltaz + deltaP * 144 * 32.174 / rho) / (0.5 - 2 * fF * L / D));%Flow velocity (ft/s)

Table 5–10 Template for Solving One Nonlinear Algebraic Equation (File **OneNLEtemplate.m**)

Line	Command, %Comment
1	function % Insert here your file name after function (Use Alphanumberic name only)
2	clear, clc, format short g, format compact
3	xguess= % Replace this line with the xguess line(s) from Polymath report.
4	disp('Variable values at the initial estimate');
5	disp([' Unknown value ' num2str(xguess) ' Function Value ' num2str(NLEfun(xguess))]);
6	xsolv=fzero(@NLEfun,xguess);
7	disp(' Variable values at the solution');
8	disp([' Unknown value ' num2str(xsolv) ' Function Value ' num2str(NLEfun(xsolv))]);
9	%- -
10	% Replace this and the following line with the function copied from the Polymath report
11	% Do not include the xguess line(s)

The creation of the MATLAB program requires that lines 1–11 of the template shown in Table 5–10 be saved as a new MATLAB m-file, The complete function NLEfun should be copied from the POLYMATH report (shown in Lines 2–21 of Table 5–9) and pasted into the same new m-file (replacing the comments in lines 10-11). The initial guess for the unknown should be copied from line 1 of the POLYMATH report and pasted into the m-file where indicated (line 3 in Table 5–10). An arbitrary function name, **Prob_5_2a2.m** for this example, should be assigned for the m-file, and it should be inserted in line 1. The same name, **Prob_5_2a2.m**, should be used to store the file. The resulting completed m-file is given in Table 5–11.

Table 5–11 MATLAB Code for Problem 5.2(a) (File **Prob_5_2a2.m**)

Line	Command, %Comment
1	function Prob_5_2a2
2	clear, clc, format short g, format compact
3	xguess = 10.5 ;
4	disp('Variable values at the initial estimate');
5	disp([' Unknown value ' num2str(xguess) ' Function Value ' num2str(NLEfun(xguess))]);
6	xsolv=fzero(@NLEfun,xguess);
7	disp(' Variable values at the solution');
8	disp([' Unknown value ' num2str(xsolv) ' Function Value ' num2str(NLEfun(xsolv))]);
9	%- - - - - - - - - - - - - - - - - - - -
10	function fv = NLEfun(v);
11	T = 60; %Temperature (deg F)
12	epsilon = 0.00015; %Surface rougness of the pipe (ft)
13	rho = 62.122 + T * (0.0122 + T * (-1.54e-4 + T * (2.65e-7 - T * 2.24e-10)));%Fluid density (lb/cu. ft.)
14	deltaz = 300; %Elevation difference (ft)
15	deltaP = -150; %Pressure difference (psi)
16	vis = exp(-11.0318 + 1057.51 / (T + 214.624)); %Fluid viscosity (lbm/ft-s)
17	L = 1000; %Effective length of pipe (ft)
18	D = 7.981 / 12; %Inside diameter of pipe (ft)
19	pi = 3.1416; %The constant pi
20	eoD = epsilon / D; %Pipe roughness to diameter ratio (dimensionless)
21	Re = D * v * rho / vis; %Reynolds number (dimensionless)
22	if (Re < 2100) %Fanning friction factor (dimensionless)
23	fF = (16 / Re) ;
24	else
25	fF = (1 / (16 * (log10(eoD / 3.7 - 5.02 * log10(eoD / 3.7 + 14.5 / Re) / Re)) ^ 2));
26	end
27	q = v * pi * D ^ 2 / 4* 7.481 * 60; %Flow rate (gpm)
28	fv = v - sqrt((32.174 * deltaz + deltaP * 144 * 32.174 / rho) / (0.5 - 2 * fF * L / D));%Flow velocity (ft/s)

The MATLAB provided function *fzero* is used to solve the nonlinear equation. This function can be called with several different sets of input and output parameters. The particular form used in Table 5–10 is

xsolv=fzero(@NLEfun,xguess);

The first input parameter is the name of the function (*NLEfun*) while the second input parameter (xguess) is an initial estimate for the value of the unknown at the solution. The parameter returned by the *fzero* function is the value of v at the solution (xsolv). Figure 5–7 shows the result of running this MATLAB code.

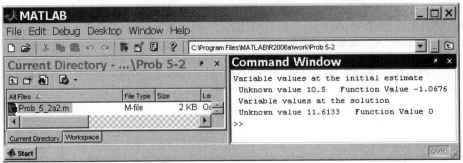

Figure 5–7 MATLAB Solution to Problem 5-2(a)

The calculated value of the flow velocity v is the same as the value provided in the problem statement.

(b) The MATLAB code can be modified so that it obtains the flow velocity v pipe diameter D, the pipe length L, and the ambient temperature T as input parameters. Thus the following changes can be made in the MATLAB file **Prob_5_2a2.m**:

1. The calculation of the flow rate q (in line 27) can be removed as this variable will have to be calculated outside of the function.
2. The definition of *pi* (shown in line 19 of Table 5–11) can be removed as MATLAB recognizes *pi* as a special value ($pi = \pi$).
3. The definition of D, L, and T (in lines 18, 17, and 11 of Table 5–11) must be removed as those variables serve as function parameters.
4. Lines 1 through 9 should be removed.
5. The variables D, L, and T must be added as input parameters to the function statement in line 1: *function* fv = NLEfun(v,D,L,T).

This MATLAB code should be saved as **NLEfun.m**. The final form of the function is shown in Table 5–12.

Table 5–12 MATLAB Function for Introducing Input Parameters (File **NLEfun.m**)

Line	Command, %Comment
1	function fv = NLEfun(v,D,L,T)
2	epsilon = 0.00015;%Surface rougness of the pipe (ft)
3	rho = 62.122 + T * (0.0122 + T * (-0.000154 + T * (0.000000265 - (T * 0.000000000224)))); %Fluid density (lb/cu. ft.)
4	deltaz = 300; %Elevation difference (ft)
5	deltaP = -150; %Pressure difference (psi)
6	vis = exp(-11.0318 + 1057.51 / (T + 214.624)); %Fluid viscosity (lbm/ft-s)
7	pi = 3.1416; %The constant pi

Line	Command, %Comment
8	eoD = epsilon / D; %Pipe roughness to diameter ratio (dimensionless)
9	Re = D * v * rho / vis; %Reynolds number (dimesionless)
10	if (Re < 2100) %Fanning friction factor (dimensionless)
11	fF = 16 / Re;
12	else
13	fF = 1 / (16 * log10(eoD / 3.7 - (5.02 * log10(eoD / 3.7 + 14.5 / Re) / Re)) ^ 2);
14	end
15	fv = v - sqrt((32.174 * deltaz + deltaP * 144 * 32.174 / rho) / (0.5 - (2 * fF * L / D))); %Flow velocity (ft/s)

This function can be used in MATLAB to make the calculation of fv after v, D, L, and T have been specified. This is shown in Figure 5–8 when $v = 10.5$, $D = 7.981 / 12$, $L = 1000$, and $T = 60$ are entered in the MATLAB Command Window. The result should and do agree with that in Figure 5–7 where $fv = -1.0676$.

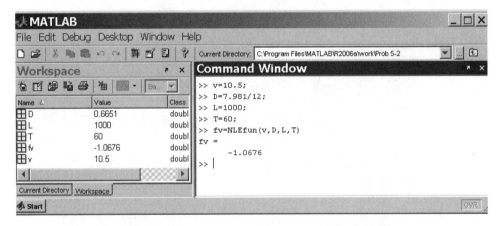

Figure 5–8 MATLAB Solution to Problem 5-2(b)

(c) Repeated calculations with different L and D values and presentation of the results in a tabular form with plotting require the preparation of a separate MATLAB script file. A representative MATLAB file is shown in Table 5–13.

Table 5–13 MATLAB *script* file for Calculating v and q Values and Presenting the Results in Tabular and Graphical Forms (File **Prob_5_2c.m**)

Line	Command, %Comment
1	function Prob_5_2c
2	clear, clc, format short g, format compact
3	D_list=[4.026/12 5.047/12 6.065/12 7.981/12]; % Inside diameter of pipe (ft)
4	T = 60; %Temperature (deg. F)
5	for i = 1:4
6	D = D_list(i);
7	j=0;
8	for L=500:500:10000

Table 5–13 (Continued) MATLAB *script* file for Calculating *v* and *q* Values and Presenting the Results in Tabular and Graphical Forms (File **Prob_5_2c.m**)

Line	Command, %Comment
9	j = j+1;
10	L_list(j)=L; % Effective length of pipe (ft)
11	[v(j,i),fval]=fzero(@NLEfun,[1 20],[],D,L,T);
12	if abs(fval)>1e-10
13	disp([' No Convergence for L = ' num2str(L) ' and D = ' num2str(D)]);
14	end
15	q(j,i) = v(j,i) * pi * D ^ 2 / 4* 7.481 * 60; %Flow rate (gpm)
16	end
17	end
18	disp(' Flow Velocity (ft/s) versus Pipe Length and Diameter');
19	disp(' Tabular Results');
20	disp('');
21	disp(' L\D D=4" D=5" D=6" D=8"');
22	Res=[L_list' v];
23	disp(Res);
24	plot(L_list,v(:,1),'-',L_list,v(:,2),'+',L_list,v(:,3),'*',L_list,v(:,4),'x');
25	legend(' D=4"',' D=5"',' D=6"',' D=8"');
26	title(' Flow Velocity')
27	xlabel('Pipe Length (ft)');
28	ylabel('Velocity (ft/s)');
29	pause
30	disp(' Flow Rate (gpm) versus Pipe Length and Diameter');
31	disp(' Tabular Results');
32	disp('');
33	disp(' L\D D=4" D=5" D=6" D=8"');
34	Res=[L_list' q(:,1) q(:,2) q(:,3) q(:,4)];
35	disp(Res);
36	plot(L_list,q(:,1),'-',L_list,q(:,2),'+',L_list,q(:,3),'*',L_list,q(:,4),'x');
37	legend(' D=4"',' D=5"',' D=6"',' D=8"');
38	title(' Flow rate')
39	xlabel('Pipe Length (ft)');
40	ylabel('Flow rate (gpm)');

Line 3 of the script file contains a row vector *(D_list)* which contains the inside diameters of the pipes (in ft) to be calculated. The ambient temperature is defined in line 4.

The flow velocity and flow rate values are calculated inside two nested *for* loops in lines 5 through 17. The variable *i* obtains the values 1, 2, 3, and 4; the current value of *D* is set as the *i*-th element of the vector *D_list*. For every value of *i*, the variable *L* obtains the values 500, 1000, ...10,000; the counter *j* is increased by one; and L is set as the *j*-th element of the vector *L_list*. The function *fzero* is called for all the *D* and *L* values. The resulting *v* values are stored in two matrices where every row is associated with one *L* value and every column is associated with a *D* value (see line 11 in Table 5–13).

Some of the parameters that are used in calling the *fzero* function are different here than in Table 5–10. The second input parameter is a vector which contains the bounds of an interval in which $f(v)$ changes sign (an error message displayed if this is not true). Instead of the third (optional) parameter, the space saver null vector: [ ] is entered. This parameter is a *structure* that contains information that controls the solution algorithm (error tolerances, maximum number of iterations allowed, etc.). If the default values are used, this parameter can be replaced by a null matrix. The additional input parameters *D*, *L*, and *T* contain values that have to be transferred to the *NLEfun* function.

The parameters returned by the *fzero* function include the value of v at the solution and the value of $f(v)$ (*fval*, an optional parameter) at the solution. In lines 11 through 13, a warning is issued in the case where the $f(v)$ value is not close enough to zero at the solution ($f(v) >= 10^{-10}$). The flow rates (*gpm*) are calculated in line 15 and stored in the matrix q, which has the same structure as the v matrix.

In lines 18 through 21, the title and the column headings of the table are displayed. In line 22, the transposed *L_list* vector is attached to the v matrix as an additional column, and the combined matrix (*Res*) is displayed in line 23.

Parts of the results obtained and displayed by this program are shown in Table 5–14. The results with MATLAB are exactly the same as obtained by Excel (see Problem 4.2).

Table 5–14 Partial Results for Flow Velocity Calculations for Various D and L Values (File **Prob_5_2c.m**)

Flow velocity (ft/s) versus Pipe Length and Diameter				
Tabular Results				
L\D	D=4"	D=5"	D=6"	D=8"
500	10.773	12.516	14.15	17.035
1000	7.4207	8.6048	9.7032	11.613
1500	5.9721	6.9243	7.8051	9.3295
2000	5.1188	5.9361	6.6912	7.9953
2500	4.5409	5.2674	5.9382	7.0953

The commands used for plotting the flow velocity versus the pipe length and diameter are shown in lines 24 through 28 of Table 5–13. The pipe length is put on the horizontal axis (x – axis), the flow velocity on the vertical axis (y – axis), and the pipe diameter is a parameter.

In line 24 of the MATLAB code, separate curves of v versus L are being plotted for the different D values. In the first triplet, the *L_list* vector is put on the x – axis, the first column of the v – vector (corresponding to the first D value) is put on the y axis, and the string '-' indicates that solid line should be used to plot this curve. There are four such triplets for the four V values. The commands *legend*, *title*, *xlabel*, and *ylabel* are self explanatory. The plot obtained using this set of commands is shown in Figure 5–9.

The pause command in line 29 causes the program to wait for a key-press before continuing to display another table and plot. The table and the plot for flow rates are obtained similarly. The plot of the flow rate versus pipe length and diameter is shown in Figure 5–10. Note that both the program file **Prob_5_2c.m** and the function file **NLEfun.m** must be in the MATLAB Workspace when this problem is solved.

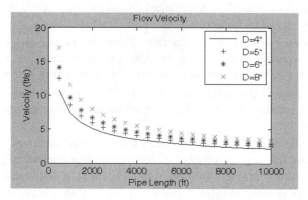

Figure 5–9 Flow Velocity versus Pipe Length and Diameter

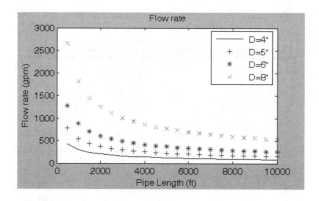

Figure 5–10 Flow Rate versus Pipe Length and Diameter

The problem solution files are found in directory Chapter 5 and designated **NLEfun.m, OneNLEtemplate.m, P5-2A.POL, Prob_5_2a1.m, Prob_5_2a2.m, Prob_5_2b.m,** and **Prob_5_2c.m.**

5.3 ADIABATIC OPERATION OF A TUBULAR REACTOR FOR CRACKING OF ACETONE

5.3.1 Concepts Demonstrated

Calculation of the conversion and temperature profile in an adiabatic tubular reactor. Demonstration of the effect of pressure and heat capacity change on the conversion in the reactor.

5.3.2 Numerical Methods Utilized

Solution of simultaneous ordinary differential equations.

5.3.3 MATLAB Options and Functions Demonstrated

Use of the MATLAB *ode45* function to solve ordinary differential equations.

5.3.4 Problem Definition

Rework Problem 4.3 (b) using MATLAB.

(a) Utilize the MATLAB code generated by POLYMATH for this problem that includes added nitrogen (part (b) in Problem 4.3). Use the MATLAB *ode45* function to calculate the final conversion and temperature values in the reactor for the case where $V = 4m^3$, $F_{A0} = 38.3$ mol/s, $P = 162$ kPa, $T_0 = 1035$, and the feed rates of the other compounds are all zero. Compare with the final values obtained by POLYMATH: $T = 907.54$ K and $X_A = 0.257$.

(b) Modify the MATLAB code to become a function called *dYfuncvecdV* that will determine the current volume V and a vector **y** containing the flow rates of compounds A, B, and C. F_{A0} and P are to be the input parameters, and the code should return a column vector of the derivatives of those variables with respect to V to the calling program.

(c) Calculate the final conversion and the final temperature in the reactor operating at constant pressure values of $P = 1.6, 1.8, \ldots 5.0$ atm, for acetone feed rates of $F_{A0} = 10, 20, 30, 35$, and 38.3 mol/s where nitrogen is fed to maintain the total feed rate 38.3 mol/s in all cases. Present the results in tabular form and prepare plots of final conversion versus P and F_{A0} and final temperature versus P and F_{A0}.

5.3.5 Solution

(a) The set of POLYMATH equations for this problem is given in Figure 4–18 for Problem 4.3(b). The MATLAB code for a problem within POLYMATH Ordinary

Differential Equations Solver can be obtained by enabling the "Show MATLAB formatted problem in report" must be enabled in the POLYMATH Settings menu under Ordinary Differential Equations. This option is accessible by opening the icon marked as: "Setup preferences and numerical parameters" as discussed in Problem 5-1. The MATLAB formatted equation set is then placed at the end of the POLYMATH report.

The generated MATLAB code for Problem 4-3(b) should be copied into the MATLAB editor, and the comments can be moved to the end of each line of code as is shown in Table 5–15.

Table 5–15 MATLAB Function Generated by POLYMATH for Problem 4.3(b) (File **Prob_5_3a1**)

Line	Command, %Comment
1	tspan = [0 4.]; % Range for the independent variable
2	y0 = [38.3; 0; 0; 1035.]; % Initial values for the dependent variables
3	function dYfuncvecdV = ODEfun(V,Yfuncvec);
4	FA = Yfuncvec(1);
5	FB = Yfuncvec(2);
6	FC = Yfuncvec(3);
7	T = Yfuncvec(4);
8	FA0 = 38.3; %Feed rate of acetone in kg-mol/s
9	k = 820000000000000 * exp(-34222 / T); %Reaction rate constant in s-1
10	XA = (FA0 - FA) / FA0; %Conversion of acetone
11	FN = 38.3 - FA0; %Feed rate of nitrogen in kg-mol/s
12	P = 162; %Pressure kPa
13	yA = FA / (FA + FB + FC + FN); %Mole fraction of acetone
14	CA = yA * P * 1000 / (8.31 * T); %Concentration of acetone in k-mol/m3
15	yB = FB / (FA + FB + FC + FN); %Mole fraction of ketene
16	yC = FC / (FA + FB + FC + FN); %Mole fraction of methane
17	rA = 0 - (k * CA); %Reaction rate in kg-mole/m3-s
18	deltaH = 80770 + 6.8 * (T - 298) - (0.00575 * (T ^ 2 - (298 ^ 2))) - (0.00000127 * (T ^ 3 - (298 ^ 3))); %Heat of reaction in J/mol-K
19	CpA = 26.6 + 0.183 * T - (0.00004586 * T ^ 2); %Heat capacity of acetone in J/mol-K
20	CpB = 20.04 + 0.0945 * T - (0.00003095 * T ^ 2); %Heat capacity of ketene in J/mol-K
21	CpC = 13.39 + 0.077 * T - (0.00001871 * T ^ 2); %Heat capacity of methane in J/mol-K
22	CpN = 6.25 + 0.00878 * T - (0.000000021 * T ^ 2); %Heat capacity of nitrogen in J/mol-K
23	dFAdV = rA; %Differential mass balance on acetone
24	dFBdV = 0 - rA; %Differential mass balance on ketene
25	dFCdV = 0 - rA; %Differential mass balance on methane
26	dTdV = (0 - deltaH) * (0 - rA) / (FA * CpA + FB * CpB + FC * CpC + FN * CpN); %Differential enthalpy balance
27	dYfuncvecdV = [dFAdV; dFBdV; dFCdV; dTdV];

The **MultipleDEQtemplate.m** file, shown in Table 5–16 (and available with this problem and from the POLYMATH Help Section), provides a basic MATLAB solution to a system of simultaneous ordinary differential equations. The instructions that are given as comments within the **MultipleDEQtemplate.m** file allow the creation of the MATLAB file **Prob_5_3a2.m** as shown in Table 5–17.

Table 5–16 MATLAB Template File **MultipleDEQtemplate.m** for Integrating
Simultaneous Ordinary Differential Equations

Line	Equation, %Comment
1	function % Insert here your file name after function (Use Alphanumberic names only)
2	clear, clc, format short g, format compact
3	tspan= % Replace this line with tspan line from Polymath report
4	y0= % Replace this line with y0 line from Polymath report
5	disp(' Variable values at the initial point ');
6	disp(['t = ' num2str(tspan(1))]);
7	disp(' y dy/dt ');
8	disp([y0 ODEfun(tspan(1),y0)]);
9	[t,y]=ode45(@ODEfun,tspan,y0);
10	for i=1:size(y,2)
11	disp([' Solution for dependent variable y' int2str(i)]);
12	disp([' t y' int2str(i)]);
13	disp([t y(:,i)]);
14	plot(t,y(:,i));
15	title([' Plot of dependent variable y' int2str(i)]);
16	xlabel(' Independent variable (t)');
17	ylabel([' Dependent variable y' int2str(i)]);
18	pause
19	end
20	%- - - - - - - - - - - - - - - - - - - -
21	% Replace this and the following line with the function copied from the Polymath report
22	% Do not include the tspan and y0 lines

Table 5–17 MATLAB Code for Problem 5.3(a) with m-File **Prob_5_3a2.m**

Line	Equation, %Comment
1	function Prob_5_3a2
2	clear, clc, format short g, format compact
3	tspan = [0 4.]; % Range for the independent variable
4	y0 = [38.3; 0; 0; 1035.]; % Initial values for the dependent variables
5	disp(' Variable values at the initial point ');
6	disp(['t = ' num2str(tspan(1))]);
7	disp(' y dy/dt ');
8	disp([y0 ODEfun(tspan(1),y0)]);
9	[t,y]=ode45(@ODEfun,tspan,y0);
10	for i=1:size(y,2)
11	disp([' Solution for dependent variable y' int2str(i)]);
12	disp([' t y' int2str(i)]);
13	disp([t y(:,i)]);
14	plot(t,y(:,i));
15	title([' Plot of dependent variable y' int2str(i)]);
16	xlabel(' Independent variable (t)');
17	ylabel([' Dependent variable y' int2str(i)]);
18	pause
19	end
20	%- - - - - - - - - - - - - - - - - - - -
21	function dYfuncvecdV = ODEfun(V,Yfuncvec);

Table 5–17 (Continued) MATLAB Code for Problem 5.3(a) with m-File **Prob_5_3a2.m**

Line	Equation, %Comment
22	FA = Yfuncvec(1);
23	FB = Yfuncvec(2);
24	FC = Yfuncvec(3);
25	T = Yfuncvec(4);
26	FA0 = 10; %Feed rate of acetone in kg-mol/s
27	k = 820000000000000 * exp(-34222 / T); %Reaction rate constant in s-1
28	XA = (FA0 - FA) / FA0; %Conversion of acetone
29	FN = 38.3 - FA0; %Feed rate of nitrogen in kg-mol/s
30	P = 162; %Pressure kPa
31	yA = FA / (FA + FB + FC + FN); %Mole fraction of acetone
32	CA = yA * P * 1000 / (8.31 * T); %Concentration of acetone in k-mol/m3
33	yB = FB / (FA + FB + FC + FN); %Mole fraction of ketene
34	yC = FC / (FA + FB + FC + FN); %Mole fraction of methane
35	rA = 0 - (k * CA); %Reaction rate in kg-mole/m3-s
36	deltaH = 80770 + 6.8 * (T - 298) - (0.00575 * (T ^ 2 - (298 ^ 2))) - (0.00000127 * (T ^ 3 - (298 ^ 3))); %Heat of reaction in J/mol-K
37	CpA = 26.6 + 0.183 * T - (0.00004586 * T ^ 2); %Heat capacity of acetone in J/mol-K
38	CpB = 20.04 + 0.0945 * T - (0.00003095 * T ^ 2); %Heat capacity of ketene in J/mol-K
39	CpC = 13.39 + 0.077 * T - (0.00001871 * T ^ 2); %Heat capacity of methane in J/mol-K
40	CpN = 6.25 + 0.00878 * T - (0.000000021 * T ^ 2); %Heat capacity of nitrogen in J/mol-K
41	dFAdV = rA; %Differential mass balance on acetone
42	dFBdV = 0 - rA; %Differential mass balance on ketene
43	dFCdV = 0 - rA; %Differential mass balance on methane
44	dTdV = (0 - deltaH) * (0 - rA) / (FA * CpA + FB * CpB + FC * CpC + FN * CpN); %Differential enthalpy balance
45	dYfuncvecdV = [dFAdV; dFBdV; dFCdV; dTdV];

Note that in m-File **Prob_5_3a2.m** shown in Table 5–17, the arrays tspan and y0 corresponding to this particular problem are found in lines 3 and 4. This template utilizes the MATLAB ode45 library function to solve the system of ordinary differential equations in line 9. This function can be called with several different sets of input and output parameters.

$$[t,y]=ode45(@ODEfun,tspan,y0);$$

The first input parameter is the name of the function (*ODEfun*) that returns the vector dy/dV for the input vector y. The second input parameter is a vector that contains the initial and the final value of the independent variable V. The third parameter is the vector of initial values for the dependent variables.

The MATLAB function *ode45* integrates the system of differential equations $y' = F(t,y)$ from the initial value to the final value that are specified in the second input parameter. The algorithm invokes the variable step-size, explicit, 5th-order Runge-Kutta (RK) method where the 4th-order RK step is used for error estimation.

The parameters returned by the *ode45* function include a vector of the independent variable values t between the specified initial and final values, and a matrix of the dependent variable values y.

The MATLAB program m-file **Prob_5_3a2.m** displays tabular results and plots separately for each dependent variable versus the volume V as shown in Figure 5–11 for the 1st variable y1 that represents the molar flow rate of acetone, F_A. The final value obtained for the temperature (the 4th variable) is $T = 907.55$ K and for the conversion (the 5th variable) is $X_A = 0.25726$. These results are essentially the same as obtained when POLYMATH is used for the solution.

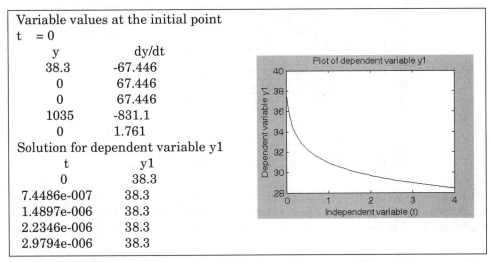

Figure 5–11 Partial Output from Execution of MATLAB file **Prob_5_3a.m**.

(b) F_{A0} and P may be added to the input parameters of the function $dYfuncvecdV$ by altering the function statement (line 21 in Table 5–17) to

```
function dYfuncvecdV = ODEfun(V,Yfuncvec,FA0,P);
```

Also, the definitions of F_{A0} in line 28 and P in line 30 in Table 5–17 must be removed from the function and placed after line 3. Additionally the following statements in lines must be modified to pass these variables to other functions.

```
disp([y0 ODEfun(tspan(1),y0,FA0,P)]);
[t,y]=ode45(@ODEfun,[0 4],y0,[],FA0,P);
```

The resulting MATLAB file is **Prob_5_3b.m** that is partially shown in Table 5–18.

Table 5–18 Partial Listing of MATLAB Program for Part (b) (File **Prob_5_3b.m**)

No.	Command, %Comment
1	function Prob_5_3b
2	clear, clc, format short g, format compact
3	tspan = [0 4.]; % Range for the independent variable
4	FA0 = 38.3; %Feed rate of acetone in kg-mol/s
5	P = 162; %Pressure kPa
6	y0 = [38.3; 0; 0; 1035.; 0]; % Initial values for the dependent variables
7	disp(' Variable values at the initial point ');

Table 5–18 (Continued) Partial Listing of MATLAB Program for Part (b) (File **Prob_5_3b.m**)

No.	Command, %Comment
8	disp([' t = ' num2str(tspan(1))]);
9	disp(' y dy/dt ');
10	disp([y0 ODEfun(tspan(1),y0,FA0,P)]);
11	[t,y]=ode45(@ODEfun,[0 4],y0,[],FA0,P);
12	for i=1:size(y,2)
13	disp([' Solution for dependent variable y' int2str(i)]);
14	disp([' t y' int2str(i)]);
15	disp([t y(:,i)]);
16	plot(t,y(:,i));
17	title([' Plot of dependent variable y' int2str(i)]);
18	xlabel(' Independent variable (t)');
19	ylabel([' Dependent variable y' int2str(i)]);
20	pause
21	end
22	%- -
23	function dYfuncvecdV = ODEfun(V,Yfuncvec,FA0,P);

(c) Repeated calculations with different values of F_{A0} and P can be accomplished by preparation of a new calling function (which replaces **MultipleDEQtemplate.m** of Table 5–17). Part of this calling function is shown in Table 5–19.

Table 5–19 MATLAB Function for Calculating Final Conversion (Xfin) and Final Temperature (Tfin) in the Tubular Reactor

No.	Command, %Comment
1	function Prob_5_3c
2	clear, clc, format short g, format compact
3	FA0set = [10 20 30 35 38.3]; %Feed rate of acetone in kg-mol/s
4	P_set(1)=1.6;
5	i=1;
6	while P_set(i)<=5;
7	i=i+1;
8	P_set(i)=P_set(i-1)+0.2; % Pressure in atm
9	end
10	n_P = size(P_set');
11	for i=1:5
12	FA0=FA0set(i);
13	for j=1:n_P
14	P=P_set(j)*101.325; % Pressure in kPa
15	y0=[FA0; 0; 0; 1035; 0];
16	[V,y]=ode45(@ODEfun,[0 4],y0,[],FA0,P);
17	Xfin(j,i)=y(end,5);
18	Tfin(j,i)=y(end,4);
19	end
20	end

A row vector *FA0set* in line 3 specifies the five desired acetone feed rate values. A row vector *P_set* is generated in lines 4 through 9 that contains all the pressure values as specified in the problem statement. The first element in this vector is specified in line 4. The remaining elements can be represented by a linear series with increment of 0.2. Those elements are generated by means of a *while* statement in lines 6 through 9.

The final conversion and final temperature values are calculated inside two nested *for* loops in lines 11 through 20. The variable *i* obtains the values 1, 2, 3, 4, and 5 and the current value of F_{A0} is set as the *i*-th element of the vector *FA0set*. For every value of *i*, the variable *j* obtains the values 1, 2, 3, ... up to *n_P*, and the current value of *P* is set as the *j*-th element of the vector *Pr_set**101.325 (to convert from atm to kPa). The initial value of the vector *y0* is specified and the function *ode45* is called for all the *i* and *j* values; the resultant final conversion and final temperature values are stored in two matrices where every row is associated with one *P* value and every column is associated with a F_{A0} value.

Note that the call to the *ode45* function is changed in order to enable passing additional parameters to the *ODEfun* (see line 16 in Table 5–19).

$$[V,y]=ode45(@ODEfun,[0\_4],y0,[],FA0,P);$$

The first three input parameters are the same as previously discussed. Note however that the fourth parameter is the *options* parameter, which is a *structure* that contains information that controls the solution algorithm (error tolerances, initial minimal and maximal integration step-sizes, etc.). If the default values are used, this parameter can be replaced by a null matrix, [], as done in this case. The fifth and sixth parameters are F_{A0} and *P* that are being passed to the *ODEfun* function.

Details for preparation of the tables and plots using the values stored in the vector *P_set* and in the matrices *Xfin* and *Tfin* are given Problems 5.1 and 5.2. The portion of the MATLAB code for the tabular and graphical output is shown in Table 5–20.

Table 5–20 MATLAB Code for Tabular and Graphical Output

Line	Command, %Comment
21	% --
22	disp(' Final Conversion versus FA0 and Pressure');
23	disp(' Tabular Results');
24	disp('');
25	disp(' Pressure FA0=10 FA0=20 FA0=30 FA0=35 FA0=38.3 ');
26	Res=[P_set' Xfin(:,1) Xfin(:,2) Xfin(:,3) Xfin(:,4) Xfin(:,5)];
27	disp(Res);
28	plot(P_set,Xfin(:,1),'-',P_set,Xfin(:,2),'+',P_set,Xfin(:,3),'*',P_set,Xfin(:,4),'x',P_set,Xfin(:,5),'o');
29	legend('FA0=10','FA0=20','FA0=30','FA0=35','FA0=38.3');
30	title(' Final Conversion versus FA0 and Pressure')
31	xlabel('Pressure (atm)');
32	ylabel('Final Conversion');
33	pause

Table 5–20 (Continued) MATLAB Code for Tabular and Graphical Output

Line	Command, %Comment
34	disp(' Final Temperature versus FA0 and Pressure');
35	disp(' Tabular Results');
36	disp('');
37	disp(' Pressure FA0=10 FA0=20 FA0=30 FA0=35 FA0=38.3 ');
38	Res=[P_set' Tfin(:,1) Tfin(:,2) Tfin(:,3) Tfin(:,4) Tfin(:,5)];
39	disp(Res);
40	plot(P_set,Tfin(:,1),'-',P_set,Tfin(:,2),'+',P_set,Tfin(:,3),'*',P_set,Tfin(:,4),'x',P_set,Tfin(:,5),'o');
41	legend('FA0=10','FA0=20','FA0=30','FA0=35','FA0=38.3');
42	title(' Final Temperature versus FA0 and Pressure')
43	xlabel('Pressure (atm)');
44	ylabel('Temperature (K)');
45	% ---

The complete MATLAB program for part (c) of this problem is assembled by placing the MATLAB function from Table 5–15 at the end of the MATLAB code given in Table 5–20 starting in line 46.

The plot of final conversion versus P and F_{A0} is presented in Figure 5–12, and the plot of the final temperature versus P and F_{A0} is presented in Figure 5–13.

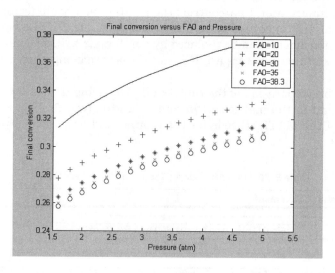

Figure 5–12 Final Conversion versus P and F_{A0} in the Tubular Reactor

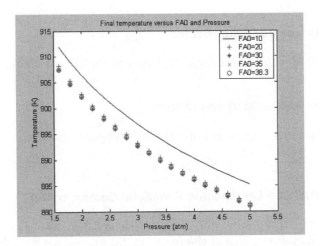

Figure 5–13 Final Temperature versus P and F_{A0} in the Tubular Reactor

The problem solution files are found in directory Chapter 5 and designated **P5-3A.POL**, **MultipleDEQtemplate.m**, **Prob_5_3a1.m**, **Prob_5_3a2.m**, **Prob_5_3b.m**, and **Prob_5_3c.m**.

5.4 CORRELATION OF THE PHYSICAL PROPERTIES OF ETHANE

5.4.1   Concepts Demonstrated

Correlations for ideal gas heat capacity, vapor pressure, and liquid viscosity.

5.4.2   Numerical Methods Utilized

Linear and nonlinear regression of data with linearization and transformation functions.

5.4.3   MATLAB Options and Functions Demonstrated

Matrix and array operations; left division and matrix inversion; the functions *size, ones, eye,* and *mean*. Use of the *load* function to import data from text files, the *input* function to input data from the command window, and the intrinsic function *fminsearch* to search minimum with several independent variables.

5.4.4   Problem Definition

Tables F–1 through F–4 of Appendix F, discussed in Problem 4.4, present values for different properties of ethane (ideal gas heat capacity, vapor pressure, and liquid viscosity) as function of temperature. Various regression models will be fitted to the properties of Appendix F using MATLAB.

(a) Construct a MATLAB function which solves the linear regression problem $\mathbf{Xb} = \mathbf{y}$, where $\mathbf{X}$ is the matrix of the independent variable values, $\mathbf{y}$ is the vector of dependent variable values, and $\mathbf{b}$ is the vector of the linear regression model parameters. The input parameters of the function are $\mathbf{X}$, $\mathbf{y}$, and a logical variable which indicates whether there is a free parameter. The returned parameters are: β and the respective confidence intervals, the calculated values of the dependent variable **ycalc**, the linear correlation coefficient R^2, and the variance. Test the function by fitting the Wagner equation to vapor pressure data of ethane from Table F–3 of Appendix F.

(b) Fit 3rd- and 5th-degree polynomials to the heat capacity of ethane for for the data given in Tables F–1 and F–2 of Appendix F by using the multiple linear regression function developed in (a). Compare the quality of the representation of the various data sets with the polynomials of different degrees.

(c) Fit the Antoine equation to liquid viscosity of ethane given in Table F–4 of Appendix F.

5.4.5 Solution

(a) A MATLAB function for carrying out multiple linear regressions is shown in Table 5–21 and available as a template regression file **MlinReg.m**. The input parameters are: X – the matrix of the independent variable values; y – the vector of dependent variable values; and *freeparm* – a logical variable which indicates whether there is a free parameter (*freeparm* = 1) or there is no free parameter (*freeparm* = 0). The returned parameters are: *Beta* – the vector of the regression model parameter values; *ConfInt* – the vector of confidence intervals; *ycalc* – the vector of calculated values of the dependent variable; *Var* – the variance; and *R2* – the linear correlation coefficient.

Table 5–21 A MATLAB Template Function for Multiple Linear Regression, m-file **MlinReg.m**

Line	Command, %Comment
1	function [Beta, ConfInt, ycalc, Var, R2]=MlinReg(X,y,freeparm)
2	[m,n]=size(X); % m-number of rows or data points, n-number of columns or regression variables
3	if freeparm
4	X=[ones(m,1) X]; % Add column of ones if there is a free parameter
5	npar=n+1;
6	else
7	npar=n;
8	end
9	Beta=X\y; % Solve X'Beta = Y using QR decomposition
10	ycalc=X*Beta; % Calculated dependent variable values
11	Var= ((y-ycalc)'*(y-ycalc))/(m-npar); % variance
12	ymean=mean(y);
13	R2=(ycalc-ymean)'*(ycalc-ymean)/((y-ymean)'*(y-ymean));%linear correlation coefficient
14	% Calculate the confidence intervals
15	A=X'*X;
16	Ainv=A\eye(size(A)); %Calculate the inverse of the X'X matrix
17	tdistr95=[12.7062 4.3027 3.1824 2.7764 2.5706 2.4469 2.3646 2.306 2.2622 2.2281...
18	2.2010 2.1788 2.1604 2.1448 2.1315 2.1199 2.1098 2.1009 2.093 2.086 2.0796...
19	2.0739 2.0687 2.0639 2.0595 2.0555 2.0518 2.0484 2.0452 2.0423 2.0395 2.0369...
20	2.0345 2.0322 2.0301 2.0281 2.0262 2.0244 2.0227 2.0211 2.0195 2.0181 2.0167...
21	2.0154 2.0141]; % 95 percent probability t-distr. values
22	if (m-npar)>45
23	t=2.07824-0.0017893*(m-npar)+0.000008089*(m-npar)^2; % t for degr. freedom > 45
24	else
25	t=tdistr95(m-npar);
26	end
27	for i=1:npar
28	ConfInt(i,1)=t*sqrt(Var*Ainv(i,i)); %confidence intervals
29	end

The function starts by using the MATLAB *size* function in line 2 to determine the number of rows or data points (m) and the number of columns or regression parameters (n) of the **X** matrix. The following *if* statement (lines 3 through 8) introduces an additional column of 1 (ones) to the **X** matrix using the *ones* function and the number of the parameters is set at $n+1$ when the *freeparm* = *true*. Otherwise the number of the parameters is set at n.

The multiple linear regression problem $\mathbf{X}\beta = \mathbf{y}$ is solved using the MATLAB *mldivide* (left matrix divide) function (line 9). This function solves the overdetermined system of equations, in the least squares sense, using QR decomposition. Consequently, the calculated β vector is used to obtain the $\mathbf{y}_{calc}$ values: $\mathbf{y}_{calc} = \mathbf{X}\beta$.

The variance is calculated (in line 11) using the equation:

$$s^2 = \frac{\sum\limits_{i=1}^{m} (y_i - \hat{y}_i)}{m - n_{\text{par}}} \tag{5-1}$$

where y_i are the measured and $\hat{y}$ are the calculated values of the dependent variable. Note that the sum of squares is calculated by vector transpose-vector multiplication.

The linear correlation coefficient is calculated in lines 12 and 13 from

$$R^2 = \frac{\sum\limits_{i=1}^{m} (\hat{y}_i - \bar{y})^2}{\sum\limits_{i=1}^{m} (y_i - \bar{y})^2} \tag{5-2}$$

where $\bar{y}$ is the sample mean of the dependent variable.

The remainder of the function statements are dedicated to the calculation of the 95% confidence intervals of the parameters using the equation

$$\beta_i - ts\sqrt{a_{ii}} \le \beta_i \le \beta_i + ts\sqrt{a_{ii}} \qquad i = 0, 1, \ldots m \tag{5-3}$$

where t is the statistical t-distribution value corresponding to the degrees of freedom and the% confidence selected, s is the standard deviation (square root of the variance) $s = \sqrt{s^2}$, and a_{ii} is the ith diagonal element of the inversed normal $(\mathbf{X}^T\mathbf{X})$ matrix.

The normal matrix is calculated in line 15 and inverted in line 16. Tabular values of the 95% t-distribution are stored in the *tdistr95* vector for up to 45 degrees of freedom. For larger number of degrees of freedom, the t-values are calculated by second-degree polynomial extrapolation (see line 23). The confidence intervals are calculated in line 28.

Use of the **MlinReg.m** function to fit the Wagner equation to vapor pressure data requires generation of the matrix of independent variables $\mathbf{X}$ and the vector of the dependent variables $\mathbf{y}$. The columns of vector $\mathbf{X}$ and the vector $\mathbf{y}$ were actually prepared in the POLYMATH solution in part (b) of Problem 4.4. In that solution, the following transformations (columns) were defined based on the original data of temperature T and vapor pressure P:

TR = T / 305.32;

InPR = In(P / 4872000);
t = (1 - TR) / TR;
t15 = (1 - TR) ^ 1.5 / TR;
t3 = (1 - TR) ^ 3/ TR and
t6 = (1 - TR) ^ 6 / TR

Defining y = InPR; X_1= t; X_2= t15; X_3 = t3, and X_4 = t6, where X_1, X_2, X_3, and X_4 are the respective columns is the **X** matrix, provides the linear representation of the Wagner equation (Equation 4-45 in Problem 4.4). The few first elements of those columns are shown in Table 5–22. The full five columns of numbers (107 rows) can be copied and pasted into a text (say, notepad) file. This text file (saved under the name VPfile.txt) can be loaded into MATLAB and the regression carried out.

Table 5–22 Part of the POLYMATH Data Table of Transformed Vapor Pressure and Temperature Data

No.	y = InPR	X1 = t	X2 = t15	X3 = t3	X4 = t6
1	−14.8684	2.318696	1.938126	1.13187	0.386034
2	−14.3694	2.248085	1.870275	1.07692	0.357059
3	−13.873	2.180417	1.805374	1.024827	0.33023
4	−13.4249	2.11551	1.743244	0.97541	0.305383
5	−13.0011	2.0532	1.683718	0.928503	0.282365

The multiple linear regression of the Wagner equation is thus represented by

$$y = a_0 + a_1 X_1 + a_2 X_2 + a_3 X_3 + a_4 X_4 \qquad (5\text{-}4)$$

where $a_0 = 0$.

The interactive use of the MlinReg function **MlinReg.m** for regression of the transformed vapor pressure data in file VPfile.txt is accomplished with the MATLAB program presented in Table 5–23.

Table 5–23 MATLAB Script File **Prob_5_4a.m** for Multiple Linear Regression of Wagner Equation

Line	Command, %Comment
1	% filename Prob_5_4a
2	clear, clc, format short g, format compact
3	prob_title = (['Vapor Pressure Correlation for Ethane']);
4	ind_var_name=['Functions of Reduced Temp.'];
5	dep_var_name=['Logarithm of Reduced Pressure'];
6	fname=input('Please enter the data file name > ');
7	xyData=load(fname);
8	X=xyData(:,2:end);
9	y=xyData(:,1);
10	[m,n]=size(X);
11	freeparm=input(' Input 1 if there is a free parameter, otherwise input 0 > ');
12	[Beta,ConfInt,ycal,Var, R2]=MlinReg(X,y,freeparm);
13	disp([' Results,' prob_title]);

Table 5–23 (Continued) MATLAB Script File **Prob_5_4a.m** for Multiple Linear Regression of Wagner Equation

Line	Command, %Comment
14	Res=[];
15	if freeparm==0, nparm = n-1; else nparm = n; end
16	for i=0:nparm
17	if freeparm==0; ii=i+1; else ii=i; end
18	Res=[Res; ii Beta(i+1) ConfInt(i+1)];
19	end
20	disp(' Parameter No. Beta Conf_int');
21	disp(Res);
22	disp([' Variance ', num2str(Var)]);
23	disp([' Correlation Coefficient ', num2str(R2)]);
24	pause
25	plot(y,y-ycal,'*') % residual plot
26	title(['Residual Plot, ' prob_title])
27	xlabel([dep_var_name '(Measured)'])
28	ylabel('Residual')
29	pause
30	%
31	%Plot of experimental and calculated data
32	%
33	for i=1:m
34	index(i)=i;
35	end
36	plot(index,ycal, 'r-',index,y,'b+')
37	title(['Calculated/Experimental Data ' prob_title])
38	xlabel(['Point No.'])
39	ylabel([dep_var_name])

During the MATLAB solution, the current directory must set to where the the files **Prob_5_4a.m**, **MlinReg.m** and **VPfile.txt** files reside. The MATLAB solution is provided in Figure 5–14.

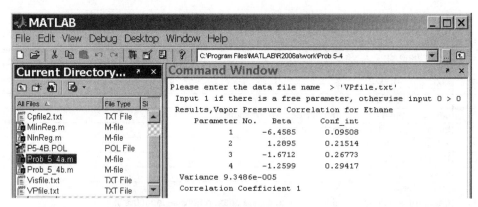

Figure 5–14 MATLAB Command Window and Current Directory for Fitting Wagner Equation to Vapor Pressure Data

A comparison of the results from MATLAB in Figure 5–14 with the Excel results shown in Figure 4–39 of Problem 4.4 shows that the results are identical. The residual plot from MATLAB (see Figure 5–15) is essentially the same as the residual plot generated in Excel (Figure 4–40 in Problem 4.4) (Note sign reversal in the residual definition.).

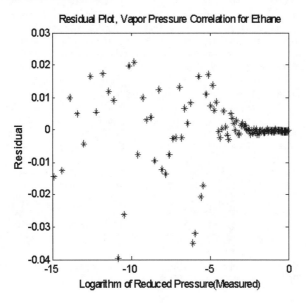

Figure 5–15 Residual Plot for Vapor Pressure Data Represented by the Wagner Equation

(b) (Partial Solution) The heat capacity data for ethane involves elevated temperatures in K. When temperatures of the order of several hundreds of K are the independent variables, the MATLAB linear equation solver (or matrix inversion) functions will issue matrix ill-conditioning related messages when fitting a third-degree polynomial. Attempts to fit a fifth-degree polynomial may be impossible as the linear equation solver function will reduce the dimension of the coefficient matrix involved so that actually a fourth-degree polynomial is fitted.

The discussion of appropriate dimensional reduction is beyond the scope of this book; however, both the error messages and the dimension reduction can be prevented by normalizing the temperature data. This normalization can be simply accomplished by dividing all the temperature values by the maximal temperature. Use of the *MlinReg* function for polynomial regression requires that the columns of the **X** matrix be prepared with the normalized temperature raised to various degrees. Thus for the fifth-degree polynomial, the following transformations should be defined in the POLYMATH Data Table.

Tnor=T_K/1500; y=Cp ;X1=Tnor; X2=Tnor^2; X3=Tnor^3; X4=Tnor^4; X5=Tnor^5

A linear representation of a fifth-degree polynomial equation   (Equation 4-42 in Problem 4.4) using these transformations can be implemented by creat-

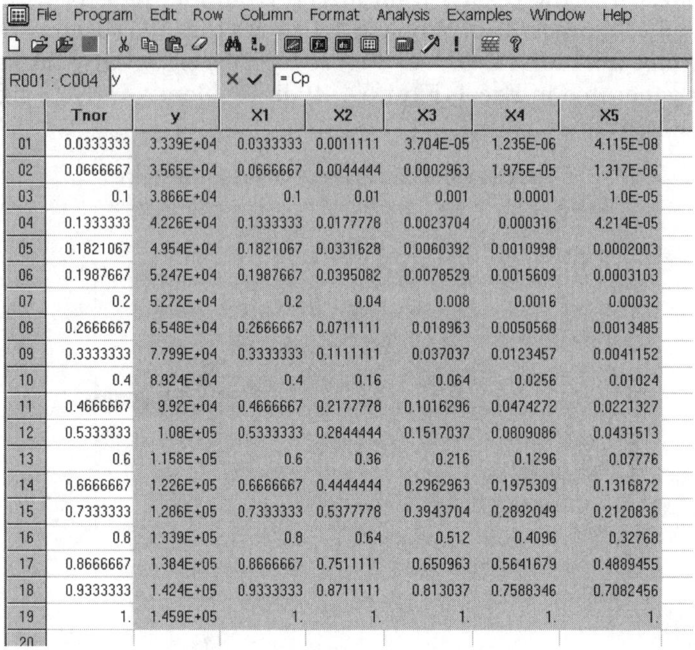

	Tnor	y	X1	X2	X3	X4	X5
01	0.0333333	3.339E+04	0.0333333	0.0011111	3.704E-05	1.235E-06	4.115E-08
02	0.0666667	3.565E+04	0.0666667	0.0044444	0.0002963	1.975E-05	1.317E-06
03	0.1	3.866E+04	0.1	0.01	0.001	0.0001	1.0E-05
04	0.1333333	4.226E+04	0.1333333	0.0177778	0.0023704	0.000316	4.214E-05
05	0.1821067	4.954E+04	0.1821067	0.0331628	0.0060392	0.0010998	0.0002003
06	0.1987667	5.247E+04	0.1987667	0.0395082	0.0078529	0.0015609	0.0003103
07	0.2	5.272E+04	0.2	0.04	0.008	0.0016	0.00032
08	0.2666667	6.548E+04	0.2666667	0.0711111	0.018963	0.0050568	0.0013485
09	0.3333333	7.799E+04	0.3333333	0.1111111	0.037037	0.0123457	0.0041152
10	0.4	8.924E+04	0.4	0.16	0.064	0.0256	0.01024
11	0.4666667	9.92E+04	0.4666667	0.2177778	0.1016296	0.0474272	0.0221327
12	0.5333333	1.08E+05	0.5333333	0.2844444	0.1517037	0.0809086	0.0431513
13	0.6	1.158E+05	0.6	0.36	0.216	0.1296	0.07776
14	0.6666667	1.226E+05	0.6666667	0.4444444	0.2962963	0.1975309	0.1316872
15	0.7333333	1.286E+05	0.7333333	0.5377778	0.3943704	0.2892049	0.2120836
16	0.8	1.339E+05	0.8	0.64	0.512	0.4096	0.32768
17	0.8666667	1.384E+05	0.8666667	0.7511111	0.650963	0.5641679	0.4889455
18	0.9333333	1.424E+05	0.9333333	0.8711111	0.813037	0.7588346	0.7082456
19	1.	1.459E+05	1.	1.	1.	1.	1.
20							

Figure 5–16 **X** Data Matrix with Normalized Temperature Data as Created in POLYMATH for MATLAB Regression

ing the **X** matrix within POLYMATH as shown in Figure 5–16. The highlighted data in POLYMATH can be copied and saved as a text file (for example, Cpfile.txt) to be used in MATLAB to carry out the regression of the fifth-degree polynomial. Note that for fitting a third-degree polynomial, only the first four data columns (without the headings) should be copied and pasted into a text file.

The MATLAB script file shown in Table 5–24, which interacts with the user to obtain the name of the input file (in line 5) and the value of the *freeparm* parameter (in line 9), calls the MlinReg function to carry out the regression and present numerical and graphical results. The program is self-explanatory for the most part. In lines 13 through 16, the results table, which includes the parameter number, parameter value, and the confidence interval value, is prepared. Depending on whether there is a free parameter, the number of the first parameter is either 0 (zero) or 1 (no free parameter).

Table 5–24 MATLAB Script-file for Carrying Out Multiple Linear Regression for Heat Capacity of Ethane and Presentation of the Results

Line	Command, %Comment
1	% filename Prob_5_4b
2	clear, clc, format short g, format compact
3	prob_title = (['Heat Capacityof Ethane']);
4	ind_var_name=['Normalized Temp'];
5	dep_var_name=['Heat Capacity J/kmol*K '];

Line	Command, %Comment
6	fname=input('Please enter the data file name > ');
7	xyData=load(fname);
8	X=xyData(:,2:end);
9	y=xyData(:,1);
10	[m,n]=size(X);
11	freeparm=input(' Input 1 if there is a free parameter, 0 otherwise > ');
12	[Beta, ConfInt,ycal, Var, R2]=MlinReg(X,y,freeparm);
13	disp([' Results, ' prob_title]);
14	Res=[];
15	if freeparm==0, nparm = n-1; else nparm = n; end
16	for i=0:nparm
17	if freeparm, ii=i+1; else ii=i; end
18	Res=[Res; ii Beta(i+1) ConfInt(i+1)];
19	end
20	disp(' Parameter No. Beta Conf_int');
21	disp(Res);
22	disp([' Variance ', num2str(Var)]);
23	disp([' Correlation Coefficient ', num2str(R2)]);
24	plot(y,y-ycal,'*')
25	title(['Residual Plot, ' prob_title]) % residual plot
26	xlabel([dep_var_name '(Measured)'])
27	ylabel('Residual')
28	pause
29	plot(X(:,2),ycal, 'r-',X(:,2),y,'b+')
30	title(['Calculated/Experimental Data ' prob_title])
31	xlabel([ind_var_name])
32	ylabel([dep_var_name])

The interaction with the MATLAB script program and the results for the regression with a fifth-order polynomial using data file **Cpfile.txt** are shown in Figure 5–17. A comparison with the Excel solution for this same problem, in Figure 4–32 of Problem 4.4, indicates that the variance, R^2, and the free parameter values are identical for up to at least four decimal digits. The remaining parameters given in Figure 5–17 can be compared to the same values as presented in the Excel solution in Figure 4–32 by division by the maximal temperature raised to the corresponding degree. For example, $a_3 = -2.1962e+006/1500^3 = -6.5073E-04$, which is in close agreement with same parameter value as in Problem 4.4, Figure 4–32.

The MATLAB driver program prepares a residual plot as well as a plot of calculated and measured values of the dependent variable versus the independent variable values. The latter plot is demonstrated Figure 5–18 for the case of the fifth-degree polynomial for the 1st set of the heat capacity data.

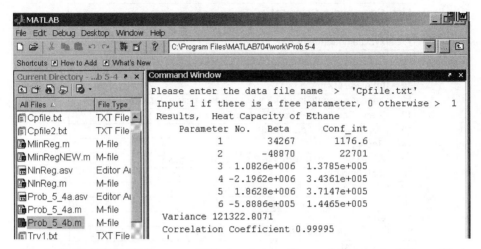

Figure 5–17 MATLAB Command Window Interaction and Output for Ethane Heat
Capacity Represented by Fifth-order Polynomial

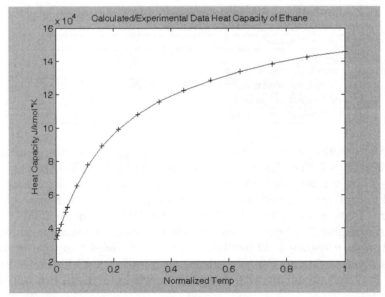

Figure 5–18 Fifth-degree Polynomial Representation of the 1st Set of Ethane Heat
Capacity Data

(c) The *fminsearch* function provided by MATLAB can be used for nonlinear
regression with several parameters. This function uses the Nelder-Mead simplex
(direct search) method to locate a local minimum close to the initial estimate provided. The determination of the parameters of the Antoine equation using the
fminsearch function in MATLAB first requires a file containing the Y and X data
(in this case, Y = ln μ, X = T). This can be accomplished with POLYMATH for this

problem by defining these columns of transformed data in POLYMATH. These two columns of data can then be pasted into a text file for use by MATLAB (file name: visfile.txt, for this example). The next step is to prepare a MATLAB function which calculates the sum of squares of the errors, $\sum (Y - Y_{calc})^2$, for a set of the parameter values. This function is shown in Table 5–25.

Table 5–25 The AntFun Function File **AntFun.m** for Calculating the Sum of Squares of Errors for the Antoine Equation for Viscosity

Line	Command, %Comment
1	function [f,Ycalc]=AntFun(parm,X,Y)
2	a=parm(1);
3	b=parm(2);
4	c=parm(3);
5	Ycalc(:,1)=(a+b./(X+c));
6	resid(:,1)=Y-Ycalc;
7	f=resid'*resid;

The *AntFun* function obtains as input parameters the vector **parm**, which contains the current value of the Antoine equation parameters: *A, B,* and *C* (see Equation 4-47 in Problem 4.4) and the vectors of the measured data **X** and **Y**. The returned parameters are the scalar *f* which contains the sum of squares of the errors and the vector **Ycalc** which contains the calculated values of the dependent variable (ln μ).

The *fminsearch* function is called to minimize the sum of squares of errors as follows.

Beta=fminsearch(@AntFun,parm,options,X,Y);

The input parameters are the name of the function that calculates the sum of squares of the errors, the vectors **parm, X,** and **Y** that were described previously, and the structure *options* which contains technical parameters of the function such as error tolerances and maximum number of iterations. The output of *fminsearch* is the set of the parameter values which give a local minimum of the sum of squares.

A MATLAB script file for nonlinear regression is given in Table 5–26 with file name **NlnReg.m**.

Table 5–26 MATLAB Function (**NlnReg.m**) for Multiple Nonlinear Regression and Presentation of the Results

Line	Command, %Comment
1	% filename NlnReg.m
2	function NlnReg
3	clear, clc, format short g, format compact
4	fname=input('Please enter the data file name between apostrophes > ');
5	xyData=load(fname); % Load data
6	X=xyData(:,1);

Table 5–26 (Continued) MATLAB Function (**NlnReg.m**) for Multiple Nonlinear Regression and Presentation of the Results

Line	Command, %Comment
7	m=size(X,1); %Determine the number of data points
8	Y=xyData(:,3);
9	prob_title = ([' Antoine Equation Parameters, Nonlinear Regression']);
10	dep_var_name=['ln of Viscosity '];
11	ind_var_name=['Temperature (K)'];
12	parm=input('Please enter vector of initial parameter estimates [a1,...] > ');
13	npar=size(parm,2); % Determine the number of the parameters
14	options=optimset('MaxFunEvals',1000); % Change the default value for MaxFunEvals
15	Beta=fminsearch(@AntFun,parm,options,X,Y); % Find optimal parameters using fminsearch
16	[f,Ycalc]=AntFun(Beta,X,Y); % Compute Y (calculated) at the optimum
17	disp([' Results,' prob_title]);
18	Res=[];
19	for i=1:npar
20	Res=[Res; i Beta(i)];
21	end
22	disp(' Parameter No. Value ');
23	disp(Res);
24	s2=sum((Y-Ycalc)'*(Y-Ycalc))/(m-npar); %variance
25	disp([' Variance ', num2str(s2)]);
26	ymean=mean(Y);
27	R2=(Ycalc-ymean)'*(Ycalc-ymean)/((Y-ymean)'*(Y-ymean));%linear correlation coefficient
28	disp([' Correlation Coefficient ', num2str(R2)])
29	plot(Y,Y-Ycalc,'*') % Residual plot
30	title(['Residual Plot, ' prob_title])
31	xlabel([dep_var_name '(Measured)'])
32	ylabel('Residual')
33	pause
34	plot(X,Ycalc,'r-',X,Y,'b+') %Plot of experimental and calculated data
35	title(['Calculated/Experimental Data ' prob_title])
36	xlabel([ind_var_name])
37	ylabel([dep_var_name])

This general program first interacts with the user to obtain the name of the input file (in line 4) and the initial estimates for the parameter values (in line 12). In line 14 the statement

options=optimset('MaxFunEvals',1000);

is used to increase the number of function evaluations allowed from the default value. (The statement: optimset('fminsearch') may be useful in the MATLAB command window to find the active function parameters and their default values).

The *fminsearch* function is used (in line 14) to find the optimal parameter values (in line 15), and the optimal results are calculated in line 16. The results table is created in lines 17 through 23, and it includes the parameter numbers and the parameter values. The variance and the linear correlation coefficient are calculated and displayed in lines 24 through 28. A residual plot is created in lines 29 through 37.

MATLAB execution of the nonlinear regression script file **NlnReg.m**, along with the **AntFun.m** function file and the data file **visfile.txt**, yields the data table shown in Figure 5–19 for fitting the Antoine equation to the viscosity data of this problem. Note that the initial estimates for the parameters are the same as those used in the POLYMATH solution shown in Figure 4–46.

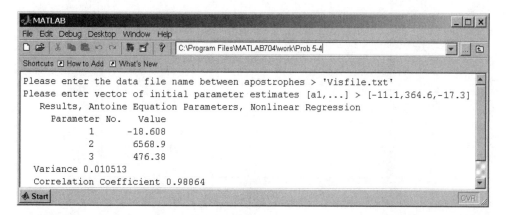

Figure 5–19 MATLAB Command Window Showing Nonlinear Regression Results

The results given in Figure 5–20 compare almost exactly with the results obtained with POLYMATH in Figure 4–45 to four significant figures. The comparison with the Excel results of Figure 4–48 is not as exact but still quite reasonable.

The generated residual plot, Figure 5–21, shows a clearly cyclic pattern as also shown in the Excel results of Figure 4–49, indicating some significant errors and a less-than-satisfying correlation. This is also indicated in the plot of calculated ln of the viscosity from the Antoine equation versus the ln of the viscosity data in Figure 5–20.

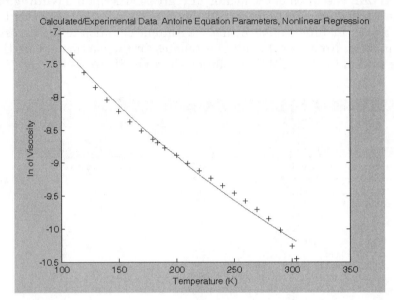

Figure 5–20 Comparison of Regressed Antone Equation with Liquid Viscosity Data

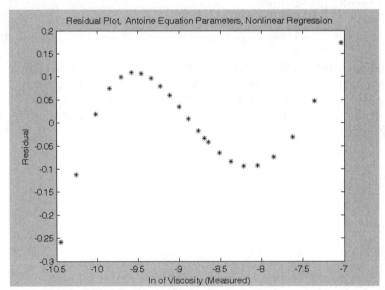

Figure 5–21 Residual Plot for Liquid Viscosity Data Represented by the Antoine Equation

The problem solution files are found in directory Chapter 5 and designated **P5-4B.POL**, **AntFun.m**, **Cpfile.txt**, **Cpfile2.txt**, **MlinReg.m**, **NlnReg.m**, **Prob_5_4a.m**, **Prob_5_4b.m**, **Visfile.txt**, and **VPfile.txt**.

5.5 COMPLEX CHEMICAL EQUILIBRIUM BY GIBBS ENERGY MINIMIZATION

5.5.1 Concepts Demonstrated

Formulation of the chemical equilibrium problem as a Gibbs energy minimization problem with atomic balance constraints. The use of Lagrange multipliers to introduce the constraints into the objective function. Conversion of the minimization problem into a system of nonlinear algebraic equations.

5.5.2 Numerical Methods Utilized

Solution of a system of nonlinear algebraic equations.

5.5.3 MATLAB Options and Functions Demonstrated

Use of the MATLAB Optimization Toolbox function *fsolve* (unconstrained) and the provided MATLAB function *conles* (constrained) for solving a system of nonlinear algebraic equations.

5.5.4 Problem Definition

Work Problem 4.5(a) using MATLAB.

(a) Obtain the MATLAB function generated by POLYMATH for this problem. Use the MATLAB Optimization Toolbox function *fsolve*[*] (from the initial estimates given in Table 4–10 of Problem 4.5) to solve the problem in order to obtain the Lagrange multiplier values and the molar amounts of the various components at equilibrium.

(b) Use the MATLAB *conles* function (a constrained nonlinear algebraic equation solver) to solve the system of equations from the same initial estimates as in part (a).

(c) Determine the objective function value of part (b) at the solution.

[*] The use of *fsolve* requires access to the MATLAB Optimization Toolbox. Please skip to part (b) if this is not available for use. Please note that a compiled MATLAB program, **conles.p**, is provided for your use in part (b) that solves constrained nonlinear algebraic equations.

5.5.5 Solution

(a) The set of POLYMATH equations for this problem is shown in Table 4–11 of Problem 4.5. In order to obtain the equation set arranged in the order calculation and formatted as a MATLAB *function,* the POLYMATH option "Show MATLAB formatted problem in report" must be turned on. This option is accessible by opening the icon marked as "Setup preferences and numerical parameters." The MATLAB formatted equation set is displayed in the POLYMATH report page together with the solution results.

The resultant MATLAB equation set from POLYMATH should be copied into the MATLAB editor. This code then requires some modifications. It is shown in Table 5–27 and discussed below.

Table 5–27 MATLAB Function Generated by POLYMATH for Problem 4.5 (Slightly Modified)

Line	Command, % Comment
1	function fx = MNLEfun(x);
2	lamda1 = x(1);
3	lamda2 = x(2);
4	lamda3 = x(3);
5	H2 = x(4);
6	H2O = x(5);
7	CO = x(6);
8	CO2 = x(7);
9	CH4 = x(8);
10	C2H6 = x(9);
11	C2H4 = x(10);
12	C2H2 = x(11);
13	O2 = x(12);
14	R = 1.9872;
15	sum = H2 + O2 + H2O + CO + CO2 + CH4 + C2H6 + C2H4 + C2H2;
16	fx(1,1) = 2 * CO2 + CO + 2 * O2 + H2O - 4; %Oxygen balance
17	fx(2,1) = 4 * CH4 + 4 * C2H4 + 2 * C2H2 + 2 * H2 + 2 * H2O + 6 * C2H6 - 14; %Hydrogen balance
18	fx(3,1) = CH4 + 2 * C2H4 + 2 * C2H2 + CO2 + CO + 2 * C2H6 - 2; %Carbon balance
19	fx(4,1) = log(H2 / sum) + 2 * lamda2;
20	fx(5,1) = -46.03 / R + log(H2O / sum) + lamda1 + 2 * lamda2;
21	fx(6,1) = -47.942 / R + log(CO / sum) + lamda1 + lamda3;
22	fx(7,1) = -94.61 / R + log(CO2 / sum) + 2 * lamda1 + lamda3;
23	fx(8,1) = 4.61 / R + log(CH4 / sum) + 4 * lamda2 + lamda3;
24	fx(9,1) = 26.13 / R + log(C2H6 / sum) + 6 * lamda2 + 2 * lamda3;
25	fx(10,1) = 28.249 / R + log(C2H4 / sum) + 4 * lamda2 + 2 * lamda3;
26	fx(11,1) = C2H2 - exp(-(40.604 / R + 2 * lamda2 + 2 * lamda3)) * sum;
27	fx(12,1) = O2 - exp(-2 * lamda1) * sum;

The following code changes have been implemented in Table 5–27.

1. The function statement: function fx=MNLEfun(x) has been added as the first line of the function.

2. The functions that are used for solving minimization problems or systems of equations require passing the input arguments inside a single vector (**x** in this case). The values stored in **x** are substituted into the local variables in a consistent manner, thus commands like lamda1 = x(1) have been added.

3. The resultant function values are stored in a single column vector (see lines 16—27 in Table 5–27).

The solution of this system of equations can utilize the MATLAB template file provided in POLYMATH's Help materials for solving a system of nonlinear equations. This **MultipleNLEtemplate.m** file is shown in Table 5–28.

Table 5–28 Template File **MultipleNLEtemplate.m** for Solving A System of Nonlinear Algebraic Equations

Line	Command, %Comment
1	function Prob_5_5a1
2	clear, clc, format short g, format compact
3	xguess = [10. 10. 10. 5.992 1. 1. 0.993 0.001 0.001 0.001 0.001 0.0001]; % initial guess vector
4	disp('Variable values at the initial estimate');
5	fguess=MNLEfun(xguess);
6	disp(' Variable Value Function Value')
7	for i=1:size(xguess,2);
8	disp([' x' int2str(i) ' ' num2str(xguess(i)) ' ' num2str(fguess(i))]);
9	end
10	options = optimset('Diagnostics',['off'],'TolFun',[1e-9],'TolX',[1e-9]);
11	xsolv=fsolve(@MNLEfun,xguess,options);
12	disp('Variable values at the solution');
13	fsolv=MNLEfun(xsolv);
14	disp(' Variable Value Function Value')
15	for i=1:size(xguess,2);
16	disp([' x' int2str(i) ' ' num2str(xsolv(i)) ' ' num2str(fsolv(i))])
17	end

The MATLAB Optimization Toolbox function *fsolve* is used to solve the system of nonlinear equations. This function can be called with several different sets of input and output parameters. The particular form used in Table 5–28 is

xsolv=fsolve(@MNLEfun,xguess,options);

The first input parameter is the name of the function (*MNLEfun*) while the second input parameter (*xguess*) is a vector of initial estimates for the values of the unknowns at the solution. The *options* input parameter is a structure which contains parameters associated with the solution method, error tolerances, etc. Several selected parameters are set by the options command shown in line 10 of Table 5–28.

The parameter returned by the *fsolve* function is the vector of the values of the unknowns at the solution (*xsolv*). The function fsolve uses minimization techniques (such as the Levenberg Marguardt method with line search) in order to solve the system of nonlinear equations.

The template file (of Table 5–28) followed by the function (in Table 5–27) are copied and pasted into one file (**Prob_4_5a1.m**) in order to complete the program (note that the vector of initial estimates *xguess* has been already introduced into the template file).

Execution of the program does not provide an acceptable solution as error messages are generated by MATLAB such as "Optimization terminated: no further progress can be made. ....". A potential cause for the termination of the optimization can be that there are errors in the conversion of the function from POLYMATH to MATLAB.

This possibility can be investigated by comparing the components of the vector $\mathbf{f}(\mathbf{x})$ calculated at the initial estimate $\mathbf{x}_0$ by MATLAB with the values obtained by POLYMATH as shown in Table 5–29. Note that in order to obtain $\mathbf{f}(\mathbf{x})$ values in POLYMATH, the option: "Calculate initial guess values" should be set at true. (This option can be found under the icon "Setup preferences and numerical parameters" and "Nonlinear equations.") It can be seen that the values are essentially the same; thus, the POLYMATH to MATLAB conversion is correct, and the "no convergence" is caused by the inability of the solution algorithm to handle this particular problem with the initial guesses provided.

Table 5–29 Initial Variable x and Function f(x) Values Obtained by POLYMATH and MATLAB

No.	Variable	Initial Value	f(x) POLYMATH	f(x) MATLAB
1	lamda1	10	-0.0138	-0.0138
2	lamda2	10	0	0
3	lamda3	10	0	0
4	H2	5.992	19.59441	19.5944
5	H2O	1	4.640743	4.6407
6	CO	1	-6.32142	-6.3214
7	CO2	0.993	-19.8127	-19.8127
8	CH4	0.001	43.21608	43.2161
9	C2H6	0.001	84.04539	84.0454
10	C2H4	0.001	65.11171	65.1117
11	C2H2	0.001	0.001	0.001
12	O2	0.0001	1.00E-04	1.00E-04

This is a challenging problem for the unconstrained optimization used by the MATLAB function *fsolve*. In this case, fsolve will perform the solution to this problem if line 10 in Table 5–28 is modified to change the tolerance and algorithm.

```
options = optimset('Diagnostics','off','TolFun',1e9,'TolX',1e-14,
        'NonlEqnAlgorithm','gn');
```

The results of this modified MATLAB program are shown in Figure 5–22 where the optimal parameter values exactly match those from POLYMATH and closely match those from Excel that are both summarized in Table 4–15. Note that the solution shown in Figure 5–22 also includes very small imaginary components that may be considered as insignificant relative to the real part of the solution.

(b) A constrained optimization usually provides a more robust solution from a variety of initial guesses. Since a constrained optimization may not be available in MATLAB, the *conles* function (Shacham[1]) is provided for use. The compiled MATLAB file conles.p utilizes the same constrained nonlinear algebraic equation solver algorithm that is used by POLYMATH to solve the constrained minimization problem.

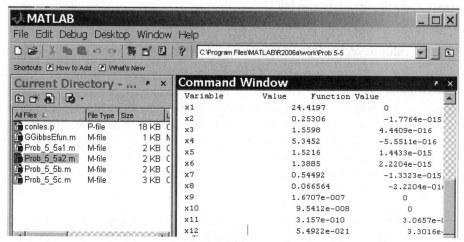

Figure 5–22 Solution to Part (a) Utilizing Modified *fsolve* MATLAB function for Unconstrained Optimization of Nonlinear Equations

The function call of *conles* is:

[xsolv,y,dy,info]=conles(fun,xguess,pote,dfun,tol,maxit,derfun,print);

Input Parameters for *conles*

fun is a function which accepts input x and returns a column vector equation values f evaluated at x.

xguess is a vector of length n. It contains an estimate of the solution of the system of equations. This initial estimate must satisfy the constraints.

pote is an vector of length n. It contains information on whether a given variable is constrained or not. The value of the j-th component of pote is set as follows: $pote(j) = 0$ — variable number j is not constrained, $pote(j) = 1$ — a solution is sought where $x(j) > 0$, and $pote(j) = 2$ — $x(j)$ must not obtain a value less than or equal to zero during the solution process.

dfun (optional) is a function which accepts input x and returns an $n \times n$ matrix of partial derivatives (Jacobian matrix).

tol is a nonnegative input variable. Convergence occurs if the Euclidean norm of the relative errors between two successive iterations is at most *tol*.

maxit is a positive integer input variable. Termination occurs if the number of iterations is at *maxit*.

derfun is an logical input variable. *Derfun* = 1 (true) indicates that analytical derivatives are being supplied by the user. *Derfun* = (0) false indicates that the Jacobian matrix has to be approximated by divided differences.

print is a logical variable which indicates whether printing of intermediate results is required *(print =1 (true))*.

Output Parameters for *conles*

xsolv is a vector of length n which contains the variable values at the solution.

y is a column vector of length n which contains the f(x) values at the solution.

dy is an $n \times n$ square matrix which contains the partial derivatives at the solution.

info is an integer variable set as follows: *info* = 0 — the Euclidean norm of the relative errors between two successive iterates is at most *tol* in magnitude; *info* = 1 — improper input parameters; *info* = 2 — initial estimate values do not satisfy the constraints; *info* = 3 — number of iterations has reached *maxit*; and *info* = 4 or 5 — iteration is not making satisfactory progress or diverging.

The function *conles* may easily replace *fsolve* for solving this problem. This is accomplished by modifying the template file (of Table 5–28) by removing the call to *fsolve* (lines 10 and 11) and inserting the commands given in Table 5–30.

Table 5–30 Modifications for Using Function *conles*

10a	pote=[0 0 0 2 2 2 2 2 2 2 2 2];
10b	tol=1e-9;
10c	maxit=100;
10d	derfun=0;
10e	print=0;
11	[xsolv,y,dy,info]=conles(@MNLEfun,xguess,pote,[],tol,maxit,derfun,print);

Lines 10a through 10e set the input parameter values (of the *conles* function). Note that all the variables, except the Lagrange multipliers, are constrained to positive values throughout the solution process (*pote* (j) = 2). No function for calculating the matrix of partial derivatives is provided (*derfun* = 0).

In line 11, the *conles* function is called to solve the system of the constrained nonlinear algebraic equations. Note that a space saver, [ ], is used instead of the 4th parameter of the function, as there is no function for calculating the matrix of partial derivatives. The MATLAB results are shown in Table 5–31. The variable and function values are essentially the same that were obtained by POLYMATH (presented in Table 4-12 of Problem 4.5).

Table 5–31 Solution Obtained Using the *conles* Function

Variable Values at the Initial Estimate			Variable Values at the Solution		
Variable	Value	Function Value	Variable	Value	Function Value
x1	10	−0.0138	x1	24.4197	0
x2	10	0	x2	0.25306	−1.78E-15
x3	10	0	x3	1.5598	0
x4 - H2	5.992	19.5944	x4 - H2	5.3452	−3.33E-16
x5 - H2O	1	4.6407	x5 - H2O	1.5216	−1.89E-15
x6 - CO	1	−6.3214	x6 - CO	1.3885	−2.22E-16

Variable Values at the Initial Estimate			Variable Values at the Solution		
Variable	Value	Function Value	Variable	Value	Function Value
x7 - CO2	0.993	-19.8127	x7 - CO2	0.54492	6.88E-15
x8 - CH4	0.001	43.2161	x8 - CH4	0.066564	4.44E-16
x9 - C2H6	0.001	84.0454	x9 - C2H6	1.67E-07	−1.68E-13
x10 - C2H4	0.001	65.1117	x10 - C2H4	9.54E-08	−2.60E-13
x11 - C2H2	0.001	0.001	x11 - C2H2	3.16E-10	−1.55E-25
x12 - O2	0.0001	1.00E-04	x12 - O2	5.46E-21	3.88E-28

(c) The calculation of the objective function (Equation 4-49 in Problem 4.5) can be obtained by adding the commands shown in Table 5–32 to the MATLAB statements following line 17 in Table 5–28. The calculated minimal objective function with MATLAB (–104.34) is the same as determined in the POLYMATH solution.

Table 5–32 MATLAB Commands for Calculating the Gibbs Energy at the Solution

Line	Command, %Comment
18	H2=xsolv(4);
19	H2O=xsolv(5);
20	CO=xsolv(6);
21	CO2=xsolv(7);
22	CH4=xsolv(8);
23	C2H6=xsolv(9);
24	C2H4=xsolv(10);
25	C2H2=xsolv(11);
26	O2=xsolv(12);
27	R = 1.9872;
28	sum = H2 + O2 + H2O + CO + CO2 + CH4 + C2H6 + C2H4 + C2H2;
29	G_O2 = O2 * log(abs(O2 / sum));
30	G_H2 = H2 * log(H2 / sum);
31	G_H2O = H2O * (-46.03 / R + log(H2O / sum));
32	G_CO = CO * (-47.942 / R + log(CO / sum));
33	G_CO2 = CO2 * (-94.61 / R + log(CO2 / sum));
34	G_CH4 = CH4 * (4.61 / R + log(abs(CH4 / sum)));
35	G_C2H6 = C2H6 * (26.13 / R + log(abs(C2H6 / sum)));
36	G_C2H4 = C2H4 * (28.249 / R + log(abs(C2H4 / sum)));
37	G_C2H2 = C2H2 * (40.604 / R + log(abs(C2H2 / sum)));
38	ObjFun = G_H2 + G_H2O + G_CO + G_O2 + G_CO2 + G_CH4 + G_C2H6 + G_C2H4 + G_C2H2

www

The problem solution files are found in directory Chapter 5 and designated **P5-5A.POL**, **conles.p**, **Prob_5_5a1.m**, **Prob_5_5a2.m**, **Prob_5_5b.m**, and **Prob_5_5c.m**.

REFERENCE

1. Shacham, M., "Numerical Solution of Constrained Non-Linear Algebraic Equations," *International Journal of Numerical Methods in Engineering*, **23**, 1455–1481 (1986).

Advanced Techniques in Problem Solving

6.1 SOLUTION OF STIFF ORDINARY DIFFERENTIAL EQUATIONS

6.1.1 Concepts Demonstrated

Simulation of chemical or biological reactions in a batch reactor process that can lead to very high reaction rates with very low reactant concentrations.

6.1.2 Numerical Methods Utilized

Solution of systems of ordinary differential equations that become stiff during the course of the integration.

6.1.3 Problem Statement

A biological process involves the growth of biomass from substrate as studied by Garritsen.[1] The material balances on this batch process yield

$$\frac{dB}{dt} = \frac{kBS}{(K+S)} \tag{6-1}$$

$$\frac{dS}{dt} = -\frac{0.75kBS}{(K+S)} \tag{6-2}$$

where B and S are the respective biomass and substrate concentrations. The reaction kinetics are such that $k = 0.3$ and $K = 10^{-6}$ in consistent units.

(a) Solve this set of differential equations starting at $t_0 = 0$, where $S = 5.0$ and $B = 0.05$ to a final time given by $t_f = 20$. Assume consistent units.
(b) Plot S and B with time for the conditions of part (a).

6.1.4 Solution

This set of equations can be entered into the POLYMATH *Simultaneous Differential Equation Solver* as previously discussed. If the equations are entered correctly and a non-stiff integration algorithm (such as the default RKF45 algorithm) is used to solve the system, the integration will proceed in the usual way up to t = 16.34. However, from this point on there will be no further progress and the integration will have to be stopped (by pressing Ctrl-C). Non-progress of the integration may indicate that the system of equations is stiff and thus requires special "stiff" solution methods.

In order to determine mathematically that a particular system is stiff, the matrix of partial derivatives of the differential equations with respect to each of the dependent variables must be calculated. If among the eigenvalues of this matrix there is at least one that is negative and has a large absolute value, then the system is referred to as stiff.

For this problem, the matrix of partial derivatives may be evaluated by first rewriting Equations (6-1) and (6-2) using simplified notation and introducing the known kinetic constants.

$$f_1 = \frac{dB}{dt} = \frac{0.3BS}{10^{-6} + S} \tag{6-3}$$

$$f_2 = \frac{dS}{dt} = -\frac{0.225BS}{10^{-6} + S} \tag{6-4}$$

Differentiation of these equations with respect to B and S yields

$$J_{11} = \frac{\partial f_1}{\partial B} = \frac{0.3S}{(10^{-6} + S)} \qquad J_{12} = \frac{\partial f_1}{\partial S} = \frac{0.3 \times 10^{-6} B}{(10^{-6} + S)^2} \tag{6-5}$$

$$J_{21} = \frac{\partial f_2}{\partial B} = -\frac{0.225S}{(10^{-6} + S)} \qquad J_{22} = \frac{\partial f_2}{\partial S} = -\frac{0.225 \times 10^{-6} B}{(10^{-6} + S)^2} \tag{6-6}$$

The eigenvalues designated by λ of the matrix of partial derivatives (the **J** matrix) must satisfy the equation $\det(\mathbf{J} - \lambda\mathbf{I}) = 0$ where **I** is an identity matrix. In the case of a 2×2 matrix, this is a quadratic equation the roots of which are given by

$$\lambda_{1,2} = \frac{J_{11} + J_{22} \pm \sqrt{(J_{11} + J_{22}) - 4(J_{11}J_{22} - J_{12}J_{21})}}{2} \tag{6-7}$$

Introduction of the problem equations, the J matrix components, and the eigenvalue expressions into the POLYMATH *Differential Equation Solver Program* yields the equation set shown in Table 6–1.

Table 6–1 Problem Equations plus Stiffness Calculations (File **P6-01-2.POL**)

Line	Equation
1	d(S)/d(t) = -k * y * B * S / (Km + S)
2	d(B)/d(t) = k * B * S / (Km + S)
3	J11 = k * S / (Km + S) # dF1/dB
4	J12 = k * Km * B / (Km + S) ^ 2 # dF1/dS
5	J21 = -0.75 * k * S / (Km + S) # dF2/dB
6	J22 =- 0.75 * k * Km * B / (Km + S) ^ 2 # dF2/dS
7	lamda1 = (J11 + J22 + sqrt((J11 + J22) ^ 2 - 4 * (J11 * J22 - J12 * J21))) / 2
8	lamda2 = (J11 + J22 - sqrt((J11 + J22) ^ 2 - 4 * (J11 * J22 - J12 * J21))) / 2

Line	Equation
9	k = 0.3
10	y = .75
11	Km = 1.e-6
12	t(0) = 0
13	S(0) = 5
14	B(0) = 0.05
15	t(f) = 20

Integrating this system of equations up to t_f =16.330 using the POLYMATH RKF45 integration algorithm yields the following eigenvalues: at $t = 0$, $\lambda_1 = 0.3$; $\lambda_2 = 0.0$ while at t_f =16.335, $\lambda_1 = 0.0$; $\lambda_2 = -1.511 \times 10^6$. Thus, indeed at $t = 16.335$, one of the eigenvalues becomes negative with very large absolute value which indicates that the system has become stiff. The default integration algorithm (fifth-order Runge-Kutta-Fehlberg) in POLYMATH and most general software packages cannot solve the problem to $t_f = 20$.

Selection of the "STIFF" algorithm in POLYMATH and integrating up to $t_f = 20$ results in the biomass and substrate concentration profiles shown in Figure 6–1.

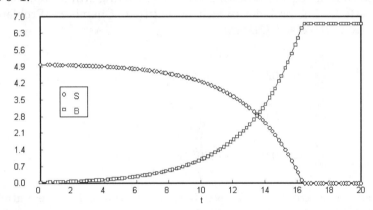

Figure 6–1 Change of Substrate and Biomass Concentration (File **P6-01-2.POL**)

This indicates that at $t \sim 16.3$, the amount of substrate becomes essentially zero, and there is no growth in the amount of the biomass. So for this example problem, the stiff integration algorithm is required in order to achieve the desired results to $t_f = 20$.

In general, it is usually advisable to use the stiff algorithm whenever a default integration algorithm either fails to solve a problem giving error messages with strange results or progresses very slowly toward a solution. If both the default and stiff integration algorithms fail to solve an ODE problem, then there are probably errors in the model equations, the values of the constants, or the initial values of some of the variables.

The problem solution files are found in directory Chapter 6 with file names **P6-1-1.POL** and **P6-1-2.POL**.

www

6.2 STIFF ORDINARY DIFFERENTIAL EQUATIONS IN CHEMICAL KINETICS

6.2.1   Concepts Demonstrated

Solution of a reaction kinetics model for an unsteady-state batch reactor process and model validation for reaction rate equations.

6.2.2   Numerical Methods Utilized

Solution of a stiff system of simultaneous ordinary differential equations.

6.2.3   Problem Statement

Gear,[2] who has developed well-known methods for solving stiff systems of ODEs, presented the following problem (which he entitled "Chemistry Problem") for testing software to solve stiff ODEs.

$$\frac{dy_1}{dt} = -0.013y_1 - 1000y_1y_3$$

$$\frac{dy_2}{dt} = -2500y_2y_3 \qquad\qquad (6\text{-}8)$$

$$\frac{dy_3}{dt} = -0.013y_1 - 1000y_1y_3 - 2500y_2y_3$$

The initial conditions are $y_1(0) = 1$, $y_2(0) = 1$, and $y_3(0) = 0$. These equations are usually integrated from $t_0 = 0$ up to $t_f = 50$.

Since its introduction, this example has been frequently used and cited in the literature (see, for example, p. 734 in Press et al.[3]).

(a)   Solve the system defined by the Equation Set (6-8) with the given initial conditions. Compare the solutions and the execution times when using the RKF45 and STIFF integration algorithms.

(b)   Assuming that y_1, y_2, and y_3 represent concentrations of different species, does the solution obtained make sense and seem feasible?

(c)   If the system defined by Equation Set (6-8) represents reaction rate equations, what typographical error could cause infeasible results? Suggest a reaction sequence that can be represented by the system, correct the typographical error, and resolve the equations in their correct form.

6.3 MULTIPLE STEADY STATES IN A SYSTEM OF ORDINARY DIFFERENTIAL EQUATIONS

6.3.1 Concepts Demonstrated

Dynamic material and energy balances for the unsteady-state model of a fluidized bed reactor leading to possible multiple steady states.

6.3.2 Numerical Methods Utilized

Determination of all solutions of a system of nonlinear algebraic equations, solution of stiff ODE systems, and effect of round-off errors in ill-conditioned algebraic and differential systems.

6.3.3 Problem Statement

Luss and Amundson[4] studied a simplified model for dynamics of a catalytic fluidized bed in which an irreversible gas phase reaction $A \rightarrow B$ was assumed to occur. The mass and energy conservation equations along with the kinetic rate constant for this system are given as Equation Set (6-9), which will be referred to as Set I.

$$\frac{dP}{d\tau} = P_e - P + H_g(P_p - P)$$

$$\frac{dT}{d\tau} = T_e - T + H_T(T_p - T) + H_W(T_W - T)$$

$$\frac{dP_p}{d\tau} = \frac{H_g}{A}[P - P_p(1 + K)] \tag{6-9}$$

$$\frac{dT_p}{d\tau} = \frac{H_T}{C}[(T - T_p) + FKP_p]$$

$$K = 0.0006 \exp(20.7 - 15000/T_P)$$

where T = absolute temperature of reactant in fluid (°R), P = partial pressure of the reactant in the fluid (atm), T_p = temperature of the reactant at the catalyst surface (°R), P_p = partial pressure of the reactant at the catalyst surface (atm), K = dimensionless reaction rate constant, τ = dimensionless time, and the subscript e indicates entrance conditions. The dimensionless constants were given as $H_g = 320$, $T_e = 600$, $H_T = 266.67$, $H_W = 1.6$, $T_W = 720$, $F = 8000$, $A = 0.17142$, $C = 205.74$, and $P_e = 0.1$.

Aiken and Lapidus[5] subsequently used Set I as a test example for a program they had developed, but they rewrote the equations by introducing the numerical values into the system and rounded some of the coefficients. The result is Equation Set (6-10) that will be referred to as Set II.

$$\frac{dP}{d\tau} = 0.1 + 320P_p - 321P$$

$$\frac{dT}{d\tau} = 1752 - 269T + 267T_p$$

$$\frac{dP_p}{d\tau} = 1.88 \times 10^3 [P - P_p(1 + K)] \qquad \textbf{(6-10)}$$

$$\frac{dT_p}{d\tau} = 1.3(T - T_p) + 1.04 \times 10^4 KP_p$$

$$K = 0.0006 \exp(20.7 - 15000/T_p)$$

(a) Introduce the values provided by Luss and Amundson into Set I and observe the differences between Set I and Set II.

(b) Find all the steady-state solutions for both Set I and Set II in the range of $500°R \leq T \leq 1300°R$.

(c) Solve both Set I and Set II using the initial values at $\tau = 0$ given by $P = 0.1$, $T = 600$, $P_p = 0$, and $T_p = 761$. The final value is to be $\tau = 1500$.

(d) Explain the differences in steady-state and dynamic solutions obtained while using the original Set I and the modified Set II.

6.3.4   Solution (Partial)

(a) The steady-state solutions are easily determined by setting the time derivatives in the four differential equations of Sets I and II equal to zero. Each set can then be reformulated to give a single implicit nonlinear algebraic equation, which should be equal to zero, while the rest of the variables can be calculated from explicit expressions. Introducing the numerical values of the constants into Set I and reformulating yields

$$f(T) = 1.296(T - T_p) + 10369KP_p$$

$$T_p = (269.267T - 1752)/266.667$$

$$P_p = 0.1/(1 + 321K) \qquad \textbf{(6-11)}$$

$$P = (320P_p + 0.1)/321$$

$$K = 0.0006 \exp(20.7 - 15000/T_p)$$

Converting a system of nonlinear algebraic equations into a form where there is only one implicit equation is the simplest way to find all of the solutions for such a system in a particular interval.

6.4 ITERATIVE SOLUTION OF AN ODE BOUNDARY VALUE PROBLEM

6.4.1 Concepts Demonstrated

Heat transfer in a one-dimensional slab, which involves both conduction with variable thermal conductivity and radiation to the surroundings at one surface.

6.4.2 Numerical Methods Utilized

Numerical solution of an ordinary differential equation where a variable describing the system must be optimized using the secant method and the method of false position.

6.4.3 Problem Statement

Heat conduction is occurring within a one-dimensional slab of variable thermal conductivity as shown in Figure 6–2. One surface of the slab is maintained at temperature T_0, and the other surface at temperature T_S has radiative heat transfer with the surroundings that act as a black body at temperature T_B. The slab thickness is given by L. There is negligible convection because a vacuum is maintained between the slab and the surroundings. Details on various modes of heat transfer can be found in Geankoplis[6] and Thomas.[7] Chapter 9 also presents the essential equations, boundary conditions, and typical units.

A steady-state energy balance on a differential element within the slab indicates that the heat flux is constant since there is no generation of heat within the slab. Application of Fourier's law in the x direction therefore gives

$$\frac{dT}{dx} = -\left(\frac{q_x}{A}\right)/k = -Q_x/k \tag{6-12}$$

where T is in K, q_x is the heat transfer in the x direction in W or J/s, A is the

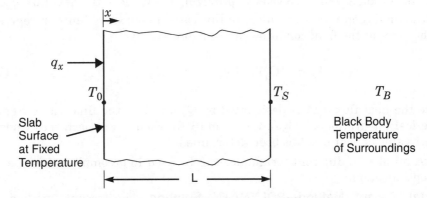

Figure 6–2 Slab with Conduction and Radiation at the Surface

cross-sectional area that is normal to the direction of heat conduction in m^2, Q_x is the heat flux in W/m^2, k is the thermal conductivity of the medium in W/m·K, and x is the distance in m.

For this problem, the thermal conductivity of the medium is temperature dependent and given by

$$k = 30(1 + 0.002T) \tag{6-13}$$

The radiation from the slab surface is given by the Stefan-Boltzmann law for a black body with an emissivity of unity and a view factor of unity. The resulting heat flux at the slab surface (or at any position within the slab) is

$$\left.\frac{q_x}{A}\right|_{x = L} = Q_x\big|_{x = L} = \left.\sigma(T_S^4 - T_B^4)\right|_{x = L} \tag{6-14}$$

where σ is the Stefan-Boltzmann constant with a value of 5.676×10^{-8} W/m^2·K^4.

> The surface of the slab is maintained at $T_0 = 290$ K and the black body temperature of the surroundings is $T_B = 1273$ K. $L = 0.2$ m.
> (a)   Calculate and plot the temperature profile within the slab using the secant method to determine the constant heat flux within the slab. What is the corresponding value of T_S?
> (b)   Repeat part (a) using the method of false position.

6.4.4   Solution

This problem requires the solution of the ODE given by Equation (6-12) during which the thermal conductivity must be calculated by Equation (6-13). The initial condition for T is T_0, which is known. The final condition is given by Equation (6-14).

This problem will be solved by optimizing the value of the heat flux, q_x/A, so that the final condition is satisfied. In this case, an objective function representing the error at the final condition can be expressed by

$$\varepsilon(q_x/A) = \varepsilon(Q_x) = \left.[Q_x - \sigma(T^4 - T_B^4)]\right|_{x = L} \tag{6-15}$$

where the heat flux q_x/A is designated by Q_x and T is the final value from the numerical integration at $x = L$. Equation (6-15) should approach zero when the correct value of heat flux has been determined.

A number of different techniques can be used to accomplish this one variable optimization.

(a) Secant Method—POLYMATH Solution The secant method (see Himmelblau[8] or Hanna and Sandall[9]) can be applied by applying the Newton formula to the heat flux Q_x.

$$Q_{x,\text{new}} = Q_x - \varepsilon(Q_x)/\varepsilon'(Q_x) \qquad (6\text{-}16)$$

with the derivative approximated by

$$\varepsilon'(Q_x) \cong \frac{\varepsilon(Q_x + \delta Q_x) - \varepsilon(Q_x)}{\delta Q_x} \qquad (6\text{-}17)$$

where δ is 0.0001 to provide a small increment in the current value of Q_x. The equations necessary to determine $Q_{x,\text{new}}$ can be calculated *simultaneously with the numerical ODE solution* for $\varepsilon(Q_x)$, thereby allowing the approximation of $\varepsilon'(Q_x)$ from Equation (6-17) and a *new estimate* for $Q_{x,\text{new}}$ from Equation (6-16) at the conclusion of each iteration.

The POLYMATH equation set for carrying out the first iteration using the secant method with an arbitrary initial estimate of $Q_x = -100,000$ is given in Table 6–2.

Table 6–2 Secant Method (File **P6-04A.POL**)

Line	Equation
1	d(T1)/d(x)=-Qx1/k1
2	d(T)/d(x)=-Qx/k
3	Qx=-100000
4	k=30*(1+0.002*T)
5	k1=30*(1+0.002*T1)
6	TB=1273
7	delta=0.0001
8	Qx1=(1+delta)*Qx
9	err=Qx-5.676e-8*(T^4-TB^4)
10	err1=Qx1-5.676e-8*(T1^4-TB^4)
11	derr=(err1-err)/(delta*Qx)
12	QxNEW=Qx-err/derr
13	x(0)=0
14	T1(0)=290
15	T(0)=290
16	x(f)=0.2

The final value for QxNEW at the end of the integration for the first iteration is −133,944 with units of $J/m^2 \cdot s$. The negative sign here indicates that the heat flux is in the negative x direction. The second iteration gives a QxNEW value of −133,014, and the third iteration yields −133,013, indicating that a converged solution has been obtained. The final value of T, which is T_S, is 729.17 K. The temperature distribution from the POLYMATH solution is given in Figure 6–3, where the nonlinear profile is a result of the temperature dependency of the thermal conductivity.

The secant method converges very rapidly for this problem, but initial estimates should be fairly accurate. The major disadvantage is the need to carry the equations for calculation of the derivative expression during the solution. Several trial solutions or some simplified calculations can be used to obtain the initial estimate of the heat flux Q_x.

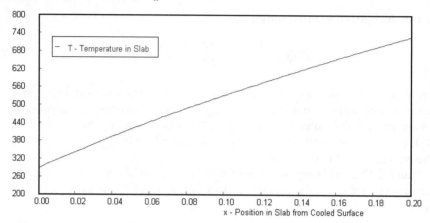

Figure 6–3 Temperature Profile in the Slab

(a) Secant Method—MATLAB Solution The secant method (Dahlquist et al.,[10] Press et al.[3]) can also be written

$$z_{k+1} = z_k - f(z_k)\frac{z_k - z_{k-1}}{f(z_k) - f(z_{k-1})} \quad f(z_k) \neq f(z_{k-1}) \quad k = 2,3... \tag{6-18}$$

where z is the independent variable (unknown) and k is the point number. Two initial estimates, z_1 and z_2, have to be provided, and Equation (6-18) is used to calculate the next estimate for the variable. The iterations are continued until the absolute value of $f(z_k)$ is less than a specified error tolerance.

In this particular case, Q_x is the unknown. The calculation of $f(Q_x) = \varepsilon(Q_x)$ for a particular value of $Q_{x,k}$ involves substitution the corresponding Q_x value into Equation (6-12), integrating the differential equations from the initial point $x = 0$ up to $x = L$ and calculating the function value using Equation (6-15). The initial estimates to be used for Q_x are $Q_{x,1} = -100,000$ and $Q_{x,2} = -150,000$. The error tolerance $| f(Q_x)| < 10^{-3}$ will be used to stop the iterations.

The secant method solution will be carried out using MATLAB. First the model equations are introduced into POLYMATH (using the first initial estimate $Q_{x,1} = -100,000$) as shown in Table 6–3.

Table 6–3 Secant Method (File **P6-04A2.POL**)

Line	Equation
1	d(T)/d(x)=-Qx/k
2	Qx=-100000
3	k=30*(1+0.002*T)

Line	Equation
4	TB=1273
5	fQx=Qx-5.676e-8*(T^4-TB^4)
6	x(0)=0
7	T(0)=290
8	x(f)=0.2

Integration of the POLYMATH equation set yields the values $T|_{x=0.2} = 636.12$ and $f(Q_x)|_{x=0.2} = 3.976 \times 10^4$.

POLYMATH can be used convert the set of equations into a MATLAB function, and a template file available in POLYMATH's Help Section can be used to integrate the differential equation and display the results. The POLYMATH to MATLAB conversion and the use of the template file are explained in detail in Problem 5.3.

The MATLAB function generated by POLYMATH for integrating the differential equation can be modified to automatically carry out the iterations involved with the secant method. This modification involves

1. Specifying Qx as one of the input parameters of the function and removing the specification of its value from the function (see line 2 in Table 6–3).

2. Determining $f(Q_x)$ at $x = 0.02$ for the iterations and thus calculating $f(Q_x)$ outside of the function. (It can be removed from the function, including the specification of TB).

The modified form of the MATLAB function is given in Table 6–4.

Table 6–4 Modified Form of the MATLAB Function

Line	Equation
1	function dYfuncvecdx = ODEfun(x,Yfuncvec,Qx);
2	T = Yfuncvec(1);
3	k = 30 * (1 + 0.002 * T);
4	dTdx = -Qx / k;
5	dYfuncvecdx = [dTdx];

Extensive modification of the template program is required in order to carry out the iteration. The part of the program where the initial values for Qx are set and the iterations are carried out is shown in Table 6–5.

Table 6–5 MATLAB Program

Line	Equation
1	function Prob_6_4a
2	clear, clc, format short g, format compact
3	tspan = [0 0.2]; % Range for the independent variable
4	y0 = [290.]; % Initial values for the dependent variables

Table 6–5 (Continued) MATLAB Program

Line	Equation
5	TB = 1273;
6	disp(' Variable values at the initial point ');
7	disp(['t = ' num2str(tspan(1))]);
8	k=1;
9	Qx(k) = -100000;
10	disp(' y dy/dt ');
11	disp([y0 ODEfun(tspan(1),y0,Qx(k))]);
12	[t,y]=ode45(@ODEfun,tspan,y0,[],Qx(k));
13	f(k) = Qx(k) - 5.676e-8 * (y(end,1) ^ 4 - TB ^ 4);
14	disp(' Point. No. Qx f(Qx) TS');
15	disp([k Qx(k) f(k) y(end,1)]);
16	k=2;
17	Qx(k) = -150000;
18	[t,y]=ode45(@ODEfun,tspan,y0,[],Qx(k));
19	f(k) = Qx(k) - 5.676e-8 * (y(end,1) ^ 4 - TB ^ 4);
20	disp([k Qx(k) f(k) y(end,1)]);
21	err=abs(f(k));
22	while (err>1e-3) & (k<20)
23	Qx(k+1)=Qx(k)-f(k)*(Qx(k)-Qx(k-1))/(f(k)-f(k-1));
24	[t,y]=ode45(@ODEfun,tspan,y0,[],Qx(k+1));
25	f(k+1) = Qx(k+1) - 5.676e-8 * (y(end,1) ^ 4 - TB ^ 4);
26	disp([k+1 Qx(k+1) f(k+1) y(end,1)]);
27	err=abs(f(k+1));
28	k=k+1;
29	end

The consecutive values of Q_x and $f(Q_x)$ are stored in the row vectors **Qx** and **f**. Calculation of $f(Q_x)$ involves three stages.

1. The value of Q_x is set (see lines 9, 17, and 23).
2. The integration is carried out for the specified value of Q_x (lines 12, 18, and 24) using the MATLAB library function *ode45*. Note that the basic form of the call to the ode45 function

    ([t,y]=ode45(@ODEfun,tspan,y0)

    has been modified to allow the passing of Q_x as an input argument.
3. The function value is calculated by the introduction of the appropriate values into Equation (6-15), (see lines 13, 19, and 25). The array **y** contains the temperature values as function of x in its first column. Thus the last value in this column (y(end, 1)) is the temperature at $x = 0.2$.

In lines 3 through 21, the initial values are set for $f(Q_{x,1})$ and $f(Q_{x,2})$, and some of the calculation results are displayed. In lines 22 through 29, the secant method iterations are carried out as long as $|f(Q_x)|>10^{-3}$ and k < 20. The point number k, and the values of $Q_{x,k}$ and $f(Q_{x,k})$ are displayed. The results of those iterations are shown in Table 6–6.

Table 6–6 Iterations for the Secant Method Using MATLAB

Point No.	Q_x	$f(Q_x)$	T_S
1	-100000.0	39764	636.12
2	-150000.0	-21355	774.4
3	-132530.0	597.58	727.86
4	-133010.0	8.7568	729.15
5	-133010.0	-0.00365	729.17
6	-133010.0	2.22E-08	729.17

The first secant method iteration (point No. 3) yields the value of $Q_x = -132,530$ with units of J/m^2·s. The negative sign here indicates that the heat flux is in the negative x direction. The second iteration (point No. 4) gives a Q_x value of $-133,010$, and the third iteration yields -133,010, indicating that a converged solution has been obtained. The convergence criterion $|f(Q_x)| < 10^{-3}$ is satisfied only in the fourth iteration indicating that the precision requested is too high and the convergence criterion can be relaxed. The final value of T, which is T_S, is 729.17 K, which is in agreement with the POLYMATH solution.

The POLYMATH and MATLAB problem solution files are found in directory Chapter 6 with file names **P6-04A.POL** and **Prob_6_4a.m**.

(b) False Position Method—POLYMATH Solution This method for the optimization of a single variable is initiated by trial solutions that determine two points that bracket the solution at which the values of the function to be minimized are of opposite sign. (See Himmelblau[8] or Carnahan et al.[11] for more details.) For this problem, let the negative functional value be given by $FQ_{xN} = \varepsilon(Q_xN)$ and the positive functional value by $FQ_{xP} = \varepsilon(Q_xP)$, where Q_{xN} and Q_{xP} lead to negative and positive values, respectively.

The new estimate of the variable for the next step is given by

$$Q_{x,\text{new}} = Q_{xN} - \frac{(Q_{xN} - Q_{xP})(FQ_{xN})}{(FQ_{xN} - FQ_{xP})} \tag{6-19}$$

and the function is determined to be FQ_{xNEW}. The new variable estimate replaces either Q_{xN} or Q_{xP} and the new function replaces either FQ_{xN} or FQ_{xP} depending on the sign of FQ_{xNEW}.

Initial trials determined that $Q_x = -100,000$ resulted in $\varepsilon(Q_xP) = 39,764$ and that $Q_x = -150,000$ resulted in $\varepsilon(Q_xN) = -21,355$. These values bracketed the solution and were used to start the iterations utilizing Equation (6-19). The POLYMATH equation set for carrying out the first iteration using the false position method is given in Table 6–7.

The first three iterations of the false position method are summarized in Table 6–8, where the converged values of iteration 3 agree with part (a) and the temperature profile is identical.

Table 6–7 False Position Method

Line	Equation
1	d(T)/d(x)=-QxNEW/k
2	QxN=-150000
3	k=30*(1+0.002*T)
4	QxP=-100000
5	TB=1273
6	FQxN=-21355
7	FQxP=39764
8	QxNEW=QxN-(QxN-QxP)*(FQxN)/(FQxN-FQxP)
9	FQxNEW=QxNEW-5.676e-8*(T^4-TB^4)
10	x(0)=0
11	T(0)=290
12	x(f)=0.2

Table 6–8 Iterations for False Position Method

Iteration	QxP	FQxP	QxN	FQxN	QxNEW	FQxNEW	T$_S$
1	−100000	39764	−150000	−21355	−132530	597.849	727.86
2	−132530	597.849	−150000	−21355	−133006	8.7448	729.15
3	−133006	8.7448	−150000	−21355	−133013	−0.1614	729.17

This method has the advantage that derivatives are not needed in the calculations and thus the POLYMATH equation set is simpler for complex problems. The method may not converge typically when the initial starting points are not close to the solution and/or the function is very nonlinear.

The POLYMATH problem solution file is found in directory Chapter 6 with file name **P6-04B.POL**.

(b) False Position Method—MATLAB Solution The results in Table 6–6 show that at $Q_x = -100,000$ FQ_x is positive and at $Q_x = -150,000$ FQ_x is negative. Those values bracket the solution and can be used to start the iterations utilizing Equation (3-23). The False Position (regula falsi) Method iterations can be carried out by the MATLAB program of the "Secant" method with the modifications starting with line 9 that are shown in Table 6–9.

Table 6–9 MATLAB Program Changes for False Position Method

Line	Command
9	QxP = -100000;
10	disp(' y dy/dt ');
11	disp([y0 ODEfun(tspan(1),y0,QxP)]);
12	[t,y]=ode45(@ODEfun,tspan,y0,[],QxP);
13	FQxP= QxP - 5.676e-8 * (y(end,1) ^ 4 - TB ^ 4);
14	disp(' Point. No. Qx Fqx TS');
15	disp([k QxP FQxP y(end,1)]);

Line	Command
16	QxN=-150000;
17	[t,y]=ode45(@ODEfun,tspan,y0,[],QxN);
18	FQxN= QxN - 5.676e-8 * (y(end,1) ^ 4 - TB ^ 4);
19	k=k+1;
20	disp([k QxN FQxN y(end,1)]);
21	err=10;
22	while (abs(err)>1e-3) & (k<20)
23	k=k+1;
24	QxNEW=QxN-(QxN-QxP)*FQxN/(FQxN-FQxP);
25	[t,y]=ode45(@ODEfun,tspan,y0,[],QxNEW);
26	FQxNEW = QxNEW - 5.676e-8 * (y(end,1) ^ 4 - TB ^ 4);
27	disp([k QxNEW FQxNEW y(end,1)]);
28	err= FQxNEW;
29	if (sign(FQxNEW)==sign(FQxN)), QxN=QxNEW; FQxN=FQxNEW;
30	else QxP=QxNEW; FQxP=FQxNEW;
31	end
32	end

The main difference between this program and the program of the secant method is that the consecutive values of Q_x and $f(Q_x)$ are not stored in arrays, but as individual variables with the same names as in Equation (6-19). Replacement of the old estimates by the new one is carried out by the commands in lines 30 through 32 using the logic explained previously.

The results of the iterations using the false position method are shown in Table 6–10.

Table 6–10 Iterations for False Position Method

Point No.	Q_x	$f(Q_x)$	T_S
1	-100000.0	39764	636.12
2	-150000.0	-21355	774.4
3	-132530.0	597.58	727.86
4	-133010.0	8.7568	729.15
5	-133010.0	0.12827	729.17
6	-133010.0	0.001879	729.17
7	-133010.0	2.75E-05	729.17

The MATLAB program needed five iterations to satisfy the convergence criterion: $|f(Q_x)|<10^{-3}$ (in comparison to the four iterations of the MATLAB program for the secant method). This demonstrates the limitation of the false position method where one of the initial points is never replaced during the solution process (in this case the initial point with the negative FQ_x value), which can slow down the convergence considerably in comparison to the secant method.

 The MATLAB problem solution file is found in directory Chapter 6 with file name **Prob_6_4b.m**.

6.5 SHOOTING METHOD FOR SOLVING TWO-POINT BOUNDARY VALUE PROBLEMS

6.5.1   Concepts Demonstrated

Methods for solving second-order ordinary differential equations with two-point boundary values typically used in transport phenomena and reaction kinetics.

6.5.2   Numerical Methods Utilized

Conversion of a second-order ordinary differential equation into a system of two first-order differential equations, a shooting method for solving ODEs, and use of the secant method to solve two-point boundary value problems.

6.5.3   Problem Statement

The diffusion and simultaneous first-order irreversible chemical reaction in a single phase containing only reactant A and product B results in a second-order ordinary differential equation given by

$$\frac{d^2 C_A}{dz^2} = \frac{k}{D_{AB}} C_A \tag{6-20}$$

where C_A is the concentration of reactant A (kg-mol/m^3), z is the distance variable (m), k is the homogeneous reaction rate constant (s^{-1}), and D_{AB} is the binary diffusion coefficient (m^2/s). A typical geometry for Equation (6-20) is that of a one-dimensional layer that has its surface exposed to a known concentration and allows no diffusion across its bottom surface. Thus the initial and boundary conditions are

$$C_A = C_{A0} \qquad \text{for } z = 0 \tag{6-21}$$

$$\frac{dC_A}{dz} = 0 \qquad \text{for } z = L \tag{6-22}$$

where C_{A0} is the constant concentration at the surface ($z = 0$) and there is no transport across the bottom surface ($z = L$), so the derivative is zero.

This differential equation has an analytical solution given by

$$C_A = C_{A0} \frac{\cosh\,[L(\sqrt{k/D_{AB}})(1 - z/L)]}{\cosh(L\sqrt{k/D_{AB}})} \tag{6-23}$$

(a) Numerically solve Equation (6-20) with the boundary conditions of (6-21) and (6-22) for the case where $C_{A0} = 0.2$ kg-mol/m^3, $k = 10^{-3}$ s^{-1}, $D_{AB} = 1.2 \ 10^{-9}$ m^2/s, and $L = 10^{-3}$ m. This solution should utilize an ODE solver with a shooting technique and employ the secant method for converging on the boundary condition given by Equation (6-22).

(b) Compare the concentration profiles over the thickness as predicted by the numerical solution of (a) with the analytical solution of Equation (6-23).

(c) Obtain a numerical solution for a second-order reaction that requires the C_A term on the right side of Equation (6-20) to become squared. The second-order rate constant is given by $k = 0.02$ m^3/(kg-mol·s).

6.5.4   Solution (Partial)

Solving Higher Order Ordinary Differential Equations

Most mathematical software packages can solve only systems of first-order ODEs. Fortunately, the solution of an nth-order ODE can be accomplished by expressing the equation as a series of simultaneous first-order differential equations each with a boundary condition. This is the approach that is typically used for the integration of higher-order ODEs.

(a) Equation (6-20) is a second-order ODE, but it can be converted into a system of first-order equations by substituting new variables for the higher order derivatives. In this particular case, a new variable y can be defined that represents the first derivative of C_A with respect to z. Thus Equation (6-20) can be written as

$$\frac{dC_A}{dz} = y$$

$$\frac{dy}{dz} = \frac{k}{D_{AB}} C_A$$

$$(6\text{-}24)$$

This set of first-order ODEs can be entered into the POLYMATH *Simultaneous Differential Equation Solver* for solution, but initial conditions for both C_A and y are needed. Since the initial condition of y is not known, an iterative method (also referred to as a shooting method) can be used to find the correct initial value for y that will yield the boundary condition given by Equation (6-22).

Shooting Method—Trial and Error

The shooting method is used to solve a boundary value problem by iterating on the solution of an initial value problem. Known initial values are utilized while unknown initial values are optimized to achieve the corresponding boundary conditions. Either trial-and-error or variable optimization techniques are used to achieve convergence on the boundary conditions.

For this problem, a first trial-and-error value for the initial condition of y (for example, $y_0 = -150$) is used to carry out the integration and calculate the error for the boundary condition designated by ε. Thus the difference between the calculated and desired final value of y at $z = L$ is given by

$$\varepsilon(y_0) = y_{f,\,\text{calc}} - y_{f,\,\text{desired}} \tag{6-25}$$

Note that for this example, $y_{f,\text{desired}} = 0$ and thus $\varepsilon(y_0) = y_{f,\text{calc}}$ only because this desired boundary condition is zero.

The equations as entered in the POLYMATH *Simultaneous Differential Equation Solver* for an initial trial-and-error solution are given in Table 6–11.

Table 6–11 POLYMATH Program (File **P6-05A1.POL**)

Line	Equation
1	d(CA)/d(z)=y
2	d(y)/d(z)=k*CA/DAB
3	k=0.001
4	DAB=1.2E-9
5	err=y
6	z(0)=0
7	CA(0)=0.2
8	y(0)=-150
9	z(f)=0.001

The calculation of `err` in the POLYMATH equation set that corresponds to Equation (6-25) is only valid at the end of the ODE solution. Repeated reruns of this POLYMATH equation set with different initial conditions for y can be used in a trial-and-error mode to converge upon the desired boundary condition for y_0, where $\varepsilon(y_0)$ or `err` $\cong 0$. Some results are summarized in Table 6–12 for various values of y_0. The desired initial value for y_0 lies between -130 and -140. This trial-and-error approach can be continued to obtain a more accurate value for y_0, or an optimization technique can be applied.

Table 6–12 Trial Boundary Conditions for Equation Set (6-24) in Problem Part (a)

$y_0 \ (z = 0)$	-120	-130	-140	-150
$y_{f,\text{calc}} \ (z = L)$	17.23	2.764	-11.70	-26.16
$\varepsilon(y_0)$	17.23	2.764	-11.70	-26.16

www

The POLYMATH problem solution file is found in directory Chapter 6 with file name **P6-05A1.POL**.

Secant Method for Boundary Condition Convergence

A very useful method for optimizing the proper initial condition is to consider this determination to be a problem in finding the zero of a function. In the notation of this problem, the variable is y_0 and the function is $\varepsilon(y_0)$. The secant method, an effective method for optimizing a single variable, has been described in Problem 6.4 and will be applied here.

According to this method, an improved estimate for y_0 can be calculated using the equation

$$y_{0,\,new} = y_0 - \varepsilon(y_0)/\varepsilon'(y_0) \tag{6-26}$$

where $\varepsilon'(y_0)$ is the derivative of ε at $y = y_0$. The derivative, $\varepsilon'(y_0)$, can be estimated using a finite difference approximation:

$$\varepsilon'(y_0) \cong \frac{\varepsilon(y_0 + \delta y_0) - \varepsilon(y_0)}{\delta y_0} \tag{6-27}$$

where δy_0 is a small increment in the value of y_0. It is very convenient that $\varepsilon(y_0 + \delta y_0)$ can be calculated *simultaneously with the numerical ODE solution* for $\varepsilon(y_0)$, thereby allowing calculation of $\varepsilon'(y_0)$ from Equation (6-27) and a *new estimate* for y_0 from Equation (6-26).

Using $\delta = 0.0001$ for this example, the POLYMATH equation set for carrying out the first step in the secant method procedure is given in Table 6–13.

Table 6–13 POLYMATH Program (File **P6-05A2.POL**)

Line	Equation
1	d(CA)/d(z) = y #
2	d(y)/d(z) = k*CA/DAB #
3	d(CA1)/d(z) = y1 #
4	d(y1)/d(z) = k*CA1/DAB #
5	k = 0.001 #
6	DAB = 1.2E-9 #
7	err = y-0 #
8	err1 = y1-0 #
9	y0 = -130 #
10	L = .001 #
11	delta = 0.0001 #
12	CAanal = 0.2*cosh(L*(k/DAB)^.5*(1-z/L))/(cosh(L*(k/DAB)^.5)) #
13	derr = (err1-err)/(delta*y0) #
14	ynew = y0-err/derr #
15	z(0)=0
16	CA(0)=0.2
17	y(0)=-130
18	CA1(0)=0.2
19	y1(0)=-130.013
20	z(f)=0.001

This set of equations yields the results summarized in Table 6–14 where the new estimate for y_0 is the final value of the POLYMATH variable ynew or −131.911.

Table 6–14 Partial Results for Selected Variables during First Secant Method Iteration

Variable	Initial Value	Minimum Value	Maximum Value	Final Value
CA	0.2	0.1404279	0.2	0.1404606
CA1	0.2	0.1404135	0.2	0.1404457
derr	1.	1.	1.446418	1.446418
err	-130.	-130.	2.764383	2.764383
err1	-130.013	-130.013	2.745579	2.745579
y	-130.	-130.	2.764383	2.764383
y1	-130.013	-130.013	2.745579	2.745579
ynew	-5.227E-11	-131.9112	-5.227E-11	-131.9112
z	0	0	0.001	0.001

Another iteration of the secant method can be obtained by starting with the new estimate and modifying the initial conditions for y and y1 and the value of y0 in the POLYMATH equation set. The second iteration indicates that the err is approximately 3.e−4 and that ynew is unchanged, indicating that convergence has been obtained.

This problem solution emphasizes the value of first obtaining an approximate solution for a split boundary value problem. Application of the secant method procedure from a reasonable starting point will usually converge very efficiently to a solution; however, unreasonable starting points can lead to numerical difficulties and often do not yield a solution.

www

The POLYMATH problem solution file is found in directory Chapter 6 with file named **P6-05A2.POL**. This problem is also solved with Excel, Maple, MathCAD, MATLAB, Mathematica, and POLYMATH as Problem 7 in the Set of Ten Problems discussed on the book web site.

6.6 EXPEDITING THE SOLUTION OF SYSTEMS OF NONLINEAR ALGEBRAIC EQUATIONS

6.6.1   Concepts Demonstrated

Complex chemical equilibrium calculations.

6.6.2   Numerical Methods Utilized

Solution of systems of nonlinear algebraic equations, and techniques useful for effective solutions and for examining possible multiple solutions of such systems.

6.6.3   Problem Statement

The following reactions are taking place in a constant volume, gas-phase batch reactor:

$$A + B \leftrightarrow C + D$$
$$B + C \leftrightarrow X + Y$$
$$A + X \leftrightarrow Z$$

A system of algebraic equations describes the equilibrium of the preceding reactions. The nonlinear equilibrium relationships utilize the thermodynamic equilibrium expressions, and the linear relationships have been obtained from the stoichiometry of the reactions.

$$K_{C1} = \frac{C_C C_D}{C_A C_B} \qquad K_{C2} = \frac{C_X C_Y}{C_B C_C} \qquad K_{C3} = \frac{C_Z}{C_A C_X}$$

$$C_A = C_{A0} - C_D - C_Z \qquad C_B = C_{B0} - C_D - C_Y \tag{6-28}$$

$$C_C = C_D - C_Y \qquad C_Y = C_X + C_Z$$

In this equation set C_A, C_B, C_C, C_D, C_X, C_Y, and C_Z are concentrations of the various species at equilibrium resulting from initial concentrations of only C_{A0} and C_{B0}. The equilibrium constants K_{C1}, K_{C2}, and K_{C3} have known values.

Solve this system of equations when $C_{A0} = C_{B0} = 1.5$, $K_{C1} = 1.06$, $K_{C2} = 2.63$, and $K_{C3} = 5$ starting from three sets of initial estimates.

(a) $C_D = C_X = C_Z = 0$

(b) $C_D = C_X = C_Z = 1$

(c) $C_D = C_X = C_Z = 10$

6.6.4   Solution and Partial Solution

The equation set (6-28) can be entered into the POLYMATH *Simultaneous Algebraic Equation Solver,* but the nonlinear equilibrium expressions must be written as functions that are equal to zero at the solution. A simple transformation of the equilibrium expressions to the required functional form yields

$$f(C_D) = \frac{C_C C_D}{C_A C_B} - K_{C1}$$

$$f(C_X) = \frac{C_X C_Y}{C_B C_C} - K_{C2}$$

(6-29)

$$f(C_Z) = \frac{C_Z}{C_A C_X} - K_{C3}$$

The POLYMATH equation set utilizing the transformed nonlinear equations is given in Table 6–15.

Table 6–15 POLYMATH Program (File **P6-06A.POL**)

Line	Equation
1	f(CD)=CC*CD/(CA*CB)-KC1
2	f(CX)=CX*CY/(CB*CC)-KC2
3	f(CZ)=CZ/(CA*CX)-KC3
4	KC1=1.06
5	CY=CX+CZ
6	KC2=2.63
7	KC3=5
8	CA0=1.5
9	CB0=1.5
10	CC=CD-CY
11	CA=CA0-CD-CZ
12	CB=CB0-CD-CY
13	CD(0)=0
14	CX(0)=0
15	CZ(0)=0

When the preceding equations are used in POLYMATH, the available algorithms will fail to solve the problem for all sets of initial estimates specified in (a), (b), and (c). An error message such as *"Error: Zero denominator not allowed"* [for part (a)] or an error message such as *"Solution did not converge"* [parts (b) and (c)] will be displayed. The failure of POLYMATH and most other programs for solving nonlinear equations is that division by unknowns makes the equations very nonlinear or sometimes undefined. The solution methods that are based on linearization (such as the Newton-Raphson method) may diverge for highly nonlinear systems or cannot continue for undefined functions.

Expediting the Solution of Nonlinear Equations

A simple transformation of the nonlinear function can make many functions much less nonlinear and easier to solve by simply eliminating division by the unknowns. In this case, the equation set (6-29) can be modified to

$$f(C_D) = C_C C_D - K_{C1} C_A C_B$$
$$f(C_X) = C_X C_Y - K_{C2} C_B C_C \qquad \textbf{(6-30)}$$
$$f(C_Z) = C_Z - K_{C3} C_A C_X$$

Using the modified nonlinear equations in POLYMATH produces the solutions summarized in Table 6–16 for the three sets of initial conditions in parts (a), (b), and (c).

Table 6–16 Multiple Solutions of the Chemical Equilibrium Problem

Variable	Part (a)	Part (b)	Part (c)
CD	0.705334	0.0555561	1.0701
CX	0.177792	0.59722	-0.322716
CZ	0.373977	1.08207	1.13053
CA	0.420689	0.36237	-0.700638
CB	0.242897	-0.234849	-0.377922
CC	0.153565	-1.62374	0.262286
CY	0.551769	1.67929	0.807818

Note that the initial conditions for problem part (a) converged to all positive concentrations. However, the initial conditions for parts (b) and (c) converged to some negative values for some of the concentrations. Thus a "reality check" on Table 6–16 for physical feasibility reveals that the negative concentrations in parts (b) and (c) are the basis for rejecting these solutions as not representing a physically valid situation.

This problem illustrates the desirability of entering nonlinear functions in a way in which the unknown variable will not lead to highly nonlinear behavior or division by zero. Another option to alleviate the solution of systems of algebraic equations is to convert them to a system where there is only one implicit equation, and the rest of the variables can be calculated from explicit expressions. This approach is demonstrated in Problem 6.3, although it cannot be applied to this chemical equilibrium problem.

Additionally, this problem shows that a correct numerical solution to a properly posed problem may be an infeasible solution, and thus it should be rejected as an unrealistic solution (for example, negative concentrations). *In all cases where the solutions of simultaneous nonlinear equations are required, it is very important to specify initial estimates inside the feasible region and as close to the ultimate solution as possible.*

The POLYMATH problem solution file for part (a) is found in the *Simultaneous Algebraic Equation Solver Library* located in directory Chapter 6 with file named **P6-06A1.POL**. This problem is also solved with Excel, Maple, MathCAD, MATLAB, Mathematica, and POLYMATH as Problem 4 in the Set of Ten Problems discussed on the book web site.

6.7 SOLVING DIFFERENTIAL ALGEBRAIC EQUATIONS—DAEs

6.7.1   Concepts Demonstrated

Batch distillation of an ideal binary mixture.

6.7.2   Numerical Methods Utilized

Solution of a system of equations comprising of ordinary differential and implicit algebraic equations using the controlled integration technique.

6.7.3   Problem Statement

For a binary batch distillation process involving two components designated 1 and 2, the moles of liquid remaining, L, as a function of the mole fraction of the component 2, x_2, can be expressed by the following equation:

$$\frac{dL}{dx_2} = \frac{L}{x_2(k_2 - 1)} \tag{6-31}$$

where k_2 is the vapor liquid equilibrium ratio for component 2. If the system may be considered ideal, the vapor liquid equilibrium ratio can be calculated from $k_i = P_i/P$, where P_i is the vapor pressure of component i and P is the total pressure.

A common vapor pressure correlation is the Antoine equation, which utilizes three parameters A, B, and C for component i as given next, where T is the temperature in °C.

$$P_i = 10^{\left(A + \frac{B}{T + C}\right)} \tag{6-32}$$

The temperature in the batch still follows the bubble point curve. The bubble point temperature is defined by the implicit algebraic equation, which can be written using the vapor liquid equilibrium ratios as

$$k_1 x_1 + k_2 x_2 = 1 \tag{6-33}$$

Consider a binary mixture of benzene (component 1) and toluene (component 2) that is to be considered as ideal. The Antoine equation constants for benzene are $A_1 = 6.90565$, $B_1 = -1211.033$, and $C_1 = 220.79$. For toluene, $A_2 = 6.95464$, $B_2 = -1344.8$, and $C_2 = 219.482$ (Dean[12]). P is the pressure in mm Hg and T the temperature in °C.

The batch distillation of benzene (component 1) and toluene (component 2) mixture is being carried out at a pressure of 1.2 atm. Initially, there are 100 mol of liquid in the still, comprised of 60% benzene and 40% toluene (mole fraction basis). Calculate the amount of liquid remaining in the still when concentration of toluene reaches 80% using the two approaches discussed in the following section and compare the results.

6.7.4 Solution (Partial)

This problem requires the simultaneous solution of Equation (6-31) while the temperature is calculated from the bubble point considerations implicit in Equation (6-33). A system of equations comprising differential and implicit algebraic equations is called differential algebraic and referred to as DAEs. There are several numerical methods for solving DAEs. Most problem-solving software packages including POLYMATH do not have the specific capability for DAEs.

Approach 1 The first approach will be to use the *controlled integration technique* proposed by Shacham et al.[13] Using this method, the nonlinear Equation (6-33) is rewritten with an error term given by

$$\varepsilon = 1 - k_1 x_1 - k_2 x_2 \qquad \text{(6-34)}$$

where the ε calculated from this equation provides the basis for keeping the temperature of the distillation at the bubble point. This is accomplished by changing the temperature in proportion to the error in a manner analogous to a proportion controller action. Thus this can be represented by another differential equation:

$$\frac{dT}{dx_2} = K_c \varepsilon \qquad \text{(6-35)}$$

where a proper choice of the proportionality constant K_c will keep the error below a desired error tolerance.

A simple trial-and-error procedure is sufficient for the calculation of K_c. At the beginning K_c is set to a small value (say $K_c = 1$), and the system is integrated. If ε is too large, then K_c must be increased and the integration repeated. This trial-and-error procedure is continued until ε becomes smaller than a desired error tolerance throughout the entire integration interval.

The temperature at the initial point is not specified in the problem, but it is necessary to start the problem solution at the bubble point of the initial mixture. This separate calculation can be carried out on Equation (6-33) for $x_1 = 0.6$ and $x_2 = 0.4$ and the Antoine equations using the POLYMATH *Simultaneous Algebraic Equation Solver*. The resulting initial temperature is found to be $T_0 = 95.5851$.

The system of equations for the batch distillation as they are introduced into the POLYMATH *Simultaneous Differential Equation Solver* using $K_c = 0.5 \times 10^6$ is as follows and the partial results from the solution are summarized in Table 6–18.

Table 6–17 POLYMATH Program (File **P6-07A1.POL**)

Line	Equation
1	d(L)/d(x2)=L/(k2*x2-x2)
2	d(T)/d(x2)=Kc*err
3	Kc=0.5e6
4	k2=10^(6.95464-1344.8/(T+219.482))/(760*1.2)
5	x1=1-x2
6	k1=10^(6.90565-1211.033/(T+220.79))/(760*1.2)
7	err=(1-k1*x1-k2*x2)
8	x2(0)=0.4
9	L(0)=100
10	T(0)=95.5851
11	x2(f)=0.8

Table 6–18 Partial Results for DAE Binary Distillation Problem (File **P6-07A1.POL**)

Variable	Initial Value	Minimal Value	Maximal Value	Final Value
err	-3.646E-07	-3.646E-07	7.747E-05	7.747E-05
k1	1.311644	1.311644	1.856602	1.856602
k2	0.5325348	0.5325348	0.7857526	0.7857526
Kc	5.0E+05	5.0E+05	5.0E+05	5.0E+05
L	100.	14.04555	100.	14.04555
T	95.5851	95.5851	108.5693	108.5693
x1	0.6	0.2	0.6	0.2
x2	0.4	0.4	0.8	0.8

The final values from Table 6–18 indicate that 14.05 mol of liquid remain in the column when the concentration of toluene reaches 80%. During the distillation the temperature increases from 95.6°C to 108.6°C. The error calculated from Equation (6-34) increases from about -3.6×10^{-7} to 7.75×10^{-5} during the numerical solution, but it is still small enough for the solution to be considered accurate.

The POLYMATH data file for Approach 1 is found in directory Chapter 6 with file named **P6-07A1.POL**. This problem is also solved with Excel, Maple, MathCAD, MATLAB, Mathematica, and POLYMATH as Problem 8 in the Set of Ten Problems discussed on the book web site.

Approach 2   A different approach for solving this problem can be used because Equation (6-33) can be differentiated with respect to x_2 to yield

$$\frac{dT}{dx_2} = \frac{(k_2 - k_1)}{\ln(10)\left[x_1 k_1 \dfrac{-B_1}{(C_1 + T)^2} + x_2 k_2 \dfrac{-B_2}{(C_2 + T)^2}\right]} \tag{6-36}$$

This equation can be integrated simultaneously with Equation (6-31) to provide the bubble point temperature during the problem solution.

6.8 METHOD OF LINES FOR PARTIAL DIFFERENTIAL EQUATIONS

6.8.1 Concepts Demonstrated

Unsteady-state heat conduction in one-dimensional slab with one face insulated and constant thermal diffusivity.

6.8.2 Numerical Methods Utilized

Application of the numerical method of lines to solve a partial differential equation that involves the solution of simultaneous ordinary differential equations and explicit algebraic equations.

6.8.3 Problem Statement[*]

Unsteady-state heat transfer in a slab in the x direction is described by the partial differential equation (see Geankoplis[6] for derivation)

$$\frac{\partial T}{\partial t} = \alpha \frac{\partial^2 T}{\partial x^2} \tag{6-37}$$

where T is the temperature in K, t is the time in s, and α is the thermal diffusivity in m^2/s given by $k/\rho c_p$. In this treatment, the thermal conductivity k in W/m·K, the density ρ in kg/m^3, and the heat capacity c_p in J/kg·K are all considered to be constant.

Consider an example problem posed by Geankoplis[6] where a slab of material with a thickness of 1.00 m is supported on nonconducting insulation. This slab is shown in Figure 6–4. For a numerical problem solution, the slab is divided into N sections with $N + 1$ node points. The slab is initially at a uniform temperature of 100°C. This gives the initial condition that all the internal node temperatures are known at time $t = 0$.

$$T_n = 100 \text{ for } n = 2 \ldots (N + 1) \text{ at } t = 0 \tag{6-38}$$

If at time zero the exposed surface is suddenly held constant at a temperature of 0°C, this gives the boundary condition at node 1:

$$T_1 = 0 \quad \text{for } t \geq 0 \tag{6-39}$$

Another boundary condition is that the insulated boundary at node $N + 1$ allows no heat conduction. Thus

$$\frac{\partial T_{N+1}}{\partial x} = 0 \quad \text{for } t \geq 0 \tag{6-40}$$

Note that this problem is equivalent to having a slab of twice the thickness exposed to the initial temperature on both faces.

[*] Adapted from Geankoplis,[6] pp. 353–56, with permission.

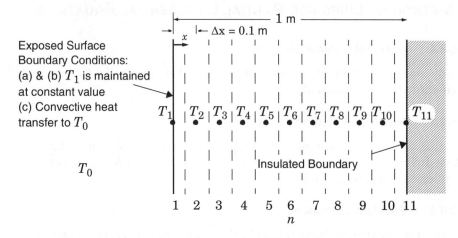

Figure 6–4 Unsteady-State Heat Conduction in a One-Dimensional Slab

When convection is considered as the only mode of heat transfer to the surface of the slab, an energy balance can be made at the interface that relates the energy input by convection to the energy output by conduction. Thus at any time, the transport normal to the slab surface in the x direction is given by

$$h(T_0 - T_1) = -k\frac{\partial T}{\partial x}\bigg|_{x=0} \qquad (6\text{-}41)$$

where h is the convective heat transfer coefficient in $\text{W/m}^2 \cdot \text{K}$ and T_0 is the ambient temperature.

(a) Numerically solve Equation (6-37) with the initial and boundary conditions of (6-38), (6-39), and (6-40) for the case where $\alpha = 2 \times 10^{-5} \text{ m}^2/\text{s}$ and the slab surface is held constant at $T_1 = 0°\text{C}$. This solution should utilize the numerical method of lines with 10 sections. Plot the temperatures T_2, T_3, T_4, and T_5 as functions of time to 6000 s.

(b) Repeat part (a) with 20 sections. Compare results with part (a) to verify that solution for part (a) is accurate.

(c) Repeat parts (a) and (b) to a time of 1500 s for the case where heat convection is present at the slab surface. The heat transfer coefficient is $h = 25.0 \text{ W/m}^2 \cdot \text{K}$, and the thermal conductivity is $k = 10.0 \text{ W/m} \cdot \text{K}$.

The Numerical Method of Lines

The method of lines is a general technique for solving partial differential equations (PDEs) by typically using finite difference relationships for the spatial derivatives and ordinary differential equations for the time derivative, as discussed by Schiesser.[14] For this problem with $N = 10$ sections of length $\Delta x = 0.1$ m, Equation (6-37) can be rewritten using a central difference formula for the second derivative [Appendix A, Equation (A-9)] as

$$\frac{\partial T_n}{\partial t} = \frac{\alpha}{(\Delta x)^2}(T_{n+1} - 2T_n + T_{n-1}) \quad \text{for } (2 \le n \le 10) \tag{6-42}$$

The boundary condition given in Equation (6-39) can be expressed as

$$T_1 = 0 \tag{6-43}$$

The second boundary condition represented by Equation (6-40) can be written using a second-order backward finite difference [Appendix A, Equation (A-7)] as

$$\frac{\partial T_{11}}{\partial x} = \frac{3T_{11} - 4T_{10} + T_9}{2\Delta x} = 0 \tag{6-44}$$

which can be solved for T_{11} to yield

$$T_{11} = \frac{4T_{10} - T_9}{3} \tag{6-45}$$

Surface Boundary Condition

The energy balance at the slab surface given by Equation (6-41) can be used to determine a relationship between the slab surface temperature T_1, the ambient temperature T_0, and the temperatures at internal node points. In this case, the second-order forward difference equation for the first derivative [Appendix A, Equation (A-5)] can be applied to T_1

$$\frac{\partial T_1}{\partial x} = \frac{(-T_3 + 4T_2 - 3T_1)}{2\Delta x} \tag{6-46}$$

and can be substituted into Equation (6-41) to yield

$$h(T_0 - T_1) = -k\frac{\partial T}{\partial x}\bigg|_{x=0} = -k\frac{(-T_3 + 4T_2 - 3T_1)}{2\Delta x} \tag{6-47}$$

The preceding equation can by solved for T_1 to give

$$T_1 = \frac{2hT_0\Delta x - kT_3 + 4kT_2}{3k + 2h\Delta x} \tag{6-48}$$

and this can be used to calculate T_1 during the method of lines solution.

6.8.4   Solution

(a) The problem then requires the solution of Equations (6-42), (6-43), and (6-45) which results in nine simultaneous ordinary differential equations and two explicit algebraic equations for the 11 temperature nodes. This set of equations can be entered into the POLYMATH *Simultaneous Differential Equation*

Solver. Note that the full screen editor in POLYMATH is very helpful for this problem. The resulting equation set is given in Table 6–19.

Table 6–19 POLYMATH Program (File **P6-08A.POL**)

Line	Equation
1	d(T2)/d(t)=alpha/deltax^2*(T3-2*T2+T1)
2	d(T3)/d(t)=alpha/deltax^2*(T4-2*T3+T2)
3	d(T4)/d(t)=alpha/deltax^2*(T5-2*T4+T3)
4	d(T5)/d(t)=alpha/deltax^2*(T6-2*T5+T4)
5	d(T6)/d(t)=alpha/deltax^2*(T7-2*T6+T5)
6	d(T7)/d(t)=alpha/deltax^2*(T8-2*T7+T6)
7	d(T8)/d(t)=alpha/deltax^2*(T9-2*T8+T7)
8	d(T9)/d(t)=alpha/deltax^2*(T10-2*T9+T8)
9	d(T10)/d(t)=alpha/deltax^2*(T11-2*T10+T9)
10	alpha=2.e-5
11	deltax=.10
12	T1=0
13	T11=(4*T10-T9)/3
14	t(0)=0
15	T2(0)=100
16	T3(0)=100
17	T4(0)=100
18	T5(0)=100
19	T6(0)=100
20	T7(0)=100
21	T8(0)=100
22	T9(0)=100
23	T10(0)=100
24	t(f)=6000

The plots of the temperatures in the first four sections, node points 2 ... 5, are shown in Figure 6–5. The transients in temperatures show an approach to steady state. The numerical results are compared to the hand calculations of a finite difference solution by Geankoplis (pp. 471–3)[6] at the time of 6000 s in Table 6–20. These results indicate that there is general agreement regarding the

Table 6–20 Results for Unsteady-State Heat Transfer in One-Dimensional Slab at t = 6000 s

Distance from Slab Surface in m	Geankoplis[6] $\Delta x = 0.20$ m		Method of Lines (a) $\Delta x = 0.10$ m		Method of Lines (b) $\Delta x = 0.05$ m	
	n	T in °C	n	T in °C	n	T in °C
0	1	0.0	1	0.0	1	0.0
0.2	2	31.25	3	31.71	5	31.68
0.4	3	58.59	5	58.49	9	58.47
0.6	4	78.13	7	77.46	13	77.49
0.8	5	89.84	9	88.22	17	88.29
1.0	6	93.75	11	91.66	21	91.72

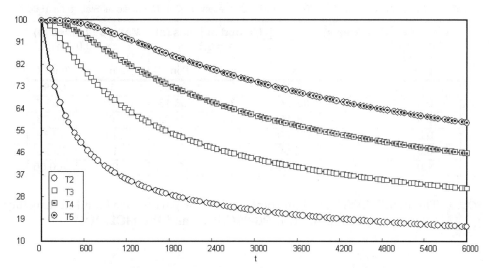

Figure 6–5 Temperature Profiles for Unsteady-State Heat Conduction in a One-Dimensional Slab (File **P6-08A.POL**)

problem solution, but differences between the temperatures at the nodes increase as the nodes approach the insulated boundary of the slab.

The POLYMATH problem solution file for part (a) is found in directory Chapter 6 with file named **P6-08A.POL**.

(b) The accuracy of the numerical solution can be investigated by doubling the number of sections for the numerical method of lines solution. This just involves adding an additional 10 equations given by the relationship in Equation (6-42) and modifying Equation (6-45) to calculate T_{21}. The results for this change in the POLYMATH equation set are also summarized in Table 6–20. Here the numerical solution is only slightly changed from the previous solution in part (a), which gives reassurance to the first choice of 10 sections for this problem. The temperature profiles are virtually unchanged.

The POLYMATH problem solution file for part (b) is found in directory Chapter 6 with file named **P6-08B.POL**.

(c) The calculation of T_1 at node 1 is required by the convection boundary condition for this case, and Equation (6-48) can be entered into the equation set used in part (a) along with an equation for the ambient temperature T_0. This equation set should indicate a somewhat slower response of the temperatures within the slab because of the additional resistance to heat transfer.

A comparison with the approximate hand calculations by Geankoplis[6] is summarized in Table 6–21. In this case, the simplified hand calculations give results that have some error relative to the numerical method of lines solutions, which are in good agreement with each other.

Table 6–21 Unsteady-State Heat Transfer with Convection in One-Dimensional Slab at t = 1500 s

Distance from Slab Surface in m	Geankoplis[6] $\Delta x = 0.20$ m		Method of Lines (a) $\Delta x = 0.10$ m		Method of Lines (b) $\Delta x = 0.05$ m	
	n	T in °C	n	T in °C	n	T in °C
0	1	64.07	1	64.40	1	64.99
0.2	2	89.07	3	88.13	5	88.77
0.4	3	98.44	5	97.38	9	97.73
0.6	4	100.00	7	99.61	13	99.72
0.8	5	100.00	9	99.96	17	99.98
1.0	6	100.00	11	100.00	21	100.00

www

The POLYMATH problem solution files for part (c) are found in directory Chapter 6 with files named **P6-08C.POL** and **P6-08C2.POL**.

6.9 ESTIMATING MODEL PARAMETERS INVOLVING ODEs USING FERMENTATION DATA

6.9.1   Concepts Demonstrated

Determination of rate expression parameters using experimental concentration data for a fermentation process which produces penicillin.

6.9.2   Numerical Methods Utilized

Integration of the differential equations for obtaining calculated concentration data in an inner loop and using a multivariable minimization program to obtain the optimal parameter values in an outer loop.

6.9.3   Problem Statement (Adapted from Constantinides[15])

When the microorganism *Penicillium Chrysogenum* is grown in a batch fermenter under carefully controlled conditions, the cells reproduce at a rate which can be modeled by the logistic law

$$\frac{dy_1}{dt} = b_1 y_1 \left(1 - \frac{y_1}{b_2}\right) \tag{6-49}$$

where y_1 is the concentration of the cells expressed as percent dry weight. In addition, the rate of production of penicillin has been mathematically quantified by the equation where y_2 is the units of penicillin per mL.

$$\frac{dy_2}{dt} = b_3 y_1 - b_4 y_2 \tag{6-50}$$

(a)   Use the experimental data in Table 6–22 to find the values of the parameters b_1, b_2, b_3 and b_4 which minimize the sum of squares of the differences between the calculated and experimental concentrations (y_1 and y_2) for all the data points. The following initial estimates can be used: $b_1 = 0.1$; $b_2 = 4.0$; $b_3 = 0.02$ and $b_4 = 0.02$.

(b)   Plot the calculated and experimental values of y_1 and y_2 using the optimal parameter values.

Table 6–22 Penicillin Growth Data in Batch Fermenter (File **P6_9.txt**)

Time h	Cell Concentration Percent Dry Weight y_1	Penicillin Concentration Units/mL y_2
0	0.18	0
10	0.12	0
22	0.48	0.0089

Table 6–22 (Continued) Penicillin Growth Data in Batch Fermenter (File **P6_9.txt**)

Time h	Cell Concentration Percent Dry Weight y_1	Penicillin Concentration Units/mL y_2
34	1.46	0.0642
46	1.56	0.2266
58	1.73	0.4373
70	1.99	0.6943
82	2.62	1.2459
94	2.88	1.4315
106	3.43	2.0402
118	3.37	1.9278
130	3.92	2.1848
142	3.96	2.4204
154	3.58	2.4615
166	3.58	2.283
178	3.34	2.7078
190	3.47	2.6542

6.9.4   Solution

The solution of the parameter estimation problem when the model is in the form of systems of ODEs can be divided into three stages.

1. Integration of the differential equations for a specified set of the parameter values in order to obtain the calculated dependent variable values at the same time (independent variable) intervals where experimental data is available.
2. Calculation of the sum of squares of the differences between the calculated and experimental values of the dependent variables.
3. Application of an optimization program which modifies the parameter values so as to obtain the minimum of the sum of squares.

These three stages can be most conveniently be carried out using MATLAB. First, the model equations are introduced into POLYMATH and integrated using the parameter values that were provided as initial estimates. The POLYMATH set of equations is given in Table 6–23.

Table 6–23 POLYMATH Equation Set (File **P6-09.POL**)

Line	Equation
1	d(y1)/d(t) = b1 * y1 * (1 - y1 / b2)
2	d(y2)/d(t) = b3 * y1 - b4 * y2
3	b1 = 0.1
4	b2 = 4
5	b3 = 0.02
6	b4 = 0.02
7	y1(0) = 0.18

Line	Equation
8	y2(0) = 0
9	t(0) = 0
10	t(f) = 190

The set of equations from POLYMATH is then converted to a MATLAB function by selection of the MATLAB output within the POLYMATH software. The MATLAB function generated by POLYMATH must be modified by removing the specification of the numerical values of the parameters and adding the vector *Beta* to the list of the input parameters of the function. The resultant MATLAB function is given in Table 6–24.

Table 6–24 MATLAB Function for Integration

No.	Commands
1	function dYfuncvecdt = ODEfun(t,Yfuncvec,Beta);
2	y1 = Yfuncvec(1);
3	y2 = Yfuncvec(2);
4	b1=Beta(1);
5	b2=Beta(2);
6	b3=Beta(3);
7	b4=Beta(4);
8	dy1dt = b1 * y1 * (1 - y1 / b2);
9	dy2dt = b3 * y1 - b4 * y2;
10	dYfuncvecdt = [dy1dt; dy2dt];

Next, the function that calculates the of the sum of squares of the differences between the calculated and experimental values of the dependent variables must be prepared. This function obtains the vector of the parameters, *Beta,* and the matrix of the experimental data, *data,* as input variables (see line 1 in Table 6–25) and returns the scalar *sum* which contains the sum of squares of the differences.

The first column of the matrix *data*, which contains the independent variable values (the time in hours in this particular case), is stored in the column vector *time* (line 4). The dependent variables measured values are stored in the matrix *ym* (line 5). The matrix *yc* will contain the calculated values of the dependent variables. In its first row, the measured values are stored (line 6).

Table 6–25 MATLAB Function for Calculation of Sum-of-Squares

No.	Command/Comment
1	function sum=SumSqr(Beta,data)
2	global Feval SumSave
3	Feval=Feval+1;
4	time=data(:,1);
5	ym=data(:,2:end);
6	yc(1,:)=ym(1,:);

Table 6–25 (Continued) MATLAB Function for Calculation of Sum-of-Squares

No.	Command/Comment
7	[nr,nc]=size(ym); % nr - umber of data points;
8	% nc - number of dependent variables
9	tspan = [time(1) time(2)]; % Range for the independent variable
10	y0 = ym(1,:); % Initial values for the dependent variables
11	for i=2:nr
12	[t,y]=ode45(@ODEfun,tspan,y0,[],Beta);
13	yc(i,:)=y(end,:);
14	if i<nr
15	tspan = [time(i) time(i+1)];
16	y0 = yc(i,:);
17	end
18	end
19	sum=0;
20	for ic=1:nc
21	sum=sum+(ym(:,ic)-yc(:,ic))'*(ym(:,ic)-yc(:,ic));
22	end
23	if sum< SumSave*0.9
24	disp([' Function evaluation No. ' num2str(Feval) ' Sum of Squares ' num2str(sum)]);
25	SumSave=sum;
26	end

The integration is carried out in a piecewise manner (in the for loop, in lines 11 through 18) in order to obtain the dependent variable calculated values at exactly the same intervals where measured values are available. Accordingly, the start of the first range of the independent variable is the first time value from the *time* vector and the end of the range is the second time value in the same vector (line 9). The initial values of the dependent variables are the values stored in the first row of the matrix *ym* (line 10). The differential equations are integrated using the *ode45* library function in line 12. Note that the vector of parameters *Beta* is added to the input parameters of the integration routine. The dependent variable values in the last row of the resultant *y* vector are stored as the next set of calculated values in the matrix *yc* (line 13).

If there are additional data points available the next time interval is substituted into the vector tspan (line 15) and the last calculated point of the dependent variables is substituted into the vector *y*0 (line 16) and the integration is repeated.

After the integration is finished for the entire range of the independent variable, the sum of squares of the differences between the experimental and calculated values are computed (in lines 19 through 22).

The function prints intermediate results (number of function evaluations and the sum of squares value) when there is 10% reduction in the sum of squares from the previously reported value. The commands in lines 2, 3, and 23 thorough 26 are related to the printing of the intermediate results.

An addition function is needed in order to create the data file, define initial estimates for the parameters, execute a minimization routine for finding the optimal values and display the results. Those tasks are accomplished in the following *Prob_6_9* function for this problem that is shown in Table 6–26.

Table 6–26 MATLAB Function for Minimization (File **Prob_6_9.m**)

No.	Command/Comment
1	function Prob_6_9
2	clear, clc, format short g, format compact
3	global Feval SumSave
4	load P6_9.txt % Load experimental data as a text file.
5	% First column - independent variable. 2nd
6	% and higher columns dependent variables.
7	% All columns must be full. Missing values
8	% are not permitted
9	Beta=[0.1 4 0.02 0.02]; % Initial estimate for parameter values
10	Feval=0;
11	SumSave=realmax;
12	[BetaOpt,Fval]=fminsearch(@SumSqr,Beta,[],P6_9);
13	disp(' ');
14	disp('Optimal Results ');
15	disp(' Parm. No. Value')
16	for i=1:size(Beta,2)
17	disp([i BetaOpt(i)]);
18	end
19	disp([' Final Sum of Squares ' num2str(Fval) ' Function evaluations ' num2str(Feval)]);
20	tspan = [P6_9(1,1) P6_9(end,1)];
21	y0=[P6_9(1,2) P6_9(1,end)];
22	[t,y]=ode45(@ODEfun,tspan,y0,[],BetaOpt);
23	for i=1:size(y,2)
24	hold off
25	plot(t,y(:,i));
26	title([' Plot of dependent variable y' int2str(i)]);
27	xlabel(' Independent variable (t)');
28	ylabel([' Dependent variable y' int2str(i)]);
29	hold on
30	plot(P6_9(:,1),P6_9(:,i+1),'*')
31	pause
32	end

In this function the experimental data matrix is loaded from a text file named *P6_9.txt* (in line 4). The initial estimates for the parameters are set in line 9 and the minimization function is called in line 12. The *fminsearch* library function of *MATLAB* is used for minimization. This function uses the simplex method of Nelder and Mead[16] to find the minimum of the sum of squares. The input parameters are: 1) the name of the function that calculates the sum of squares, 2) the vector of the initial estimates of the parameter values, 3) the structure *Options* (in this case it is replaced by a null matrix [ ]) and 4), the experimental data matrix. The output parameters are 1) the vector of the optimal parameter values, and 2) the sum of squares of the differences between the experimental and the calculated values at the optimum.

The results are displayed in lines 13 through 19. The plots of the calculated and experimental values of the dependent variables versus the independent variable are prepared and displayed in lines 20 through 32.

The complete MATLAB program file can be assembled by incorporating the MATLAB Function for Calculation of Sum-of-Squares of Table 6–25 and the MATLAB Function for Integration of Table 6–24 with the MATLAB Function for Minimization of Table 6–26 (File **Prob_6_9.m**).

The final results and some of the intermediate results are shown in Table 6–27. The initial sum of squares, 36.13, was reduced to the value 1.558 after 63 function evaluations. About 210 additional function evaluations were required for reducing the sum of squares by an additional 8% to 1.436.

Table 6–27 Intermediate and Final Results of the Sum of Squares Minimization

Function evaluation No. 1 Sum of Squares 36.1271	
Function evaluation No. 63 Sum of Squares 1.5582	
Optimal Results	
Parameter Number	Value
1	0.049875
2	3.634
3	0.020459
4	0.02652
Final Sum of Squares 1.4358 Function evaluations 268	

The optimal values of b_1, b_2, b_3, and b_4 (shown in Table 6–27) can be introduced into the model equations in order to obtain the calculated values of y_1 and y_2. The plot of the calculated and experimental values versus time for the cell concentration (y_1) is shown in Figure 6–6, and for the concentration of penicillin (y_2) is shown in Figure 6–7. The data representation is very good for the concentration profiles in both cases.

The problem solution files for the problem are found in directory Chapter 6 and designated **P6-09.POL**, **Prob_6_9.m**, and **P6_9.txt**.

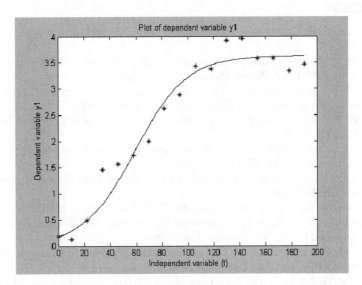

Figure 6–6 Calculated and Measured Cell Concentrations versus Time

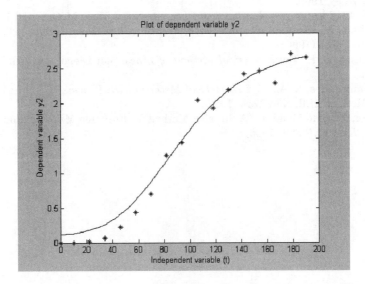

Figure 6–7 Calculated and Measured Penicillin Concentrations versus Time

REFERENCES

1. Garritsen, A.W., University of Technology, Delft, Netherlands, personal communication, 1992.
2. Gear, C. W., "The Automatic Integration of Stiff Ordinary Differential Equations," *Proc. of the IP68 Conf.*, North-Holland, Amsterdam, 1969.
3. Press, W. H., Teukolsky, S. A., Vetterling, W. T., and Flannery, B. P. *Numerical Recipes,* 2nd ed. Cambridge, MA: Cambridge Univ. Press., 1992.
4. Luss, D., and Amundson, N. R. "Stability of Batch Catalytic Fluidized Beds," *AIChE J., 14* (2), 211 (1968).
5. Aiken, R. C., and Lapidus, L. "An Effective Integration Method for Typical Stiff Systems," *AIChE J., 20* (2), 368 (1974).
6. Geankoplis, C. J. *Transport Processes and Unit Operations*, 3rd ed. Englewood Cliffs, NJ: Prentice Hall, 1993.
7. Thomas, L. C. *Heat Transfer*, Englewood Cliffs, NJ: Prentice Hall, 1992.
8. Himmelblau, D. M. *Basic Principles and Calculations in Chemical Engineering*, 6th ed., Englewood Cliffs, NJ: Prentice Hall, 1996.
9. Hanna, O. T., and Sandall, O. C. *Computational Methods in Chemical Engineering*, Englewood Cliffs, NJ: Prentice Hall, 1995.
10. Dahlquist, G, Björk, A. and Anderson, N. *Numerical Methods*, Englewood Cliffs, NJ: Prentice Hall, 1974.
11. Carnahan, B., Luther, H. A., and Wilkes, J. O. *Applied Numerical Methods*, New York: Wiley, 1969.
12. Dean, A. (Ed.) *Lange's Handbook of Chemistry*, New York: McGraw-Hill, 1973.
13. Shacham, M., Brauner, N., and Pozin, M. *Computers Chem Engng.*, *20*, Suppl., pp. S1329–S1334 (1996).
14. Schiesser, W. E. *The Numerical Method of Lines*, San Diego, CA: Academic Press, 1991.
15. Constantinides, A. *Applied Numerical Methods with Personal Computers*, Example 7-1, McGraw Hill, New York, 1987.
16. Nelder, J.A. and Mead, R. "A Simplex Method for Function Minimization," *The Computer Journal*, 7:308 (1965).

CHAPTER 7

Thermodynamics

7.1 COMPRESSIBILITY FACTOR VARIATION FROM VAN DER WAALS EQUATION

7.1.1 Concepts Demonstrated

Compressibility factor variation with reduced pressure from the van der Waals equation of state at various reduced temperatures.

7.1.2 Numerical Methods Utilized

Solution of a single nonlinear equation and integration of an ordinary differential equation to permit continuous plotting of various functions.

7.1.3 Problem Statement

The van der Waals equation of state can be used to calculate the compressibility factors for gases, as has been discussed in Problem 2.1. Thermodynamic textbooks often provide charts of compressibility factors for various reduced temperatures, T_r, as a function of reduced pressure, P_r. An example is given by Kyle.[1]

Consider the calculations necessary for a general compressibility factor chart for CO_2 whose critical temperature is $T_C = 304$ K and whose critical pressure is $P_C = 72.9$ atm.

(a) Plot the compressibility factor for CO_2 over the reduced pressure range of $0.1 \leq P_r \leq 10$ for a constant reduced temperature of $T_r = 1.1$.
(b) Repeat part (a) at $T_r = 1.3$.
(c) Repeat part (a) at $T_r = 2.0$.
(d) Plot the results of (a), (b), and (c) on a single figure and compare your results with those from a generalized compressibility chart, as given in textbooks such as Kyle[1] and Sandler.[2]

7.1.4   Solution

(a)–(d) Approach 1 The first approach is a direct solution of the van der Waals equation for a variety of reduced pressures at the desired reduced temperature. Thus the POLYMATH *Simultaneous Algebraic Equation Solver* can be used, as discussed in Problem 2.1. For desired values of T_r and P_r, the molar volume V and the compressibility factor Z can be calculated. This approach requires repetitive solution of a nonlinear equation in order to generate continuous curves of Z. While this approach is straightforward, it requires a number of calculations plus the creation of a figure from the results.

 The POLYMATH problem solution files for parts (a), (b), and (c) are found in directory Chapter 7 with files named **P7-01A1.POL**, **P7-01B1.POL**, and **P7-01C1.POL**.

(a)–(d) Approach 2   Another more efficient approach to generating the continuous curves of compressibility factor is to convert the van der Waals equation into a differential equation to describe how the specific volume changes with reduced pressure. This differential equation, along with the definitions of reduced pressure and compressibility factor, can be used to generate a continuous curve for compressibility, as outlined next.

The van der Waals equation (see Problem 2.1) can be rewritten as

$$P = \frac{RT}{V-b} - \frac{a}{V^2} \tag{7-1}$$

Differentiating Equation (7-1) with respect to V yields

$$\frac{dP}{dV} = -\frac{RT}{(V-b)^2} + \frac{2a}{V^3} \tag{7-2}$$

The variation of V with P_r can be determined using

$$\frac{dV}{dP_r} = \frac{dV}{dP}\frac{dP}{dP_r} \tag{7-3}$$

By definition, $P_r = P/P_c$; therefore,

$$\frac{dP}{dP_r} = P_c \tag{7-4}$$

The derivative from the preceding equation and the inverse of Equation (7-2) can be introduced into Equation (7-3) to yield the variation of V with P_r. Thus

$$\frac{dV}{dP_r} = \left[-\frac{RT}{(V-b)^2} + \frac{2a}{V^3}\right]^{-1} P_c \tag{7-5}$$

Differential Equation (7-5) can be solved simultaneously using the POLY-MATH *Simultaneous Differential Equation Solver* for a single value of T_r while the independent variable P_r is changed over the desired range $0.1 \leq P_r \leq 10$. In order to obtain the initial value (at $P_r = 0.1$) for Equation (7-5), the POLYMATH *Simultaneous Algebraic Equation Solver* can be used to solve the van der Waals equation for V when P_r is set to 0.1 and T_r is known. For example, when P_r is 0.1 and $T_r = 1.1$, the solution is $V = 3.71998$ liter/g-mol.

The equations for solution of the van der Waals equation for a range of P_r values can be entered into the POLYMATH *Simultaneous Differential Equation Solver* as shown in Table 7–1.

Table 7–1 POLYMATH Problem Code (File **P7-01A2.POL**)

Line	Equation
1	d(V)/d(Pr)=Pc/(-R*T/(V-b)^2+2*a/V^3)
2	Tc=304.2
3	Pc=72.9
4	R=0.08206
5	Tr=1.1
6	P=Pc*Pr
7	T=Tc*Tr
8	b=R*Tc/(8*Pc)
9	a=27*R^2*Tc^2/(Pc*64)
10	Z=P*V/(R*T)
11	Pr(0)=0.1
12	V(0)=3.71998
13	Pr(f)=10

Solution of this set of equations generates the plot of Z versus P_r for $T_r = 1.1$ that is given in Figure 7–1.

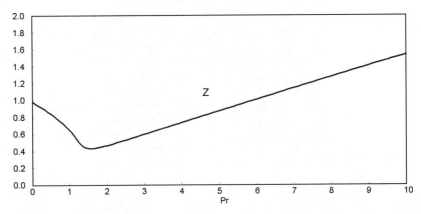

Figure 7–1   Compressibility Factor Variation as Function of P_r at $T_r = 1.1$, from the van der Waals Equation

This plot can be compared with similar curves generated from experimental data as presented in Kyle[1] and Sandler.[2] Deviations are found at higher reduced temperatures and pressures where the van der Waals equation becomes less accurate. Similar plots for other T_r values can be made for other reduced temperatures when the initial value for V is changed as a result of solving the van der Waals equation using the POLYMATH *Simultaneous Algebraic Equation Solver*.

 The POLYMATH problem solution files for parts (a), (b), and (c) are found in directory Chapter 7 with files named **P7-01A2.POL**, **P7-01B2.POL**, and **P7-01C2.POL**.

In part (d) it is desirable to have the compressibility for all three T_r's on the same plot. This can be accomplished in POLYMATH by modifying the equation set with indices 1, 2, and 3 for reduced temperatures of 1.1, 1.3, and 2.0, respectively. The respective initial conditions as determined from separate solutions of the van der Waals equations are V1 = 3.71998, V2 = 4.4344, and V3 = 6.89988. The POLYMATH equation set is given in Table 7–2.

Table 7–2 POLYMATH Problem Code (File **P7-01D.POL**)

Line	Equation
1	d(V1)/d(Pr)=Pc/(-R*T1/(V1-b)^2+2*a/V1^3)
2	d(V2)/d(Pr)=Pc/(-R*T2/(V2-b)^2+2*a/V2^3)
3	d(V3)/d(Pr)=Pc/(-R*T3/(V3-b)^2+2*a/V3^3)
4	Pc=72.9
5	R=0.08206
6	Tc=304.2
7	Tr1=1.1
8	Tr2=1.3
9	Tr3=2.0
10	P=Pc*Pr
11	T2=Tc*Tr2
12	T1=Tc*Tr1
13	T3=Tc*Tr3
14	b=R*Tc/(8*Pc)
15	a=27*R^2*Tc^2/(Pc*64)
16	Z1=P*V1/(R*T1)
17	Z2=P*V2/(R*T2)
18	Z3=P*V3/(R*T3)
19	Pr(0)=0.1
20	V1(0)=3.71998
21	V2(0)=4.4344
22	V3(0)=6.89988
23	Pr(f)=10

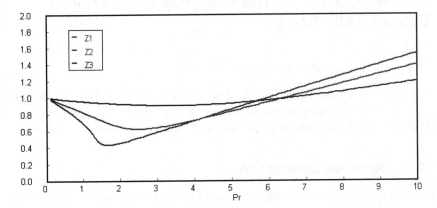

Figure 7–2 Compressibility Factors from the van der Waals Equation at $T_r = 1.1$ (Z1), 1.3 (Z2), and 2.0 (Z3)

The plot for all three T_r's is shown in Figure 7–2.

The POLYMATH problem solution file for part (d) is found in directory Chapter 7 with file named **P4-01D.POL**.

7.2 COMPRESSIBILITY FACTOR VARIATION FROM VARIOUS EQUATIONS OF STATE

7.2.1   Concepts Demonstrated

Use of Virial, Soave-Redlich-Kwong, Peng-Robinson, and Beattie-Bridgeman equations of state for calculation of compressibility factors, and plotting compressibility factors versus reduced pressure at constant reduced temperatures.

7.2.2   Numerical Methods Utilized

Solution of a single nonlinear algebraic equation and conversion of a nonlinear equation to a differential equation for generation of a continuous solution.

7.2.3   Problem Statement

(a) Select a component from Table 7–3 and plot the compressibility factor over the reduced pressure range of $0.5 \leq P_r \leq 11$ for a constant reduced temperature of $T_r = 1.3$. Calculations should be based on one of the following equations of state: Soave-Redlich-Kwong, Peng-Robinson, Beattie-Bridgeman or Virial.

(b) Repeat part (a) at $T_r = 2.0$.

(c) Repeat part (a) at $T_r = 10.0$.

(d) Plot the results of parts (a), (b), and (c) on a single figure.

Additional Information and Data

The Soave-Redlich-Kwong, Peng-Robinson, and Beattie-Bridgeman equations of state are described in Problem 2.9.

The Virial Equation of State The Virial equation of state[3] is normally expressed as

$$P = \frac{RT}{V - B} \tag{7-6}$$

where

P = pressure in atm

V = molar volume in liters/g-mol

T = temperature in K

R = gas constant   (R = 0.08206 atm·liter/g-mol·K)

B = the second Virial coefficient in liter/g-mol

Various techniques have been suggested to calculate B. The method outlined by Smith and Van Ness[4] utilizes

$$B = \frac{RT_c}{P_c}[B^{(0)} + \omega B^{(1)}] \tag{7-7}$$

where

T_c = the critical temperature (K)

P_c = the critical pressure (atm)

ω = the acentric factor

and

$$B^{(0)} = 0.083 - \frac{0.422}{T_r^{1.6}} \tag{7-8}$$

$$B^{(1)} = 0.139 - \frac{0.172}{T_r^{4.2}} \tag{7-9}$$

The preceding treatment of the virial equation of state is only accurate for $T_r > 4$.

Table 7–3 summarizes values of T_c, P_c, and ω for selected substances. The constants for the Beattie-Bridgeman equation for selected substances are given in Table 7–4.

Table 7–3 Critical Constants of Selected Substances[a]

Substance		Critical Temperature T_c (K)	Critical Pressure P_c (atm)	Acentric Factor ω
Hydrogen	H_2	33.3	12.8	−0.218
Nitrogen	N_2	126.2	33.5	0.039
Oxygen	O_2	154.8	50.1	0.025
Carbon Dioxide	CO_2	304.2	72.9	0.239
Carbon Monoxide	CO	133	34.5	0.066
Nitrous Oxide	N_2O	309.7	71.7	0.165
Water	H_2O	647.4	218.3	0.344
Ammonia	NH_3	405.5	111.3	0.250
Methane	CH_4	191.1	45.8	0.011
Ethane	C_2H_6	305.5	48.2	0.099
Propane	C_3H_8	370.0	42.0	0.153
n-Butane	C_4H_{10}	425.2	37.5	0.199

[a]A more extensive table is given by Kyle[1]

Table 7–4 Constants for the Beattie-Bridgeman Equation

Gas	A_0	a	B_0	b	$c \times 10^{-4}$
Hydrogen	0.1975	−0.00506	0.02096	−0.04359	0.0504
Nitrogen	1.3445	0.02617	0.05046	−0.00691	4.20
Oxygen	1.4911	0.02562	0.04624	0.004208	4.80
Air	1.3012	0.01931	0.04611	−0.001101	4.34
Carbon Dioxide	5.0065	0.07132	0.10476	0.07235	66.00
Ammonia	2.3930	0.17031	0.03415	0.19112	476.87
Carbon Monoxide	1.3445	0.02617	0.05046	−0.00691	4.20
Nitrous Oxide	5.0065	0.07132	0.10476	0.07235	66.00
Methane	2.2769	0.01855	0.05587	−0.01587	12.83
Ethane	5.8800	0.05861	0.09400	0.01915	90.00
Propane	11.9200	0.07321	0.18100	0.04293	120.00
n-Butane	17.7940	0.12161	0.24620	0.094620	350.00

7.2.4   Solution (Suggestions)

See Section 7.1.4 for methods of solution.

7.3 ISOTHERMAL COMPRESSION OF GAS USING REDLICH-KWONG EQUATION OF STATE

7.3.1 Concepts Demonstrated

Work and volume change during isothermal compression of an ideal and real gas.

7.3.2 Numerical Methods Utilized

Solution of an ordinary differential equation.

7.3.3 Problem Statement

Carbon dioxide is being compressed isothermally at 50 °C to 1/100th of its initial volume. The initial pressure is 1.0 atm.

(a) Calculate the work per g-mol of gas required using the ideal gas assumption and employing the Redlich-Kwong equation of state.

(b) Check whether the variation of the pressure P and volume V during compression follows the expected path of PV = constant for the ideal and nonideal cases of part (a).

(c) Repeat (a) and (b) for the cases when the gas to be compressed is (1) air and (2) methane. Explain the differences in the behavior of the three gases.

Additional Information and Data

Redlich-Kwong Equation of State The Redlich-Kwong equation of state is given by

$$P = \frac{RT}{(V-b)} - \frac{a}{V(V+b)\sqrt{T}} \qquad (7\text{-}10)$$

where

$$a = 0.42747\left(\frac{R^2 T_c^{5/2}}{P_c}\right) \qquad (7\text{-}11)$$

$$b = 0.08664\left(\frac{RT_c}{P_c}\right) \qquad (7\text{-}12)$$

P = pressure in atm
V = molar volume in liters/g-mol
T = temperature in K
R = gas constant (R = 0.08206 (atm·liter/g-mol·K))
T_c = critical temperature in K
P_c = critical pressure in atm

Critical properties of CO_2 and methane can be found in Table 7–3.

7.3.4   Solution (Partial)

(a)   For a closed system in a reversible isothermal compression from V_1 to V_2, the work W done on the system can be calculated from

$$\frac{dW}{dV} = -P \tag{7-13}$$

For an ideal gas, this equation can be integrated from V_1 to V_2, yielding

$$W = -RT \ln\frac{V_2}{V_1} \tag{7-14}$$

For the nonideal case, Equation (7-13) should be numerically integrated using Equation (7-10) to calculate the pressure as a function of the volume. The initial volume, V_1, should be calculated using the Redlich-Kwong equation of state given in Equation (7-10) by solving this nonlinear equation for V_1 at 1.0 atm and 50 °C. This solution for V_1 can be obtained using the POLYMATH *Simultaneous Algebraic Equation Solver*. The result is $V_1 = 26.4134$ liters so that $V_2 = 0.2641$ liters since the problem states that V_2 is simply $0.01\ V_1$.

 The POLYMATH problem solution file for the initial volume estimate of part (a) is found in directory Chapter 7 with file named **P4-03A1.POL**.

Change of Variable to Allow Decreasing Independent Variable in ODEs
Most numerical integration programs, including POLYMATH, proceed with increasing values of the independent variable. In problems where this is not the case, a convenient new independent variable can be introduced. In this case, a simple linear relationship of this new variable Y to variable V can be employed:

$$V = 26.4134 - Y \tag{7-15}$$

so that as this new (dummy) variable Y progresses from an initial condition of $Y = 0.0$ to a final value of $Y = 26.1493$, then V goes from $V_1 = 26.4134$ to $V_2 = 0.2641$. The only major change to the original problem equations is that Equation (7-13) must be rewritten using the independent variable Y as

$$\frac{dW}{dY} = P \tag{7-16}$$

since $dV = -dY$ from Equation (7-15). Also, algebraic Equation (7-15) must be introduced into the ODE solution.

(b)   If the rule, PV = constant, holds for the process, then a plot of $\ln(P)$ versus $\ln(V)$ should give a straight line with slope of -1.0.

The necessary calculations can be obtained during a single numerical solution by using Equations (7-10) to (7-12) and Equations (7-14) to (7-16) with the initial/final conditions as discussed for Y to solve part (a) of this problem. Part (b) can also be solved during the same numerical ODE solution by simply creating the necessary plotting variables. The equation set for the POLYMATH *Simultaneous Differential Equation Solver* is given in Table 7–5.

Table 7–5 POLYMATH Problem Code (**File P7-03A2.POL**)

Line	Equation
1	d(W)/d(Y)=P
2	R=0.08206
3	T=50+273.15
4	Tc=304.2
5	Pc=72.9
6	V=26.4134-Y
7	lnV=ln(V)
8	b=0.08664*R*Tc/Pc
9	a=0.42747*R^2*Tc^(5/2)/Pc
10	Pideal=R*T/V
11	Wideal=-R*T*ln(V/26.4134)
12	lnPideal=ln(Pideal)
13	P=R*T/(V-b)-a/(V*(V+b)*sqrt(T))
14	lnP=ln(P)
15	Y(0)=0
16	W(0)=0
17	Y(f)=26.1493

Numerical results for the calculated values of P, Pideal, W, and Wideal for different volumes during the compression are summarized in Table 7–6. Note that the compressions, summarized in Table 7–6, result from separate POLYMATH integrations to various percentages of the final compression volume (i.e., 0%, 10%, ..., 100%).

Table 7–6 Partial Tabular Results for Isothermal Compression of Carbon Dioxide

V	%Vf	P	Pideal	W	Wideal
26.4134	0	1.0000015	1.0039483	0	0
23.79847	10	1.1093993	1.1142602	2.7530198	2.7644734
21.18354	20	1.2456727	1.2518063	5.8253164	5.8510483
18.56861	30	1.420112	1.4280923	9.3007788	9.3448054
15.95368	40	1.6513613	1.6621675	13.30143	13.36974
13.33875	50	1.9725694	1.988019	18.014738	18.116835
10.72382	60	2.4489014	2.4727839	23.750825	23.903152
8.10889	70	3.228489	3.2701996	31.080157	31.315016
5.49396	80	4.7360817	4.8266986	41.24304	41.638663
2.87903	90	8.8831912	9.2106331	57.928225	58.774284
0.2641	100	67.419813	100.40776	112.66436	122.12188

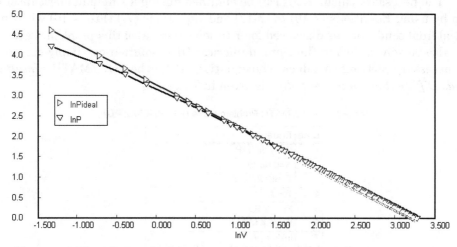

Figure 7–3 Plot of lnP and lnPideal versus lnV for Isothermal Compression of Carbon
Dioxide

Here the suffix "ideal" refers to the value calculated using the ideal gas assumption. The results in Table 7–6 indicate that during the initial compression, the ideal and the real work are nearly equivalent as the system is nearly ideal. However, as the compression proceeds toward very small volume and high pressure, the differences between the ideal and real pressure and the ideal and real work become very substantial as the system becomes very nonideal.

Figure 7–3 presents the plot of lnP and lnPideal versus lnV. This figure provides a clear indication that lnPideal versus lnV for the ideal gas indeed gives a straight line with slope of negative one as expected, whereas the slope of lnP versus lnV for the real CO_2 calculations changes considerably for pressures higher than $P \sim 12$ atm as the system becomes nonideal.

Similar calculations using the Redlich-Kwong equation of state for air and methane require only modifications to Tc, Pc, and the initial/final conditions in the POLYMATH program; however, the initial/final conditions will first require solving the equation of state as a nonlinear equation for the starting temperature and pressure.

The problem solution file for the second row in Table 7–6 of part (b) is found in directory Chapter 7 with file named **P7-03A2.POL**. A MATLAB file that creates all of Table 7–6 is named **TABLE7_6.m**.

7.4 THERMODYNAMIC PROPERTIES OF STEAM FROM REDLICH-KWONG EQUATION

7.4.1 Concepts Demonstrated

Calculation of compressibility factor, molar volume, liquid enthalpy, and vapor enthalpy from an equation of state and ideal gas enthalpy change.

7.4.2 Numerical Methods Utilized

Solution of a nonlinear algebraic equation.

7.4.3 Problem Statement

If you have ever used a thermodynamic diagram (such as pressure enthalpy diagram) to try to find the specific volume, compressibility factor, or enthalpy of a pure substance, you probably noticed how difficult it is to get an accurate value from the diagram. You probably concluded that with access to a computer "there ought to be a better way."

This problem requires the development of a simple numerical computation that will provide thermodynamic properties of steam to approximately three digits accuracy with calculations utilizing the Redlich-Kwong equation of state.

(a) Outline a general calculational procedure to use the Redlich-Kwong equation of state to calculate the molar volume (m^3/kg-mol), specific volume (m^3/kg), compressibility factor, and enthalpy of water vapor at a specified temperature and pressure.

(b) Compare your results with data found in the steam tables or thermodynamics textbooks for saturated steam at 100, 200, and 300°C. Please pay particular attention to the possibility of multiple solutions to the R-K equation in some regions. Typically the highest molar volume represents the correct value, while other solutions are extraneous.

(c) Observe the behavior of the R-K equation in three different regions: (1) below the critical point, (2) close to the critical point, and (3) above the critical point.

Additional Information and Data

The Redlich-Kwong equation of state is described in Problem 7.3, and the critical properties for water can be found in Table 7–3.

The molar specific heats of gases in the ideal gas state are typically given in the literature as a polynomial in temperature:

$$C_p^o = a_0 + a_1 T + a_2 T^2 + a_3 T^3 \tag{7-17}$$

where

T = temperature in K

C_p = molar specific heat in cal/g-mol·K

a_0, a_1, a_2, and a_3 = constants specific for the substance. For water vapor, a_0 = 7.700, $a_1 = 0.04594 \times 10^{-2}$, $a_2 = 0.2521 \times 10^{-5}$, and $a_3 = -0.8587 \times 10^{-9}$.

The usual reference state for the enthalpy of liquid water is H_l^o = 0 kJ/kg at T_0 = 0 °C = 273.15 K, where the enthalpy of vaporization is ΔH_{vap}^o = 2501.3 kJ/kg. Thus the enthalpy of water vapor as an ideal gas at temperature T can be calculated from

$$H_v^o = H_l^o + \Delta H_{vap}^o + \int_{T_0}^{T} C_p^o \, dT \tag{7-18}$$

where

$$\int_{T_0}^{T} C_p^o \, dT = a_0(T - T_0) + \frac{a_1}{2}(T^2 - T_0^2) + \frac{a_2}{3}(T^3 - T_0^3) + \frac{a_3}{4}(T^4 - T_0^4) \tag{7-19}$$

and the temperatures are in K.

The isothermal enthalpy departure can be expressed in the integral equation

$$(H_v - H_v^o) = \int_{\infty}^{V}\left[T\left(\frac{\partial P}{\partial T}\right)_V + V\left(\frac{\partial P}{\partial V}\right)_T \right] dV \tag{7-20}$$

The Redlich-Kwong equation of state, which accounts for nonidealities, can be introduced into Equation (7-20) and the integration can be carried out analytically. The integrated form of the equation can be found in Edmister.[6]

$$(H_v - H_v^o) = RT\left[Z - 1 - \frac{1.5a}{bRT^{1.5}} \ln\left(1 + \frac{b}{V}\right) \right] \tag{7-21}$$

Thus the enthalpy of steam (water in the vapor phase) at T and P can be calculated from

$$H_v = H_l^o + \Delta H_{vap}^o + \int_{T_0}^{T} C_p^o \, dT + RT\left[Z - 1 - \frac{1.5a}{bRT^{1.5}} \ln\left(1 + \frac{b}{V}\right) \right] \tag{7-22}$$

using the usual reference states and any conversion factors to maintain consistent units.

7.4.4   Solution (Partial)

(a) For specified values of T and P, the Redlich-Kwong equation can be solved for the molar volume using the POLYMATH *Simultaneous Algebraic Equation Solver*. In the two-phase region this equation may have up to three roots; the smallest one is an approximate solution for liquid, the largest is accurate for vapor, and the intermediate has no physical meaning.

After calculation of the molar volume and then specific volume, the compressibility factor can be calculated from its definition:

$$Z = \frac{PV}{RT} \tag{7-23}$$

Additional calculations follow from the previous discussion and lead to the final calculation of H_v by evaluation of Equation (7-22). All of the preceding equations can be solved with the POLYMATH *Simultaneous Algebraic Equation Solver.*

(a) The POLYMATH code for this problem when T in °C and P in atm are provided is given in Table 7–7.

Table 7–7 POLYMATH Problem Code (**File P7-04A.POL**)

Line	Equation
1	f(V)=P-(R*T/(V-b)-a/(V*(V+b)*sqrt(T)))
2	P=1
3	R=0.08206
4	Tc=647.4
5	Pc=218.3
6	Vsp=V/18
7	TC=100
8	T0=273.15
9	T=273.15+TC
10	b=0.08664*R*Tc/Pc
11	a=0.42747*R^2*Tc^(5/2)/Pc
12	Hv0=7.7*(T-T0)+0.04594e-2*(T^2-T0^2)/2+0.2521e-5*(T^3-T0^3)/3-0.8587e-9*(T^4-T0^4)/4
13	Z=P*V/(R*T)
14	Hdep=24.218*R*T*(Z-1-1.5*a*ln(1+b/V)/(b*R*T^1.5))
15	Hv=2501.3+4.1868*(Hv0+Hdep)/18
16	V(min)=20
17	V(max)=40

Note that the conversion factor 24.218 has been added to the equation for Hdep to convert from liter·atm to cal. The conversion factor 4.1868/18 is added to the equation for Hv to convert cal/g-mol to kJ/kg.

(b) The case of saturated steam at $T = 100$°C where $P = 1$ atm has been entered in the equation set. The solution yields $V = 30.4027$ liter/g-mol, $v = 1.68904$ m³/kg, $Z = 0.99288$, and $H_v = 2686.09$ kJ/kg. Steam tables give $v = 1.6729$ m³/kg and $H_v = 2676.1$ kJ/kg for saturated steam at the same condition.

Only the temperature and pressure in the POLYMATH equation set need to be changed in order to obtain the specific volume and saturated steam enthalpy. Modification for other pure components can easily be made.

The POLYMATH problem solution files for parts (a) and (b) are found in directory Chapter 7 with files named **P7-04A.POL** and **P7-04B.POL**.

7.5 ENTHALPY AND ENTROPY DEPARTURE USING THE REDLICH-KWONG EQUATION

7.5.1   Concepts Demonstrated

Use of the Redlich-Kwong equation to calculate isothermal enthalpy and entropy departure versus reduced temperature and pressure.

7.5.2   Numerical Methods Utilized

Solution of a nonlinear algebraic equation and integration of an ordinary differential equation.

7.5.3   Problem Statement

(a)   Select a component from Table 7–3 and plot the enthalpy departure and entropy departure functions over the reduced pressure range of $0.5 \leq P_r \leq 30$ for constant reduced temperatures of $T_r = 1.2, 1.4,$ and 3.0. Please place all results on a single plot.

(b)   Modify the output of part (a) to allow a direct comparison of your plots with generalized plots available in the literature. Typically these plots require $(\Delta H^*)/T_c$ and ΔS^* versus $\ln(P_r)$.

(c)   Compare your calculated results with generalized correlations and experimental data wherever available (data from Mollier diagrams, etc.).

Additional Information and Data

The enthalpy and entropy departure functions for the Redlich-Kwong equation are presented by Edmister[6] as

$$\frac{\Delta H^*}{RT} = \frac{3a}{2bRT^{1.5}}\ln\left(1 + \frac{b}{V}\right) - (Z - 1) \qquad \text{(7-24)}$$

and

$$\frac{\Delta S^*}{R} = \frac{a}{2bRT^{1.5}}\ln\left(1 + \frac{b}{V}\right) - \ln\left(Z - \frac{Pb}{RT}\right) \qquad \text{(7-25)}$$

where

$$\Delta H^* = H^0 - H \qquad \Delta S^* = S^0 - S$$

H^0 = ideal gas enthalpy at temperature T in cal/g-mol

H = gas enthalpy at temperature T in cal/g-mol

S^0 = ideal gas entropy at temperature T in cal/g-mol·K

S = gas entropy at temperature T in cal/g-mol·K

P = pressure in atm

V = molar volume in liters/g-mol

T = temperature in K

R = gas constant (R = 0.08206 liter·atm/g-mol·K) (R = 1.9872 cal/g-mol·K)

Z = compressibility factor

and a and b are the Redlich-Kwong equation constants calculated from critical pressure and temperature as discussed in Problem 7.3.

7.5.4   Solution (Partial)

Calculational Approach Since continuous curves for enthalpy and entropy departure are desired, the approach discussed in Section 7.1.4 will be used to develop a differential equation for the Redlich-Kwong equation. The enthalpy and entropy departure functions will then be calculated and plotted during the numerical solution.

The Redlich-Kwong equation, which is discussed in Section 7.3.3,

$$P = \frac{RT}{V-b} - \frac{a}{V(V+b)\sqrt{T}} \tag{7-26}$$

can be differentiated with respect to V to yield

$$\frac{dP}{dV} = -\frac{RT}{(V-b)^2} + \frac{a}{\sqrt{T}}\left[\frac{2V+b}{V^2(V+b)^2}\right] \tag{7-27}$$

The introduction of Equations (7-27) and (7-4) into Equation (7-3) yields

$$\frac{dV}{dP_r} = \left\{-\frac{RT}{(V-b)^2} + \frac{a}{\sqrt{T}}\left[\frac{2V+b}{V^2(V+b)^2}\right]\right\}^{-1} P_c \tag{7-28}$$

Equation (7-28) can be numerically integrated using the POLYMATH *Simultaneous Differential Equation Solver* for a desired value of T_r while the independent variable P_r is changed over the desired range $(0.5 \leq P_r \leq 30)$. During the integration, the enthalpy and entropy departure functions given by Equations (7-24) and (7-25) can be calculated for plotting.

(a) The component selected for this example is water so that a comparison can be made with data from steam tables. The initial condition for V must be separately calculated using the POLYMATH *Simultaneous Algebraic Equation Solver* and the Redlich-Kwong equation of state. This calculation, which is similar to that presented in Section 7.4.4, gives $V = 0.523726$ liters for the initial conditions where $P_r = 0.5$ and $T_r = 1.2$.

The POLYMATH equation set for both the enthalpy departure function (Hdep) and the entropy departure function (Sdep) is given in Table 7–8.

Table 7–8 POLYMATH Problem Code (File **P7-05A1.POL**)

Line	Equation
1	d(V)/d(Pr)=Pc/(-R*T/((V-b)^2)+a*(2*V+b)/((V^2*(V+b)^2)*sqrt(T)))
2	Tc=647.4
3	Pc=218.3
4	R=0.08206
5	Tr=1.2
6	P=Pc*Pr
7	T=Tc*Tr
8	b=0.08664*R*Tc/Pc
9	a=0.42747*R^2*Tc^(5/2)/Pc
10	Z=P*V/(R*T)
11	HdepFUN=(3*a/(2*b*R*T^1.5))*ln(1+b/V)-(Z-1)
12	SdepFUN=(a/(2*b*R*T^1.5))*ln((1+b/V))-ln(Z-P*b/(R*T))
13	Pr(0)=0.5
14	V(0)=0.523726
15	Pr(f)=30

A plot of POLYMATH variables Hdep and Sdep versus Pr for Tr = 1.2 is shown in Figure 7–4, where both functions go through a maximum with Pr.

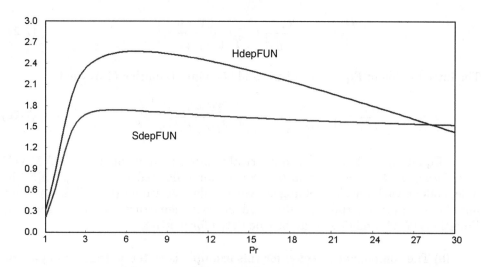

Figure 7–4 Enthalpy and Entropy Departure Functions for Water Vapor at $T_r = 1.2$ (File **P7-05A1.POL**)

Since it is desirable to have the Hdep plots for all three Tr's on the same plot, the POLYMATH equation set can be modified to solve all three cases simultaneously. The Tr's can be coded with indices 1, 2, and 3 for reduced temperatures of 1.2, 1.4, and 3.0, respectively, in a similar manner to the method discussed in Section 7.1.4. The respective initial conditions as determined from the Redlich-Kwong equation solutions are V1 = 0.523726, V2 = 0.639885, and V3 = 1.46182. The results for all three reduced temperatures are given in Figure 7–5.

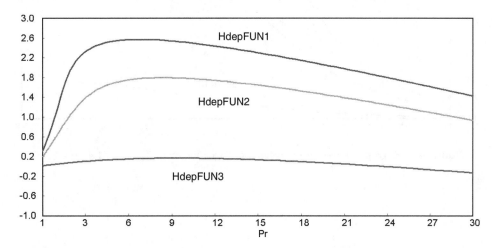

Figure 7–5 Enthalpy Departure Expressions for Water Vapor at T_r = 1.2, 1.4, and 3.0 (File **P7-05A2.POL**)

The POLYMATH problem solution files for part (a) are found in directory Chapter 7 with files named **P7-05A1.POL** and **P7-05A2.POL**.

(b)   Most of the plots in the literature require $(\Delta H^*)/T_c$ and ΔS^* versus P_r on a logarithmic scale. Thus these variables can also be calculated during the numerical solution. Typical results are presented in Figure 7–6.

(c)   Most thermodynamic textbooks (for example, Kyle[1] and Sandler[2]) present figures for generalized enthalpy and entropy departure for ideal gas behavior. The numerical results for T_r = 1.2 and P_r = 10 are calculated to be $(\Delta H^*)/T_c$ = 6.031 cal/g-mol·K and ΔS^* = 3.374 cal/g-mol·K. These values are slightly lower than what is shown in the generalized figures (100–101 of Kyle)[1], but the generalized figures are for substances with Z_c = 0.27, while for water Z_c = 0.23.

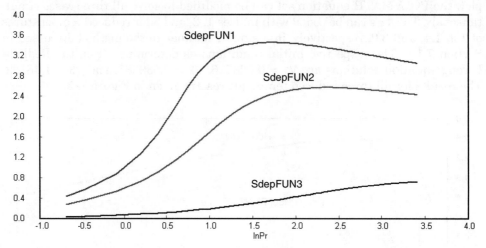

Figure 7–6 Entropy Departure Curves for Water Vapor at T_r = 1.2, 1.4, and 3.0

7.6 FUGACITY COEFFICIENTS OF PURE FLUIDS FROM VARIOUS EQUATIONS OF STATE

7.6.1   Concepts Demonstrated

Use of various equations of state to calculate fugacity coefficients for pure substances in the liquid or gaseous state.

7.6.2   Numerical Methods Utilized

Solution of a single nonlinear algebraic equation and conversion of a nonlinear equation to a differential equation.

7.6.3   Problem Statement

(a)   Select a component from Table 7–3 and use the van der Waals equation of state to plot the calculated fugacity coefficient over the reduced pressure range of $0.5 \le P_r \le 30$ for constant reduced temperatures of $T_r =$ 1.2, 1.4 and 3.0. Please place all results on a single plot, and compare your results with those from a generalized compressibility chart, as given in textbooks such as Kyle[1] and Sandler.[2]

(b)   Repeat part (a) using the Redlich-Kwong equation of state.

(c)   Repeat part (a) using the Peng-Robinson equation of state.

(d)   Compare your calculated results with experimental data wherever available (data from Mollier diagrams, etc.).

Additional Information and Data

The fugacity coefficient is defined by $\phi = f/P$, where f is the fugacity of the component with units of pressure and P is the pressure. The fugacity coefficients of pure substances can be calculated from a number of equations of state which are summarized by Walas.[7]

van der Waals

$$\ln \phi = Z - 1 - \frac{a}{RTV} - \ln\left[Z\left(1 - \frac{b}{V}\right)\right] \tag{7-29}$$

Redlich-Kwong

$$\ln \phi = Z - 1 - \ln\left[Z\left(1 - \frac{b}{V}\right)\right] - \frac{a}{bRT^{1.5}}\ln\left(1 + \frac{b}{V}\right) \tag{7-30}$$

Peng-Robinson

$$\ln \phi = Z - 1 - \ln(Z - B) + \left(-\frac{A}{2\sqrt{2}B}\right) \ln \frac{(Z + 2.414B)}{(Z - 0.414B)} \qquad \text{(7-31)}$$

where

$$A = 0.45724 \alpha P_r / T_r^2 \qquad \text{(7-32)}$$

$$\alpha = [1 + (0.37464 + 1.54226\omega - 0.26992\omega^2)(1 - T_r^{0.5})]^2 \qquad \text{(7-33)}$$

$$B = 0.07780 P_r / T_r \qquad \text{(7-34)}$$

Equations (7-29) to (7-34) are applicable to both the vapor and liquid phases, according to the root selected in solving the equation of state for Z (or for V). The smallest root for Z is for the liquid phase, the largest for the vapor phase.

7.6.4   Solution (Comments)

The equation of state of interest can be solved while continuously changing the pressure along an isotherm for the compressibility factor as demonstrated in Problem 7.5. After the molar volume and compressibility factor have been calculated, the fugacity coefficient can be determined.

Critical properties and the acentric factor for water can be found in Table 7–3. The van der Waals equation and the appropriate constants are discussed in Problem 2.1. The Redlich-Kwong equation of state is described in Problem 7.3.

7.7 FUGACITY COEFFICIENTS FOR AMMONIA—EXPERIMENTAL AND PREDICTED

7.7.1   Concepts Demonstrated

Determination of fugacity from experimental data and calculation of fugacity coefficients from various equations of state.

7.7.2   Numerical Methods Utilized

Integration of experimental data and solution of nonlinear algebraic equations.

7.7.3   Problem Statement

Experimental measurements have yielded the compressibility factors at various pressures for ammonia at 100°C summarized in Table 7–9.

(a)   Calculate the fugacity coefficients from the experimental data given in Table 7–9 and summarize these values in a table for $100 \leq P \leq 1100$ in increments of 100 atm.

(b)   Calculate the fugacity coefficients from the van der Waals equation of state and enter these values in the table created in part (a).

(c)   Repeat (b) for the Redlich-Kwong equation of state.

(d)   Repeat (b) for the Peng-Robinson equation of state.

(e)   Discuss any conclusions you can make about the accuracy of the equations of state that you have considered.

Table 7–9 Experimental Data for Ammonia[a]

Pressure atm	Z	Pressure atm	Z	Pressure atm	Z
1.374	0.9928	30.47	0.8471	300	0.3212
3.537	0.9828	33.21	0.831	400	0.4145
5.832	0.9728	36.47	0.8111	500	0.506
8.632	0.9599	40.41	0.7864	600	0.5955
11.352	0.9468	45.19	0.7538	700	0.6828
14.567	0.9315	51.09	0.7102	800	0.7684
19.109	0.9085	58.28	0.6481	900	0.8507
22.84	0.889	100	0.1158	1000	0.9333
26.12	0.8714	200	0.2221	1100	1.014

[a]Data are from Walas[7] where the original reference is given as Gmelin, *Handbook der anorganischen Chemie*, 5, 426 (1935).

Additional Information and Data

The fugacity coefficient, either from experimental data or from an equation of state, may be calculated from

$$\ln \phi = \int_0^P \frac{Z-1}{P} dP \qquad (7\text{-}35)$$

The use of a cubic spline is recommended for evaluation of the integral in Equation (7-35) when experimental data are available for Z as a function of P.

7.7.4   Solution (Suggestions)

The formulas for the calculation of fugacity coefficients from various equations of state are given by Equations (7-29) to (7-31).

The POLYMATH data file for Table 7–9 is found in directory Chapter 7 with file named **P7-07.POL**.

7.8 FLASH EVAPORATION OF AN IDEAL MULTICOMPONENT MIXTURE

7.8.1 Concepts Demonstrated

Calculation of bubble point and dew point temperatures and associated vapor, and liquid compositions for flash evaporation of an ideal multicomponent mixture.

7.8.2 Numerical Methods Utilized

Solution of a single nonlinear algebraic equation.

7.8.3 Problem Statement

A flash evaporator must separate ethylene and ethane from a feed stream which also contains propane and n-butane. A diagram of the evaporator is given in Figure 7–7.

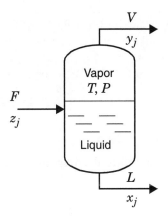

Figure 7–7 Flash Evaporator

where

> F = total feed flow rate in lb-mol/hr
>
> V = total vapor flow rate in lb-mol/hr
>
> L = total liquid flow rate in lb-mol/hr
>
> z_j = mole fraction of component j in the feed stream
>
> y_j = mole fraction of component j in the vapor stream
>
> x_j = mol fraction of component j in the liquid stream
>
> n_c = total number of components

The feed composition entering the evaporator is given in Table 7–10. The evaporator will operate under high pressure, between 15 and 25 atm, with a feed stream at 50 °C.

Table 7–10 Liquid Composition and Antoine Equation Constants [a]

Component	Mole Fraction	A	B	C
Ethylene	0.1	6.64380	395.74	266.681
Ethane	0.25	6.82915	663.72	256.681
Propane	0.5	6.80338	804.00	247.04
n-Butane	0.15	6.80776	935.77	238.789

[a]Thermodynamics Research Center API44 Hydrocarbon Project, *Selected Values of Properties of Hydrocarbon and Related Compounds*, Texas A&M University, College Station, Texas (1978), with permission. (available from NIST, Gaithersburg, MD)

(a) Calculate the percent of the total feed at 50°C that is evaporated and the corresponding mole fractions in the liquid and vapor streams for the following pressures: $P = 15, 17, 19, 21, 23$, and 25 atm.
(b) Determine the dew and the bubble point temperatures of the feed stream.

Additional Information and Data

The vapor pressure of the individual components can be calculated from the Antoine equation:

$$P_j = 10^{\left[A_j - \left(\frac{B_j}{C_j + T}\right)\right]} \tag{7-36}$$

where P_j is the vapor pressure of component j. The constants A_j, B_j, and C_j are specific for component j, and T is the temperature in °C. The Antoine equation constants for the hydrocarbons of this problem are presented in Table 7–10, where P_j has units of mm Hg.

7.8.4   Solution (Partial)

The material balance and phase equilibrium equations for a flash evaporator can be formulated into a single nonlinear algebraic equation (Henley and Rosen,[8] 341) given by

$$f(\alpha) = \sum_{j=1}^{n_c} (x_j - y_j) = \sum_{j=1}^{n_c} \frac{z_j(1-k_j)}{1+\alpha(k_j-1)} = 0 \qquad (7\text{-}37)$$

where α is the vapor to feed ratio, $\alpha = V/F$, and k_j is the vapor liquid equilibrium ratio for component j. Once α has been determined, the vapor and liquid mole fractions can be calculated from

$$x_j = \frac{z_j}{1+\alpha(k_j-1)} \qquad (7\text{-}38)$$

and

$$y_j = k_j x_j \qquad (7\text{-}39)$$

For ideal systems, the vapor liquid equilibrium ratio can be calculated from

$$k_j = \frac{P_j}{P} \qquad (7\text{-}40)$$

where P_j is the vapor pressure of component j and P is the total pressure in the evaporator. The k_j's needed in this problem can be calculated at a particular pressure and temperature by using the Antoine correlation, Equation (7-36), for each vapor pressure P_j in Equation (7-40).

(a) This problem requires the solution of the single nonlinear Equation (7-37) along with Equations (7-36) and (7-38) through (7-40) for each component j. This set of equations can be solved for α, the vapor to feed ratio, and the resulting mole fractions. The interval for the solution is $0 \le \alpha \le 1$, since there is no physical meaning to $V/F > 1$ or $V/F < 0$. The problem solution as entered into POLYMATH *Simultaneous Algebraic Equation Solver* for the case of $P = 20$ atm and $T = 50°C$ is given in Table 7–11.

Table 7–11 POLYMATH Problem Code (File **P7-08A.POL**)

Line	Equation
1	f(alpha)=x1*(1-k1)+x2*(1-k2)+x3*(1-k3)+x4*(1-k4)
2	P=20*760
3	TC=50
4	k1=10^(6.6438-395.74/(266.681+TC))/P
5	k2=10^(6.82915-663.72/(256.681+TC))/P
6	k3=10^(6.80338-804/(247.04+TC))/P
7	k4=10^(6.80776-935.77/(238.789+TC))/P
8	x1=0.1/(1+alpha*(k1-1))
9	x2=0.25/(1+alpha*(k2-1))
10	x3=0.5/(1+alpha*(k3-1))
11	x4=0.15/(1+alpha*(k4-1))
12	y1=k1*x1

Table 7–11 (Continued) POLYMATH Problem Code (File **P7-08A.POL**)

Line	Equation
13	y2=k2*x2
14	y3=k3*x3
15	y4=k4*x4
16	alpha(min)=0
17	alpha(max)=1

The solution obtained for this case is $\alpha = 0.6967$, indicating that at this temperature 69.67% of the feed has evaporated at a pressure of 20 atm. The corresponding mole fractions in the vapor and in the liquid streams are as shown in Table 7–12.

Table 7–12 Summary of Mole Fractions

Component Mole Fractions	Ethylene	Ethane	Propane	*n*-Butane
Feed	0.1	0.25	0.5	0.15
Vapor	0.1398	0.3139	0.4692	0.07713
Liquid	0.008574	0.1032	0.5708	0.3174

The POLYMATH problem solution file for part (a) ($P = 20$ atm and $T = 50°C$) is found in the directory Chapter 7 with file named **P7-08A.POL**.

(b) For $\alpha = 0$ and constant pressure, Equation (7-37) reduces to

$$f(T) = 1 - \sum_{j=1}^{n_c} z_j k_j \tag{7-41}$$

Solution of this equation for T gives the bubble point temperature.

For $\alpha = 1$ and constant pressure, Equation (7-37) reduces to

$$f(T) = \sum_{j=1}^{n_c} \frac{z_j}{k_j} - 1 \tag{7-42}$$

Solution of Equation (7-42) for T yields the dew point temperature.

7.9 FLASH EVAPORATION OF VARIOUS HYDROCARBON MIXTURES

7.9.1 Concepts Demonstrated

Calculation of bubble point and dew point temperatures and associated vapor and liquid compositions for flash evaporation of an ideal multicomponent mixture.

7.9.2 Numerical Methods Utilized

Solution of a single nonlinear algebraic equation.

7.9.3 Problem Statement

Complete Problem 7.8 for a different three- or four-component mixture of hydrocarbons. Select the components from Table 7–13 and assign mole fractions in the feed that sum to one.

Table 7–13 Antoine Equation Constants for Various Hydrocarbons[a]

Substance	A	B	C
Methane	6.64380	395.74	266.681
Ethane	6.82915	663.72	256.681
Propane	6.80338	804.00	247.04
n-Butane	6.80776	935.77	238.789
n-Pentane	6.85296	1064.84	232.012
n-Hexane	6.87601	1171.17	224.408
n-Heptane	6.89677	1264.90	216.544
n-Octane	6.91868	1351.99	209.155
n-Nonane	6.93893	1431.82	202.011
n-Decane	6.94363	1495.17	193.858

[a]Thermodynamics Research Center API44 Hydrocarbon Project, *Selected Values of Properties of Hydrocarbon and Related Compounds*, Texas A&M University, College Station, Texas (1978), with permission. (available from NIST, Gaithersburg, MD)

7.9.4 Solution (Suggestion)

The pertinent equations are given in Problem 7.8.

7.10 CORRELATION OF ACTIVITY COEFFICIENTS WITH THE VAN LAAR EQUATIONS

7.10.1   Concepts Demonstrated

Estimation of parameters in the Van Laar equations for the correlation of binary activity coefficients.

7.10.2   Numerical Methods Utilized

Linear and nonlinear regression, transformation of data for regression, calculation and comparisons of confidence intervals, residual plots, and sum of squares.

7.10.3   Problem Statement

The Van Laar equations for correlation of binary activity coefficients are

$$\gamma_1 = \exp\left\{ A / [1 + (x_1/x_2)(A/B)]^2 \right\} \tag{7-43}$$

$$\gamma_2 = \exp\left\{ B / [1 + (x_2/x_1)(B/A)]^2 \right\} \tag{7-44}$$

where x_1 and x_2 are the mole fractions of components 1 and 2, respectively, and γ_1 and γ_2 are the activity coefficients. Parameters A and B are constant for a particular binary mixture.

Equations (7-43) and (7-44) can be combined to give the excess Gibbs energy expression:

$$g = G_E/RT = x_1 \ln \gamma 1 + x_2 \ln \gamma 2 = ABx_1x_2/(Ax_1 + Bx_2) \tag{7-45}$$

Activity coefficients at various mole fractions are available from Table 3–7 for the benzene and n-heptane system from which the g in Equation (7-45) can be calculated. Linear regression can be used to estimate the parameter values of A and B. An alternate method is to sum Equations (7-43) and (7-44) and to utilize nonlinear regression on this sum to determine the values of both A and B.

(a) Use linear regression on Equation (7-45) with the data of Table 3–7 to determine A and B in the Van Laar equations for the benzene and n-heptane binary system.

(b) Estimate A and B by employing nonlinear regression on Equation (7-45) and a single equation that is the sum of Equations (7-43) and (7-44).

(c) Compare the results of the regressions in (a) and (b) using parameter confidence intervals, residual plots, and sums of squares of errors (least-squares summations calculated with both activity coefficients).

(d) Comment on the best choice to correlate this data set between the Margules equations of Problem 3.8 and the Van Laar equations of this problem.

7.10.4   Solution (Suggestions)

The approaches to this problem are similar to those that were used in Problem 3.8 to determine the Margules equations parameters.

Equation (7-45) can be rewritten in a linearized form for the determination of A and B using linear regression as

$$\frac{x_1}{x_1 \ln \gamma_1 + x_2 \ln \gamma_2} = \frac{1}{A} + \frac{1}{B}\frac{x_1}{x_2} \tag{7-46}$$

 The POLYMATH problem data file is found in directory Chapter 3 with file named **P3-08A.POL**.

7.11 VAPOR LIQUID EQUILIBRIUM DATA FROM TOTAL PRESSURE MEASUREMENTS I

7.11.1   Concepts Demonstrated

Calculation of vapor composition from liquid composition and total pressure data using the Gibbs-Duhem equation, and consistency testing for activity coefficients.

7.11.2   Numerical Methods Utilized

Differentiation and integration of tabular data, solution of ordinary differential equations, and use of l'Hôpital's rule for functions undefined at the initial point of the solution interval.

7.11.3   Problem Statement

Data of liquid composition versus total pressure for the system benzene (1) and acetic acid (2) are presented in Table 7–14.

(a)   Calculate the composition of the vapor phase as a function of liquid phase composition using the Gibbs-Duhem equation.

(b)   Calculate the activity coefficients for benzene and acetic acid in the liquid phase.

(c)   Check the consistency of the activity coefficients using the Gibbs-Duhem equation.

Table 7–14 Total Pressure Data at 50°C for the System Benzene (1) Acetic Acid (2)[a]

x_1	P (mm Hg)	x_1	P (mm Hg)
0.0	57.52	0.8286	250.20
0.0069	58.2	0.8862	259.00
0.1565	126.00	0.9165	261.11
0.3396	175.30	0.9561	264.45
0.4666	189.50	0.9840	266.53
0.6004	224.30	1.0	271.00
0.7021	236.00'		

[a]*International Critical Tables*, 1st ed., Vol. III, McGraw Hill, New York, 1928, p. 287.

Additional Information and Data

Vapor liquid equilibrium data for binary systems are very important for computing multicomponent and multiphase equilibrium. A simple and economical experimental method for obtaining such data is to measure the total vapor pressure of the binary mixture as a function of composition at constant temperature. The total pressure and liquid composition data can then be used to calculate the vapor phase composition and activity coefficients in the liquid phase.

The Gibbs-Duhem equation is used as the basis for calculating the vapor phase composition and the activity coefficients. This equation can be written for a binary mixture as

$$x_1\left(\frac{\partial \ln \gamma_1}{\partial x_1}\right)_{P,\,T} + x_2\left(\frac{\partial \ln \gamma_2}{\partial x_1}\right)_{P,\,T} = 0 \tag{7-47}$$

where x_i is the mole fraction of the component i in the liquid phase and γ_i is the liquid phase activity coefficient of component i.

The activity coefficients are defined as

$$\gamma_i = \frac{y_i P}{x_i P_i^v} \tag{7-48}$$

where y_i is the mole fraction of component i in the vapor phase, P is the total pressure, and P_i^v is the vapor pressure of component i.

Introduction of Equation (7-48) into Equation (7-47) with some manipulations (Balzhiser et al., 448)[9] leads to

$$\frac{dy_1}{dx_1} = \frac{y_1(1-y_1)}{y_1 - x_1}\frac{d \ln P}{dx_1} \tag{7-49}$$

If P versus x data are available, Equation (7-49) can be used to calculate y_1 versus x_1. This calculation requires an expression for the derivative $d \ln P/dx_1$ as a function of x_1. The generation of y_1 then requires the numerical solution of Equation (7-49) integrated from $x_1 = 0$, where $y_1 = 0$ to $x_1 = 1.0$.

7.11.4   Solution

For this problem, the POLYMATH *Polynomial, Multiple Linear, and Nonlinear Regression Program* can be used to obtain a continuous representation of the total pressure versus composition data from Table 7–14. The third-order polynomial that results is given by

$$P = 57.6218 + 463.137x_1 - 419.736x_1^2 + 169.871x_1^3 \tag{7-50}$$

This is the highest-degree polynomial in which all the coefficients are significantly different from zero. Consequently, this polynomial can be used to represent the variation of P as a function of x_1.

The POLYMATH program solution file for Table 7–14 is found in directory Chapter 7 with file named **P7-11A.POL**.

Application of Equation (7-49) requires values for $d\ln P/dx_1$, and these can be obtained from Equation (7-50) as

$$\frac{d\ln(P)}{dx_1} = \frac{1}{P}\frac{dP}{dx_1} = \frac{1}{P}[463.137 - 2(419.736)x_1 + 3(169.871)x_1^2] \tag{7-51}$$

The initial condition on the integration of Equation (7-49) poses a difficulty because the equation is undefined when $x_1 = 0$ and $y_1 = 0$. The application of l'Hôpital's rule can be used to determine the initial value of the derivative as

$$\frac{dy_1}{dx_1} = 1 + \frac{d\ln(P)}{dx_1}\bigg|_{x_1 = 0} \tag{7-52}$$

The use of a finite difference expression for the left side of Equation (7-52) allows a value of y_1 to be calculated for a small value of x_1. Thus for $x_1 = 0.00001$

$$y_1\bigg|_{x_1 = 0.00001} = \left(1 + \frac{d\ln(P)}{dx_1}\bigg|_{x_1 = 0}\right)10^{-5} = (1 + 8.0375)10^{-5} = 9.0375 \times 10^{-5} \tag{7-53}$$

From these initial values of x_1 and y_1, Equation (7-49) can be integrated, but it becomes undefined once more for $x_1 = 1$ as $y_1 = 1$. The practical solution to this difficulty is to stop the integration when x_1 becomes 0.99999, giving reasonable results.

During the course of the integration, the activity coefficients can be calculated from

$$\gamma_1 = \frac{y_1 P}{x_1 P_1^v} \qquad \text{and} \qquad \gamma_2 = \frac{(1 - y_1)P}{(1 - x_1)P_2^v} \tag{7-54}$$

and P can be evaluated from Equation (7-50). The pure component vapor pressures are known from Table 7–14 to be $P_1^v = 271.00$ mm Hg and $P_2^v = 57.52$ mm Hg.

The Gibbs-Duhem equation can also be used to check the quality and the consistency of the results based on experimental data. For a binary mixture, the Gibbs-Duhem equation requires that the integral given by

$$I = \int_0^1 \ln\frac{\gamma_1}{\gamma_2}dx_1 \tag{7-55}$$

is equal to zero.

Equation (7-55) can be written as

$$\frac{dI}{dx_1} = \ln\frac{\gamma_1}{\gamma_2} \tag{7-56}$$

and this differential equation can be evaluated during the problem solution since γ_1 and γ_2 are calculated during the numerical integration. The initial condition on I is zero.

Thus the problem consists of ordinary differential Equations (7-49) and (7-56) to be solved simultaneously with algebraic Equations (7-50), (7-51), and (7-54).

This set of equations can be entered into the POLYMATH *Simultaneous Differential Equation Solver* as presented in Table 7–15.

Table 7–15 POLYMATH Problem Code (File **P7-11ABC.POL**)

Line	Equation
1	d(I)/d(x1)=ln(gamma1/gamma2)
2	d(y1)/d(x1)=dy1dx1
3	P=57.6218+463.137*x1-419.736*x1^2+169.871*x1^3
4	gamma1=y1*P/(x1*271)
5	gamma2=(1-y1)*P/((1-x1)*57.52)
6	dlnpdx=(463.137-2*419.736*x1+3*169.871*x1^2)/P
7	dy1dx1=y1*(1-y1)*dlnpdx/(y1-x1)
8	x1(0)=1e-05
9	I(0)=0
10	y1(0)=9.029e-05
11	x1(f)=0.999999

The calculated mole fractions and activity coefficient values are summarized in Table 7–16 at approximate equal intervals of x1. Note that these results require separate POLYMATH integrations to the various final values of x1.

Table 7–16 Calculated y1, γ_1, γ_2, and P Values for the Benzene Acetic Acid System

x1	y1	gamma1	gamma2	P
1e–05	9.03e–05	1.920	1.002	57.63
0.1000	0.4785	1.764	1.006	99.91
0.2000	0.6513	1.620	1.022	134.8
0.3000	0.7410	1.489	1.051	163.4
0.4000	0.7968	1.371	1.099	186.6
0.5000	0.8359	1.268	1.172	205.5
0.6000	0.8666	1.178	1.282	221.1
0.7000	0.8934	1.104	1.448	234.4
0.8000	0.9206	1.047	1.701	246.5
0.9000	0.9536	1.010	2.085	258.3
0.999999	0.999999	0.99961	2.3924	270.89

Figure 7–8 shows a plot of the integral given by $I = \int_0^{x_1} \ln\frac{\gamma_1}{\gamma_2} dx_1$ versus x_1 from $x_1 = 0.00001$ up to $x_1 = 0.999999$.

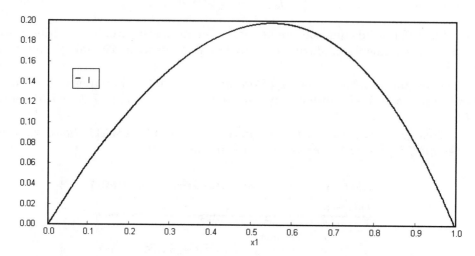

Figure 7–8 Integral of $\ln(\gamma_1/\gamma_2)$ from $x_1 = 0.00001$ to $x_1 = 0.99999$

While the integration cannot be carried out up to exactly $x_1 = 1$ because Equation (7-49) is undefined at this point, it is clearly seen that the integral gets very close to zero (value is -2.2×10^{-3}) for $x_1 = 0.99999$, indicating that the activity coefficient calculations are consistent.

The determination of the polynomial that represents the pressure P has a critical influence on this problem. This has a particularly major effect on the activity coefficients. Numerical treatment of experimental data to obtain derivative information is always challenging.

The POLYMATH problem solution file is located in directory Chapter 7 with file named **P7-11ABC.POL**. A MATLAB file that creates all of Table 7–16 is named **TABLE7_16.m**.

7.12 VAPOR LIQUID EQUILIBRIUM DATA FROM TOTAL PRESSURE MEASUREMENTS II

7.12.1 Concepts Demonstrated

Calculation of vapor composition from liquid composition and total pressure data using the Gibbs-Duhem equation, and consistency testing for activity coefficients.

7.12.2 Numerical Methods Utilized

Differentiation and integration of tabular data, solution of ordinary differential equations, and use of l'Hôpital's rule for functions undefined at the initial point of the solution interval.

7.12.3 Problem Statement

Data from the *International Critical Tables*[10] (286–290) of total pressure versus liquid composition are available for four different systems in the Appendix: water-methyl alcohol (Table C–1), ethyl ether-chloroform (Table C–2), toluene-acetic acid (Table C–3), and chloroform-acetone (Table C–4).

Select one of the four systems with total pressure versus liquid composition data.

(a) Calculate the composition of the vapor phase as function of liquid phase composition using the Gibbs-Duhem equation.
(b) Calculate the activity coefficients for both components in the liquid phase.
(c) Check the consistency of the activity coefficients using the Gibbs-Duhem equation.

7.12.4 Solution (Suggestions)

See Section 7.11.4 for method of solution. Note that Equation (7-49) must be integrated in the direction of increasing pressure (see Van Ness[5]). Note that systems that form azeotropes (such as toluene-acetic acid) must be divided into two parts at the azeotropic composition. The integration should be carried out in the two parts in opposite directions.

7.13 COMPLEX CHEMICAL EQUILIBRIUM

7.13.1   Concepts Demonstrated

Calculation of equilibrium concentrations for a complex reacting system in a constant volume batch reactor and use of the van't Hoff equation to estimate changes in equilibrium constants with temperature.

7.13.2   Numerical Methods Utilized

Effective solution of systems of nonlinear algebraic equations, and useful techniques for examining the possible multiple solutions of such systems.

7.13.3   Problem Statement

Problem 6.6 involves a gas mixture of 50% A and 50% B charged to a constant volume batch reactor in which equilibrium is quickly achieved. The initial total concentration is 3.0 g-mol/dm^3. Three independent reactions are known to occur.

Reaction 1: $A + B \leftrightarrow C + D$ $\qquad K_{C1}(350 \text{ K}) = 1.06$

Reaction 2: $C + B \leftrightarrow X + Y$ $\qquad K_{C2}(350 \text{ K}) = 2.63$

Reaction 3: $A + X \leftrightarrow Z$ $\qquad K_{C3}(350 \text{ K}) = 5.0 \text{ dm}^3/\text{g-mol}$

At 330 K, the equilibrium constants based on concentrations are $K_{C1}(330 \text{ K}) = 0.7$, $K_{C2}(330 \text{ K}) = 4.0$, and $K_{C3}(330 \text{ K}) = 5.0 \text{ dm}^3/\text{g-mol}$.

(a) Calculate the equilibrium concentrations of all reaction components at 330 K.

(b) Repeat part (a) at 370 K.

7.13.4   Solution (Suggestions)

(a) See Section 6.6.4 for solution assistance.

(b) The equilibrium constants at 370 K can be estimated using the van't Hoff equation with the assumption of constant heat of reaction, ΔH_R.

$$\frac{d \ln K_P}{dT} = \frac{\Delta H_R}{RT^2} \qquad\qquad \textbf{(7-57)}$$

The relationship between K_C and K_P is discussed by Fogler.[11]

7.14 REACTION EQUILIBRIUM AT CONSTANT PRESSURE OR CONSTANT VOLUME

7.14.1 Concepts Demonstrated

Calculation of equilibrium concentrations for both constant pressure and constant volume conditions using a combination of elemental balances and equilibrium expressions.

7.14.2 Numerical Methods Utilized

Solution of simultaneous nonlinear and linear algebraic equations.

7.14.3 Problem Statement

At high temperatures and low pressures, hydrogen sulfide and sulfur dioxide undergo the following reversible reactions:

$$H_2S \leftrightarrow H_2 + \frac{1}{2}S_2 \qquad K_{P1} = 0.45 \text{ atm}^{1/2} \qquad \text{(7-58)}$$

$$2H_2S + SO_2 \leftrightarrow 2H_2O + \frac{3}{2}S_2 \qquad K_{P2} = 28.5 \text{ atm}^{1/2} \qquad \text{(7-59)}$$

An initial gas mixture contains 45 mol % H_2S, 25 mol % SO_2, and the balance inert N_2 at a pressure of 1.2 atm.

(a) Calculate the equilibrium mole fractions of all reaction components when the preceding reactions are at equilibrium at a constant pressure of 1.2 atm.

(b) Repeat (a) for equilibrium at constant volume when the initial pressure is 1.2 atm.

7.14.4 Solution (Suggestions)

Material balances on the various species involved in this system (H, S, O, N) can yield linear algebraic equations that can be solved with Equations (7-58) and (7-59). The POLYMATH *Simultaneous Algebraic Equation Solver* can be used to solve all the equations. It is advisable to enter all equations as nonlinear equations for simultaneous solution. Please apply the suggestions given in Section 6.6.4 regarding the expediting of solutions of nonlinear equations.

REFERENCES

1. Kyle, B. G. *Chemical and Process Thermodynamics*, 2nd ed., Englewood Cliffs, NJ: Prentice Hall, 1992.
2. Sandler, S. I. *Chemical and Engineering Thermodynamics*, New York: Wiley, 1989.
3. Modell, M., Reid, C. *Thermodynamics and its Applications*, 2nd ed., Englewood Cliffs, NJ: Prentice Hall, 1983.
4. Smith, J. M., and Van Ness, H. C. *Introduction to Chemical Engineering Thermodynamics*, 3rd ed., New York: McGraw-Hill, 1975.
5. Van Ness, H. C. *AIChE J.* **16** (1), 18 (1970).
6. Edmister, W. C. *Hydrocarb. Process.*, **47** (10), 145–149 (1968)..
7. Walas, S. M. *Phase Equilibrium in Chemical Engineering*, Stoneham, MA: Butterworth, 1985.
8. Henley, E. J., and Rosen, E. M. *Material and Energy Balance Computation.* New York: Wiley, 1969.
9. Balzhiser, R. E., Samuels, M. R., and Eliassen, J. D. *Chemical Engineering Thermodynamics*, Englewood Cliffs, NJ: Prentice Hall, 1972.
10. *International Critical Tables*, 1st ed., Vol III, New York: McGraw-Hill, 1928.
11. Fogler, H. S. *Elements of Chemical Reaction Engineering*, 2nd ed., Englewood Cliffs, NJ: Prentice Hall, 1992

Fluid Mechanics

8.1 LAMINAR FLOW OF A NEWTONIAN FLUID IN A HORIZONTAL PIPE

8.1.1 Concepts Demonstrated

Solution of momentum balance to obtain shear stress and velocity profiles for a Newtonian fluid in laminar flow in a horizontal pipe and comparison of numerical and analytical solutions.

8.1.2 Numerical Methods Utilized

Solution of simultaneous first order ordinary differential equations employing a shooting technique to converge on the desired boundary conditions, use of combined variables, and avoidance of division by zero in calculating expressions.

8.1.3 Problem Statement

An incompressible Newtonian fluid is flowing at steady state inside a horizontal circular pipe at constant temperature. This flow is fully developed so that the velocity profile does not vary in the direction of flow. Figure 8–1 presents a schematic of this laminar flow for the coordinate system where the flow is in the x direction and the inside pipe radius is R in m.

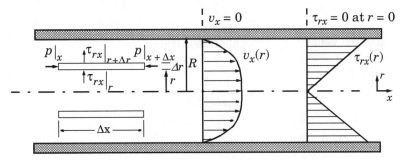

Figure 8–1 Differential Element for Laminar Flow Treatment in a Horizontal Pipe

A shell momentum balance of a control volume within the pipe gives the differential equation

$$\frac{d}{dr}(r\tau_{rx}) = \left(\frac{\Delta p}{L}\right)r \tag{8-1}$$

where r is the radius in m, τ_{rx} is the shear stress at radius r in kg/m·s^2 or Pa, and Δp is the pressure drop in kg/m·s^2 or Pa over length L in m.

For a Newtonian fluid, the shear stress (or momentum flux) is linearly related to the velocity gradient by

$$\tau_{rx} = -\mu\frac{dv_x}{dr} \tag{8-2}$$

where μ = the viscosity in kg/m·s or Pa·s.

The boundary conditions for Equations (8-1) and (8-2) are that

$$\tau_{rx} = 0 \qquad \text{at } r = 0 \tag{8-3}$$

$$v_x = 0 \qquad \text{at } r = R \tag{8-4}$$

The analytical solution with the two boundary conditions yields

$$\tau_{rx} = \left(\frac{\Delta p}{2L}\right)r \tag{8-5}$$

$$v_x = \frac{\Delta p}{4\mu L}R^2\left[1 - \left(\frac{r}{R}\right)^2\right] \tag{8-6}$$

where $\Delta p = p_0 - p_L$.

The average velocity $v_{x,\text{av}}$ is calculated from

$$v_{x,\text{av}} = \frac{1}{\pi R^2}\int_0^R v_x 2\pi r\, dr \tag{8-7}$$

to give the analytical solution

$$v_{x,\text{av}} = \frac{(p_0 - p_L)R^2}{8\mu L} = \frac{(p_0 - p_L)D^2}{32\mu L} \tag{8-8}$$

(a) Numerically solve Equations (8-1) and (8-2) with the boundary conditions given by Equations (8-3) and (8-4) for water at 25 °C with μ =
8.937×10^{-4} kg/m·s, Δp = 500 Pa, L = 10 m, and R = 0.009295 m. This
solution should utilize an ODE solver with a shooting technique and
should employ some technique for converging on the boundary condition given by Equation (8-4).
(b) Compare the calculated shear stress and velocity profiles with the analytical solutions given by Equations (8-5) and (8-6).
(c) Modify your solution to part (a) to include calculation of the average
velocity given by Equation (8-7) and compare your solution with the
analytical solution of Equation (8-8).

8.1.4 Solution

This set of differential equations is similar to the set solved in Problem 6.5, and
the same techniques can be used.

(a) The differential equations to be solved in this problem consist of Equations (8-1) and (8-2). Equation (8-2) can be rearranged as

$$\frac{dv_x}{dr} = -\frac{\tau_{rx}}{\mu} \tag{8-9}$$

There is no need to use the chain rule from differential calculus to work on
the derivative in Equation (8-1) to separate out the derivative $d\tau_{rx}/dr$. Instead, in
this case, and in most problems, it is highly recommended that the combined
variable such as $r\tau_{rx}$ should be retained, and an algebraic equation should be
used to calculate the individual variables. In this case the algebraic equation is

$$\tau_{rx} = \frac{r\tau_{rx}}{r} \tag{8-10}$$

The simultaneous solution of the two ordinary differential equations, Equations (8-1) and (8-10), requires that an initial condition for v_x be assumed at $r = 0$
so that the boundary condition given by $v_x = 0$ will be satisfied at $r = R$. The
shooting method discussed in Section 6.5.4 will be used to determine this initial
condition. A first trial initial condition for v_x, say $v_{x0} = 2.0$, can be used to integrate the equations and calculate the error in the boundary condition for v_x at
$r = R$. This error can be defined as ε, which is the difference between the numerical solution and the desired final value of v_x at $r = R$.

$$\varepsilon(v_{x0}) = v_{x,\,\text{calc}} - v_{x,\,\text{desired}} \tag{8-11}$$

In this problem, $v_{x,\text{desired}} = 0$ indicates no velocity at the wall of the pipe.

The equation set as entered into the POLYMATH *Simultaneous Differential Equation Solver* for the first step of the shooting method is given in Table 8–1.

Table 8–1 POLYMATH Problem Code (File **P8-01A.POL**)

Line	Equation
1	d(Vx)/d(r)=-TAUrx/mu
2	d(rTAUrx)/d(r)=deltaP*r/L
3	deltaP=500
4	L=10
5	TAUrx=if(r>0)then(rTAUrx/r)else(0)
6	mu=8.937e-4
7	R=.009295
8	err=Vx-0
9	r(0)=0
10	Vx(0)=1.20842
11	rTAUrx(0)=0
12	r(f)=0.009295

The numerical solution of the preceding equation set cannot be carried out because of the division by r in the calculation of TAUrx, because r is zero when the integration is initiated.

Division by Zero

A simple way to avoid division by zero is to use the "if … then … else … " capability in POLYMATH. In this case the algebraic equation for calculation of τ_{rx} in POYMATH can be modified to

TAUrx=if(r>0)then(rTAUrx/r)else(0)

and the numerical solution of the initial trial solution can be achieved.

Boundary Condition Convergence

Once the initial trial solution has been obtained, additional integrations can be made in order to minimize the error as calculated by Equation (8-11). This can utilize either trial and error or some simple convergence logic, such as the secant or false position methods, which are discussed in Problem 6.4.

The final solution of this two-point boundary value problem for part (a) is presented in Table 8–2.

Table 8–2 Partial Results for Selected Variables

Variable	Initial Value	Maximum Value	Minimum Value	Final Value
r	0	0.009295	0	0.009295
rTAUrx	0	0.00215993	0	0.00215993
Vx	1.20842	1.20842	2.39622e-06	2.39622e-06

Table 8–2 (Continued) Partial Results for Selected Variables

Variable	Initial Value	Maximum Value	Minimum Value	Final Value
TAUrx	0	0.232375	0	0.232375
mu	0.0008937	0.0008937	0.0008937	0.0008937
deltaP	500	500	500	500
L	10	10	10	10
err	1.20842	1.20842	2.39622e-06	2.39622e-06
R	0.009295	0.009295	0.009295	0.009295

The POLYMATH problem solution file for part (a) is found in directory Chapter 8 with file named **P8-01A.POL**.

(b) The analytical solutions for the shear stress and velocity profiles given in Equations (8-5) and (8-6) can be entered directly into the POLYMATH equation set for evaluation as

TAUrxANAL=(deltaP/(2*L))*r
VxANAL=(deltaP*R^2/(4*mu*L))*(1-(r/R)^2)

Evaluation of the preceding equations can easily be compared with the numerical solution by requesting POLYMATH to print a table for results at 10 intervals during the integration. This is shown in Table 8–3, where the numerical and analytical solutions are shown to agree to approximately six significant figures.

Table 8–3 Comparisons of Numerical and Analytical Solutions for τ_{rx} and v_x (File **TABLE8_3.m**)

r	TAUrx	TAUrxANAL	Vx	VxANAL
0	0	0	1.20842	1.2084176
0.0009295	0.0232375	0.0232375	1.1963358	1.1963334
0.001859	0.046475	0.046475	1.1600833	1.1600809
0.0027885	0.0697125	0.0697125	1.0996624	1.09966
0.003718	0.09295	0.09295	1.0150732	1.0150708
0.0046475	0.1161875	0.1161875	0.9063156	0.9063132
0.005577	0.139425	0.139425	0.77338966	0.77338727
0.0065065	0.1626625	0.1626625	0.61629537	0.61629298
0.007436	0.1859	0.1859	0.43503273	0.43503034
0.0083655	0.2091375	0.2091375	0.22960174	0.22959934
0.009295	0.232375	0.232375	2.396218e-06	0

(c) The average velocity in the pipe is calculated from Equation (8-7) which is an integral equation. This equation can be differentiated with respect to r to yield the differential equation

$$\frac{d}{dr}(v_{x,\,av}) = \frac{v_x 2r}{R^2}$$

(8-12)

whose initial condition is zero. This equation can be entered as another differential equation in the POLYMATH equation set and integrated to $r = R$. Note that $v_{x,av}$ is determined only at the end of the integration. The analytical solution for average velocity is given by Equation (8-8), which can be calculated during the numerical solution. The additional statements for POLYMATH are

```
d(Vxav)/d(r)=Vx*2*r/R^2
VxavANAL=deltaP*R^2/(8*mu*L)
```

The numerical and analytical solutions for the average v_x are 0.6042112 m/s and 0.6042088 m/s, respectively, which are in good agreement to five significant figures.

 The POLYMATH problem solution file for this complete problem is found in directory Chapter 8 with file named **P8-01ABC.POL**. This directory also contains a MATLAB file named **TABLE8_3.m** that is programmed to generate Table 8–3.

8.2 LAMINAR FLOW OF NON-NEWTONIAN FLUIDS IN A HORIZONTAL PIPE

8.2.1   Concepts Demonstrated

Solution of momentum balance to obtain shear stress and velocity profiles for power law fluids in laminar flow in a horizontal pipe, and comparison of numerical and analytical solutions.

8.2.2   Numerical Methods Utilized

Solution of simultaneous first order ordinary differential equations employing a shooting technique to converge on the desired boundary conditions and use of combined variables in problem solution.

8.2.3   Problem Statement

The shear stress for a number of non-Newtonian fluids can be described by

$$\tau_{rx} = -K\left|\frac{dv_x}{dr}\right|^{n-1}\frac{dv_x}{dr} \tag{8-13}$$

where parameter K has units of $\mathrm{N \cdot s^n/m^2}$ and exponent n is the flow index. For $n < 1$, the fluid is *pseudoplastic* and for $n > 1$, the fluid is *dilatant*. When $n = 1$, the fluid is Newtonian.

Application of Equation (8-13) in numerical solutions typically requires the differential equation for velocity.

$$\frac{dv_x}{dr} = -\left(\frac{\tau_{rx}}{K}\right)^{1/n} \quad (\text{if } \tau_{rx} > 0) \tag{8-14}$$

$$\frac{dv_x}{dr} = \left(\frac{-\tau_{rx}}{K}\right)^{1/n} \quad (\text{if } \tau_{rx} \leq 0) \tag{8-15}$$

The analytical velocity profile for these fluids flowing in a horizontal cylindrical pipe is given by

$$v_x = \frac{n}{n+1}\left(\frac{\Delta p}{2KL}\right)^{1/n}R^{(n+1)/n}\left[1 - \left(\frac{r}{R}\right)^{(n+1)/n}\right] \tag{8-16}$$

(a)   A dilatant fluid is flowing in a horizontal pipe where $K = 1.0 \times 10^{-6}$, $n = 2$, $\Delta p = 100$ Pa, $L = 10$ m, and $R = 0.009295$ m. The shear stress τ_{rx} and velocity profile for v_x should be plotted versus the radius r of the pipe. Calculate the average velocity in the pipe $v_{x,\text{av}}$.

(b)   Compare the calculated velocity profile with the analytical solution given by Equation (8-16).

(c)   Repeat part (a) for a pseudoplastic fluid with $K = 0.01$ and $n = 0.5$.

(d)   Repeat part (b) for the pseudoplastic fluid of part (c).

8.2.4   Solution (Suggestions)

(a) The numerical solution requires ordinary differential equations for both τ_{rx} and v_x. The equation for τ_{rx} in the pipe is obtained from a shell momentum balance in the pipe. The result is Equation (8-1), which is independent of the fluid properties. Note that use of the combined variable $r\tau_{rx}$ is strongly suggested in Problem 8.1. This requires a separate algebraic expression, Equation (8-10), to yield τ_{rx} during the numerical solution. The POLYMATH "if ... then ... else ... " statement can be used to select the correct option given by Equations (8-14) and (8-15). Convergence to the boundary conditions is discussed in Section 8.1.4.

8.3 VERTICAL LAMINAR FLOW OF A LIQUID FILM

8.3.1   Concepts Demonstrated

Shear stress, velocity profiles, and average velocity calculations for laminar flow of Newtonian and Bingham fluids down a vertical surface.

8.3.2   Numerical Methods Utilized

Solution of simultaneous first order ordinary differential equations employing a shooting technique to converge on the desired boundary conditions.

8.3.3   Problem Statement

A liquid is flowing down a vertical surface as a fully established film where the velocity profile in the film does not vary in the direction of flow. This is illustrated in Figure 8–2. This situation is discussed in more detail by Geankoplis[1] and Bird et al.[2]

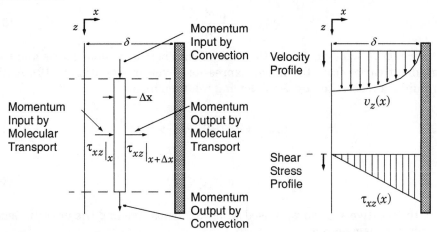

Figure 8–2 Differential Element for Vertical Flow of a Liquid Film

A shell momentum balance yields a differential equation for the shear stress (momentum flux)

$$\frac{d\tau_{xz}}{dx} = \rho g \tag{8-17}$$

where the boundary condition for the free surface is

$$\tau_{xz} = 0 \quad \text{at } x = 0 \tag{8-18}$$

For a Newtonian fluid, the shear stress is related to the velocity gradient by

$$\tau_{xz} = -\mu \frac{dv_z}{dx} \qquad \text{(8-19)}$$

where the boundary condition at the wall is

$$v_z = 0 \qquad \text{at } x = \delta \qquad \text{(8-20)}$$

The average velocity in the film can be calculated from

$$v_{z,\,av} = \frac{1}{\delta} \int_0^{\delta} v_z \, dx \qquad \text{(8-21)}$$

The analytical solution for the velocity distribution is

$$v_z = \frac{\rho g \delta^2}{2\mu} \left[1 - \left(\frac{x}{\delta}\right)^2 \right] \qquad \text{(8-22)}$$

and the analytical solution for the average velocity is

$$v_{z,\,av} = \frac{\rho g \delta^2}{3\mu} \qquad \text{(8-23)}$$

For non-Newtonian fluids, the relationship between shear stress and velocity gradient varies from the linear expression given in Equation (8-19). A Bingham fluid is represented by the following relationships:

$$\tau_{xz} = -\mu_0 \frac{dv_z}{dx} \pm \tau_0 \qquad (\text{if } |\tau_{xz}| > \tau_0) \qquad \text{(8-24)}$$

$$\frac{dv_z}{dx} = 0 \qquad (\text{if } |\tau_{xz}| \le \tau_0) \qquad \text{(8-25)}$$

where the positive sign on τ_0 is used when τ_{xz} is positive and the negative sign is used when τ_{xz} is negative.

For numerical solutions where the differential equation for velocity is needed, Equations (8-24) and (8-25) can be expressed as

$$\frac{dv_z}{dx} = \frac{(\tau_0 - \tau_{xz})}{\mu_0} \qquad \text{if } \tau_{xz} > \tau_0$$

$$\frac{dv_z}{dx} = 0 \qquad \text{if } |\tau_{xz}| \le \tau_0 \qquad \textbf{(8-26)}$$

$$\frac{dv_z}{dx} = \frac{(\tau_0 + \tau_{xz})}{-\mu_0} \qquad \text{if } \tau_{xz} < (-\tau_0)$$

with constants τ_0 and μ_0 having known values for the particular fluid.

(a) Numerically solve Equations (8-17) and (8-19) with the boundary conditions given by Equations (8-18) and (8-20) for an oil that is a Newtonian fluid with $\mu = 0.15$ kg/m·s and $\rho = 840.0$ kg/m^3. The film thickness is $\delta = 0.002$ m. The shear stress τ_{xz} and velocity profile for v_z should be plotted versus the location x in the film. Calculate the average velocity in the film $v_{z,av}$.

(b) Compare the calculated velocity profile with the analytical solutions given by Equation (8-22).

(c) Repeat part (a) for a Bingham plastic fluid by solving Equations (8-17) and (8-26), where $\tau_0 = 6.0$ kg/m·s^2, $\mu_0 = 0.20$ kg/m·s, and $\rho = 800$ kg/m^3. What is unique about the velocity profile for this fluid?

8.3.4 Solution (Suggestions)

(a) and (b) Similar differential equations are solved in Problem 6.5.

(c) Equation (8-26) along with the logic as discussed for a Bingham plastic fluid must be used instead of Equation (8-19).

8.4 LAMINAR FLOW OF NON-NEWTONIAN FLUIDS IN A HORIZONTAL ANNULUS

8.4.1   Concepts Demonstrated

Shear stress, velocity profiles, and average velocity calculations for laminar flow of Newtonian, dilatant, and pseudoplastic fluids in a horizontal annulus.

8.4.2   Numerical Methods Utilized

Solution of simultaneous first-order ordinary differential equations employing a shooting technique to converge on the desired boundary conditions.

8.4.3   Problem Statement

An incompressible fluid is flowing within an annulus between two concentric horizontal pipes, as shown in Figure 8–3 where the flow is fully developed. The shell momentum balance discussed in Section 8.1.3 applies.

$$\frac{d}{dr}(r\tau_{rx}) = \left(\frac{\Delta p}{L}\right)r \tag{8-27}$$

The particular fluid will determine the relationship between the shear rate and the shear stress. For a Newtonian fluid, this can be written as

$$\frac{dv_x}{dr} = -\frac{\tau_{rx}}{\mu} \tag{8-28}$$

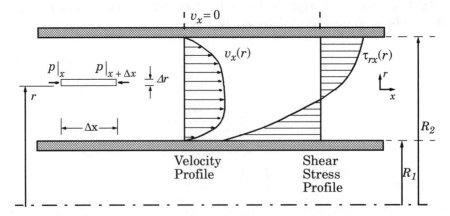

Figure 8–3 Differential Element for Laminar Flow in a Horizontal Annulus

Equations (8-27) and (8-28) describe the flow in the annulus, but the boundary conditions are that the velocity at each wall is zero. Thus

$$v_x = 0 \quad \text{at } r = R_1 \tag{8-29}$$

$$v_x = 0 \quad \text{at } r = R_2 \tag{8-30}$$

An analytical solution for the velocity profile in the annulus with a Newtonian fluid is

$$v_x = \frac{\Delta p}{4\mu L}\left[R_2^2 - r^2 + \frac{(R_2^2 - R_1^2)}{\ln(R_2/R_1)}\left(\ln\frac{r}{R_2}\right)\right] \tag{8-31}$$

where the pressure drop is given by $\Delta p = p_0 - p_L$. The average velocity $v_{x,\text{av}}$ is calculated from

$$v_{x,\text{av}} = \frac{1}{\pi(R_2^2 - R_1^2)}\int_{R_1}^{R_2} v_x 2\pi r \, dr \tag{8-32}$$

to give the analytical solution

$$v_{x,\text{av}} = \frac{\Delta p}{8\mu L}\left[R_1^2 + R_2^2 - \frac{(R_2^2 - R_1^2)}{\ln(R_2/R_1)}\right] \tag{8-33}$$

For a Newtonian fluid such as water, Equation (8-9) relates the derivative of the velocity to shear stress. A number of non-Newtonian fluids can be described by

$$\tau_{rx} = -K\left|\frac{dv_x}{dr}\right|^{n-1}\frac{dv_x}{dr} \tag{8-34}$$

which is discussed in Section 8.2.3 for both pseudoplastic and dilatant fluids. Relationships between the velocity derivative and the shear stress are given in Equations (8-14) and (8-15).

(a)   Numerically solve Equations (8-27) and (8-28) with the boundary conditions given by Equations (8-29) and (8-30) for water, a Newtonian fluid, at 25°C with $\mu = 8.937 \times 10^{-4}$ kg/m·s, $\Delta p = 100$ Pa, $L = 10$ m, $R_1 = 0.02223$ m, and $R_2 = 0.03129$ m. The shear stress τ_{rx} and velocity v_x should be plotted versus the radius r in the annulus from R_1 to R_2. Calculate the average velocity in the annulus from the differential equation form of Equation (8-32), as detailed in Problem 8.1.

(b)   Compare the calculated velocity profile with the analytical solutions given by Equation (8-31). Compare the calculated average velocity with the analytical solution given by Equation (8-32).

(c)   Repeat part (a) for a dilatant fluid with $K = 1.0 \times 10^{-6}$ and $n = 2$.

(d)   Repeat part (a) for a pseudoplastic fluid with $K = 0.01$ and $n = 0.5$.

8.4.4   Solution (Suggestion)

(a)–(c) A good initial estimate for τ_{rx} at R_1 is −0.33 for all fluids.

8.5 TEMPERATURE DEPENDENCY OF DENSITY AND VISCOSITY OF VARIOUS LIQUIDS

8.5.1   Concepts Demonstrated

General correlation of density and viscosity for liquids as a function of temperature in both English and SI units.

8.5.2   Numerical Methods Utilized

Regression of experimental data to different functional forms.

8.5.3   Problem Statement

Fluid flow calculations typically require correlations of the density and viscosity of liquids at various temperatures. In this section, various equations representing these properties as a function of temperature will be determined for water as a representative case.

Tables D–1 to D–2 in Appendix D present density and absolute viscosity of various liquids as a function of temperature in both English and SI units.

(a)   Determine appropriate correlations for the temperature dependency of density and viscosity of liquid water in both English and SI units using the data of Tables D–1 and D–2.

(b)   Select one of the components in Table D–3 and provide similar correlations that give the results for density and viscosity in SI units.

8.5.4   Solution (Partial)

The techniques for finding the most appropriate correlations are discussed in Chapter 2. Using those methods and the data in Table D–2 and Table D–2 yields the following results for water.

(a) In English units, the data of Table D–2 for water yield

$$\rho = 62.122 + 0.0122T - 1.54 \times 10^{-4}T^2 + 2.65 \times 10^{-7}T^3 - 2.24 \times 10^{-10}T^4 \qquad \textbf{(8-35)}$$

$$\ln \mu = -11.0318 + \frac{1057.51}{T + 214.624} \qquad \textbf{(8-36)}$$

where T is in °F, ρ is in lb_m/ft^3, and μ is in $lb_m/ft \cdot s$.

In SI units, the data of Table D–1 for water yield

$$\rho = 46.048 + 9.418T - 0.0329T^2 + 4.882\times10^{-5}T^3 - 2.895\times10^{-8}T^4 \qquad \textbf{(8-37)}$$

$$\ln \mu = -10.547 + \frac{541.69}{T - 144.53} \qquad \textbf{(8-38)}$$

where T is in K, ρ is in kg/m^3, and μ is in Pa·s or kg/m·s.

(b) Table D–3 provides only data in English units. Use appropriate conversion factors to first convert the data to SI units and then regress equations similar to Equations (8-35) to (8-38) in SI units.

The POLYMATH data files for Tables D–1, D–2 and D–3 are found directory Tables with files named **D-01.POL**, **D-02.POL**, and **D-03.POL**.

8.6 TERMINAL VELOCITY OF FALLING PARTICLES

8.6.1 Concepts Demonstrated

Calculation of terminal velocity of solid particles falling in liquid or gas under gravity and additional forces.

8.6.2 Numerical Methods Utilized

Solution of single nonlinear equation.

8.6.3 Problem Statement

(a) Calculate the terminal velocity for 65 mesh particles of coal $\rho_p = 1800$ kg/m^3 falling in one of the liquids listed in Tables D–2 and D–3 at 25°C.

(b) Estimate the terminal velocity of the coal particles in a centrifugal separator where the acceleration is 30g.

Additional Information and Data

Assuming that the coal particles are spherical, a force balance on a particle yields

$$v_t = \sqrt{\frac{4g(\rho_p - \rho)D_p}{3C_D\rho}} \tag{8-39}$$

where v_t is the terminal velocity in m/s, g is the acceleration of gravity given by $g = 9.80665$ m/s^2, ρ_p is the particle's density in kg/m^3, ρ is the fluid density in kg/m^3, D_p is the diameter of the spherical particle in m, and C_D is a dimensionless drag coefficient.

The drag coefficient on a spherical particle at terminal velocity varies with the Reynolds number (Re) as follows (Perry et al.[3], 5-63 and 5-64):

$$C_D = 24/Re \quad \text{for} \quad Re < 0.1 \tag{8-40}$$

$$C_D = (24/Re)(1 + 0.14Re^{0.7}) \quad \text{for} \quad 0.1 \le Re \le 1{,}000 \tag{8-41}$$

$$C_D = 0.44 \quad \text{for} \quad 1{,}000 < Re \le 350{,}000 \tag{8-42}$$

$$C_D = 0.19 - 8 \times 10^4/Re \quad \text{for} \quad 350{,}000 < Re \tag{8-43}$$

where $Re = D_p v_t \rho/\mu$ and μ is the viscosity in Pa·s or kg/m·s.

8.6.4 Solution for Water

(a) For conditions similar to those of this problem, the Reynolds number will not exceed 1,000 so that only Equations (8-40) and (8-41) need to be applied. The logic that selects the proper equation based on the value of Re is shown in

Equation (8-44) using the "if … then … else … " capability within the POLY-MATH *Simultaneous Algebraic Equation Solver.*

$$C_D = \text{if } (Re < 0.1) \text{ then } (24/Re) \text{ else } (24\langle 1 + 0.14Re^{0.7}\rangle/Re) \qquad \textbf{(8-44)}$$

The numerical values of the additional variables are $D_p = 0.208$ mm (p. 941 in McCabe et al.[4]) and for water the $\rho = 994.6$ kg/m^3 and $\mu = 8.931 \times 10^{-4}$ kg/m·s, as calculated using Equations (8-37) and (8-38) at $T = 298.15$ K.

Equation (8-39) should be rearranged in order to avoid possible division by zero and negative square roots as it is entered into the form of a nonlinear equation for POLYMATH.

$$f(v_t) = v_t^2(3C_D\rho) - 4g(\rho_p - \rho)D_p \qquad \textbf{(8-45)}$$

The equation set for POLYMATH is given in Table 8–4.

Table 8–4 POLYMATH Problem Code (File **P8-06A.POL**)

Line	Equation
1	f(vt)=vt^2*(3*CD*rho)-4*g*(rhop-rho)*Dp
2	rho=994.6
3	g=9.80665
4	rhop=1800
5	Dp=0.208e-3
6	vis=8.931e-4
7	Re=Dp*vt*rho/vis
8	CD=if(Re<0.1)then(24/Re)else(24*(1+0.14*Re^0.7)/Re)
9	vt(min)=0.0001
10	vt(max)=0.05

Specification of POLYMATH variables vt(min) = 0.0001 and vt(max) = 0.05 leads to the results summarized in Table 8–5.

Table 8–5 Terminal Velocity Solution

Variable	Value	$f(\ )$
vt	0.0157816	−8.882e−16
Re	3.65564	
CD	8.84266	

(b) The terminal velocity in the centrifugal separator can be calculated by replacing the g in Equation (8-45) by $30g$. Introduction of this change to the equation set gives the results that $v_t = 0.2060$ m/s, $Re = 47.72$, and $C_D = 1.557$. Only the viscosity and the density of the liquid should be changed to carry out the calculations for other fluids.

The POLYMATH problem solution files are found in directory Chapter 8 and named **P8-06A.POL** and **P8-06B.POL**. This problem is also solved with Excel, Maple, MathCAD, MATLAB, Mathematica, and POLYMATH as Problem 5 in the Set of Ten Problems discussed on the book web site.

8.7 COMPARISON OF FRICTION FACTOR CORRELATIONS FOR TURBULENT PIPE FLOW

8.7.1 Concepts Demonstrated

Various correlations for calculating friction factors of turbulent fluid flow in pipes.

8.7.2 Numerical Methods Utilized

Solution of nonlinear algebraic equations.

8.7.3 Problem Statement

The Fanning friction factor can be used to calculate the friction loss for isothermal liquid flow. This friction factor is dependent upon the Reynolds number and the surface roughness factor ε in m. The Reynolds number is $Re = Dv\rho/\mu$, where ρ is the density of the fluid in kg/m^3 and μ is the viscosity in kg/m·s. A widely used chart gives experimental values of the friction factor as a function of Re and ε. (See, for example, Geankoplis[1] or Perry et al.[3])

There are also a number of implicit and explicit correlations for the friction factor in turbulent flow where $Re > 3000$. For hydraulically smooth pipes where $\varepsilon/D = 0$, an implicit equation is

$$\frac{1}{\sqrt{f_F}} = 4.0\log(Re\sqrt{f_F}) - 0.4 \quad \text{(Nikuradse[5] equation)} \qquad \textbf{(8-46)}$$

and an explicit equation is

$$f_F = 0.0791 Re^{-1/4} \quad \text{(Blasius equation)} \qquad \textbf{(8-47)}$$

For rough pipes, where surface roughness characterized by ε/D ratios is important, a widely used implicit equation is

$$\frac{1}{\sqrt{f_F}} = -4.0\log\left(\frac{\varepsilon}{D} + \frac{4.67}{Re\sqrt{f_F}}\right) + 2.28 \quad \text{(Colebrook and White[6] equation)} \qquad \textbf{(8-48)}$$

and explicit equations are given by

$$f_F = \frac{1}{16\left\{\log\left[\dfrac{\varepsilon/D}{3.7} - \dfrac{5.02}{Re}\log\left(\dfrac{\varepsilon/D}{3.7} + \dfrac{14.5}{Re}\right)\right]\right\}^2} \quad \text{(Shacham[7] equation)} \qquad \textbf{(8-49)}$$

and

$$f_F = \left\{-3.6 \ \log\left[\frac{6.9}{Re} + \left(\frac{\varepsilon/D}{3.7}\right)^{10/9}\right]\right\}^{-2} \qquad \text{(Haaland[8] equation)} \qquad \textbf{(8-50)}$$

(a) Summarize calculated friction factors from Equations (8-46) through (8-50) at $Re = 10^4$ and 10^7 for smooth pipes where $\varepsilon/D = 0$. Comment on the differences between the implicit and explicit equations.

(b) Summarize calculated friction factors from Equations (8-48) through (8-50) at $Re = 10^4$ and 10^7 for rough pipes where $\varepsilon/D = 0.0001$ and 0.01. Comment on the differences between the implicit and explicit equations.

(c) Compare the results of (a) and (b) with the friction factors obtained from a general graphical presentation of the Fanning friction factor.

8.7.4   Solution (Suggestions)

All of the explicit and implicit equations can be solved simultaneously for a particular set of conditions by using the POLYMATH *Simultaneous Algebraic Equation Solver*. Results can easily be compared in a table.

8.8 CALCULATIONS INVOLVING FRICTION FACTORS FOR FLOW IN PIPES

8.8.1 Concepts Demonstrated

Calculation of friction factor and pressure drop for both laminar and turbulent liquid flow in pipes.

8.8.2 Numerical Methods Utilized

Solution of a system of simultaneous nonlinear algebraic equations.

8.8.3 Problem Statement

The Fanning friction factor can be used to calculate the friction loss for isothermal liquid flow in uniform circular pipes by the formula

$$F_f = 2f_F \frac{\Delta L v^2}{D} \tag{8-51}$$

where F_f is the friction loss in N·m/kg or J/kg, f_F is the Fanning friction factor (dimensionless), ΔL is the length of the pipe in m, v is the fluid velocity in m/s, and D is the pipe diameter in m.

The friction factor can be used to predict the pressure drop due to friction loss from

$$\Delta p_f = \rho F_f = 2f_F \rho \frac{\Delta L v^2}{D} \tag{8-52}$$

where Δp_f is the pressure drop in Pa and ρ is the density in kg/m^3. The average fluid velocity can be calculated from

$$v = \frac{q}{\pi D^2/4} \tag{8-53}$$

where q is the volumetric flow rate in m^3/s.

If the flow is in the laminar region when $Re < 2100$, the Fanning friction factor can be calculated from

$$f_F = \frac{16}{Re} \tag{8-54}$$

Otherwise for turbulent flow in a smooth tube when $Re > 2100$, the Nikuradse[5] correlation represented by Equation (8-46) can be utilized to calculate the Fanning friction factor.

(a)   A heat exchanger is required that will be able to handle $Q = 2.5$ liter/s
of water through a smooth pipe with an equivalent length of $L = 100$ m.
The maximum total pressure drop is to be $\Delta P = 103$ kPa at a temperature of 25°C. Calculate the diameter D of the pipe required for this
application and select the appropriate tube from Table D–4.
(b)   What will be the pressure drop for the pipe selected in part (a) if the
temperature drops to 5°C?
(c)   Repeat parts (a) and (b) for another liquid from Table D–3.

8.8.4   Solution (Partial)

(a) The calculations in this problem require a friction factor correlation,
which in turn depends upon the Reynolds number Re. Both correlations can be
combined in the POLYMATH *Simultaneous Algebraic Equation Solver* by using
the "if … then … else … " statement in the nonlinear equation, which represents
both laminar and turbulent flow.

For turbulent flow, the Nikuradse correlation of Equation (8-46) can be used
and rearranged as

$$f_F = \frac{1}{[4.0 \log{(Re\,\sqrt{f_F})} - 0.4]^2} \qquad \text{(8-55)}$$

In POLYMATH, this nonlinear equation can be entered as

$$f(fF)=fF-1/(4*\log(Re*sqrt(fF))-0.4)^2 \qquad \text{(8-56)}$$

For laminar flow, Equation (8-54), which is a linear equation, can also be
expressed in POLYMATH as a nonlinear equation as

$$f(fF)=fF-16/R \qquad \text{(8-57)}$$

Equations (8-56) and (8-57) can be used in a single POLYMATH "if … then
… else … " expression to achieve a solution for the friction factor which depends
upon the value of the Reynolds number Re.

Equation (8-52) is also a nonlinear equation that can be entered into POLY-
MATH as a function of the diameter D. Thus two nonlinear equations can be
solved simultaneously by entering the equation set shown in Table 8–6 where
the density and viscosity of water at 25°C are calculated from equations developed in Problem 8.5.

Table 8–6 POLYMATH Problem Code (File **P8-08A.POL**)

Line	Equation
1	(D)=dp-2*fF*rho*v*v*L/D
2	f(fF)=if(Re<2100)then(fF-16/Re)else(fF-1/(4*log(Re*sqrt(fF))-0.4)^2)
3	dp=103000

Line	Equation
4	L=100
5	T=25+273.15
6	pi=3.1416
7	q=0.0025
8	rho=46.048+T*(9.418+T*(-0.0329+T*(4.882e-5-T*2.895e-8)))
9	vis=exp(-10.547+541.69/(T-144.53))
10	v=q/(pi*D^2/4)
11	Re=D*v*rho/vis
12	D(0)=0.01
13	fF(0)=0.001

This set of equations yields the results summarized in Table 8–7, where the required pipe diameter is determined to be D = 0.039 m or 39 mm. A comparison of this diameter with the data of available heat exchanger tube diameters in Table D–4 indicates that the minimal available tube is a 2-inch OD BWG No. 10 tube with an inside diameter of 43.99 mm.

Table 8–7 Solution for Problem 8.8 (File **P8-08A.POL**)

Variable	Value	f( )
D	0.0389653	1.12E-9
fF	0.00459053	-8.674E-19
dp	103000	
L	100	
pi	3.1416	'
q	0.0025	
rho	994.572	
Re	90973.6	
T	298.15	
v	2.09649	
vis	0.0008931	

(b) Changing the temperature to 278.15 K (5°C), setting D = 0.04399 m, and solving for Δp with an initial guess of 103,000 Pa results in a calculated pressure drop of Δp = 64.82 kPa.

(c) The calculations for other liquids can be carried out by modification of the correlation equations for density and viscosity of the particular liquid.

 The POLYMATH problem solution files are found in directory Chapter 8 with files named **P8-08A.POL** and **P8-08B.POL**.

8.9 AVERAGE VELOCITY IN TURBULENT SMOOTH PIPE FLOW FROM MAXIMUM VELOCITY

8.9.1   Concepts Demonstrated

Conversion of maximum velocity into average velocity using the universal velocity distribution and comparison with experimental values.

8.9.2   Numerical Methods Utilized

Solution of a single nonlinear equation and regression of general expressions to experimental data.

8.9.3   Problem Statement

It is often necessary to convert maximum flow velocity in a pipe v_{max} to average velocity v_{av}. A common example is the use of a pitot tube, which measures v_{max}. The ratio of v_{av}/v_{max} can be calculated for a smooth tube from the following equation given by McCabe et al.:[4]

$$v_{av} = \frac{v_{max}}{1 + (3.75\sqrt{f_F/2})} \tag{8-58}$$

where f_F is the Fanning friction factor. For smooth tubes, the friction factor can be calculated from the Nikuradse[5] Equation (8-46) using the Reynolds number based on the average velocity.

Water is flowing at 25°C inside a 1-inch OD BWG–10 tube. (See Table D–4 for tube specifications.) A technician needs a simple relationship that can be used to estimate the average velocity from a measured maximum velocity.

(a) Calculate the average velocities for the maximum velocities given in Table 8–8 and enter these values in the table.

(b) Obtain a simple but accurate correlation of the predicted velocity as a function of the maximum velocity that a technician could use.

(c) Enter the correlation values in the Table 8–8 for comparison purposes.

Table 8–8 Comparison of Measured and Calculated Water Velocities

No.	Measured v_{max} (m/s)	Calculated v_{av} (m/s)	Correlated v_{av} (m/s)	No.	Measured v_{max} (m/s)	Calculated v_{av} (m/s)	Correlated v_{av} (m/s)
1	0.2			6	6.4		
2	0.4			7	12.8		
3	0.8			8	25.6		
4	1.6			9	51.2		
5	3.2			10	102.4		

8.10 CALCULATION OF THE FLOW RATE IN A PIPELINE

8.10.1 Concepts Demonstrated

Application of the general mechanical energy balance for incompressible fluids, and calculation of flow rate in a pipeline for various pressure drops.

8.10.2 Numerical Methods Utilized

Solution of a single nonlinear algebraic equation.

8.10.3 Problem Statement

Figure 8–4 shows a pipeline which delivers liquid at constant temperature T from point 1, where the pressure is p_1 and the elevation is z_1 to point 2, where the pressure is p_2 and the elevation is z_2. The effective length of the pipe line, including fittings and expansion losses, is L and its diameter is D.

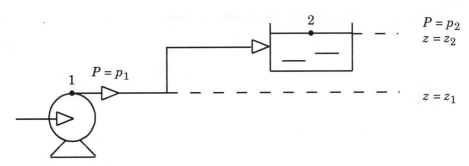

Figure 8–4 Liquid Flow in a Pipeline

(a) Calculate the flow rate q (in gal/min) in the pipeline for water at 60 °F. The pipeline is nominal 6-inch-diameter schedule 40 commercial steel pipe with L = 5000 ft, P_1= 150 psig, P_2 = 0 psig, z_1 = 0 ft, and z_2 = 300 ft.

(b) Plot the calculated flow rate as a function of pressure difference (200 psig max). What is the minimum pressure difference needed to start flow?

(c) Add curves to the plot created in part (b) for both 4- and 8-inch-diameter schedule 40 commercial steel pipe and summarize any observed general trends.

Additional Information and Data

The general mechanical energy balance on an incompressible liquid in turbulent flow is given by Geankoplis[1] as

$$\frac{1}{2}(v_2^2 - v_1^2) + g(z_2 - z_1) + \frac{P_2 - P_1}{\rho} + \sum F + W_S = 0 \qquad \text{(8-59)}$$

where $\sum F$ is the summation of frictional losses given by Equation (8-51) and W_S is the shaft work done by the system in consistent units. Other terms in Equation (8-59) are defined in Section 8.8.3. For English units, the conversion factor $g_c = 32.174 \text{ ft} \cdot \text{lb}_m/\text{lb}_f \cdot \text{s}^2$ must be used as necessary to maintain consistent units in Equation (8-59).

8.10.4   Solution (Suggestions)

The inside diameter of the 6-inch steel pipe is shown in Table D–5, and its surface roughness can be found in Table D–5. The explicit Shacham Equation (8-49) is convenient for calculation of the Fanning friction factor in order to use Equation (8-51) to evaluate the $\sum F$ term in Equation (8-59). Equations (8-35) and (8-36) can be used to determine the viscosity and density of water in English units. It is important that each terms in Equation (8-59) have the same units.

(a) The calculated flow rate is $q = 368.7$ gal/min.

8.11 FLOW DISTRIBUTION IN A PIPELINE NETWORK

8.11.1 Concepts Demonstrated

Calculation of flow rates and pressure drops in a pipeline network using Fanning friction factors.

8.11.2 Numerical Methods Utilized

Solution of systems of nonlinear algebraic equations.

8.11.3 Problem Statement

Water at 25°C is flowing in the pipeline network given in Figure 8–5. The pressure at the exit of the pump is 15 bar (15×10^5 Pa) above atmospheric, and the water is discharged at atmospheric pressure at the end of the pipeline. All the pipes are 6-inch schedule 40 steel with an inside diameter of 0.154 m. The equivalent lengths of the pipes connecting different nodes are the following: L_{01} = 100 m, $L_{12} = L_{23} = L_{45}$ = 300 m, and $L_{13} = L_{24} = L_{34}$ = 1200 m.

(a) Calculate all the flow rates and pressures at nodes 1, 2, 3, and 4 for the pipeline network shown in Figure 8–5. The Fanning friction factor can be assumed to be constant at f_F = 0.005 for all pipelines.

(b) Investigate how the flow rates and pressure drops change when one of the pipelines becomes blocked. Select one of the flow rates (q_{12}, q_{23}, q_{24}, or q_{13}) equal to zero and repeat part (a).

(c) Rework a more accurate solution for part (a) by using Equation (8-49) to calculate each Fanning friction factor where $\varepsilon = 4.6 \times 10^{-5}$ m. Which calculated flow rate from part (a) has the greatest error?

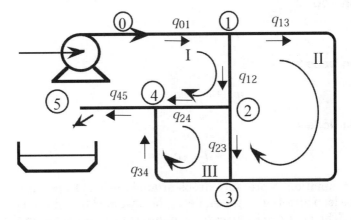

Figure 8–5 Pipeline Network

Additional Information and Data

For the solution of this problem it is convenient to express the pressure drop from node i to node j as

$$\Delta P_{ij} = k_{ij}(q_{ij})^2 \qquad \text{(8-60)}$$

where ΔP_{ij} is the pressure drop and q_{ij} is the volumetric flow rate between nodes i and j. The k_{ij} terms in Equation (8-60) are related to the Fanning friction factors and average fluid velocities, as described by Equations (8-52) and (8-53). Thus

$$k_{ij} = \frac{2f_F\rho\dfrac{\Delta L_{ij}}{D}}{(\pi D^2/4)^2} = \frac{32f_F\rho\Delta L_{ij}}{\pi^2 D^5} \qquad \text{(8-61)}$$

There are two relationships that govern the steady-state flow rate in pipeline networks. First, *the algebraic sum of the flow rates at each node must be zero.* Second, *the algebraic sum of all pressure drops in a closed loop must be zero.*

The summation of flow rates at each node is just an application of the conservation of mass balance, which can be expressed as

$$\text{INPUT} - \text{OUTPUT} = \text{ZERO}$$

The balance equations are written based on assumed flow directions. If the solution yields some negative flow rates, then the actual flow direction is opposite to the assumed flow direction.

Balance on Node 1

$$q_{01} - q_{12} - q_{13} = 0 \qquad \text{(8-62)}$$

Balance on Node 2

$$q_{12} - q_{24} - q_{23} = 0 \qquad \text{(8-63)}$$

Balance on Node 3

$$q_{23} + q_{13} - q_{34} = 0 \qquad \text{(8-64)}$$

Balance on Node 4

$$q_{24} + q_{34} - q_{45} = 0 \qquad \text{(8-65)}$$

The summation of pressure drops around various loops utilizes ΔP_{ij} with a + sign if the loop direction is the same as the assumed flow direction; otherwise it is negative. The pressure drop across a pump is negative when in the direction of flow.

Summation on Loop I

$$\Delta P_{01} + \Delta P_{12} + \Delta P_{24} + \Delta P_{45} + \Delta P_{PUMP} = 0 \qquad \text{(8-66)}$$

Summation on Loop II

$$\Delta P_{13} - \Delta P_{23} - \Delta P_{12} = 0 \qquad \text{(8-67)}$$

Summation on Loop III

$$\Delta P_{23} + \Delta P_{34} - \Delta P_{24} = 0 \qquad \text{(8-68)}$$

(a) The pressure drops can be expressed as functions of q_{ij} using Equation (8-60). This substitution leads to seven equations with seven unknown flow rates designated as q_{01}, q_{12}, q_{13}, q_{24}, q_{23}, q_{34}, and q_{45}. Introduction of this system of equations (including the numerical values of the different k_{ij}'s) into the POLYMATH *Simultaneous Algebraic Equation Solver* yields the code given in Table 8–11.

Table 8–9 POLYMATH Problem Code (File **P8-11A.POL**)

Line	Equation
1	f(q01)=q01-q12-q13
2	f(q12)=q12-q24-q23
3	f(q13)=q23+q13-q34
4	f(q24)=q24+q34-q45
5	f(q23)=k01*q01^2+k12*q12^2+k24*q24^2+k45*q45^2+deltaPUMP
6	f(q34)=k13*q13^2-k23*q23^2-k12*q12^2
7	f(q45)=k23*q23^2+k34*q34^2-k24*q24^2
8	deltaPUMP=-15.e5
9	fF=0.005
10	rho=997.08
11	D=0.154
12	pi=3.1416
13	k01=32*fF*rho*100/(pi^2*D^5)
14	k12=32*fF*rho*300/(pi^2*D^5)
15	k24=32*fF*rho*1200/(pi^2*D^5)
16	k45=32*fF*rho*300/(pi^2*D^5)
17	k13=32*fF*rho*1200/(pi^2*D^5)
18	k23=32*fF*rho*300/(pi^2*D^5)
19	k34=32*fF*rho*1200/(pi^2*D^5)
20	q01(0)=0.1
21	q12(0)=0.1
22	q13(0)=0.1
23	q24(0)=0.1
24	q23(0)=0.1
25	q34(0)=0.1
26	q45(0)=0.1

The initial estimates for all the volumetric flow rates can be set at 0.1 m^3/s.

The POLYMATH data file for part (a) is found in directory Chapter 8 with file named **P8-11A.POL**.

(b) Note that if one of the flow rates is set to zero, the number of unknowns is reduced by one. Consequently, the number of equations must be reduced by one. Remember, with POLYMATH, the variable within a nonlinear equation entry on the left side in the parentheses does not need to be a part of the expression on the right side. However, all nonlinear expression variables need to be within the parentheses on the left to identify all the problem variables to POLYMATH.

8.12 WATER DISTRIBUTION NETWORK

8.12.1 Concepts Demonstrated

Calculation of flow rates and pressure drops in a pipeline network using Fanning friction factors.

8.12.2 Numerical Methods Utilized

Solution of systems of nonlinear algebraic equations.

8.12.3 Problem Statement

Ingels and Powers[9] investigated the following water distribution network shown in Figure 8–6. There is one input source to node 1, which can deliver 1400 gpm. The demand at nodes 2, 3, and 5 is 420 gpm and at node 6 is 140.0 gpm.

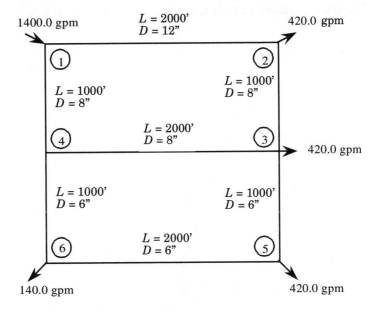

Figure 8–6 Pipeline Network for Problem 8.12

(a)   Calculate the flow rates through the different pipes and the head differences between the supply node and the different customers for water at 60 °F. The pipe lengths and diameters are shown in Figure 8–6. A constant Fanning friction factor of $f_F = 0.005$ can be assumed for all calculations.

(b)   What pressure in feet of water head must be developed by the supply pump at node 1 to maintain all flow rates if the line between nodes 1 and 2 must be shut down for repairs?

8.12.4   Solution (Suggestions)

(a) Background information can be found in Problems 8.8 through 8.11. Only five of the possible six node balances for flow rates give independent relationships. Thus pressure drop considerations around the two loops will be needed. Conversion factors including g_c may be needed to maintain consistent units.

8.13 PIPE AND PUMP NETWORK

8.13.1 Concepts Demonstrated

Calculations of flow rates in a pipe and pump network using an overall mechanical energy balance and accounting for various frictional losses.

8.13.2 Numerical Methods Utilized

Solution of systems of nonlinear algebraic equations.

8.13.3 Problem Statement

Water at 60°F and one atmosphere is being transferred from tank 1 to tank 2 with a 2-hp pump that is 75% efficient, as shown in Figure 8–7. All the piping is 4-inch schedule 40 steel pipe except for the last section, which is 2-inch schedule 40 steel pipe. All elbows are 4-inch diameter, and a reducer is used to connect to the 2-inch pipe. The change in elevation between points 1 and 2 is $z_2 - z_1 = 60$ ft.

(a) Calculate the expected flow rate in gal/min when all frictional losses are considered.
(b) Repeat part (a) but only consider the frictional losses in the straight pipes.
(c) What is the % error in flow rate for part (b) relative to part (a)?
(d) Repeat parts (a), (b), and (c) for one of the liquids given in Table D–3.

Additional Information and Data

The various frictional losses and the mechanical energy balance are discussed by Geankoplis[1] and Perry et al.[3]

8.13.4 Solution (Suggestions)

(a) Frictional losses for this problem include contraction loss at tank 1 exit, friction in three 4-inch elbows, contraction loss from 4-inch to 2-inch pipe, friction in both the 2-inch and 4-inch pipes, and expansion loss at the tank 2 entrance. The explicit equation given by Equation (8-49) can be used to calculate the friction factor for both sizes of pipe. Equations (8-35) and (8-36) can be used to determine the viscosity and density of water in English units. Additional data are found in Appendix D.

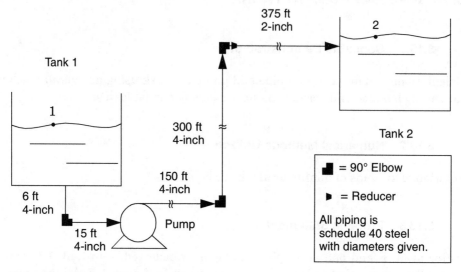

Figure 8–7 Pipe and Pump Network

8.14 Optimal Pipe Length for Draining a Cylindrical Tank in Turbulent Flow

8.14.1 Concepts Demonstrated

Use of the overall mechanical energy balance and unsteady-state material balance in fluid flow calculations.

8.14.2 Numerical Methods Utilized

Solution of nonlinear equations and differential algebraic equations.

8.14.3 Problem Statement

There is a need to design a piping system for the rapid draining of a tank in an emergency, and you are called upon to settle a disagreement. A simplified view of the tank is shown in Figure 8–8. Some engineers say that the shorter the length of the pipe, the less time it will take to drain the tank. Others argue that a longer pipe will lead to shorter drain times. Still others say that there is an optimal length. You will have to calculate the draining time for different lengths of the pipe and determine the length L that will minimize the draining time t_f for water, which is expected to remain in *turbulent flow* during the draining time.

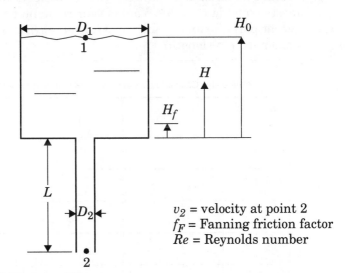

v_2 = velocity at point 2
f_F = Fanning friction factor
Re = Reynolds number

Figure 8–8 Tank with Draining Pipe

The drain pipe is a nominal 1/2-inch schedule 40 steel pipe with ε = 0.00015. The tank diameter D_1 is 3 ft, and the initial liquid level is always H_0 = 6 ft. The liquid in the tank is water at 60°F. The tank is considered to be drained when the final level H_f reaches 1 inch. The variables are limited to the ranges 1 in $\leq H \leq$ 6 ft and 1 in $\leq L \leq$ 10 ft.

(a)   Investigate the effects of various values of L and H on v_2, f_F, and Re by completing Table 8–10 using the steady-state mechanical energy balance and assuming friction losses only in the pipe. Comment on the assumption of a constant Fanning friction factor for an approximate solution of this problem, and verify that the flow is turbulent.

(b)   Show that an approximate solution for the draining time assuming a constant value of the friction factor is given by

$$t_f = \frac{D_1^2}{D_2^2}(\sqrt{H_0 + L} - \sqrt{H_f + L})\sqrt{\frac{2}{g}\left(1 + \frac{4f_F L}{D_2}\right)} \qquad \textbf{(8-69)}$$

where t_f is the time (in seconds) that it takes to reach a final level where $H_f = 1$ inch. Here the f_F is to be the average of the friction factors calculated from Table 8–10 for the case where H_0 is 6 ft. Perform calculations for $L = 0.0833$ ft to 10 ft until you find an optimum or a clear trend in the change of the draining time that will support your recommended pipe length and estimated draining time.

(c)   Solve part (b) numerically where the entrance and flow constriction effects can be neglected but the friction factor can vary. The steady-state mechanical energy equation can be used in this case. Are there meaningful differences between the results obtained in this part of the question and in part (b)? What is your recommended pipe length L?

(d)   Repeat parts (b) and (c) for the case where H_0 is 3 ft. What is your recommended pipe length?

Table 8–10 Calculated Values for Tank Draining of Water at 60°F

L ft	H ft	f_F	v_2 ft/s	Re
1/12	1/12			
1/12	6			
10	1/12			
10	6			
5	3			

Additional Information and Data

The unsteady-state mechanical energy balance can be applied between points 1 and 2 shown in Figure 8–8. The assumption of an incompressible fluid leads to the isothermal mechanical energy balance given in Equation (8-59). Note that this equation is only an approximation as the time-dependent terms have been neglected, which is reasonable in this case. Application to this problem yields an expression for the outlet velocity v_2

$$v_2^2 = \frac{2g(H+L)}{1 + 4f_F\left(\dfrac{L}{D_2}\right)} \tag{8-70}$$

where the Fanning friction factor can be evaluated from explicit Equation (8-49). An unsteady-state material balance on the fluid in the tank yields

$$\frac{dH}{dt} = -\frac{D_2^2 v_2}{D_1^2} \tag{8-71}$$

8.14.4   Solution (Suggestions)

(a) This problem becomes one of solving nonlinear Equation (8-70) for the exit velocity, where the friction factor is calculated from explicit Equation (8-49). Equations (8-35) and (8-36) can be used to determine the viscosity and density of water in English units.

(b) The approximate solution can be derived by substituting Equation (8-70) into differential Equation (8-71) and integrating from H_0 to H_f. The calculations involved only require repeated evaluations of Equation (8-69) with the average friction factor for various L's.

(c) This part involves the solution of a differential equation and a nonlinear algebraic equation, which requires a DAE problem solution (discussed in Problem 6.7).

8.15 OPTIMAL PIPE LENGTH FOR DRAINING A CYLINDRICAL TANK IN LAMINAR FLOW

8.15.1 Concepts Demonstrated

Use of the overall mechanical energy balance and unsteady-state material balance in fluid flow calculations.

8.15.2 Numerical Methods Utilized

Solution of nonlinear equations and differential algebraic equations.

8.15.3 Problem Statement

The liquid to be placed in the tank discussed in Problem 8.14 is hydraulic fluid MIL-M-5606 at 30°F whose properties are given in Table D–3. As in Problem 8.14, you are to determine the length of the pipe that will minimize the draining time for this hydraulic fluid, which is expected to remain in *laminar flow* during the draining time. The variables are limited to the ranges 1 in $\leq H \leq$ 6 ft and 1 in $\leq L \leq$ 10 ft.

(a) Investigate the effects of various values of L and H on v_2, f_F, and Re by completing Table 8–11 using the steady-state mechanical energy balance and assuming friction losses only in the pipe. Comment on the assumption of a constant Fanning friction factor for an approximate solution of this problem and verify that the flow is laminar.

(b) Solve this problem numerically where the entrance and flow constriction effects can be neglected but the friction factor can vary. The steady-state mechanical energy equation can be used in this case, and H_0 is 2.0 feet. What is your recommended pipe length L?

(c) Repeat part (b) for H_0 of 1.0 ft.

(d) Determine an analytical solution to part (b) and compare the results with the numerical solution of part (b).

Table 8–11 Calculated Values for Tank Draining of Hydraulic Fluid at 30°F

L ft	H ft	f_F	v_2 ft/s	Re
1/12	1/12			
1/12	6			
10	1/12			

Table 8–11 Calculated Values for Tank Draining of Hydraulic Fluid at 30°F

L ft	H ft	f_F	v_2 ft/s	Re
10	6			
5	3			

8.15.4 Solution (Suggestions)

The Fanning friction factor for laminar flow in pipes is given by Equation (8-54).

8.16 BASEBALL TRAJECTORIES AS A FUNCTION OF ELEVATION

8.16.1 Concepts Demonstrated

Effect of drag and fluid density on a solid object moving through a fluid.

8.16.2 Numerical Methods Utilized

Conversion of higher order differential equations to simultaneous first-order differential equations and solution of a two-point boundary value problem.

8.16.3 Problem Statement

Enthusiastic New York Mets baseball fans maintain that the Colorado Rockies baseball team has an unfair advantage because balls travel further in the thin air of Denver, Colorado. It is a good engineering assumption that the major difference between the conditions in New York and Denver is the decrease of the air density because of the much higher elevation in Denver.

(a)   A good batter can impart an initial velocity of 120 ft/s to a baseball. What distance will the ball travel in Shea Stadium in New York when the initial angle is 30° with the horizontal?

(b)   Use the data in Table 8–12 to calculate the distance that the ball will travel in Coors Stadium in Denver for the same initial velocity and angle as given in part (a).

(c)   What would be the best angle to optimize the resulting distance for the initial velocity of 120 ft/s in Shea Stadium? Plot the results for y versus x.

Additional Information and Data

Table 8–12 Data for Baseball Trajectories

Description	Variable	Value
Mass of the baseball	m	$0.313 \, \text{lb}_\text{m}$
Projected area	A	$0.046 \, \text{ft}^2$
Drag coefficient (assuming turbulent flow)	C_D	0.44
The acceleration of gravity (the conversion factor g_c has the same numerical value as g)	g	$32.174 \, \text{ft/s}^2$
Height of the ball with initial velocity imparted	y_0	5 ft
The angle with the horizontal in which the ball is being contacted	θ	30° for part (a)

Table 8–12 Data for Baseball Trajectories

Description	Variable	Value
The elevation of the field above sea level at Shea Stadium in New York	Z_1	100 ft
The elevation of the field above sea level in Coors Stadium in Denver	Z_2	5280 ft

It can be shown (Riggs[10]) that the trajectory of a baseball can be represented by the following pair of second-order differential equations:

$$m\frac{d^2x}{dt^2} = -k\frac{dx}{dt}\sqrt{\left(\frac{dx}{dt}\right)^2 + \left(\frac{dy}{dt}\right)^2}$$

$$m\frac{d^2y}{dt^2} = -k\frac{dy}{dt}\sqrt{\left(\frac{dx}{dt}\right)^2 + \left(\frac{dy}{dt}\right)^2} - mg \tag{8-72}$$

where x and y are the horizontal and vertical distances and k is the drag coefficient given by

$$k = \frac{C_D A \rho}{2g_c} \tag{8-73}$$

In the preceding equation, C_D is the drag coefficient, ρ is the density of the air, and g_c is a conversion factor. The air density changes with elevation above sea level given by Z according to

$$\rho = 0.07647 \, \exp\left(\frac{-Z}{3.33 \times 10^4}\right) \tag{8-74}$$

where Z is in ft and the units of ρ are lb_m/ft^3.

8.16.4 Solution (Suggestions)

The POLYMATH *Simultaneous Differential Equation Solver* program can be used to solve this problem. First, the set of two second-order differential equations can be converted into a set of four first-order equations. The technique was discussed in Problem 6.5.

For this problem, it is convenient to define

$$v_x = \frac{dx}{dt}$$

$$v_y = \frac{dy}{dt} \tag{8-75}$$

so that the pair of Equation (8-72) becomes four first-order ordinary differential equations.

$$\frac{dx}{dt} = v_x$$

$$\frac{dy}{dt} = v_y$$

$$\frac{d(v_x)}{dt} = -\frac{k}{m}v_x\sqrt{v_x^2 + v_y^2} \qquad \textbf{(8-76)}$$

$$\frac{d(v_y)}{dt} = -\frac{k}{m}v_y\sqrt{v_x^2 + v_y^2} - g$$

The corresponding initial values at $t = 0$ are given by

$$x = 0$$

$$y = y_0$$

$$v_x = V_0\cos\theta \qquad \textbf{(8-77)}$$

$$v_y = V_0\sin\theta$$

where V_0 is the initial velocity of the ball.

(a) The problem is defined by the differential equation set (8-76) and the initial conditions (8-77). Equations (8-73) and (8-74) must be evaluated during the numerical solution. For Shea Stadium, $Z = 100 + y$. The boundary condition that indicates that the ball has impacted the ground is that $y = 0$, and the corresponding x is the distance.

Retaining a Value when a Condition Is Satisfied

When using POLYMATH, a convenient way to save the value of x when y = 0 for this problem is to use the "if ... then ... else ... " capability, as illustrated by

d(x)/d(t)=if (y>0) then (Vx) else (0)

The preceding statement conveniently holds the value of x when y becomes negative.

(b) The solution for Coors Stadium requires the modification that $Z = 5280 + y$. The equations must again be integrated to $y = 0$ to determine the x distance that the ball will travel.

(c) This solution uses the equations of part (a), but the angle θ can be varied so as to maximize the value of x when y becomes zero.

8.17 VELOCITY PROFILES FOR A WALL SUDDENLY SET IN MOTION—LAMINAR FLOW

8.17.1 Concepts Demonstrated

Unsteady-state one-dimensional boundary layer velocity profiles for a Newtonian fluid in laminar flow near a wall, which is suddenly set in motion at a constant velocity.

8.17.2 Numerical Methods Utilized

Application of the numerical method of lines to solve a partial differential equation and solution of simultaneous ordinary differential equations.

8.17.3 Problem Statement

A large volume of water at 25°C with $\mu = 8.931 \times 10^{-4}$ kg/m·s and $\rho = 994.6$ kg/m^3 is bounded on one side by a plane surface, which is initially at rest with v_x = 0. At time $t = 0$, the plane is suddenly set in motion with a constant velocity V. The flow is expected to be laminar, and there is no gravity force or pressure gradients. Water may be considered a Newtonian fluid. This situation is shown in Figure 8–9 where the velocity profile is shown at a particular time t.

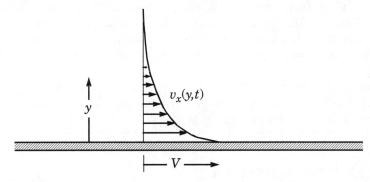

Figure 8–9 Velocity Profile of a Newtonian Fluid Near a Wall Suddenly Set in Motion.

For this situation, the equation of continuity and the equation of motion for one-dimensional flow of a Newtonian fluid (see Bird et al.[2]) yield

$$\frac{\partial v_x}{\partial t} = \frac{\mu}{\rho}\frac{\partial^2 v_x}{\partial y^2} \tag{8-78}$$

where the initial and boundary conditions are given by

$$v_x = 0 \qquad \text{at any } y \qquad t \le 0$$
$$v_x = V \qquad \text{at } y = 0 \qquad t > 0 \qquad \text{(8-79)}$$
$$v_x = 0 \qquad \text{at } y = \infty \qquad t > 0$$

The analytical solution to this problem by Bird et al.[2] uses the error function, which is available in many mathematical tables and given in Table 3 of Appendix A.

$$\frac{v_x}{V} = 1 - \frac{2}{\sqrt{\pi}} \int_0^\eta e^{-\eta^2} d\eta = 1 - \text{erf}(\eta) \qquad \text{(8-80)}$$

where

$$\eta = \frac{y}{\sqrt{\dfrac{4\mu t}{\rho}}} \qquad \text{(8-81)}$$

(a) Use the numerical method of lines to solve Equation (8-78) with the boundary conditions given by Equation (8-79). The velocity of the wall is given as $V = 0.2$ m/s. The y distance to be considered is 0.1 m and the recommended Δy distance between nodes is 0.01 m. Generate the numerical solution to a time of 500 s. Plot the velocities with time for $y = 0.01, 0.02, 0.03,$ and 0.04 m as a function of time to 500 s.

(b) Repeat the calculations of part (a) with $\Delta y = 0.005$ m to verify a reasonable solution. Make a comparison of the results of parts (a) and (b) at 500 s for the y distances used in the plots for part (a).

(c) Make a separate plot similar to Figure 8–9 at $t = 500$ s using the results of the part (b) calculations.

(d) Compare the results with the error function solution given by Equations (8-80) and (8-81) with the results generated for the four locations investigated in parts (a) and (b) at 500 s.

8.17.4 Solution (Suggestions)

(a) The numerical method of lines is discussed in Problem 6.8. In this application, the number of nodes must be adequate to describe the fluid, which is actually in movement. Typically, 10 to 20 nodes should be adequate if the Δy spacing is properly chosen to be within the boundary layer as it is being established. The boundary layer is the region near the wall where the fluid velocity v_x is greater than 1% of the velocity V.

The set of equations can be entered into the POLYMATH *Simultaneous Differential Equation Solver*. The capability of the editor to duplicate equations is convenient for entry of the ordinary differential equations for the various nodes.

(c) The error function is summarized in many textbooks and compilations of mathematical tables. It can also easily be generated by numerically solving the ordinary differential equation given by

$$\frac{d}{dz}[\operatorname{erf}(z)] = \frac{2e^{-z^2}}{\sqrt{\pi}} \tag{8-82}$$

with an initial condition of $\operatorname{erf}(z) = 0$ and the integration continued to z yielding $\operatorname{erf}(z)$. Thus a plot or table of $\operatorname{erf}(z)$ versus x can easily be calculated, as has been accomplished in Table A-3 of Appendix A.

 The POLYMATH file for calculation of the error function is found in directory Chapter 8 with file named **P8-ERROR.POL**.

www

8.18 BOUNDARY LAYER FLOW OF A NEWTONIAN FLUID ON A FLAT PLATE

8.18.1   Concepts Demonstrated

Numerical solution of the equations of motion and continuity for a laminar boundary layer flow of a Newtonian fluid over a flat plate. Prediction of boundary layer thickness and drag force.

8.18.2   Numerical Methods Utilized

Transformation of partial differential equations into an ordinary differential equation, reduction of a higher order differential equation to a system of first-order differential equations, and numerical solution of simultaneous ordinary differential equations employing a shooting technique.

8.18.3   Problem Statement

An important case in laminar boundary layer theory involves the flow of a Newtonian fluid on a very thin flat plate, as shown in Figure 8–10.

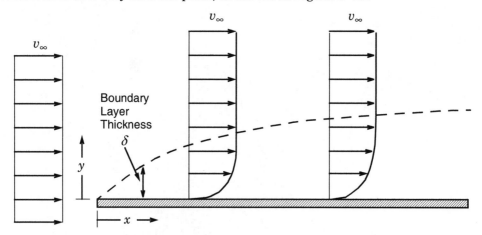

Figure 8–10 Laminar Boundary Layer for Flow Past a Flat Plate

For this situation, the equation of motion and the continuity equation become (Geankoplis[1])

$$v_x \frac{\partial v_x}{\partial x} + v_y \frac{\partial v_x}{\partial y} = \frac{\mu}{\rho} \frac{\partial^2 v_x}{\partial y^2} \tag{8-83}$$

$$\frac{\partial v_x}{\partial x} + \frac{\partial v_y}{\partial y} = 0 \tag{8-84}$$

where the boundary conditions are $v_x = v_y = 0$ at $y = 0$ (y is the distance from the plate) and $v_x = v_\infty$ at $y = \infty$.

These partial differential equations can be transformed into an ordinary differential equation (see Schlichting[11] for details) by considering the boundary layer thickness δ to be proportional to $\sqrt{\dfrac{\mu x}{\rho v_\infty}}$.

A dimensionless coordinate can be defined by $\eta = y/\delta$ so that

$$\eta = y\sqrt{\frac{\rho v_\infty}{\mu x}} \tag{8-85}$$

Also, a stream function that satisfies the continuity equation, Equation (8-84), may be defined by

$$\psi = \sqrt{\frac{\mu x v_\infty}{\rho}} f(\eta) \tag{8-86}$$

where

$$v_x = \frac{\partial}{\partial y}(\psi) \text{ and } v_y = -\frac{\partial}{\partial x}(\psi) \tag{8-87}$$

Thus the velocities become a function of the dimensionless stream function $f(\eta)$ given by

$$v_x = v_\infty f'(\eta) \tag{8-88}$$

$$v_y = \frac{1}{2}\sqrt{\frac{\mu v_\infty}{\rho x}}(\eta f' - f) \tag{8-89}$$

where the prime indicates differentiation with respect to η.

After introduction of Equations (8-88) and (8-89) into Equation (8-83), the partial differential equations are transformed into the ordinary differential equation in the dimensionless stream function given by

$$ff'' + 2f''' = 0 \tag{8-90}$$

with the initial conditions

$$f = 0 \text{ and } f' = 0 \text{ at } \eta = 0 \tag{8-91}$$

and the final condition

$$f' = 1 \text{ at } \eta = \infty \tag{8-92}$$

Boundary Layer Thickness

The boundary layer thickness is usually described as where the velocity is less than $0.99v_\infty$, and this is where $f' \cong 0.99$. Thus Equation (8-85) can be used to determine the boundary layer thickness by

$$\delta = (\eta|_{f' = 0.99})\sqrt{\frac{\mu x}{\rho v_{\infty}}} \qquad \text{(8-93)}$$

Drag Force Due to Skin Friction

The total drag force acting upon the plate with length L and width W is given by (Geankoplis[1])

$$F_D = W \int_0^L \tau_0\,dx \qquad \text{(8-94)}$$

where the shear stress at the surface of the plate can be calculated from

$$\tau_0 = \mu\left(\frac{dv_x}{dy}\Big|_{y=0}\right) = \mu v_{\infty}\sqrt{\frac{\rho v_{\infty}}{\mu x}}(f''|_{\eta=0}) \qquad \text{(8-95)}$$

Thus evaluation of Equation (8-94) utilizing Equation (8-95) yields

$$F_D = 2(f''|_{\eta=0})\sqrt{\mu \rho v_{\infty} L} \qquad \text{(8-96)}$$

(a)   Solve the differential equation given by Equation (8-90) with initial conditions and boundary conditions given by Equations (8-91) and (8-92) for $0 \le \eta \le 10$.

(b)   Present the boundary layer thickness result of Equation (8-93) utilizing the result of part (a).

(c)   Present the drag force result of Equation (8-96) utilizing the result of part (a).

(d)   Compare the results of parts (b) and (c) with those presented by Geankoplis[1] or Schlichting[11] or a fluid dynamics textbook.

8.18.4   Solution (Partial)

(a) The third-order differential equation given by Equation (8-90) can be solved by transformation to a system of three simultaneous ordinary differential equations which can be solved by the *POLYMATH Simultaneous Ordinary Differential Equation Solver*. This can be accomplished by defining additional functions given by $g_1 = f$, $g_2 = f'$, $g_3 = f''$ (as per Carnahan et al.[12]), resulting in the differential equations

$$\frac{dg_1}{d\eta} = g_2 \qquad\qquad\qquad (8\text{-}97)$$

$$\frac{dg_2}{d\eta} = g_3 \qquad\qquad\qquad (8\text{-}98)$$

$$\frac{dg_3}{d\eta} = -\frac{1}{2}g_1 g_3 \qquad\qquad\qquad (8\text{-}99)$$

with the initial conditions

$$g_1 = 0 \text{ and } g_2 = 0 \text{ at } \eta = 0 \qquad\qquad\qquad (8\text{-}100)$$

and final condition

$$g_2 = 1 \text{ at } \eta = \infty \qquad\qquad\qquad (8\text{-}101)$$

A shooting technique can be used to determine (optimize) the initial condition for g_3 that will satisfy the final condition when η becomes large. Problem 6.5 discusses this solution technique and single-variable optimization in detail.

REFERENCES

1. Geankoplis, C. J. *Transport Processes and Unit Operations*, 3rd ed., Englewood Cliffs, NJ: Prentice Hall, 1993.
2. Bird, R. B., Stewart, W. E., and Lightfoot, E. N. *Transport Phenomena*, New York: Wiley, 1960.
3. Perry, R.H., Green, D. W., and Malorey, J. D., Eds. *Perry's Chemical Engineers Handbook*, 7th ed, New York: McGraw-Hill, 1997.
4. McCabe, W. L., Smith, J. C., and Harriot, P. *Unit Operations of Chemical Engineering*, 5th ed., New York: McGraw-Hill, 1993.
5. Nikuradse, J. *VDI-Forschungsheft*, 356 (1932).
6. Colebrook, C. F., and White, C. M. *J. Inst. Civil Eng., 10* (1) 99–118 (1937–38).
7. Shacham, M. *Ind. Eng. Chem. Fund., 19*, 228–229 (1980).
8. Haaland, S. E. *Trans. ASME*, JFE, *105*, 89 (1983).
9. Ingels, D. M., and Powers, J. E. *Chem Eng. Progr., 60* (2), 65 (1964).
10. Riggs, J. B. *An Introduction to Numerical Methods for Chemical Engineers*, 2nd ed., Lubbock, TX: Texas Tech University Press, 1994.
11. Schlichting, H., *Boundary Layer Theory*, 6th ed., New York: McGraw-Hill, 1969.
12. Carnahan, B., Luther, H. A., and Wilkes, J. O. *Applied Numerical Methods*, New York: Wiley, 1969.

C H A P T E R **9**

Heat Transfer

9.1 ONE-DIMENSIONAL HEAT TRANSFER THROUGH A MULTILAYERED WALL

9.1.1 Concepts Demonstrated

Calculation of one-dimensional heat flux and temperature distributions for heat conduction including variation of thermal conductivity with temperature and convection as a boundary condition.

9.1.2 Numerical Methods Utilized

Solution of ordinary differential equations in sequence which require matching of boundary conditions.

9.1.3 Problem Statement

Heat transfer by conduction in solids and fluids follows Fourier's law. One-dimensional heat transfer by conduction is described by

$$\frac{q_x}{A} = -k\frac{dT}{dx} \qquad (9\text{-}1)$$

where q_x is the heat transfer in the x direction in W or J/s, A is the cross-sectional area that is normal to the direction of heat conduction in m^2, k is the thermal conductivity of the medium in W/m·K, and x is the distance in m.

Convective heat transfer between solids and fluids is described by

$$\frac{q_x}{A} = h(T_w - T_f) \qquad (9\text{-}2)$$

where h is the heat transfer coefficient in W/m^2·K, T_w is the temperature of the solid surface in K, and T_f is the temperature of the fluid in K.

A common boundary condition at the solid/fluid interface is continuity of

333

heat flux given by

$$\frac{q_x}{A}\bigg|_S = -k\frac{dT}{dx}\bigg|_S = h(T_w - T_f) \qquad (9\text{-}3)$$

where S represents the solid surface whose temperature is T_w. At interfaces between two different solids, the temperature is continuous:

$$T = T_1\big|_{x=I} = T_2\big|_{x=I} \qquad (9\text{-}4)$$

where I represents the interface between the two different solids, and the heat flux continuity is given by

$$\frac{q_x}{A} = -k_1\frac{dT_1}{dx}\bigg|_I = -k_2\frac{dT_2}{dx}\bigg|_I \qquad (9\text{-}5)$$

For steady-state heat transfer by conduction, an energy balance yields

$$\frac{dq_x}{dx} = \dot{q}A \qquad (9\text{-}6)$$

where $\dot{q}$ is the rate of heat generated per unit volume in W/m³. When the generation term in Equation (9-6) is zero, then the rate of heat transfer is constant at any value of x.

Consider a problem similar to that given by Geankoplis[1] (adapted with permission) in which the wall of a cold-storage room is constructed of layers of pine, cork board, and concrete. The thicknesses of the layers are 15.0 mm, 100.0 mm, and 75.0 mm, respectively. The corresponding thermal conductivities are 0.151, 0.0433, and 0.762 in W/m·K. Figure 9–1 shows a diagram of the wall, where the pine, cork board, and concrete layers are labeled A, B, and C. The temperatures at each interface are shown in the diagram.

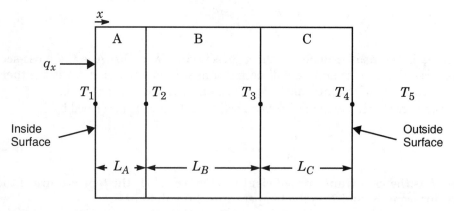

Figure 9–1 Heat Transfer through a Multilayered Wall

(a)   Calculate the heat flux through the wall if the interior surface is at 255 K and the exterior surface is at 298 K.

(b)   It is proposed to reduce the heat loss by 50% by increasing the thickness of the cork board. What total thickness of cork board would be required?

(c)   A new material is to be used instead of the cork board in part (a). Its thermal conductivity is given by $k = 2.5 \times \exp(-1225/T)$ with T in K. Repeat part (a) for this material and plot the temperature profile within the wall.

(d)   The natural convection heat transfer coefficient for a vertical plane representing the outside wall is given by $h = 1.37 \left| \dfrac{T_5 - T_4}{6} \right|^{1/4}$, where the T's are in K. Note that the absolute value is to be used in calculating h. Repeat part (c) using this information and consider that only the interior surface temperature is known to be 255 K. In this case, the exterior surface temperature must be calculated when the ambient temperature T_5 is 298 K. Plot the temperature profile within the wall.

9.1.4   Solution (Partial)

(a) Since there is no heat generation within the multi-layered wall, the heat transfer in the x direction is constant, as inferred from Equation (9-6). Equation (9-1) can therefore be rearranged and written in turn for each of the wall layers giving the differential equation set as

$$\frac{dT}{dx} = -\left(\frac{q_x}{A}\right)/k_A \qquad 0 \leq x \leq L_A$$

$$\frac{dT}{dx} = -\left(\frac{q_x}{A}\right)/k_B \qquad L_A < x \leq (L_A + L_B) \tag{9-7}$$

$$\frac{dT}{dx} = -\left(\frac{q_x}{A}\right)/k_C \qquad (L_A + L_B) < x \leq (L_A + L_B + L_C)$$

where the initial condition is given by

$$T = T_1 \qquad \text{when } x = 0 \tag{9-8}$$

and the interface temperatures are T_2, T_3, and T_4.

For constant thermal conductivities, the differential equations of Equation Set (9-7) can be integrated and the results incorporated into a single equation for

q_x /A and the overall temperature change (details given by Geankoplis[1]), yielding

$$q_x/A = \frac{T_1 - T_4}{L_A/k_A + L_B/k_B + L_C/k_C}$$ (9-9)

Thus q_x/A can be directly calculated to be −17.15 W/m^2. Note that the negative sign indicates that the heat flux is actually in the negative x direction as the energy is actually conducted into the cold-storage room. *A useful convention is to set up the heat transfer equations with the heat transfer assumed to be in the positive x direction and then use the sign of the result to indicate actual heat flow.*

(b) This part may be solved by setting the heat flux at the desired value of $0.5 \times (-17.15$ W/m$^2)$ and solving for L_B from Equation (9-9).

(c) When the thermal conductivity is a function of temperature, the differential equation for that material may be difficult to solve analytically. One approach for this problem is to solve the differential equation given in Equation Set (9-7), where the thermal conductivity k_B is expressed as a function of temperature T. This differential equation can be solved by the POLYMATH *Simultaneous Differential Equation Solver* with the use of the "if ... then ... else ..." statement to control the differential equation for temperature T depending upon the position within the wall. The shooting technique can be employed to determine a value of the heat flux so that the final boundary condition is satisfied. The initial value is $T = 255$ at $x = 0$, and the boundary condition to be satisfied is that $T = 298$ when $x = L_A + L_B + L_C = 0.19$. The POLYMATH equation set where variable Q_x is used for the heat flux (q_x/A) is given in Table 9–1.

Table 9–1 POLYMATH Program (File **P9-01C.POL**)

Line	Equation
1	d(T)/d(x) = if(x<=LA)then(-Qx/kA)else(if(x<=(LA+LB)and(x>LA))then(-Qx/kB)else(-Qx/kC))
2	err = T-298
3	LA = .015
4	Qx = -12.2
5	kA = .151
6	LB = .1
7	kB = 2.5*exp(-1225/T)
8	kC = .762
9	x(0)=0
10	T(0)=255
11	x(f)=0.19

A single-variable optimization technique as discussed in Problem 6.4 and a shooting technique as discussed in Problem 6.5 can be used to determine the heat flux Q_x.

This solution also allows convenient plotting of the temperature profile, as shown in Figure 9–2. Note that the T is not a linear function of distance in the new material because of the temperature variation of the thermal conductivity.

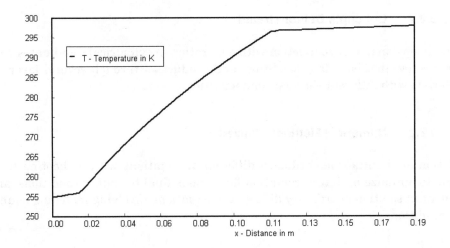

Figure 9–2 Temperature Profile through Multilayered Wall

The POLYMATH problem solution file for part (c) is found in directory Chapter 9 with file named **P9-01C.POL**.

(d) This solution should utilize Equation (9-3) to calculate T_5 since T_4 and Q_x are known at the completion of an integration. Again, a technique must be used to converge upon a value for Q_x that gives $T_5 = 298$.

9.2 HEAT CONDUCTION IN A WIRE WITH ELECTRICAL HEAT SOURCE AND INSULATION

9.2.1   Concepts Demonstrated

One-dimensional heat conduction with generation in cylindrical coordinates with a free convection boundary condition, and conduction through several layers of materials with different thermal conductivities.

9.2.2   Numerical Methods Utilized

Solution of simultaneous ordinary differential equations with a shooting technique to optimize an initial condition for various final boundary conditions, and sequential solution of ordinary differential equations involving matching boundary conditions.

9.2.3   Problem Statement

An insulated wire is carrying an electrical current. The passage of the current generates heat (thermal energy) according to

$$\dot{q} = \frac{I^2}{k_e} \tag{9-10}$$

where $\dot{q}$ has units of W/m^3, I is the current density in amps/m^2, and the electrical conductivity is given by k_e in ohm^{-1}m^{-1}. The wire has a uniform radius R_1 in m. A reasonable assumption is that $\dot{q}$ is constant and does not vary with position.

An energy balance on a cylindrical shell of thickness Δr and length L, as shown in Figure 9–3, leads to the differential equation

$$\frac{d}{dr}(rQ_r) = \dot{q}r \tag{9-11}$$

where the heat flux Q_r can be calculated from Fourier's law given by

$$Q_r = -k\frac{dT}{dr} \tag{9-12}$$

Conduction and convection are discussed in Problem 9.1.

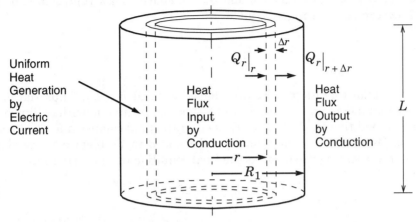

Figure 9–3 Differential Volume for Shell Balance

(a) Calculate and plot the temperature and heat flux within the wire if the wire surface is maintained at $T_1 = 15$ °C = 288.15 K, and the electrical and thermal conductivities are given by $k_e = 1.4 \times 10^5 \exp(0.0035T)$ ohm^{-1}m^{-1} and $k = 5$ W/m·K. The wire radius is $R_1 = 0.004$ m and the *total* current is maintained at $I = 400$ amps.

(b) The wire surface has a heat transfer coefficient due to natural convection given by $h = 1.32(|\Delta T|/D)^{1/4}$, where h is in W/m^2·K, ΔT is in K, and D is in m. Calculate the heat flux at the wire surface and plot the temperature distribution within the wire. The temperature of the surroundings is $T_b = 15$ °C = 288.15 K. The wire radius is $R_1 = 0.004$ m and the *total* current is maintained at $I = 50$ amps.

(c) The wire with $I = 50$ amps is covered by an insulating layer whose outer radius is $R_2 = 0.015$ m and whose constant thermal conductivity is $k_I = 0.2$ W/m·K. Calculate the heat flux at the exterior surface of the insulation and plot the temperature distribution within the wire and the insulation layer. Comment on the use of insulation in this application.

9.2.4   Solution (Partial)

(a) The energy generation expression of Equation (9-10) can be substituted into Equation (9-11) to give

$$\frac{d}{dr}(rQ_r) = \frac{I^2 r}{k_e} \tag{9-13}$$

where the initial condition is that the combined variable $(rQ_r) = 0$ at the centerline for the wire where both $r = 0$ and $Q_r = 0$. Fourier's law represented by Equation (9-12) can be rearranged to yield

$$\frac{dT}{dr} = -\frac{Q_r}{k} \tag{9-14}$$

where the final condition is known that $T = T_1$ at $r = R_1$. Thus the problem involves split boundary conditions and a shooting technique with optimization of the initial condition for T at $r = 0$, T_0, so that the final condition for T at $r = R_1$ is satisfied. Using the notation and logic developed in Problem 6.4, the objective function representing the error at the final condition can be written as

$$\varepsilon(T_0) = T|_{r = R_1} - 288.15 \tag{9-15}$$

Thus the initial condition on temperature T must be optimized to yield a final value of T, which is the known temperature of 288.15 K. When the optimized initial condition is determined, the value of the right-hand side of Equation (9-15) will be approximately zero at the final condition.

This problem formulation also requires use of the algebraic equation

$$Q_r = \frac{(rQ_r)}{r} \tag{9-16}$$

which is necessary to determine Q_r for Equation (9-14) from the combined variable (rQ_r). Division by zero at $r = 0$ in Equation (9-16) is avoided by setting $Q_r = 0$ at $r = 0$ through use of the "if ... then ... else ..." statement in the POLYMATH *Simultaneous Ordinary Differential Equation Solver.*

Thus this problem requires the simultaneous solution of Equations (9-13) through (9-16), as follows. This equation set in Table 9–2 shows the final result of a single-variable search that determines the initial condition for T at $r = 0$.

Table 9–2 POLYMATH Program (File **P9-02A.POL**)

Line	Equation
1	d(T)/d(r)=-Qr/k
2	d(rQr)/d(r)=qdot*r
3	Qr=if(r>0)then(rQr/r)else(0)
4	k=5.
5	ke=1.4e5*exp(0.0035*T)
6	R1=.004
7	err=T-288.15
8	I=400/(3.1416*R1^2)
9	qdot=I^2/ke
10	r(0)=0
11	T(0)=389.25
12	rQr(0)=0
13	r(f)=0.004

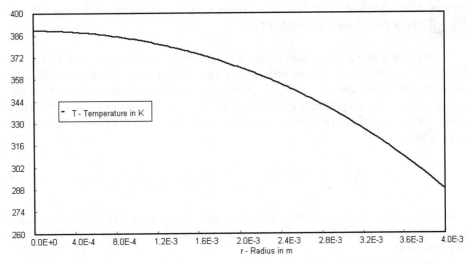

Figure 9–4 Temperature Profile for Wire with Surface Held at 288.15 K (15 °C)

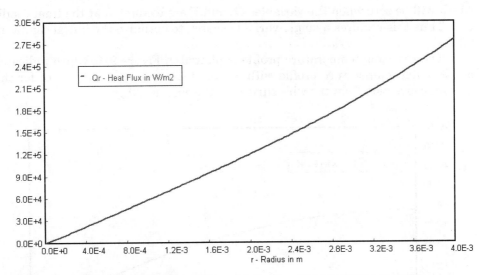

Figure 9–5 Heat Flux Profile for Wire with Surface Held at 288.15 K (15°C)

A plot of the temperature profile is shown in Figure 9–4. The calculated temperature profile exhibits the zero slope condition at $r = 0$ and matches the desired boundary condition $T = 288.15$ K (15 °C) at $r = 0.004$ m. The heat flux Q_r in W/m^2 is nonlinearly related to the radius, as indicated in Figure 9–5. Note that this numerical solution is similar to that of Problem 8.1, demonstrating the analogy between viscous flow in a pipe and this heated wire problem, as emphasized in the approach by Bird et al.[2]

The POLYMATH data file for part (a) is found in directory Chapter 9 with file named **P9-02A.POL**.

(b) The same differential equations apply as used in part (a), but the final boundary condition must account for the convection from the wire surface to the ambient temperature at $r = R_1$. The energy balance at the wire surface can be expressed, as developed in Equation (9-3), by setting the conduction (flux) to the wire surface equal to the convection (flux) from the wire surface to the surroundings. Thus

$$Q_r\big|_{r = R_1} = -k\frac{dT}{dr}\bigg|_{r = R_1} = h(T\big|_{r = R_1} - T_b) \qquad \textbf{(9-17)}$$

where the temperature of the surroundings is represented by T_b. An objective function based on the error in the energy balance of Equation (9-17) can be expressed as

$$\varepsilon(T_0) = Q_r\big|_{r = R_1} - h(T\big|_{r = R_1} - T_b) \qquad \textbf{(9-18)}$$

which will be zero when the variables Q_r and T are evaluated at the final condition. Thus this requires a single-variable search to optimize the initial condition for T.

The resulting temperature profile is plotted in Figure 9–6, which indicates a rather flat temperature profile with a large temperature driving force for the natural convection from the wire surface to the surroundings.

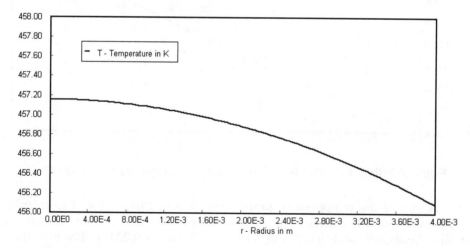

Figure 9–6 Temperature Profile in Wire with Natural Convection Boundary Condition

(c) The equation set for conduction without the generation terms must be used for the insulation layer, and a different thermal conductivity must be used. Thus the equation set within the insulation is given by

$$\frac{d}{dr}(rQ_r) = 0 \qquad R_1 < r \le R_2$$

$$\frac{dT}{dr} = -\frac{Q_r}{k_I} \qquad R_1 < r \le R_2 \qquad \textbf{(9-19)}$$

$$Q_r = \frac{(Q_r r)}{r} \qquad R_1 < r \le R_2$$

The preceding changes can be accomplished within the POLYMATH *Simultaneous Differential Equation Solver* by using the "if ... then ... else ..." capability, as discussed in Problem 9.1. The boundary condition of part (b) is then applied to the preceding equations at R_2.

9.3 RADIAL HEAT TRANSFER BY CONDUCTION WITH CONVECTION AT BOUNDARIES

9.3.1   Concepts Demonstrated

Heat transfer in the radial direction with constant thermal conductivities with convection at both boundaries.

9.3.2   Numerical Methods Utilized

Solution of a nonlinear algebraic equation with some additional explicit algebraic equations.

9.3.3   Problem Statement

A low-pressure steam system at 60 psia (292.73 °F) has saturated steam that is contained in a 2-inch schedule 40 steel pipe with a thermal conductivity of $k = 26$ btu/h·ft·°F. The pipe insulation has a thermal conductivity of k = 0.05 btu/h·ft·°F. The steam side heat transfer coefficient is $h_j = 2000$ btu/h·ft^2·°F and the air side heat transfer coefficient is $h_0 = 4$ btu/h·ft^2·°F. The ambient air temperature is 70°F.

(a)   Calculate the heat loss per foot if the insulation is 1 inch thick.
(b)   What is the outer radius of insulation in inches that is required to keep the heat loss at 50 btu/h per foot of pipe?

Additional Information and Data

The overall radial heat transfer through a cylindrical geometry with various resistances and two materials in series can be expressed (see Geankoplis[1]) as

$$q = \frac{T_1 - T_5}{1/(h_i A_i) + (r_1 - r_i)/k_A A_{A\,\mathrm{lm}} + (r_o - r_1)/k_B A_{B\,\mathrm{lm}} + 1/h_o A_o} \qquad (9\text{-}20)$$

for the case that is illustrated in Figure 9–7. Here the inside surface area is $A_i = 2\pi L r_i$, the outside surface area is $A_o = 2\pi L r_o$, and the area between the two materials is $A_1 = 2\pi L r_1$. The analytical solution when the thermal conductivities are constant requires the use of log mean areas for material A (the steel pipe) and material B (the insulation), given by

$$A_{A\,\mathrm{lm}} = \frac{A_1 - A_i}{\ln(A_1/A_i)} \qquad A_{B\,\mathrm{lm}} = \frac{A_o - A_1}{\ln(A_o/A_1)} \qquad (9\text{-}21)$$

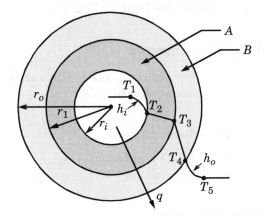

Figure 9–7 General Radial Temperature Profiles for Conduction through Two Materials with Convection at the Boundaries

9.3.4   Solution (Suggestions)

(a) The heat loss can be directly calculated from Equation (9-20) and the appropriate areas. This can conveniently be accomplished in the POLYMATH *Simultaneous Algebraic Equation Solver* since all of the equations are explicit.

(b) The value of r_o can be determined using Equation (9-20) as a nonlinear equation with the additional calculations as explicit algebraic equations. Simple modifications to the POLYMATH equation for the solution to part (a) are required.

9.4 ENERGY LOSS FROM AN INSULATED PIPE

9.4.1   Concepts Demonstrated

Radial heat conduction within pipe insulation with variable thermal conductivity and a free convection boundary condition.

9.4.2   Numerical Methods Utilized

Solution of simultaneous ordinary differential equations with a shooting technique to optimize a boundary condition.

9.4.3   Problem Statement

A horizontal steel pipe carrying steam at 450 K is to be insulated with a newly developed material whose thermal conductivity varies with temperature, as summarized in Table 9–3. The pipe can be assumed to be at the steam temperature due to an effective internal heat transfer coefficient and the large thermal conductivity of steel. The external diameter of the steel pipe is 0.033 m, and the ambient temperature is 300 K.

The external heat transfer coefficient from the surface of a cylinder due to natural convection in air can be approximated by (see Geankoplis[1])

$$h = 1.32(|\Delta T|/D)^{1/4} \tag{9-22}$$

where h is the heat transfer coefficient in $W/m^2 \cdot K$, ΔT is the temperature difference between the surface and the air in K, and D is the cylinder diameter in m.

(a)   Calculate the heat loss per meter from the uninsulated pipe.
(b)   Calculate the heat loss per meter from the pipe when the insulation thickness is 0.04 m and plot the temperature distribution in the insulation.

Table 9–3 Thermal Conductivity Data

T K	k W/m · K
250	0.042
300	0.061
350	0.077
400	0.087
450	0.094
500	0.098

9.5 HEAT LOSS THROUGH PIPE FLANGES

9.5.1   Concepts Demonstrated

Radial heat conduction in one dimension with simultaneous heat convection. Effects of various ambient temperatures on convective losses, and calculation of heat loss through pipe flanges made of different metals.

9.5.2   Numerical Methods Utilized

Solution of two point boundary value problem involving ordinary differential equations using a shooting technique and single-variable optimization.

9.5.3   Problem Statement

Metal pipes may be connected by the bolting together of two flanges on the pipe ends, as shown in Figure 9–8, without the bolts. Often these unions are not insulated so that the pipes can be quickly disconnected. For heat transfer purposes, this union can be considered as a cylinder of solid metal made from two pipe flanges. Consider the uninsulated union of Figure 9–8 where the fluid in the pipe is at 260°F and the heat transfer coefficient to the surroundings is constant at $h = 3$ btu/h$\cdot$ft$^2\cdot$°F.

(a)   Select one of the metals listed in Table 9–4 and calculate the total heat loss from a single flange for an average day when the average ambient temperature $T_\infty = 60$°F. Assume that the thermal conductivity is constant at the value reported for 212 F in Table 9–4.

(b)   Plot the heat transfer rate and temperature versus radius for part (a). What percentage of the total heat loss is from the end of the flange?

(c)   Repeat the total heat loss calculations of part (a) for the same flange material for a cold winter day ($T_\infty = 10$°F) and a very hot summer day ($T_\infty = 100$°F).

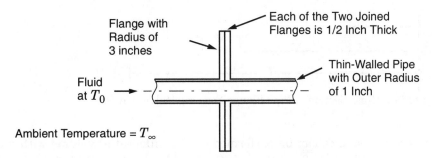

Figure 9–8 Cross Section of Union Consisting of Two Pipe Flanges

Table 9–4 Thermal Conductivity of Selected Metals (Welty,[3] with permission.)

Metal	Thermal Conductivity (btu/h · ft · °F)		
	68°F	212°F	572°F
Aluminum	132	133	133
Copper	223	219	213
Iron	42.3	39	31.6
Nickel	53.7	47.7	36.9
Stainless Steel	9.4	10.0	13
Steel (1% C)	24.8	24.8	22.9

9.5.4   Solution (Partial)

The union formed by two flanges is symmetrical, and thus only one flange has to be considered with heat loss from one exposed circular face and an exposed rim as shown in Figure 9–9. The other side of the circular face, which is the line of symmetry, is treated as insulated due to symmetry with no heat transfer across this surface. The problem becomes one dimensional, with the heat transfer being a function of the radius. The boundary condition at the inner surface where $r = R_1$ is the temperature of the fluid. The thickness of the pipe is neglected as it is very thin. The boundary condition at the outer surface where $r = R_2$ is determined by heat convection from the surface to the ambient temperature T_∞.

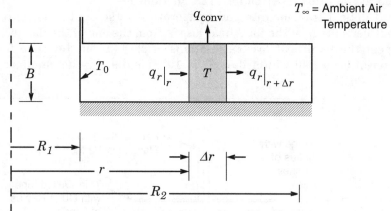

Figure 9–9 Cross Section of Pipe Flange Forming Half of Union

An energy balance can be performed on the differential volume within the flange shown in Figure 9–9, where q_r is the total heat rate in the r direction due to convection in btu/h.

INPUT + GENERATION = OUTPUT + ACCUMULATION

$$q_r\big|_r + 0 = q_r\big|_{r+\Delta r} + h(2\pi r)\Delta r(T - T_\infty) + 0 \qquad \textbf{(9-23)}$$

Rearrangement of the preceding equation and utilization of the definition of the derivative for q_r by taking the limit as $\Delta r \to 0$ yields

$$\frac{dq_r}{dr} = -h(2\pi r)(T - T_\infty) \qquad \textbf{(9-24)}$$

The relationship between heat rate q_r and heat flux Q_r at radius r requires the cross-sectional area for heat conduction at radius r, giving

$$q_r = Q_r(2\pi r)B \qquad \textbf{(9-25)}$$

Substitution of Equation (9-25) into Equation (9-24) yields

$$\frac{d}{dr}(rQ_r) = \frac{-hr(T - T_\infty)}{B} \qquad \textbf{(9-26)}$$

where the heat flux Q_r in btu/ft$^2 \cdot$h can be calculated from Fourier's law, given by

$$Q_r = -k\frac{dT}{dr} \qquad \textbf{(9-27)}$$

which can be rearranged to

$$\frac{dT}{dr} = (-Q_r)/k \qquad \textbf{(9-28)}$$

In the preceding equations, h is the heat transfer coefficient in btu/h$\cdot$ft$^2\cdot$°F, k is the thermal conductivity of the metal in btu/h$\cdot$ft$\cdot$°F, and B is the thickness of the flange in ft. The algebraic equation is given by

$$Q_r = \frac{rQ_r}{r} \qquad \textbf{(9-29)}$$

Thus the differential equations that describe the heat transfer processes consist of Equations (9-26) and (9-28) along with the algebraic relationship given Equation (9-29), which is needed to calculate Q_r since the product of two-variable rQ_r is used as a problem variable.

The boundary (final) condition on Equation (9-26) is obtained from an energy balance on the rim, which equates the heat flux by conduction to the rim to the heat that is transported from the rim by convection. Thus

$$Q_r\big|_{r=R_2} = h(T - T_\infty)\big|_{r=R_2} \qquad \textbf{(9-30)}$$

and thus the variable rQ_r is calculated from

$$(rQ_r)\big|_{r = R_2} = R_2h(T - T_\infty)\big|_{r = R_2} \qquad\qquad \textbf{(9-31)}$$

The initial condition for Equation (9-28) is that the temperature is the temperature of the fluid in the pipe designated by T_0 at radius $r = R_1$.

$$T\big|_{r = R_1} = T_0 \qquad\qquad \textbf{(9-32)}$$

(a) For the case of the flange and pipe being made of aluminum, the parameter values are given by k = 133 btu/h·ft·°F, T_0 = 260 °F, T_∞ = 60 °F, and h = 3 btu/h·ft^2·°F. The solution can utilize a shooting technique to solve the differential Equations (9-26) and (9-28), which is discussed in Problems 6.5, 9.2, and 9.4. The initial condition for Equation (9-26) must be determined for the variable rQ_r so that the boundary condition (final condition) of Equation (9-31) is met. This can be accomplished by defining an objective function as

$$\varepsilon(rQ_r) = (rQ_r)\big|_{r = R_2} - R_2h(T - T_\infty)\big|_{r = R_2} \qquad\qquad \textbf{(9-33)}$$

Thus the initial value of (rQ_r) must be optimized using a single-variable search to minimize the objective function calculated from Equation (9-33).

An equation set with the converged solution to this problem using the POLYMATH *Ordinary Differential Equation Solver* is as given in Table 9–3

Table 9–5 POLYMATH Program (File **P9-05A.POL**)

Line	Equation
1	d(rQ)/d(r)=-h*r*(T-Tinf)/B
2	d(T)/d(r)=-Q/k
3	Q=rQ/r
4	k=133
5	h=3
6	Tinf=60
7	B=0.5/12
8	err=rQ-r*h*(T-Tinf)
9	q=Q*2*3.1416*r*B
10	r(0)=0.08333
11	rQ(0)=542.4
12	T(0)=260
13	r(f)=0.25

The POLYMATH problem solution file for part (a) is found in the *Simultaneous Differential Equation Solver Library* located in directory Chapter 9 with file named **P9-05A.POL**.

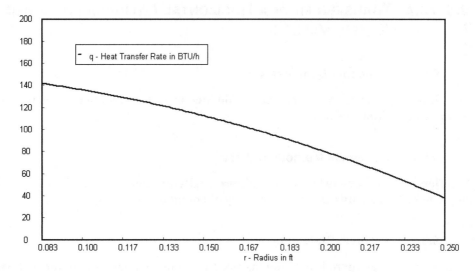

Figure 9–10 Heat Transfer Rate in Flange as a Function of Radius

(b) The actual values of the heat rate q at various values of r can be calculated from the converged solution and Equation (9-25), which is included in the preceding equation set. The results are shown in Figure 9–10. The heat rate at the end of the flange is 38.6 btu/h compared to the total heat rate at R_1 of 142 btu/h. Thus the percentage of the total lost from the end is 27.2%.

9.6 HEAT TRANSFER FROM A HORIZONTAL CYLINDER ATTACHED TO A HEATED WALL

9.6.1   Concepts Demonstrated

One-dimensional heat transfer due to conduction in a rod subject to convection and radiation from the surface.

9.6.2   Numerical Methods Utilized

Solution of simultaneous ordinary differential equations with a shooting technique in order to converge upon a variety of boundary conditions.

9.6.3   Problem Statement

A horizontal circular rod (cylinder) is used to transfer heat from a vertical wall that is held at a constant temperature $T_0 = 275°F$ to the surrounding air at temperature T_∞, as illustrated in Figure 9–11. The natural convection from the surface of the rod is given by

$$h_s = 0.27|\Delta T/D|^{1/4} \tag{9-34}$$

where h_s is the heat transfer coefficient in btu/h·ft^2·°F, ΔT is the temperature difference between the rod surface temperature T and the ambient air temperature T_∞ in °F, and D is the diameter of the rod in feet.

At the end of the rod that is exposed to the air, the heat transfer coefficient is given by

$$h_e = 0.18|\Delta T|^{1/3} \tag{9-35}$$

where the units are the same as for Equation (9-34).

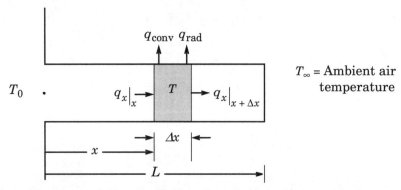

Figure 9–11 Differential Element for Heat Transfer for a Horizontal Rod

The radiative heat transfer from the rod surfaces is described by the Stefan-Boltzmann equation with an effective emissivity of ε to the surroundings at temperature T_∞

$$q_{rad} = \varepsilon \sigma A (T^4 - T_\infty^4) \tag{9-36}$$

where $\sigma = 0.1714 \times 10^{-8}$ btu/h $\cdot$ ft$^2 \cdot$ °R^4 and A is the surface area in ft^2.

For the usual case where the diameter of the rod is small relative to the length L of the rod, there will be little temperature change in the radial direction but there will be a significant temperature change in the x direction. Thus a mathematical description involves one-dimensional conduction within the rod subject to convection and radiation from the rod surface to the ambient air and surroundings.

(a) Calculate the temperature profile down the length of the horizontal rod and the total rate of heat removed from the heated wall when *only* convection from *all* rod surfaces is considered. The wall temperature is $T_0 = 275$°F, the surroundings are at $T_\infty = 70$°F, the rod diameter $D = 0.1$ ft, the rod length $L = 0.5$ ft, and the rod thermal conductivity is $k = 30$ btu/h $\cdot$ ft $\cdot$ °F.

(b) Repeat part (a) with the assumption that there is no convection from the exposed end of the rod.

(c) Repeat part (a) if radiation from all rod surfaces with an effective emissivity of $\varepsilon = 0.85$ is included with the convection.

(d) What conclusions can be reached about the boundary condition assumption made in part (b) and the inclusion of radiation in part (c)?

(e) Repeat parts (a) through (d) for $T_0 = 575$°F.

Additional Information and Data

Modeling Equations for Heat Transfer The differential equation describing the temperature in the rod can be obtained by an energy balance on the differential element depicted in Figure 9–11.

INPUT + GENERATION = OUTPUT + ACCUMULATION

$$q_x \big|_x + 0 = q_{conv} + q_{rad} + q_x \big|_{x + \Delta x} + 0 \tag{9-37}$$

Fourier's law [Equation (9-1)] applies to the conduction within the rod and can be expressed as

$$\frac{dT}{dx} = (-Q_x)/k \tag{9-38}$$

where the heat flux is represented by $Q_x = q_x/A$. The initial condition for Equation (9-38) is that $T = T_0$ at $x = 0$.

The expressions for heat flux, convection, and radiation with the corresponding areas can be introduced into Equation (9-37):

$$Q_x\left(\frac{(\pi D^2)}{4}\right)\bigg|_x = h(\pi D \Delta x)(T - T_\infty) + \varepsilon\sigma(\pi D \Delta x)(T^4 - T_\infty^4) + Q_x\left(\frac{(\pi D^2)}{4}\right)\bigg|_{x + \Delta x} \quad \textbf{(9-39)}$$

Taking the limit of the preceding equation as $\Delta x \to 0$ yields

$$\frac{dQ_x}{dx} = -\left(\frac{4}{D}\right)[h_s(T - T_\infty) + \varepsilon\sigma(T^4 - T_\infty^4)] \quad \textbf{(9-40)}$$

The boundary condition at the end of the rod is determined from a steady-state energy balance at the end surface where the area is $A = \pi D^2/4$.

$$\text{INPUT + GENERATION = OUTPUT + ACCUMULATION}$$

$$q_x\big|_{x = L} + 0 = q_{conv} + q_{rad} + 0 \quad \textbf{(9-41)}$$

Insertion of the various terms into Equation (9-41) and solving for the heat flux yields

$$Q_x\big|_{x = L} = [h(T - T_\infty) + \varepsilon\sigma(T^4 - T_\infty^4)]\big|_{x = L} \quad \textbf{(9-42)}$$

Equation for Heat Rate into Rod The total heat lost from the rod due to convection and radiation is equal to the heat conducted into the rod at the base where $x = 0$. This can be expressed by

$$q_x\big|_{x = 0} = -k\left(\frac{(\pi D^2)}{4}\right)\frac{dT}{dx}\bigg|_{x = 0} \quad \textbf{(9-43)}$$

and evaluated once the numerical solution to Equation (9-38) has been achieved.

9.6.4   Solution (Suggestions)

The numerical solution requires the simultaneous integration of Equations (9-38) and (9-40). A shooting technique, as outlined in Problem 6.5, can be used to converge upon the initial condition for $(dQ_x)/dx\big|_{x = 0}$, which results in the desired final condition for Q_x. The corresponding objective function for the final condition, which should converge to zero, can be written from Equation (9-42) as

$$\text{err} = Q_x\big|_{x = L} - [h_e(T - T_\infty) + \varepsilon\sigma(T^4 - T_\infty^4)]\big|_{x = L} \quad \textbf{(9-44)}$$

The heat transfer rate into the rod can then be obtained from Equation (9-43). The various modes of heat transfer can be modeled by adjusting the terms in the equations, and the boundary conditions can be similarly modified.

9.7 HEAT TRANSFER FROM A TRIANGULAR FIN

9.7.1 Concepts Demonstrated

One dimensional heat transfer due to conduction in a triangular fin subject to convection from the surface.

9.7.2 Numerical Methods Utilized

Solution of simultaneous ordinary differential equations with a shooting technique in order to converge upon split boundary conditions.

9.7.3 Problem Statement

Triangular fins are widely used to increase heat transfer with a minimal amount of fin material. Such a fin is shown in Figure 9–12. A simplified treatment of this fin can utilize the assumptions that the temperature T varies only in the x direction, no energy is lost from either the sides or the end of the fin, and simple convection with a constant heat transfer coefficient describes the convection from the fin surfaces.

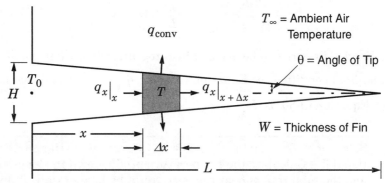

Figure 9–12 Differential Element for Heat Transfer in a Triangular Fin

Thus an energy balance on the differential volume of the fin as shown in Figure 9–12 yields

$$\text{INPUT} + \text{GENERATION} = \text{OUTPUT} + \text{ACCUMULATION}$$

$$q_x\big|_x + 0 = q_{conv} + q_x\big|_{x + \Delta x} + 0 \tag{9-45}$$

Utilization of the convective heat transfer from the upper and lower fine surfaces gives

$$q_{conv} = 2W\Delta x \sec\theta\, h_s(T - T_\infty) \tag{9-46}$$

where H is the height of the fin, W is the width of the fin, and θ is angle at the tip of the fin whose tangent is $H/2L$. Definitions and units are the same as in Problem 9.6.

Equation (9-45) can be divided by Δx and rearranged to the definition of a derivative. As the limit is taken where $\Delta x \to 0$, the following differential equation results:

$$\frac{dq}{dx} = -2W\sec\theta h_s(T - T_\infty) \tag{9-47}$$

Fourier's law for heat conduction in the fin can be written as

$$q = -kA\frac{dT}{dx} \tag{9-48}$$

where the cross-sectional area given by A can be expressed as a function of x by

$$A = \left(1 - \frac{x}{L}\right)HW \tag{9-49}$$

Thus Equation (9-48) can be rewritten using Equation (9-49) for A as

$$\frac{dT}{dx} = \frac{q}{-k\left(1 - \frac{x}{L}\right)HW} \tag{9-50}$$

where division by zero can be avoided by integration to $x = 0.999999L$

The final condition for Equation (9-47) is $q|_{x = L} = 0$ and the initial condition for Equation (9-50) is $T|_{x = 0} = T_0$.

(a) Select one of the metals in Table 9–4 for use in a triangular cross-sectional fin. Calculate the temperature profile down the length of the fin and the total rate of heat removed from the heated wall. The wall temperature is $T_0 = 212°F$, the fin length is $L = 1.0$ ft, the fin width is $W = 1.0$ ft, and the fin height is $H = 0.1$ ft. The surroundings are at temperature $T_\infty = 60°F$ and $h_s = 2.5$ btu/h $\cdot$ ft$^2 \cdot$ °F. Assume that the metal thermal conductivity remains constant at the value given in Table 9–4 for 212°F.

(b) What is the efficiency factor η_f for the fin in part (a) that is defined as the ratio of the heat rate actually transferred from the base to the heat rate if the entire fin surface was at the base temperature T_0?

(c) How much error is introduced into the result for part (b) by assuming a constant thermal conductivity at 212°F for the metal used in the fin?

(d) What second-order ordinary differential equation and boundary conditions can alternatively be used to describe the combined energy balance and Fourier's law for the constant thermal conductivity case?

9.8 SINGLE-PASS HEAT EXCHANGER WITH CONVECTIVE HEAT TRANSFER ON TUBE SIDE

9.8.1   Concepts Demonstrated

Simple heat exchanger with average, log mean, and point temperature driving forces, and convective heat transfer involving turbulent flow where physical properties vary with temperature, and overall and differential energy balances applied to simple heat exchangers.

9.8.2   Numerical Methods Utilized

Solutions of nonlinear equations and ordinary differential equations.

9.8.3   Problem Statement

A simple heat exchanger consists of a single tube with an inside diameter of $D = 0.01033$ m and an equivalent length $L = 8$ m in which $m = 1$ kg/s of a light hydrocarbon oil is heated by condensing steam at $T_S = 170°C$. Because of the high heat transfer coefficient from the steam to the tube surface and the high thermal conductivity of the tube wall, the inside temperature of the tube wall can be assumed to be the steam temperature. The oil has a viscosity which varies significantly with temperature and is given by

$$\mu = \exp[-12.86 + 1436/(T-153)] \qquad (9\text{-}51)$$

in kg/m·s with temperature T in K. Other physical properties are approximately constant with temperature and are given by $\rho = 850$ kg/m^3, $C_p = 2000$ J/kg·K, and $k = 0.140$ W/m·K. The inlet oil temperature is $T_1 = 40°C$.

(a)   Calculate the outlet temperature of the oil stream in °C using the bulk mean oil temperature for physical properties and average ΔT for the temperature driving force for the heat transfer. What is the total amount of heat transferred? The Sieder-Tate expression given by Equation (3-21) should be used to determine h_i for the oil side in turbulent flow.

(b)   Repeat part (a) utilizing the log mean temperature difference for the heat transfer.

(c)   Repeat part (a) with a differential equation approach in which the local energy balance is solved down the length of the heat exchanger, with the Sieder-Tate correlation of Equation (3-21) providing the local heat transfer coefficient from local physical properties.

Additional Information and Data

The local or point energy balance utilizing the overall heat transfer coefficient can be expressed (see Geankoplis[1]) as

$$q = h_i A \Delta T \tag{9-52}$$

where h_i is the convective heat transfer coefficient based on the inside area of the tube in $\text{W/m}^2 \cdot \text{K}$ and the inside surface area is A in m^2.

An overall energy balance on the tube side fluid yields

$$q_T = m C_p (T_2 - T_1) \tag{9-53}$$

where q_T is the total heat transfer rate and $(T_2 - T_1)$ represents the temperature increase in the fluid as it passes through the heat exchanger.

Average Temperature Driving Force When the driving force does not vary greatly down the heat exchanger, the mean temperature difference at the inlet and outlet can be used in Equation (9-52):

$$\Delta T = \Delta T_m = [(T_1' - T_1) + (T_2' - T_2)]/2 \tag{9-54}$$

where the subscript 1 represents the tube entrance location and subscript 2 represents the tube outlet location as diagrammed for the general cocurrent case in Figure 9–13. The prime ($'$) represents the hotter stream, which for this problem is the constant steam temperature T_S.

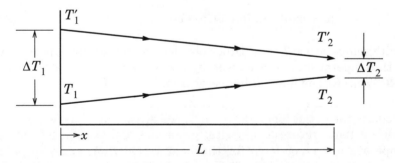

Figure 9–13 General Temperature Distribution for One-Pass Cocurrent Heat Exchanger

Log Mean Temperature Driving Force When the driving force varies significantly down the heat exchanger, the log mean temperature difference at the inlet and outlet should be used in Equation (9-52):

$$\Delta T = \Delta T_{lm} = \frac{[(T_1' - T_1) - (T_2' - T_2)]}{\ln[(T_1' - T_1)/(T_2' - T_2)]} \tag{9-55}$$

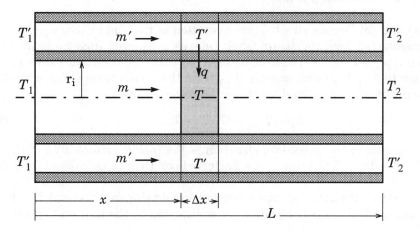

Figure 9–14 Differential Element for Energy Balance in One-Pass Cocurrent Heat Exchanger

Local Temperature Driving Force A differential energy balance can be made at a point that is located at distance x from the entrance, as shown in Figure 9–14. The steady-state energy balance on the differential element of the tube yields

INPUT + GENERATION = OUTPUT + ACCUMULATION

$$mC_pT\big|_x + q + 0 = mC_pT\big|_{x+\Delta x} + 0 \tag{9-56}$$

where m is the mass flow rate in the tube in kg/s. The q term, which represents the input from the hot stream to the tube stream in the differential element, can be expressed using Equation (9-52) and the differential area for heat transfer as

$$q = h_i(2\pi r_i \Delta x)(T' - T) \tag{9-57}$$

Rearranging and taking the limit as Δx goes to zero yields the differential equation

$$\frac{d}{dx}(mC_pT) = h_i(2\pi r_i)(T' - T) \tag{9-58}$$

For constant m and C_p, the differential energy balance simplifies to

$$\frac{dT}{dx} = \frac{h_i(2\pi r_i)(T' - T)}{mC_p} \tag{9-59}$$

This ordinary differential equation can be used to predict the temperature in the tube side of the heat exchanger as a function of distance down the exchanger. The initial condition is that $T = T_1$ at $x = 0$. Integration of this equation to $x = L$ yields the exit tube fluid temperature.

9.8.4   Solution (Partial)

(a) and (b) During steady-state operation of this heat exchanger, the total heat transferred from the steam (shell side) as calculated from Equation (9-52), where $q = q_T$, must equal the heat transferred to the light oil (tube side) given by Equation (9-53). Thus elimination of q_T from these equations yields a nonlinear equation which can be solved for the outlet temperature T_2.

$$h_i A \Delta T = m C_p (T_2 - T_1) \qquad (9\text{-}60)$$

The area term in the preceding equation is the inside surface area of the tube given by $A = 2\pi r_i L$. The total heat transferred can then be calculated from Equation (9-53).

Note that when the log mean temperature driving force is used, it is helpful to rearrange the nonlinear equation by multiplying by the ln term to give

$$h_i[(T_1' - T_1) - (T_2' - T_2)] = m C_p (T_2 - T_1) \ln[(T_1' - T_1)/(T_2' - T_2)] \qquad (9\text{-}61)$$

9.9 DOUBLE-PIPE HEAT EXCHANGER

9.9.1   Concepts Demonstrated

Calculations of cocurrent and countercurrent heat exchangers using differential energy balances that allow for local variations in physical properties and heat transfer coefficients.

9.9.2   Numerical Methods Utilized

Solution of coupled first-order ordinary differential equations with known initial conditions (cocurrent flow) and split boundary conditions (countercurrent flow).

9.9.3   Problem Statement

One mode of operation of a double-pipe heat exchanger, which is diagramed in Figure 9–15 for cocurrent operation, is to cool a steady stream of m' lb_m/h of fluid in the tube from inlet temperature T_1 to an exit temperature T_2. The fluid in the shell side with a flow rate of m lb_m/h is correspondingly heated from inlet temperature T'_1 to exit temperature T'_2.

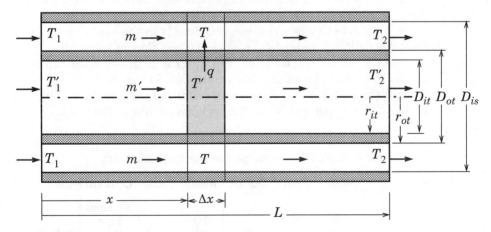

Figure 9–15 Double-Pipe Heat Exchanger for Cocurrent (Parallel) Flow

Select one of the fluids listed in Table 9–6 to be cooled by water in a double-pipe heat exchanger. The mass flow rate of cooling water is always to be 300% or three times that of the fluid mass flow rate for liquids and thirty times that of the fluid mass flow rate for gases. The cooling water is continuously available at 65°F. The local convective heat transfer coefficients on both the tube and shell sides should be calculated from the Dittus-Boelter[4] correlation, given by

$$Nu = 0.023 Re^{0.8} Pr^n \tag{9-62}$$

or

$$\frac{hD}{k} = 0.023 \left(\frac{Dv\rho}{\mu}\right)^{0.8} \left(\frac{C_p\mu}{k}\right)^n \tag{9-63}$$

where $n = 0.4$ for heating and $n = 0.3$ for cooling. The various terms are discussed in Problem 3.5. Note that the $v\rho$ term in the Re can also be conveniently expressed as the mass flux or mass velocity where $v\rho = m/A_c$, with m representing the mass flow rate and A_c representing the cross-sectional area for flow.

The center heat exchanger tube is made of copper with a constant thermal conductivity of 220 btu/ft·h·°F, and the exterior of the steel pipe shell is very well insulated. Scale formation is not a problem. The physical properties of the different fluids at various temperature are given in Tables E–2 and E–3 of Appendix E.

Select a fluid from Table 9–6 for all calculations.
(a) Calculate the length required to reach the desired exit temperature when the selected fluid is in the heat exchanger tube and the cooling water is in cocurrent flow in the shell.
(b) Calculate the length when the selected fluid is in the heat exchanger tube and the cooling water is in countercurrent flow in the shell.
(c) Calculate the length when the cooling water is in the heat exchanger tube and the selected fluid is in cocurrent flow in the shell.
(d) Calculate the length when the cooling water is in the heat exchanger tube and the selected fluid is in countercurrent flow in the shell.

Table 9–6 Data on Heat Exchanger Operation for Various Fluids

Fluid	Flow	Inlet	Outlet	Inside Tube		Outside Pipe	
	m	T	T	OD		Nominal Pipe	
	lb_m/h	°F	°F	inch	BWG	inch	Schedule
Benzene (liquid)	3000	150	80	7/8	16	$1\frac{1}{2}$	40
Kerosene (liquid)	4000	140	85	1	14	2	80
Ammonia (liquid)	2000	120	82	3/4	12	$1\frac{1}{4}$	40
Carbon Dioxide (gas)	25	250	85	5/8	16	1	40
Sulfur Dioxide (gas)	35	300	90	3/4	14	$1\frac{1}{4}$	80

Additional Information and Data

The heat exchanger can be operated with cocurrent flow in both the tube and shell, as shown in Figure 9–15, or with countercurrent flow, where the flow in the shell is opposite to that in the tube.

Cocurrent or Parallel Flow A differential energy balance on the tube fluid in concurrent flow in a manner similar to that detailed in Problem 9.8, with the prime indicating the fluid temperature or property of the higher temperature fluid, leads to

$$\frac{d}{dx}(T') = -\frac{U_i(\pi D_i)(T' - T)}{m'C_p'}$$
(9-64)

where the overall heat transfer coefficient U_i in btu/h·ft^2·°F is based on the inside area given by πD_i with the inside tube diameter D_i in ft. The initial condition for Equation (9-64) is just the inlet temperature T_1' to the tube, and the final condition is the outlet temperature T_2' from the tube.

Similarly, for the well insulated shell, a differential energy balance on the shell fluid in concurrent flow and the inside heat transfer area yields

$$\frac{dT}{dx} = \frac{U_i(\pi D_i)(T' - T)}{mC_p}$$
(9-65)

where the convective heat transfer coefficient for the shell is also based on the inside diameter D_i of the inner pipe in ft. The inlet and outlet shell temperatures are T_1 and T_2, respectively. Figure 9–13 illustrates the general temperature distributions for cocurrent flow using this notation.

Overall Heat Transfer Coefficients The local overall heat transfer coefficient based on the inside area is calculated from (Geankoplis[1])

$$U_i = \frac{1}{1/h_i + ((r_{ot} - r_{it})D_{it})/(k_t D_{tlm}) + D_{it}/(D_{ot}h_o)}$$
(9-66)

where t represents the tube wall material and dimensions. The D_{tlm} term is the log mean diameter (or area) for the tube wall.

Heat Transfer within the Shell The heat transfer coefficient for the fluid in the shell annulus is calculated by using an equivalent diameter that is determined by taking the outer diameter of the annulus and subtracting the inner diameter of the annulus. This equivalent diameter is then used in calculating the Reynolds and Nusselt numbers in the heat transfer correlations.

Countercurrent Flow The differential energy balance on the tube fluid in countercurrent flow remains the same as Equation (9-64); however, the reversal of the flow direction in the shell yields a differential energy balance given by

$$\frac{dT}{dx} = -\frac{U_i(\pi D_i)(T' - T)}{mC_p}$$
(9-67)

which is just the negative of the concurrent equation. The initial condition, however, is now the outlet shell temperature T_1, and the final condition is the inlet shell temperature T_2. Thus in this situation, the simultaneous solution of Equations (9-64) and (9-67) is a split boundary problem, which typically involves assuming an outlet temperature T_1 for the shell followed by integration to determine the inlet temperature T_2 of at the desired heat exchanger length. The solution of two-point boundary value problems is discussed in Problem 6.5.

9.9.4   Solution (Suggestions)

The properties of liquid water and the other fluids are found in Appendix E. These properties can be represented as a function of the temperature in °F by using the techniques and suggested correlation equations introduced in Problems 3.3 and 8.5.

The POLYMATH table data files for the different fluids are found in directory Tables with files named **E-02A.POL**, **E-02B.POL**, and **E-03A.POL** through **E-03C.POL**.

9.10 HEAT LOSSES FROM AN UNINSULATED TANK DUE TO CONVECTION

9.10.1 Concepts Demonstrated

Calculation of temperature in a well-mixed and uninsulated tank with convective heat losses to the surroundings.

9.10.2 Numerical Methods Utilized

Solution of a nonlinear algebraic equation and explicit algebraic equations.

9.10.3 Problem Statement

Cooling water exits a heat exchanger at a rate of $Q = 0.5$ m^3/h and temperature of $T = 80$°C. The water is collected in a cylindrical tank with a diameter $D = 1$ m and a height of $H = 2$ m. It can be assumed that the contents of the tank are well mixed and that all the heat loss is due to natural convection to the surrounding air. Water is being withdrawn from the tank at the same rate as it enters. The external surface of the tank can be assumed to be at the tank temperature, and the bottom of the tank can be considered to be very well insulated.

The properties of air at atmospheric pressure and for various temperatures are given in Table E–1 of Appendix E.

(a)   Calculate the energy lost from the tank due to heat transfer in W and the temperature of the water in °C within the tank for different surrounding air temperatures of –10, 15, and 40°C in calm air.

(b)   Repeat part (a) for windy conditions with a constant velocity of 30 mph.

Additional Information and Data

Natural Convection For natural convection from the top surface under calm conditions, the following correlations (Geankoplis[1]) can be used:

$$Nu = 0.54Ra^{1/4} \quad \text{for} \quad 10^5 < Ra < 2 \times 10^7$$
$$Nu = 0.14Ra^{1/3} \quad \text{for} \quad 2 \times 10^7 < Ra < 3 \times 10^{10}$$

(9-68)

where

$$Nu = \frac{hL}{k} \text{ is the Nusselt number}$$

$$Ra = GrPr \text{ is the Rayleigh number}$$

$$Gr = \frac{\beta g \rho^2 L^3 \Delta T}{\mu^2} \text{ is the Grashof number}$$

$$Pr = \frac{\mu C_p}{k} \text{ is the Prandtl number}$$

In the preceding equations, h is the heat transfer coefficient in $W/m^2 \cdot K$, L is a characteristic length in m, k is the air thermal conductivity in $W/m \cdot K$, β is the air coefficient of thermal expansion in $1/K$, g is the gravitational acceleration given by 9.80665 m/s^2, ρ is air density in kg/m^3, ΔT is the positive temperature difference between the tank surface and the air in K, μ is the air viscosity in $kg/m \cdot s$, and C_p is the heat capacity of air in $J/kg \cdot K$. In this problem, $L = 0.8862D$, as this is the length of a square that has the same area as the top of the tank with diameter D.

For natural convection over a wide range of Ra, the Churchill and Chu[5] correlation can be used for the vertical side of the cylindrical tank.

$$Nu = \left(0.825 + \frac{0.387Ra^{1/6}}{[1 + (0.492/Pr)^{9/16}]^{8/27}}\right)^2 \qquad \textbf{(9-69)}$$

Equations (9-68) and (9-69) should be evaluated with the properties of air at the film temperature T_f, which is given by

$$T_f = \frac{T + T_{\text{air}}}{2} \qquad \textbf{(9-70)}$$

Forced Convection For forced convection the heat transfer coefficients can be approximated for both the top and the side of the cylindrical tank by using

$$Nu = 0.0366Re_L^{0.8}Pr^{1/3} \qquad \textbf{(9-71)}$$

where the Reynolds number is defined as $Re_L = (Lv\rho)/\mu$. Note that the characteristic length for the circular flat plate, which is the top of the cylindrical tank, is estimated by $L = 0.8862D$, while the length for convection from the side of the cylindrical tank is given by $L = D$. The velocity v is defined as the wind velocity in m/s. The air properties in Equations (9-71) are evaluated at the film temperature given by Equation (9-70).

9.10.4   Solution (Suggestions)

An enthalpy balance on the tank contents yields

$$Q\rho_w C_{pw}(80 - T) = h_1 A_1(T - T_{air}) + h_2 A_2(T - T_{air}) \qquad \textbf{(9-72)}$$

where ρ_w is the density of water, C_{pw} is the heat capacity of the water, T is the outlet temperature from the tank, h_1 is the horizontal heat transfer coefficient, A_1 is the horizontal surface area for heat transfer, h_2 is the vertical heat transfer coefficient, and A_2 is the vertical surface area for heat transfer.

The properties for air can be represented as functions of temperature by applying the expressions and techniques suggested in Chapter 3 to the data of Table E–1.

(a) Calm Air The problem solution involves solving nonlinear Equation (9-72) along with algebraic Equation (9-70), while the appropriate values of h_1 and h_2 can be calculated from Equations (9-68) and (9-69).

(b) Windy Conditions Both horizontal and vertical heat transfer coefficients, h_1 and h_2, can be calculated from Equation (9-71) with care to define the Reynolds number properly for each case.

The POLYMATH table data file for air is found in directory Tables with file named **E-01.POL**.

9.11 UNSTEADY-STATE RADIATION TO A THIN PLATE

9.11.1   Concepts Demonstrated

Unsteady-state heat transfer via radiation to an object with high thermal conductivity whose temperature can be considered uniform throughout, and determination of transient and steady state behavior of temperature and heat flux.

9.11.2   Numerical Methods Utilized

Solution of an ordinary differential equation.

9.11.3   Problem Statement

A thin metal plate is to be heat treated for a period of time in a high-temperature vacuum furnace. The metal plate is 0.5 m by 0.5 m with a thickness of 0.0015 m. The initial temperature of the plate is 20°C, and the furnace temperature is 1000°C. The plate is to be suspended in the furnace so that radiation from the furnace walls provides rapid heating to both plate surfaces. The interior of the furnace and the surfaces of the plate can be considered to radiate as black bodies.

Select one of the metals whose properties are summarized in Table 9–7 and can be assumed to be constant with temperature.
(a)   Plot the temperature of the plate as a function of time to steady state.
(b)   Plot the heat flux to the plate as a function of time to steady state.
(c)   What is the time required to reach steady state defined as 99% of the possible change in plate temperature?

Table 9–7 Properties of Selected Metals (from Thomas[6])

Metal	Density ρ (kg/m^3)	Heat Capacity C_p (kJ/kg $\cdot$ K)
Copper	8,950	0.383
Iron	7,870	0.452
Nickel	8,900	0.446
Silver	10,500	0.234
Stainless Steel	8,238	0.468
Steel (1% C)	7,801	0.473
Zirconium	6,750	0.272

Additional Information and Data

This is an unsteady-state problem in which the radiative heat transfer between the furnace wall and the plate causes the thin metal plate to heat. The plate itself might be subjected to temperature variations within its thickness if it were thick and/or had a low thermal conductivity. In this case for a thin plate with high thermal conductivity, it is a good assumption that the conduction is rapid throughout the plate and that the temperature is uniform throughout.

A simple unsteady-state energy balance on the plate can be made over a time increment Δt.

$$\text{INPUT} + \text{GENERATION} = \text{OUTPUT} + \text{ACCUMULATION}$$

$$\sigma A_P F_{12}(T_F^4 - T^4)\Delta t + 0 = 0 + V_P \rho C_P(T|_{t+\Delta t} - T|_t) \qquad \textbf{(9-73)}$$

Here σ is the Stefan-Boltzmann constant with a value of 5.676×10^{-8} W/m$^2 \cdot$K^4, A_P is the surface area of both sides of the plate in m^2, V_P is the volume of the plate in m^3, F_{12} is the view factor, which can be assumed to be unity, and T_F is the temperature of the furnace in K. All other variables refer to the properties of the metal plate.

Equation (9-73) can be rearranged and the limit taken as Δt goes to zero to give the ordinary differential equation

$$\frac{dT}{dt} = \frac{\sigma A_P F_{12}(T_F^4 - T^4)}{V_P \rho C_P} \qquad \textbf{(9-74)}$$

in which the properties of the metal are assumed constant.

The heat flux Q to the plate in W/m^2 is given by

$$Q = \sigma F_{12}(T_F^4 - T^4) \qquad \textbf{(9-75)}$$

9.12 UNSTEADY-STATE CONDUCTION WITHIN A SEMI-INFINITE SLAB

9.12.1 Concepts Demonstrated

Unsteady-state heat conduction in a one-dimensional semi-infinite slab with constant properties and subjected to time-dependent boundary conditions.

9.12.2 Numerical Methods Utilized

Application of the numerical method of lines to solve a partial differential equation, and solution of simultaneous ordinary differential equations and explicit algebraic equations.

9.12.3 Problem Statement

Unsteady-state heat transfer in one dimension, when the physical properties are constant, follows the partial differential equation

$$\frac{\partial T}{\partial t} = \alpha \frac{\partial^2 T}{\partial x^2} \qquad (9\text{-}76)$$

which has been discussed in Problem 6.8.

An interesting application of this type of heat transfer involves the variation of the temperature of soil at various depths, which has been considered by Thomas.[6] Let us consider the annual variations in the surface temperature of the soil and how this will affect the temperature at various soil depths due to heat conduction. The variation of the surface soil temperature in °F during a typical year is reported by Thomas[6] to be represented by

$$T_s(t) = T_M - \Delta T_s \cos\left[\frac{2\pi}{\tau}(t - t_0)\right] \qquad (9\text{-}77)$$

where t is time in days of the year, T_M is the annual mean earth temperature in °F, ΔT_s is the amplitude of the annual variation in surface soil temperature, τ is the period that is 365 days, and t_0 is the phase constant in days. Data for the parameters are available for various locations, and typical values are given in Table 9–8.

Table 9–8 Soil Temperature Parameters for Selected U. S. Cities (from Thomas[6])

City	T_M (°F)	ΔT_s (°F)	t_0 (days)
Bismarck, ND	44	31	33
Burlington, VT	46	26	37
Chicago, IL	51	25	37
Las Vegas, NV	69	23	32
Phoenix, AZ	73	23	33

An example of a plot of annual soil surface temperature variation for Chicago is given in Figure 9–16 by utilizing the data of Table 9–8 in Equation (9-77) for $(0 \leq t \leq 365)$.

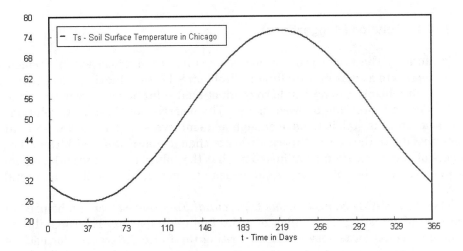

Figure 9–16 Annual Surface Temperature Variation for Chicago

Select one of the cities whose surface temperature data are given in Table 9–8. Assume that the soil's thermal diffusivity $\alpha = 0.9$ ft^2/day.

(a)   Use the numerical method of lines to solve Equation (9-76) for the temperatures at various levels beneath the soil surface by applying the surface temperature of Equation (9-77) as a boundary condition. Plot the temperatures at values of x corresponding to 8, 16, 24, and 72 ft over a four-year period. It is suggested that an initial simulation with 11 nodes and 10 evenly spaced intervals of 8 ft should be adequate. The initial condition is suggested to be the annual mean earth temperature, T_M.

(b)   Verify that the intervals you used in part (a) are adequate by doubling the number of intervals and halving the interval spacing. Compare results with part (a).

(c)   Why is there a phase change in the temperatures between the soil surface and the soils levels below? Construct some plots that clearly show this effect, and indicate why a number of years of solution time is desirable for this problem.

9.12.4   Solution (Suggestions)

(a) and (b) The numerical method of lines has been discussed in Problem 6.8 and applied to a semi-infinite fluid in Problem 8.17. In achieving a solution to this type of problem, it is desirable to conduct trial solutions with various number of nodes and intervals between nodes. The objective of these trial solutions for the semi-infinite slab is to have enough node intervals so that the variables at the greatest depth (large x in this case) do not change appreciably with time. The numbers of nodes can then be adjusted so that the solution is not greatly dependent upon the number of nodes as the nodes are increased over the same total interval.

The POLYMATH *Simultaneous Differential Equation Solver* can be used in the process of adjusting the number of nodes and their spacing. The capability to duplicate an equation is very useful in writing the finite differences for each of the nodes. Some large problems may require very large numbers of differential equations to be solved simultaneously for accurate solutions.

9.13 COOLING OF A SOLID SPHERE IN A FINITE WATER BATH

9.13.1 Concepts Demonstrated

Analysis of heat transfer via both lumped and distributed systems—formulation of the lumped and distributed models for unsteady-state conduction within a sphere and lumped model for a water bath.

9.13.2 Numerical Methods Utilized

Solution of partial differential equations with the numerical method of lines and integration of simultaneous ordinary differential equations.

9.13.3 Problem Statement

A metal fabrication company has a need for rapid cooling of solid spheres of uniform diameters made from various metals. This cooling is to be accomplished by a well-mixed water bath into which individual spheres are dropped after manufacture from an elevated temperature. A sketch of this operation is shown in the upper left of Figure 9–17.

(a) Consider a lumped analysis in which the conduction within the sphere is so rapid that the sphere maintains a uniform temperature as it cools. The sphere has a diameter $D = 0.1$ m with a constant thermal conductivity given by $k = 10$ W/m $\cdot$ °C, a density of $\rho = 8200$ kg/m^3, and a heat capacity $C_p = 0.41$ kJ/kg $\cdot$ K. The heat transfer coefficient between the sphere and the bath is $h = 220$ W/m$^2 \cdot$ °C. The initial temperature of the sphere is 300°C, and the water bath is initially at 20°C. Assume that the water bath maintains a constant temperature, and neglect heat transfer from the sphere until it enters the water bath. Calculate and make separate plots of the sphere's temperature T and heat rate q as a function of time to steady state.

(b) Repeat part (a) but take into account the heat conduction within the sphere. Plot the temperatures within the sphere at radii of 0 m, 0.05 m, and 0.1 m to steady state.

(c) Repeat part (b) but additionally consider the dynamic heating of the water bath. The volume of the bath is $V = 0.1$ m^3, and the properties of water can be considered constant with density $\rho_W = 965$ kg/m^3 and heat capacity $C_{pW} = 4.199$ kJ/kg $\cdot$ K. Plot the temperatures T_W and the temperatures at radii of 0 m, 0.05 m, and 0.1 m to steady state.

9.13.4 Solution (Partial)

This is an unsteady-state problem in which the solid sphere will cool down upon entry into the water bath. The conduction within the sphere will be rapid if the

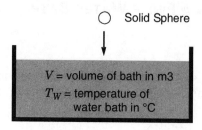

○ Solid Sphere

V = volume of bath in m3
T_W = temperature of
        water bath in °C

T_S = temperature of sphere's surface

T = temperature in differential element

$q|_r$ = heat rate at radius r

$q|_{r+\Delta r}$ = heat rate at radius $r + \Delta r$

h = convective heat transfer coefficient at
     sphere's surface

Enlarged view of solid sphere in
water bath showing differential element

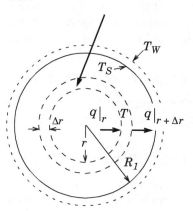

Figure 9–17 Cooling of a Spherical Solid in a Water Bath

thermal conductivity is high and/or the sphere diameter is small. The water bath
will be heated by the energy transferred from the cooling sphere.

(a) The "lumped" treatment will allow a general unsteady-state energy balance to be performed on the entire sphere, where the temperature within the
sphere is assumed to be uniform throughout at any time. In this case, there is no
input to the sphere nor any generation within the sphere. The output is via heat
transfer to the water bath, and the accumulation is related to the energy change
within the sphere. Thus an unsteady-state balance over a Δt time interval yields

$$INPUT + GENERATION = OUTPUT + ACCUMULATION$$

$$0 + 0 = h(4\pi R_1^2)(T - T_W)\Delta t + \left(\frac{4\pi R_1^3}{3}\right)\rho C_p(T|_{t+\Delta t} - T|_t) \qquad (9\text{-}78)$$

Taking the limit as Δt goes to zero followed by rearrangement yields

$$\frac{dT}{dt} = \frac{-h(4\pi R_1^2)(T - T_W)}{\left(\frac{4\pi R_1^3}{3}\right)\rho C_p} = \frac{-3h(T - T_W)}{R_1 \rho C_p} \qquad (9\text{-}79)$$

where the initial condition is $T = 300°C$ at time $t = 0$.

The rate of heat lost from the sphere can be calculated from

$$q = h(4\pi R_1^2)(T - T_W) \qquad (9\text{-}80)$$

which is just the output term of Equation (9-78).

(b) The heat conduction within the sphere can be described by making an unsteady-state energy balance on the differential element with incremental radius Δr shown in Figure 9–17 over an incremental time interval Δt.

$$\text{INPUT} + \text{GENERATION} = \text{OUTPUT} + \text{ACCUMULATION}$$

$$q|_r \Delta t + 0 = q|_{r+\Delta r} \Delta t + 4\pi r^2 \Delta r \rho C_p (T|_{t+\Delta t} - T|_t) \tag{9-81}$$

Rearrangement of Equation (9-81) followed by taking the limits as both Δr and Δt go to zero gives

$$\frac{\partial T}{\partial t} = -\left(\frac{1}{4\pi r^2 \rho C_p}\right)\frac{\partial q}{\partial r} \tag{9-82}$$

Fourier's law, expressed as

$$q = -k 4\pi r^2 \frac{\partial T}{\partial r} \tag{9-83}$$

can be substituted into Equation (9-82) to yield the second-order partial differential equation given by

$$\frac{\partial T}{\partial t} = \frac{1}{r^2 \rho C_p}\frac{\partial}{\partial r}\left(k r^2 \frac{\partial T}{\partial r}\right) \tag{9-84}$$

If the thermal conductivity k is constant, it can be removed from the partial derivative on the right-hand side of the preceding equation.

A convenient numerical solution to Equation (9-84) utilizes the numerical method of lines, which is discussed in Problem 6.8. Instead of directly solving the second-order Equation (9-84), it is convenient numerically to solve the two equations that lead to it—namely, Equations (9-82) and (9-83). This can be accomplished by using a series of ordinary derivatives for $\partial T/\partial t$ and finite difference formulas for both $\partial q/\partial r$ and $\partial T/\partial r$. This solution will be illustrated by setting up 11 nodes for 10 sections of the radius, as shown in Figure 9–18. Thus, using

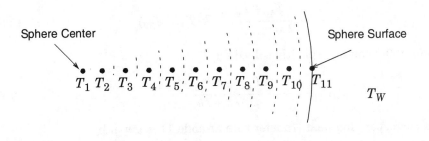

Figure 9–18 Nodes within the Sphere for Numerical Method of Lines Solution

the second-order central difference formula given by Equation (A-6) of Appendix A for the first derivatives with respect to r yields

$$\frac{\partial q_n}{\partial r} = \frac{(q_{n+1} - q_{n-1})}{2\Delta r} \quad \text{for } (2 \leq n \leq 10) \tag{9-85}$$

$$\frac{\partial T_n}{\partial r} = \frac{(T_{n+1} - T_{n-1})}{2\Delta r} \quad \text{for } (2 \leq n \leq 10) \tag{9-86}$$

Thus application of Equation (9-82) at each interior node point with Equation (9-85) results in

$$\frac{dT_n}{dt} = -\frac{1}{4\pi r_n^2 \rho C_p}\left(\frac{q_{n+1} - q_{n-1}}{2\Delta r}\right) \quad \text{for } (2 \leq n \leq 10) \tag{9-87}$$

Fourier's law of Equation (9-83) can be written for each interior node utilizing Equation (9-86) to give

$$q_n = -k4\pi r_n^2\left(\frac{T_{n+1} - T_{n-1}}{2\Delta r}\right) \quad \text{for } (2 \leq n \leq 10) \tag{9-88}$$

In the preceding equations, the arbitrary use of 11 nodes requires that $\Delta r = R_1/10$, and the radius to a particular node is given by $r_n = 0.1R_1(n-1)$.

Boundary Condition at Surface of the Sphere An energy balance at the surface of the sphere in the r direction relates that the rate of heat transferred to the surface by conduction within the sphere is equal to the convection from the sphere surface to the water bath. Thus

$$-k\frac{\partial T}{\partial r}\bigg|_{r=R_1} = h(T_{11} - T_W) \tag{9-89}$$

A second-order backward finite difference formula given by Equation (A-7) of Appendix A can be used for the partial derivative $\dfrac{\partial T}{\partial r}\bigg|_{r=R_1}$ to yield

$$-k\left(\frac{3T_{11} - 4T_{10} + T_9}{2\Delta r}\right) = h(T_{11} - T_W) \tag{9-90}$$

Equation (9-90) can be explicitly solved for T_{11}:

$$T_{11} = \frac{2\Delta rhT_W + 4kT_{10} - kT_9}{2\Delta rh + 3k} \tag{9-91}$$

and the corresponding heat transfer rate at node 11 is given by

$$q_{11} = h(4\pi R_1^2)(T_{11} - T_W) \tag{9-92}$$

Boundary Condition at Center of the Sphere At the center of the sphere, the heat transfer rate is zero; therefore,

$$q|_{r=0} = q_1 = 0 \tag{9-93}$$

which from Fourier's law also requires

$$\left.\frac{\partial T}{\partial r}\right|_{r=0} = 0 \tag{9-94}$$

A second-order forward difference formula, Equation (A-5) of Appendix A, can be used in the preceding equation:

$$\frac{\partial T_1}{\partial r} = \frac{(-T_3 + 4T_2 - 3T_1)}{2\Delta r} \tag{9-95}$$

and T_1 can be determined explicitly at any time to be given by

$$T_1 = \frac{4T_2 - T_3}{3} \tag{9-96}$$

Numerical Solution The numerical solution requires the simultaneous solution of Equations (9-87) and (9-88) for the interior nodes along with Equations (9-91) and (9-92) for the boundary condition at the sphere surface and Equations (9-93) and (9-96) for the boundary condition at the center of the sphere. The initial temperature throughout the sphere is 300°C, and the output heat rate from the sphere is given by q_{11}.

(c) The description of the water bath heating will require an unsteady-state energy balance on the well-mixed water within the tank. The tank will be assumed to be adiabatic, with the only heat transfer coming from the cooling sphere. Thus an unsteady-state balance over a Δt time interval yields

INPUT + GENERATION = OUTPUT + ACCUMULATION

$$q_{11} + 0 = 0 + V\rho_w C_{pW}(T_W|_{t+\Delta t} - T_W|_t) \tag{9-97}$$

where q_{11} as calculated in Equation (9-92) is the rate of heat transfer from the sphere at the outer surface to the water bath. Taking the limit as Δt goes to zero followed by rearrangement yields

$$\frac{dT_W}{dt} = \frac{q_{11}}{V\rho_W C_{pW}} \tag{9-98}$$

where the initial condition is $T_W = 20°C$ at time $t = 0$.

Thus the numerical solution only requires the addition of Equation (9-98) and the initial condition to the equation set used in part (b).

9.14 Unsteady-State Conduction in Two Dimensions

9.14.1   Concepts Demonstrated

Unsteady-state conduction in two dimensions with faces at known temperatures and constant thermal diffusivity.

9.14.2   Numerical Methods Utilized

Application of the numerical method of lines to solve a two-dimensional partial differential equation.

9.14.3   Problem Statement

Unsteady-state heat transfer in the x and y directions is described by the partial differential equation

$$\frac{\partial T}{\partial t} = \alpha \left(\frac{\partial^2 T}{\partial x^2} + \frac{\partial^2 T}{\partial y^2} \right) \tag{9-99}$$

where T is the temperature in K, t is the time in s, and α is the thermal diffusivity in m^2/s given by $k/\rho c_p$. In this treatment, the thermal conductivity k in W/m, the density ρ in kg/m^3, and the heat capacity c_p in J/kg are considered to be constant.

A hollow square chamber has the inside walls held at a temperature of 700 K while the outside walls are maintained at 300 K. The inside dimensions of the chamber are 1 m × 1 m and the outside dimensions are 2 m × 2 m, as shown in Figure 9–19. Grid spacing for a finite difference treatment of a symmetrical section of this chamber is also presented in Figure 9–19, where $\Delta x = \Delta y = 0.125$ m. Note that eight sections are required to describe the cross section of the chamber.

(a)   Use the numerical method of lines to solve Equation (9-99) to determine the temperatures at the node points of the chamber shown in Figure 9–19 as a function of time until steady state is reached. Prepare a plot of temperatures $T_{1,2}$, $T_{2,2}$, $T_{3,2}$, and $T_{4,2}$ versus time. The thermal diffusivity of the chamber wall is $\alpha = 5 \times 10^{-5}$ m^2/s, the thermal conductivity is $k = 1.2$ W/m·K, and the initial temperature of all chamber material is 300 K.

(b)   Calculate and plot the heat flux q per meter of chamber length through the interior wall of the chamber as a function of time until steady state is reached.

(c)   Discuss how you could increase the accuracy of your calculations for parts (a) and (b).

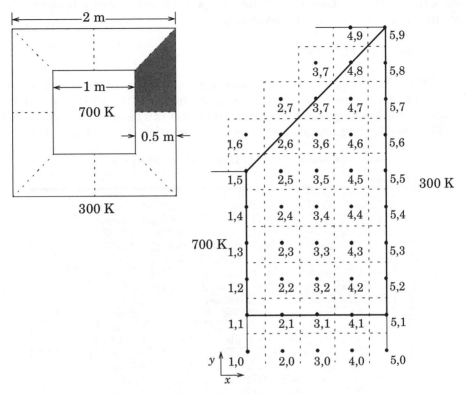

Figure 9–19 Grid Pattern for Hollow Square Chamber

The Numerical Method of Lines

The solution of a one-dimensional partial differential equation involving heat transfer utilizing the method of lines has been discussed in Problems 6.8, 9.12, and 9.13. For this problem, the time derivative can be replaced by ordinary derivatives and the two spacial derivatives can be written using finite difference approximations. Equation (9-99) can be rewritten using central difference formulas [Equation (A-3) of Appendix A] for each of the interior nodes in the grid spacing shown in Figure 9–19 as

$$\frac{dT_{n,m}}{dt} = \alpha \left[\frac{(T_{n+1,m} - 2T_{n,m} + T_{n-1,m})}{(\Delta x)^2} + \frac{(T_{n,m+1} - 2T_{n,m} + T_{n,m-1})}{(\Delta y)^2} \right] \quad \textbf{(9-100)}$$

where n represents a general node in the x direction and m represents a general node in the y direction (see Geankoplis[1]).

Due to the symmetry of this problem, the following temperature relations are true at any time

$$T_{1,6} = T_{2,5} \qquad T_{2,7} = T_{3,6} \qquad T_{3,7} = T_{4,7}$$
$$T_{2,2} = T_{2,0} \qquad T_{3,2} = T_{3,0} \qquad T_{4,2} = T_{4,0} \qquad \textbf{(9-101)}$$

and can be used in Equation (9-100) to perform calculations limited to only points within the primary grid boundary.

9.14.4 Solution (Partial)

(a) The equations for the particular interior nodes with the boundary conditions and the symmetry relationships given by Equation (9-101) result in 21 simultaneous ordinary differential equations that can be entered into the POLYMATH *Simultaneous Differential Solver*. The equation set is given in Table 9–9.

Table 9–9 POLYMATH Program (File **P9-14A.POL**)

Line	Equation
1	d(T21)/d(t) = alpha*((T31-2*T21+700)/deltax^2+(T22-2*T21+T22)/deltay^2)
2	d(T31)/d(t) = alpha*((T41-2*T31+T21)/deltax^2+(T32-2*T31+T32)/deltay^2)
3	d(T22)/d(t) = alpha*((T32-2*T22+700)/deltax^2+(T23-2*T22+T21)/deltay^2)
4	d(T41)/d(t) = alpha*((300-2*T41+T31)/deltax^2+(T42-2*T41+T42)/deltay^2)
5	d(T32)/d(t) = alpha*((T42-2*T32+T22)/deltax^2+(T33-2*T32+T31)/deltay^2)
6	d(T23)/d(t) = alpha*((T33-2*T23+700)/deltax^2+(T24-2*T23+T22)/deltay^2)
7	d(T42)/d(t) = alpha*((300-2*T42+T32)/deltax^2+(T43-2*T42+T41)/deltay^2)
8	d(T33)/d(t) = alpha*((T43-2*T33+T23)/deltax^2+(T34-2*T33+T32)/deltay^2)
9	d(T24)/d(t) = alpha*((T34-2*T24+700)/deltax^2+(T25-2*T24+T23)/deltay^2)
10	d(T43)/d(t) = alpha*((300-2*T43+T33)/deltax^2+(T44-2*T43+T42)/deltay^2)
11	d(T34)/d(t) = alpha*((T44-2*T34+T24)/deltax^2+(T35-2*T34+T33)/deltay^2)
12	d(T25)/d(t) = alpha*((T35-2*T25+700)/deltax^2+(T26-2*T25+T24)/deltay^2)
13	d(T44)/d(t) = alpha*((300-2*T44+T34)/deltax^2+(T45-2*T44+T43)/deltay^2)
14	d(T35)/d(t) = alpha*((T45-2*T35+T25)/deltax^2+(T36-2*T35+T34)/deltay^2)
15	d(T26)/d(t) = alpha*((T36-2*T26+T25)/deltax^2+(T36-2*T26+T25)/deltay^2)
16	d(T45)/d(t) = alpha*((300-2*T45+T35)/deltax^2+(T46-2*T45+T44)/deltay^2)
17	d(T36)/d(t) = alpha*((T46-2*T36+T26)/deltax^2+(T37-2*T36+T35)/deltay^2)
18	d(T46)/d(t) = alpha*((300-2*T46+T36)/deltax^2+(T47-2*T46+T45)/deltay^2)
19	d(T37)/d(t) = alpha*((T47-2*T37+T36)/deltax^2+(T47-2*T37+T36)/deltay^2)
20	d(T47)/d(t) = alpha*((300-2*T47+T37)/deltax^2+(T48-2*T47+T46)/deltay^2)
21	d(T48)/d(t) = alpha*((300-2*T48+T47)/deltax^2+(300-2*T48+T47)/deltay^2)
22	alpha = 5e-5
23	deltax = 0.125
24	deltay = 0.125
25	t(0)=0
26	T21(0)=300
27	T31(0)=300
28	T22(0)=300
29	T41(0)=300
30	T32(0)=300
31	T23(0)=300
32	T42(0)=300
33	T33(0)=300
34	T24(0)=300
35	T43(0)=300
36	T34(0)=300
37	T25(0)=300
38	T44(0)=300
39	T35(0)=300
40	T26(0)=300
41	T45(0)=300

Line	Equation
42	T36(0)=300
43	T46(0)=300
44	T37(0)=300
45	T47(0)=300
46	T48(0)=300
47	t(f)=300

(b) Once the temperature distribution has been calculated, the total heat loss at the interior wall of the chamber can be obtained by summing the heat fluxes at the various nodes due to the local temperature gradient. Fourier's law, given by

$$q_x = -kA\frac{dT}{dx}\bigg|_{x = \text{inner surface}} \tag{9-102}$$

can be applied with due consideration for the appropriate area A and the overall symmetry of the problem. The derivatives in Equation (9-102) can be obtained using the second-order forward finite difference expression whose general formula is given by [see Equation (A-5) of Appendix A]

$$\frac{dT_n}{dx} = \frac{-3T_n + 4T_{n+1} - T_{n+2}}{2\Delta x} \tag{9-103}$$

for the x direction. Thus the total heat flux through the interior wall surface (in positive x direction) for the inner chamber is given by

$$q = -8k\Delta x\left[\left(\frac{1}{2}\right)\frac{dT_{1,5}}{dx} + \frac{dT_{1,4}}{dx} + \frac{dT_{1,3}}{dx} + \frac{dT_{1,2}}{dx} + \left(\frac{1}{2}\right)\frac{dT_{1,1}}{dx}\right] \tag{9-104}$$

where the factor of 8 is for the eight sections with similar symmetry, and the factors of (1/2) are necessary because of the appropriate areas. This expression, with the derivatives calculated from Equation (9-103) for x, can be evaluated *at any time* to estimate the total heat flux.

Table 9–10 provides the needed additions to the equation set for the POLYMATH solution of part (b).

Table 9–10 Equation Set for Calculation of Total Heat Flux

Line	Equation
	dT11dx=(-T31+4*T21-3*700)/(2*deltax)
	dT12dx=(-T32+4*T22-3*700)/(2*deltax)
	dT13dx=(-T33+4*T23-3*700)/(2*deltax)
	dT14dx=(-T34+4*T24-3*700)/(2*deltax)
	dT15dx=(-T35+4*T25-3*700)/(2*deltax)
	q=-8*1.2*deltax*((1/2)*dT15dx+dT14dx+dT13dx+dT12dx+(1/2)*dT11dx)

The POLYMATH problem solution file for both parts (a) and (b) is found in directory Chapter 9 with files named **P9-14A.POL** and **P9-14AB.POL**.

REFERENCES

1. Geankoplis, C. J., *Transport Processes and Unit Operations*, 3rd ed., Englewood Cliffs, NJ: Prentice Hall, 1993.
2. Bird, R. B., Stewart, W. E., and Lightfoot, E. N., *Transport Phenomena*, New York: Wiley, 1960.
3. Welty J. R., Wicks, C. E., and Wilson, R.E., *Fundamentals of Momentum, Heat and Mass Transfer*, 3rd ed., New York: Wiley, 1984.
4. Dittus, F. W., and L. M. K. Boelter, University of California, Berkeley, *Pub. Engr.*, *2*, 443 (1930).
5. Churchill, S. W., and Chu, H. H. S., *Int. J Heat & Mass Transfer*, *18*, 1323 (1975).
6. Thomas, L. C., *Heat Transfer*, Englewood Cliffs, NJ: Prentice Hall, 1992.

CHAPTER **10**

Mass Transfer

10.1 ONE-DIMENSIONAL BINARY MASS TRANSFER IN A STEFAN TUBE

10.1.1 Concepts Demonstrated

Binary gas phase diffusion of A through stagnant B during evaporation of a pure liquid in a simple diffusion tube.

10.1.2 Numerical Methods Utilized

Numerical integration of simultaneous ordinary differential equations by a shooting technique which must satisfy split boundary conditions.

10.1.3 Problem Statement

Liquid A is evaporating into a gas mixture of A and B from a liquid layer of pure A near the bottom of a cylindrical Stefan tube, as shown in Figure 10–1. The rate of evaporation is relatively slow, so it is a good assumption that the level of the liquid surface is constant. A gas mixture is passing over the upper surface of the Stefan tube. Thus the partial pressure of A, p_{A2}, and the mole fraction of A at point 2, x_{A2}, are both known at z_2. The surface of the liquid A contains no dissolved B since B is insoluble in liquid A; therefore, liquid A exerts its vapor pressure, p_{A1}, at location z_1. The mole fraction of A at the liquid surface is given by

$$x_{A1} = \frac{P_{A0}}{P} \tag{10-1}$$

where P_{A0} is the vapor pressure of component A and P is the total pressure.

The simplest assumptions for this system would be that the temperature and pressure are constant and that the gases A and B are ideal. Thus, gas A is diffusing from the surface into the bulk stream above the surface of the Stefan tube though gas B which is stationary within the tube. This is the case of single component diffusion through a stagnant gas film.

383

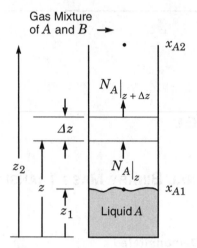

Figure 10-1 Gas Phase Diffusion of A through Stagnant B

Mass Balance on Component A within Diffusion Path

Consider a differential element between points z_1 and z_2 with a differential length of Δz. Since there is no reaction in this case, a steady-state mass balance in the positive z direction yields

$$\frac{dN_A}{dz} = 0 \tag{10-2}$$

where N_A is the flux of A relative to stationary coordinates in kg-mol/m^2·s.

Fick's Law for Binary Diffusion

Also for stationary coordinates, the general expression of Fick's law for the flux of A can be written as

$$N_A = -D_{AB}C\frac{dx_A}{dz} + \frac{C_A}{C}(N_A + N_B) \tag{10-3}$$

flux = diffusion + bulk flow (convection)

where C is the total concentration in kg-mol/m^3, D_{AB} is the molecular diffusivity of A in B in m^2/s, C_A is the concentration of A in kg-mol/m^3, and N_B is the flux of B in kg-mol/m^2·s. For this problem, the total gas concentration C is constant, and component B is stagnant. Thus N_B is zero. The mole fraction of A in the gas mixture, x_A, can be used to replace C_A/C. The modified expression for Fick's law can be written as

$$N_A = -D_{AB}\frac{dC_A}{dz} + x_A N_A \tag{10-4}$$

Solving for N_A and rearranging Equation (10-3) yields

$$\frac{dx_A}{dz} = \frac{-(1-x_A)N_A}{D_{AB}C} \tag{10-5}$$

Final Equations and Boundary Conditions

The diffusion with convection for this problem can be described by the simultaneous solution of Equations (10-2) and (10-5) with the initial condition for x_A at z_1, as given by Equation (10-1). The final value of x_{A2} at z_2 is the mole fraction of A in the gas mixture that is flowing across the top of the Stefan tube.

Analytical Solution

At constant temperature and pressure for ideal gases, the total concentration C and binary diffusivity D_{AB} may be considered constant. Thus Equation (10-5) can be solved for N_A and entered into Equation (10-2) and integrated twice using the boundary conditions given previously to give the analytical solution (see Bird et al.[1]) for the concentration profile as

$$\left(\frac{1-x_A}{1-x_{A1}}\right) = \left(\frac{1-x_{A2}}{1-x_{A1}}\right)^{\frac{(z-z_1)}{(z_2-z_1)}} \tag{10-6}$$

The analytical solution for the flux at the liquid-gas interface gives

$$N_{Az}\big|_{z=z_1} = \frac{D_{AB}C}{(z_2-z_1)(x_B)_{\text{lm}}}(x_{A1}-x_{A2}) \tag{10-7}$$

where

$$(x_B)_{\text{lm}} = \frac{(x_{B2}-x_{B1})}{\ln(x_{B2}/x_{B1})} = \frac{(x_{A1}-x_{A2})}{\ln[(1-x_{A2})/(1-x_{A1})]} \tag{10-8}$$

The evaporation of methanol into a stream of dry air is being studied in a Stefan tube apparatus with a diffusion path of 0.238 m. The measurements are made at a temperature of 328.5 K, where the vapor pressure of methanol is 68.4 kPa. The total pressure is 99.4 kPa. The binary molecular diffusion coefficient of methanol in air under these conditions is $D_{AB} = 1.991 \times 10^{-5}$ m^2/s, as estimated by Taylor and Krishna.[2]

(a)   Calculate the constant molar flux of methanol within the Stefan tube at steady state using a numerical technique.

(b)   Plot the mole fraction of methanol from the liquid methanol surface to the flowing air stream.

(c)   Compare the result of part (a) with the result calculated from Equation (10-7).

(d)   Verify several points on the numerical solution for the mole fraction profile with calculations from the analytical profile of Equation (10-6).

(e)   Repeat parts (a) through (d) for a temperature of 298.15 K, where the vapor pressure of methanol is 16.0 kPa.

(f)   Assume that the temperature within the Stefan tube varies linearly from 328.5 K at the methanol surface to 295 K in the air stream at the tube surface. Complete parts (a) and (b) for this condition.

Additional Information and Data

Geankoplis[3] (p. 396) recommends that the effect of temperature on binary diffusivities for gases vary according to the absolute temperature to the 1.75 power given by

$$D_{AB} = D_{AB}\big|_{T_1} \left(\frac{T}{T_1}\right)^{1.75} \tag{10-9}$$

where the diffusivity is known at temperature T_1.

10.1.4   Solution (Partial with Suggestions)

(a), (b), and (c) The methanol flux and mole fraction profile require the numerical integration of Equations (10-2) and (10-5). For this case, the solution of Equation (10-2) is simply that N_A is a constant. Thus the initial and final conditions for Equation (10-2) are known for the methanol mole fraction x_A at both ends of the diffusion path. At $z = 0$, $x_A = 0.688$ [using Equation (10-1)] and at $z = 0.238$, $x_A = 0$.

The shooting technique as described in Problem 6.5 can be used to determine the constant value of N_A. Thus the initial value for x_A is known to be 0.688 at $z = 0$, and N_A can be optimized using the techniques of Problem 6.4 to satisfy the boundary condition, where $x_A = 0$ at $z = 0.238$. The value of C can be calculated from the perfect gas law.

The POLYMATH *Simultaneous Differential Equation Solver* can be used to solve Equation (10-5), and the POLYMATH equation set is given in Table 10–1.

In the POLYMATH solution, the numerical solution for $N_A = 3.5461 \times 10^{-6}$ agrees with the analytical solution exactly to the five significant figures used. This is demonstrated by the results of various solutions with slightly different

Table 10–1 POLYMATH Program (File **P10-01ABC.POL**)

Line	Equation
1	d(xA)/d(z)=-(1-xA)*NA/(DAB*C)
2	P=99.4
3	NA=3.5461e-6
4	DAB=1.991e-5
5	R=8.31434
6	T=328.5
7	xA1=0.688
8	xBlm=(xA1-0)/ln(1/(1-xA1))
9	C=P/(R*T)
10	NACALC=DAB*C*(xA1-0)/((0.238-0)*xBlm)
11	z(0)=0
12	xA(0)=0.688
13	z(f)=0.238

values of N_A summarized in Table 10–2. In solutions 1 and 3, the final value for x_A is further away from zero than for the best solution 2.

Table 10–2 Comparison of Final Values for Different Initial Conditions

Solution	N_A	$x_A\|_{z\,=\,0.238}$
#1	3.5460×10^{-6}	3.29891×10^{-5}
#2	3.5461×10^{-6}	1.43643×10^{-7}
#3	3.5462×10^{-6}	-3.27029×10^{-5}

The numerical solution allows convenient plotting of the methanol mole fraction x_A along the diffusion path. This is shown in Figure 10–2.

The POLYMATH problem solution file for parts (a), (b), and (c) is found in directory Chapter 10 with file named **P10-01ABC.POL**.

(d) The analytical solution given by Equation (10-6) is a nonlinear equation that can be solved for the distance z if the mole fraction x_A is known or the equation can be solved for the mole fraction x_A if the distance z is known. Verification of the numerical solution will be carried out at three arbitrary values of z as given in Table 10–3, where the correspondence is exact to at least five significant figures.

(f) The temperature profile can be imposed on the numerical solution by adding an algebraic equation for T as a function of z. The effect of temperature on the diffusivity of methanol in air must be included by utilizing Equation (10-9).

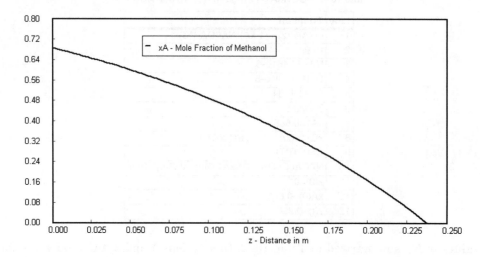

Figure 10–2 Mole Fraction of Methanol along Diffusion Path

Table 10–3 Comparison of Analytical and Numerical Methanol Mole Fraction

z	x_A Analytical	x_A Numerical
0.0714	0.557505	0.55750517
0.1428	0.37243	0.37243054
0.2142	0.109948	0.10994795

10.2 MASS TRANSFER IN A PACKED BED WITH KNOWN MASS TRANSFER COEFFICIENT

10.2.1   Concepts Demonstrated

Simple convective mass transfer from a surface to the bulk stream for a system involving transport of A through inert B with the fluid mass transfer coefficient dependent on concentration.

10.2.2   Numerical Methods Utilized

Numerical integration of an ordinary differential equation with a known initial condition and simultaneous explicit algebraic equations.

10.2.3   Problem Statement[*]

Pure water at 26.1 °C is slowly passing through a bed of benzoic acid spheres at a rate of 0.0701 ft^3/h. The spheres have a diameter of 0.251 inches, and the total surface area within the bed is 0.129 ft^2. The mass transfer coefficient for equimolar counterdiffusion varies with composition and is given by

$$k'_L = K_1 + K_2 x_A \tag{10-10}$$

where k'_L is the mass transfer coefficient for equimolar counterdiffusion (see Geankoplis,[3] p. 435) in m/s, $K_1 = 0.0551$ ft/h, $K_2 = 185.5$ ft/h, and x_A is the mole fraction of benzoic acid in the liquid phase. The saturation solubility of benzoic acid in water is 0.00184 lb-mol/ft^3 of solution at 26.1°C. The total concentration of the liquid water phase is $C = 3.461$ lb-mol/ft^3, and the volumetric flow rate is $V = 0.0701$ ft^3/h.

> (a)   Calculate and plot the concentration of benzoic acid within the bed as a function of the bed surface area to 0.129 ft^2.
> (b)   Calculate the surface area necessary to achieve a liquid phase concentration of benzoic acid at the exit of the bed that is 50% of the saturation solubility.

Additional Information and Data

A general material balance on the differential surface area of the packed bed shown in Figure 10–3 yields

$$\frac{dC_A}{d(\text{Area})} = \frac{N_A|_i}{V} \tag{10-11}$$

[*] Adapted from Geankoplis[3] with permission.

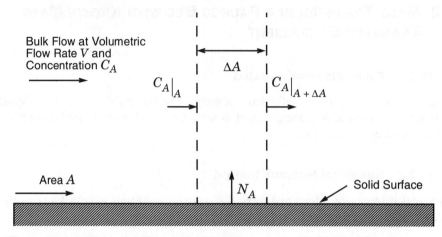

Figure 10–3 Differential Volume for Mass Transfer in Packed Bed Based on External Surface Area

The mass transfer flux for the diffusion of A (benzoic acid) through stagnant B (water) at the interface between the benzoic acid and the flowing stream is given by (Geankoplis,[3] p. 435)

$$N_A\big|_i = \frac{k'_L\, C}{x_{BM}}(x_{Ai} - x_{Ab})$$

(10-12)

where N_A has units of lb-mol/h·ft^2 and x_{BM} is given by

$$x_{BM} = \frac{x_{B2} - x_{B1}}{\ln(x_{B2}/x_{B1})}$$

(10-13)

between points 1 and 2, which in this case are the surface i and the bulk stream b respectively.

10.2.4   Solution (Suggestions)

This problem can be described by inserting Equation (10-12) into Equation (10-11) and using $C_A = x_A C$ to obtain an ordinary differential equation given by

$$\frac{dx_A}{d(\text{Area})} = \frac{k'_L}{V x_{BM}}(x_{Ai} - x_{Ab})$$

(10-14)

with the initial condition that at the entrance to the packed bed, the mole fraction of benzoic acid is zero. Thus $x_A = 0$ when Area = 0. The POLYMATH *Simultaneous Ordinary Differential Equation Solver* can be used to integrate Equation (10-14) with the algebraic equations for k'_L and x_{BM} given by Equations (10-10) and (10-13), respectively.

10.3 SLOW SUBLIMATION OF A SOLID SPHERE

10.3.1 Concepts Demonstrated

Sublimation of a solid sphere by diffusion in still gas and by a mass transfer coefficient in a moving gas.

10.3.2 Numerical Methods Utilized

Solution of simultaneous ordinary differential equations while optimizing a single parameter to achieve split boundary conditions.

10.3.3 Problem Statement

Consider the sublimation of solid dichlorobenzene, designated by A, which is suspended in still air, designated by B, at 25 °C and atmospheric pressure. The particle is spherical with a radius of 3×10^{-3} m. The vapor pressure of A at this temperature is 1 mm Hg, and the diffusivity in air is 7.39×10^{-6} m^2/s. The density of A is 1458 kg/m^3, and the molecular weight is 147.

(a) Estimate the initial rate of sublimation (flux) from the particle surface by using an approximate analytical solution to this diffusion problem. (See the following discussion for more information.)

(b) Calculate the rate of sublimation (flux) from the surface of a sphere of solid dichlorobenzene in still air with a radius of 3×10^{-3} m with a numerical technique employing a shooting technique, with ordinary differential equations that describe the problem. Compare the result with part (a).

(c) Show that expression for the rate of sublimation (flux) from the particle as predicted in part (a) is the same as that predicted by the external mass transfer coefficient for a still gas.

(d) Calculate the time necessary for the complete sublimation of a single particle of dichlorobenzene if the particle is enclosed in a volume of 0.05 m^3.

Additional Information and Data

Diffusion The diffusion of A through stagnant B from the surface of a sphere is shown in Figure 10–4. A material balance on A in a differential volume between radius r and $r + \Delta r$ in a Δt time interval yields

$$\text{INPUT} + \text{GENERATION} = \text{OUTPUT} + \text{ACCUMULATION}$$

$$(N_A 4\pi r^2)\big|_r \, \Delta t + 0 = (N_A 4\pi r^2)\big|_{r + \Delta r} \, \Delta t + 0 \qquad \text{(10-15)}$$

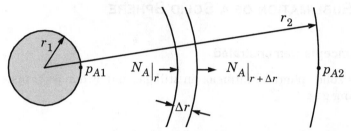

Figure 10–4 Shell Balance for Diffusion from the Surface of a Sphere

where N_A is the flux in kg-mol/m$^2 \cdot$s at radius r in m. The $4\pi r^2$ is the surface area of the sphere with radius r. Division by Δt and rearrangement of this equation while taking the limit as $\Delta r \to 0$ yields

$$\frac{d(N_A r^2)}{dr} = 0 \qquad (10\text{-}16)$$

Fick's law for the diffusion of A through stagnant B in terms of partial pressures is expressed as

$$N_A = -\frac{D_{AB}}{RT}\frac{dp_A}{dr} + \frac{p_A}{P}N_A \qquad (10\text{-}17)$$

where D_{AB} is the molecular diffusivity of A in B in m^2/s, R is the gas constant with a value of 8314.34 m$^3 \cdot$Pa/kg-mol$\cdot$K, T is the absolute temperature in K, and P is the total pressure in Pa.

Rearrangement of Equation (10-17) yields

$$\frac{dp_A}{dr} = -\frac{RTN_A\left(1 - \frac{p_A}{P}\right)}{D_{AB}} \qquad (10\text{-}18)$$

where the initial condition is that $p_A = (1/760)1.01325 \times 10^5$ Pa $= 133.32$ Pa, which is the vapor pressure of the solid A at $r = 3 \times 10^{-3}$ m. The final value is that $p_A = 0$ at some relatively large radius r.

The analytical solution to this problem can be obtained by integrating Equation (10-16) and introducing the result for N_A into Equation (10-18). The final solution as given by Geankoplis[3] (p. 391) is

$$N_{A1} = \frac{D_{AB}P}{RTr_1}\frac{(p_{A1} - p_{A2})}{p_{BM}} \qquad (10\text{-}19)$$

where subscripts 1 and 2 indicate locations and p_{BM} is given by

$$p_{BM} = \frac{p_{B2} - p_{B1}}{\ln(p_{B2}/p_{B1})} = \frac{p_{A1} - p_{A2}}{\ln((P - p_{A2})/(P - p_{A1}))} \qquad (10\text{-}20)$$

Mass Transfer Coefficient The transfer of A from the surface of the spherical particle to the surrounding gas can also be described by a mass transfer coefficient for transport of A through stagnant B.

A general relationship for gases that can be used to calculate the mass transfer coefficient for gases is presented by Geankoplis[3] (p. 446) as

$$N_{Sh} = 2 + 0.552 N_{Re}^{0.53} N_{Sc}^{1/3} \qquad \text{(10-21)}$$

Note that for a quiescent gas, the limiting value of N_{Sh} is 2 because the N_{Re} is zero.

The Sherwood number is defined as

$$N_{Sh} = k_c' \frac{D_p}{D_{AB}}$$

where k_c' is the mass transfer coefficient in m/s based on concentration and equimolar counterdiffusion, and D_p is the particle diameter in m. The particle Reynolds number is defined as

$$N_{Re} = \frac{D_p v \rho}{\mu}$$

where D_p is the particle diameter in m, v is the gas velocity in m/s, and μ is the gas viscosity in Pa·s. The Schmidt number is given by

$$N_{Sc} = \frac{\mu}{\rho D_{AB}}$$

with ρ representing the density of the gas in kg/m^3.

The mass transfer coefficient (see Geankoplis,[3] p. 435) can be used to describe the flux N_A from the surface of the sphere for transport through stagnant B by utilizing

$$N_A = \frac{k_c' P(p_{A1} - p_{A2})}{RT \quad p_{BM}} \qquad \text{(10-22)}$$

10.3.4 Solution (Partial with Suggestions)

(a) The analytical solution is given by Equations (10-19) and (10-20) which can be easily evaluated. Since comparisons will be made with the numbered solutions, these explicit equations can be entered into the particular POLYMATH Program that will be used for the numerical solution.

(b) The numerical solution involves the simultaneous solution of Equations (10-16) and (10-18) along with the following algebraic equation that is needed to calculate the flux N_A from the quantity $(N_A r^2)$:

$$N_A = \frac{(N_A r^2)}{r^2} \tag{10-23}$$

The POLYMATH *Simultaneous Differential Equation Solver* can be used to integrate these differential equations for the boundary conditions. The initial radius is the known radius of the sphere (initial condition), and the final value of the radius is much greater than the initial radius, so that the value of the partial pressure of A is effectively zero. The shooting technique for accomplishing this type of a solution is discussed in Problems 6.5 and 10.1. A POLYMATH equation set that provides the solution is given in Table 10–4.

Table 10–4 POLYMATH Program (File **P10-03AB.POL**)

Line	Equation
1	d(NAr2)/d(r)=0
2	d(pA)/d(r)=-R*T*NA*(1-pA/P)/DAB
3	R=8314.34
4	T=298.15
5	NA=NAr2/(r^2)
6	P=1.01325e5
7	DAB=7.39e-6
8	PBM=(133.32-0)/ln((P-0)/(P-133.32))
9	NACALCatr1=DAB*P*(133.32-0)/(R*T*0.003*PBM)
10	NACALCr12=NACALCatr1*0.003^2
11	r(0)=0.003
12	NAr2(0)=1.19335e-12
13	pA(0)=133.32
14	r(f)=16

The final value in this numerical solution is such that the desired value of the partial pressure of A is very nearly zero ($-7.86e-4$ Pa for the preceding equation set) at the very "large" final radius of $r = 16$ m. This "large" final radius is really somewhat arbitrary as long as the boundary condition that $p_A = 0$ is achieved, which is independent of the "large" final radius value.

A comparison of the calculated N_A with the analytical N_A at the surface of the sphere is quite satisfactory in that there is agreement to four significant places for the result of 1.326×10^{-7} kg-mol/m$^2 \cdot$s.

 The POLYMATH problem solution file for parts (a) and (b) is found in directory Chapter 10 with file named **P10-03AB.POL**.

www

(c) The resulting analytical solutions should be identical with each other.

(d) This is an unsteady-state problem that can be solved utilizing the mass transfer coefficient by making the pseudo-steady-state assumption that the mass transfer can be described by Equation (10-22) at any time. The mass transfer coefficient increases as the particle diameter decreases because for a still gas,

$$N_{Sh} = k'_c \frac{D_p}{D_{AB}} = 2 \tag{10-24}$$

as indicated by Equation (10-22); thus, in terms of the particle radius,

$$k'_c = \frac{D_{AB}}{r} \tag{10-25}$$

A material balance on component A in the well-mixed gas phase volume V with the only input due to the sublimation of A yields

INPUT + GENERATION = OUTPUT + ACCUMULATION

$$(N_A 4\pi r^2)\Big|_r \Delta t + 0 = 0 + \left(\frac{V p_A}{RT}\right)\Big|_{t+\Delta t} - \left(\frac{V p_A}{RT}\right)\Big|_t \tag{10-26}$$

The limit as $\Delta t \to 0$ and the use of Equation (10-22) for N_A give

$$\frac{dp_A}{dt} = \frac{4\pi r^2 k'_c P(p_{A1} - p_{A2})}{V} \frac{}{p_{BM}} \tag{10-27}$$

with p_{A1} being the vapor pressure of A at the solid surface and p_{A2} being the partial pressure of A in the gas volume.

A material balance on A within the solid sphere of radius r gives

INPUT + GENERATION = OUTPUT + ACCUMULATION

$$0 + 0 = (N_A 4\pi r^2)\Big|_r \Delta t + \left(\frac{4}{3}\pi r^3 \frac{\rho_A}{M_A}\right)\Big|_{t+\Delta t} - \left(\frac{4}{3}\pi r^3 \frac{\rho_A}{M_A}\right)\Big|_t \tag{10-28}$$

where ρ_A is the density of the solid and M_A is the molecular weight of the solid. Rearranging Equation (10-28) and taking the limit as $\Delta t \to 0$ yields

$$\frac{d(r^3)}{dt} = \frac{3r^2 dr}{dt} = -\frac{3 N_A M_A r^2}{\rho_A} \tag{10-29}$$

Simplifying and introducing Equation (10-22) for N_A gives

$$\frac{dr}{dt} = -\frac{M_A k'_c P(p_{A1} - p_{A2})}{\rho_A RT} \frac{}{p_{BM}} \tag{10-30}$$

The complete sublimation of A is described by the simultaneous solution of differential Equations (10-27) and (10-30) along with Equation (10-25).

10.4 CONTROLLED DRUG DELIVERY BY DISSOLUTION OF PILL COATING

10.4.1   Concepts Demonstrated

Unsteady-state dissolution of a solid into a liquid with transport described by a mass transfer coefficient with subsequent reaction.

10.4.2   Numerical Methods Utilized

Solution of simultaneous ordinary differential equations with known initial conditions that have conditional alterations during the numerical solution.

10.4.3   Problem Statement (Adapted from Fogler,[4] p. 600)

The pill to deliver a particular drug has a solid spherical inner core of pure drug D and is surrounded by a spherical outer coating of A that makes the pill palatable and helps to control the drug release. The outer coating and the drug dissolve at different rates in the stomach due to their difference in solubilities. Let C_{AS} = concentration of coating in the stomach in mg/cm^3, C_{DS} = concentration of drug in the stomach in mg/cm^3, and C_{DB} = concentration of drug in the body in mg/kg.

Three different pill formulations are available:

Pill 1 — Diameter of A = 5 mm, Diameter of D = 3 mm

Pill 2 — Diameter of A = 4 mm, Diameter of D = 3 mm

Pill 3 — Diameter of A = 3.5 mm, Diameter of D = 3 mm

Additional Information and Data

Amount of drug in inner core of each pill = 20 mg

Density of inner and outer layers = 1414.7 mg/cm^3

Solubility of outer pill layer at stomach conditions = S_A = 1.0 mg/cm^3

Solubility of inner drug core at stomach conditions = S_D = 0.4 mg/cm^3

Volume of fluid in stomach = V = 1.2 liters

Residence time in stomach = $\tau = V/v_0$ = 4 hours, where v_0 is the volumetric flow rate

Typical body weight W = 75 kg

Sherwood number = $N_{Sh} = k_L \dfrac{D_p}{D_{AB}} = 2$ (see Problem 10.3 for details)

Effective Diffusivities of A and D in stomach $D_A = D_D = 0.6$ cm^2/min

A person takes all three pills at the same time. Assume that the stomach is well mixed and that the pills remain in the stomach while they are dissolving.

(a) Plot C_{AS} and C_{DS} as a function of time for up to 12 hours after the pills are taken.

(b) If the drug is absorbed into the body (from solution in the stomach into the blood stream) by an effective mass transfer rate of $10 \times C_{DS}$ in mg/min, plot C_{AS} and C_{DS} under these conditions.

(c) If the body metabolizes the drug with an effective first-order reaction rate of $1.0 \times C_{DB}$ in mg/kg.min, plot C_{DB} in the body under the conditions of (b).

(d) Under the conditions of (c), what is the estimated time that C_{DB} is greater than 2.0×10^{-3} mg/kg, which is the minimum effective concentration level for the drug?

(e) Please suggest a more effective layering of the three pills with the drug placed only in the center to increase the time for the effective level concentration of 2.0×10^{-3} mg/kg under the same conditions.

10.4.4 Solution (Suggestions)

(a) Differential equations will need to be solved simultaneously for the drug resulting from each of the three pills. No drug will be released until the outer coating is dissolved. See Problem 10.3 for help with the drug dissolution, which is analogous to sublimation for gases.

A material balance on the volume of pill 1 yields

$$\frac{dD_1}{dt} = -\frac{2k_{L1}}{\rho}(S_A - C_{AS}) \quad \text{I. C.} \quad D_1 = 0.5 \text{ cm at } t = 0 \text{ min} \qquad \textbf{(10-31)}$$

which becomes

$$\frac{dD_1}{dt} = -\frac{2k_{L1}}{\rho}(S_D - C_{DS}) \quad \text{for } 10^{-5} \leq D_1 \leq 0.3 \text{ cm} \qquad \textbf{(10-32)}$$

and

$$\frac{dD_1}{dt} = 0 \quad \text{for } D_1 \leq 10^{-5} \text{ cm} \qquad \textbf{(10-33)}$$

where the mass transfer coefficient for pill 1, k_{L1}, depends upon diameter D_1.

$$k_{L1} = \frac{2(0.6)}{D_1} \qquad \textbf{(10-34)}$$

Similar equations can be derived for pills 2 and 3 that describe pill diameters D_2 and D_3. A single differential equation can be constructed from Equations (10-31) through (10-33) using the logic capability of nested "if ... then ... else ..." statements in the POLYMATH *Simultaneous Ordinary Differential Equation Solver.*

The differential equation describing the material balance on A within the stomach is given by

$$\frac{dC_{AS}}{dt} = \frac{1}{V}[S_{W1}k_{L1}(S_A - C_{AS})\pi D_1^2 + S_{W2}k_{L2}(S_A - C_{AS})\pi D_2^2$$

$$+ S_{W3}k_{L3}(S_A - C_{AS})\pi D_3^2] - \frac{C_{AS}}{\tau}$$

(10-35)

each "switch" is unity when each diameter D is greater that 0.3 and zero at other times. This "switch" then provides the appropriate input of the pill coating to the stomach.

A similar material balance on the drug D yields

$$\frac{dC_{DS}}{dt} = \frac{1}{V}[(1 - S_{W1})k_{L1}(S_D - C_{DS})\pi D_1^2 + (1 - S_{W2})k_{L2}(S_D - C_{DS})\pi D_2^2$$

$$+ (1 - S_{W3})k_{L3}(S_D - C_{DS})\pi D_3^2] - \frac{C_{DS}}{\tau}$$

(10-36)

in which the "switch" defined previously provides the input of drug from each pill.

A POLYMATH equation set that describes the dissolution processes for the three pills is provided in Table 10–5.

Table 10–5 POLYMATH Program (File **P10-04A.POL**)

Line	Equation
1	d(D1)/d(t) = if(D1>0.3)then(-2*kL1*(SA-CAS)/rho)else(if(D1>1e-5)then(-2*kL1*(SD-CDS)/rho)else(0.0))
2	d(CAS)/d(t) = (1/V)*(SW1*kL1*(SA-CAS)*3.1416*D1^2+SW2*kL2*(SA-CAS)*3.1416*D2^2+SW3*kL3*(SA-CAS)*3.1416*D3^2)-CAS/tau
3	d(CDS)/d(t) = (1/V)*((1-SW1)*kL1*(SD-CDS)*3.1416*D1^2+(1-SW2)*kL2*(SD-CDS)*3.1416*D2^2+(1-SW3)*kL3*(SD-CDS)*3.1416*D3^2)-CDS/tau
4	d(D2)/d(t) = if(D2>0.3)then(-2*kL2*(SA-CAS)/rho)else(if(D2>1e-5)then(-2*kL2*(SD-CDS)/rho)else(0.0))
5	d(D3)/d(t) = if(D3>0.3)then(-2*kL3*(SA-CAS)/rho)else(if(D3>1e-5)then(-2*kL3*(SD-CDS)/rho)else(0.0))
6	kL1 = 2*0.6/D1
7	SA = 1.0
8	rho = 1414.7
9	SD = 0.4
10	V = 1200
11	SW1 = if(D1>0.3)then(1.0)else(0.0)
12	SW2 = if(D2>0.3)then(1.0)else(0.0)
13	kL2 = 2*0.6/D2
14	SW3 = if(D3>0.3)then(1.0)else(0.0)
15	kL3 = 2*0.6/D3
16	tau = 240

Table 10–5 POLYMATH Program (File **P10-04A.POL**)

Line	Equation
17	t(0)=0
18	D1(0)=0.5
19	CAS(0)=0
20	CDS(0)=0
21	D2(0)=0.4
22	D3(0)=0.35
23	t(f)=150

The partial results of this solution are shown for the three pill diameters in Figure 10–5.

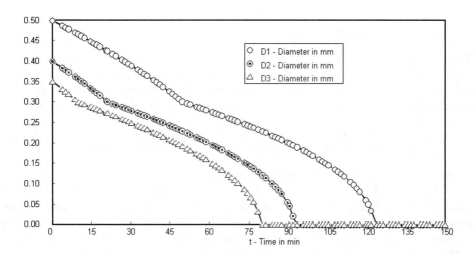

Figure 10–5 Diameter Variations for Dissolution of Three Pills

The POLYMATH problem solution file for part (a) is found in directory Chapter 10 with file named **P10-04A.POL**.

(b) The rate of absorption of the drug in terms of the mass transfer rate can be included in Equation (10-36), which was derived for part (a). Note that this absorption will tend to reduce the buildup of D within the stomach.

(c) A differential balance on the drug within the body utilizing C_{DB} needs to be made on the volume of the body, where the input is the rate of absorption is calculated as in part (b). The first-order reaction must also be included in the differential equation development. Assume the body volume to be completely mixed at all times.

(d) One technique to determine the dose time is to set up a differential equation for this time that has a derivative of unity when above the minimum effective concentration and that has a derivative of zero at all other times.

10.5 DIFFUSION WITH SIMULTANEOUS REACTION IN ISOTHERMAL CATALYST PARTICLES

10.5.1   Concepts Demonstrated

Determination of effectiveness factors for porous catalyst particles with cylindrical, and spherical geometries and various reaction orders under isothermal conditions.

10.5.2   Numerical Methods Utilized

Solution of simultaneous ordinary differential equations with split boundary values.

10.5.3   Problem Statement

The mathematical solution of simultaneous diffusion and reaction inside porous catalytic particles at constant temperature is typically formulated as an isothermal internal effectiveness factor problem. The differential equations that describe the diffusion with reaction are solved with the result expressed as

$$\eta = \frac{\text{average reaction rate within the particle}}{\text{reaction rate at the concentrations of the particle surface}} \tag{10-37}$$

where η is the isothermal internal effectiveness factor.

The general solution involves the derivation of the ordinary differential equations that describe the material balance within the particle geometry. This treatment also requires the use of Fick's law for diffusion, which is usually assumed to involve only the diffusion of the reactant with an effective diffusivity for the catalyst particle. Details are found in Fogler,[4] pp. 610–20.

The numerical solution for a spherical particle involves the material balance on a differential volume within the catalyst sphere, which yields

$$\frac{d}{dr}(N_A r^2) = -k'' a C_A r^2 \tag{10-38}$$

where N_A is the flux of reactant A, r is the radius of the spherical particle, k'' is the first-order rate constant based on particle volume, a is the surface area per unit volume of particle, and C_A is the concentration of reactant. The initial condition for Equation (10-38) is that there is no flux at the particle center; therefore, N_A or the combined variable $N_A r$ are both zero at $r = 0$.

Fick's law for the diffusion of reactant A can be written as

$$\frac{dC_A}{dr} = \frac{N_A}{-D_e} \tag{10-39}$$

where D_e is the effective diffusivity for the diffusion of reactant A in the porous particle. The boundary condition for this equation is that the concentration of A

at the particle surface is given by $C_A = C_{As}$ when $r = R$ (the radius of the spherical particle).

Since the combined variable $(N_A r^2)$ is used in Equation (10-38), an algebraic equation must be included with this problem formulation to provide N_A for Equation (10-39). Thus

$$N_A = \frac{N_A r^2}{r^2} \tag{10-40}$$

The effectiveness factor can be calculated from

$$\eta = \frac{\int_0^R k'' a C_A (4\pi r^2) dr}{k'' a C_{As} \left(\frac{4}{3}\pi R^3\right)} = \frac{3}{C_{As} R^3} \int_0^R C_A r^2 dr \tag{10-41}$$

which is the mathematical equivalent to Equation (10-37).

For convenience in calculation of the effectiveness factor during the solution of Equations (10-38) to (10-40), the effectiveness factor of Equation (10-41) can be differentiated with respect to r to obtain

$$\frac{d\eta}{dr} = \frac{3 C_A r^2}{C_{As} R^3} \tag{10-42}$$

whose initial condition is $\eta = 0$ and $r = 0$. This differential equation can be solved simultaneously with Equations (10-38) and (10-39). The final value of the effectiveness factor is given by the value of η at $r = R$ provided that the boundary conditions of Equations (10-38) and (10-39) are satisfied.

Similar problem solutions can be obtained for the slab and cylindrical geometries for a variety of reaction rate expressions. The numerical solution of these problems can provide the concentration profiles and the effectiveness factor when the boundary conditions are satisfied.

Analytical Solution
The analytical solution to the diffusion with first-order reaction is given by (Fogler,[4] p. 617, or Bird et al.,[1] p. 545)

$$\eta = \frac{3}{\phi^2} \{\phi[\coth(\phi)] - 1\} \tag{10-43}$$

where coth( ) is the hyperbolic cotangent and ϕ is the Thiele modulus defined by

$$\phi = R \sqrt{\frac{k'' a C_{As}^{n-1}}{D_e}} \tag{10-44}$$

with n representing the order of the reaction.

Consider the simultaneous diffusion and reaction inside porous catalyst particle at constant temperature.

(a) Calculate the concentration profile numerically for C_A and simultaneously determine the effectiveness factor η for a first-order irreversible reaction in a spherical particle, where $R = 0.5$ cm, $D_e = 0.1$ cm^2/s, $C_{As} = 0.2$ g-mol/cm^3, and $k''a = 6.4$ s^{-1}.

(b) Compare the result in part (a) for the effectiveness factor η with the analytical solution for η given by Equations (10-43) and (10-44).

(c) Repeat part (a) for a catalyst particle that is a long cylinder with radius $R = 0.5$ cm, $D_e = 0.1$ cm^2/s, $C_{As} = 0.2$ g-mol/cm^3, and $k''a = 6.4$ s^{-1}.

(d) Repeat part (a) for a second-order irreversible reaction with $C_{As} = 0.2$ g-mol/cm^3, $D_e = 0.1$ cm^2/s, and $k''a = 32$ cm^3/g-mol$\cdot$s. [Note that the Thiele modulus is the same as for part (a).]

(e) Repeat part (a) for a catalyst particle that is a long cylinder with radius $R = 0.5$ cm. The reaction is second order and irreversible with $C_{As} = 0.2$ g-mol/cm^3 and $k''a = 32$ cm^3/g-mol$\cdot$s. [Note that the Thiele modulus is the same as for part (a).]

10.5.4 Solution (Partial)

(a) and (b) Sphere This problem requires the numerical integration of ordinary differential equations given by Equations (10-38), (10-39), and (10-41) along with algebraic equations (10-40), (10-43), and (10-44). A suitable equation set for the POLYMATH *Simultaneous Ordinary Differential Equation Solver* is given in Table 10–6.

Table 10–6 POLYMATH Program (File **P10-05AB.POL**)

Line	Equation
1	d(eta)/d(r)=3*CA*r^2/(CAs*R^3)
2	d(CA)/d(r)=NA/(-De)
3	d(NArr)/d(r)=-kppa*CA*r^2
4	NA=if(r==0)then(0.0)else(NArr/r^2)
5	CAs=0.2
6	R=0.5
7	De=0.1
8	kppa=6.4
9	err=CA-CAs
10	phi=R*sqrt(kppa/De)
11	etacalc=(3/phi^2)*(phi*(1/tanh(phi))-1)
12	r(0)=0
13	eta(0)=0
14	CA(0)=0.029315
15	NArr(0)=0
16	r(f)=0.5

Division by Zero Note that the possible division by zero in the POLY-MATH equation set for Equation (10-40) is handled with "if ... then ... else ..." logic, as follows:

NA=if(r==0)then(0.0)else(NArr/r^2)

Boundary Condition Convergence The shooting technique can be used to determine the value of the initial concentration C_A at $r = 0$ that gives $C_A = C_{As}$ = 0.2 g-mol/cm^3. The POLYMATH equation set statement that gives the error is

err=CA-CAs

Convergence to a low value of err can be accomplished by simple trial-and-error iterations or by more sophisticated methods, as discussed in Problem 6.6.

The POLYMATH problem solution file for parts (a) and (b) is found in directory Chapter 10 with file named **P10-05AB.POL**.

(c) Cylinder The corresponding equations for the cylinder are as follows:

Material Balance Differential Equation

$$\frac{d}{dr}(N_A r) = -k'' a C_A r \tag{10-45}$$

Fick's Law Differential Equation

$$\frac{dC_A}{dr} = \frac{N_A}{-D_e} \tag{10-46}$$

Algebraic Equation

$$N_A = \frac{N_A r}{r} \tag{10-47}$$

Effectiveness Factor Integral Equation

$$\eta = \frac{\int_0^R k'' a C_A (2\pi r) dr}{k'' a C_{As}(\pi R^2)} = \frac{2}{C_{As} R^2} \int_0^R C_A r \, dr \tag{10-48}$$

Effectiveness Factor in Differential Equation Form

$$\frac{d\eta}{dr} = \frac{2 C_A r}{C_{As} R^2} \tag{10-49}$$

10.6 GENERAL EFFECTIVENESS FACTOR CALCULATIONS FOR FIRST-ORDER REACTIONS

10.6.1   Concepts Demonstrated

Demonstration of the similarity of effectiveness factor solutions for diffusion with first-order reaction in a slab, cylinder, and sphere using a modified Thiele modulus with the same volume-to-surface-area ratios.

10.6.2   Numerical Methods Utilized

Numerical solution of simultaneous ordinary differential equations with split boundary conditions.

10.6.3   Problem Statement

The cases of diffusion with first-order reaction in isothermal catalyst particles in several geometries have been shown by Aris[5] to yield similar effectiveness factors when a modified Thiele modulus is utilized. (Effectiveness factors and Thiele modulus are discussed in Problem 10.5.)

The modified Thiele modulus is defined by

$$\phi_m = \left(\frac{V}{S}\right)\sqrt{\frac{k''a}{D_e}} \tag{10-50}$$

where $\frac{V}{S}$ represents the volume to surface ratio for the particle. The ratios are summarized in Table 10–7. A plot of this modified Thiele modulus is given by Fogler,[4] p. 618.

Table 10–7 Volume-to-Surface-Area Ratios for Particles

Particle Geometry	$\dfrac{V}{S}$
Slab of thickness L with reaction on both surfaces	$\dfrac{L}{2}$
Cylinder of radius R_c	$\dfrac{R_c}{2}$
Sphere of radius R_s	$\dfrac{R_s}{3}$

Consider the simultaneous diffusion and first-order reaction inside porous catalyst particle at constant temperature. This problem will help to verify that the calculated effectiveness factors are very similar when the volume to surface ratio of the particles given in Table 10–7 are the same. A complete numerical solution will be required for each particle to enable this comparison.

(a) Determine the effectiveness factor for the base case, a spherical particle where $R = 0.3$ cm, $D_e = 0.08$ cm^2/s, $C_{As} = 0.4$ g-mol/cm^3, and $k''a = 8$ s^{-1}.

(b) Determine the effectiveness factor for a slab whose thickness gives the same modified Thiele modules as the sphere in part (a). (This is equivalent to having the same volume-to-surface-area ratio as the base case sphere.)

(c) Determine the effectiveness factor for a long cylinder whose radius gives the same modified Thiele modules as the sphere in part (a). (This is equivalent to having the same volume-to-surface-area ratio as the base case sphere.)

(d) Repeat the effectiveness factor calculations for a sphere, slab, and cylinder whose dimensions are increased to give two times the volume to surface ratio (V/S) of the original spherical catalyst particle in part (a). Note that this doubles the modified Thiele modulus.

(e) Summarize the various effectiveness factor calculations made in parts (a) through (d) for the two values of the modified Thiele modulus. Do your results confirm that the effectiveness factor values are approximately equivalent at the same modified Thiele modulus?

10.7 SIMULTANEOUS DIFFUSION AND REVERSIBLE REACTION IN A CATALYTIC LAYER

10.7.1 Concepts Demonstrated

Modeling of binary gaseous diffusion with simultaneous, isothermal, reversible reaction in a porous catalyst layer.

10.7.2 Numerical Methods Utilized

Integration of simultaneous ordinary differential equations with split boundary conditions.

10.7.3 Problem Statement

The catalytic gas phase reaction between components A and B is occurring reversibly in a catalyst layer at a particular point in a monolithic reactor.

$$2A \leftrightarrow B \tag{10-51}$$

The catalytic reaction rate expression for reactant A is given by

$$r'_A = -k\left(C_A^2 - \frac{C_B}{K_C}\right) \tag{10-52}$$

where the rate is in g-mol/cm^3·s, the rate constant k has been determined to be 8×10^4 in cm^3/s·g-mol, and the equilibrium constant is known to be $K_C = 6 \times 10^5$ cm^3/g-mol. The catalytic layer has a thickness of $L = 0.2$ cm, and the effective diffusivity of A in B for this layer is $D_e = 0.01$ cm^2/s. The reactant mixture contains only gases A and B, so only binary gas diffusion need be considered. The total concentration of A and B is $C_T = 4 \times 10^{-5}$ g-mol/cm^3.

Calculation of the effectiveness factor (see Problem 10.5 for definition) requires consideration of the simultaneous diffusion of both components along with a material balance involving the reversible reaction rate expression.

Material Balances on A and B within the Porous Layer

The material balance on component A in the z direction from the top of the porous layer results in an ordinary differential equation involving the flux N_A:

$$\frac{dN_A}{dz} = -k\left(C_A^2 - \frac{C_B}{K_C}\right) \tag{10-53}$$

with the boundary condition that there is no mass transfer flux at the bottom of the porous layer, expressed as

$$N_A\big|_{z=L} = 0 \tag{10-54}$$

A similar differential equation could be derived for component B, but it is simpler to use the algebraic relationship between the molar fluxes of A and B due to the reaction stoichiometry.

$$N_B = -\frac{1}{2}N_A \qquad \text{(10-55)}$$

Note that this equation also satisfied the boundary condition at the bottom of the porous layer, where

$$N_B\big|_{z=L} = 0 \qquad \text{(10-56)}$$

Fick's Law for Binary Diffusion

The diffusion of A in this binary system is described by

$$N_A = -D_e\frac{dC_A}{dz} + x_A(N_A + N_B) \qquad \text{(10-57)}$$

where the effective diffusivity D_e of the catalyst layer is given by

$$D_e = \frac{D_{AB}\varepsilon_p\sigma}{\tilde{\tau}} \qquad \text{(10-58)}$$

where D_{AB} is the binary diffusivity, ε_p is the catalyst porosity, σ is the constriction factor, and $\tilde{\tau}$ is the tortuosity (see Fogler,[4] p. 608, for details)

Equation (10-57) can be rearranged by incorporating Equation (10-55) and the definition of x_A in terms of concentrations to yield

$$\frac{dC_A}{dz} = \frac{\dfrac{C_A}{C_T}\left(\dfrac{N_A}{2}\right) - N_A}{D_e} \qquad \text{(10-59)}$$

with the initial condition

$$C_A\big|_{z=0} = C_{As} \qquad \text{(10-60)}$$

A similar differential equation can be derived for component B; however, it is much easier to use the overall mass balance for this gas phase system to calculate C_B. Thus

$$C_B = C_T - C_A \qquad \text{(10-61)}$$

and Equation (10-53) can therefore be expressed in terms of only C_A as

$$\frac{dN_A}{dz} = -k\left[C_A^2 - \frac{(C_T - C_A)}{K_C}\right] \qquad \text{(10-62)}$$

(a)   Use an implicit finite difference technique to calculate the effective-ness factor for the given reaction and summarize C_A and N_A in a table at 10 equally spaced intervals within the catalytic layer. The reactant concentrations at the surface of the layer are known to be $C_{As} = 3 \times 10^{-5}$ g-mol/cm^3 and $C_{Bs} = 1 \times 10^{-5}$ g-mol/cm^3.

(b)   Repeat part(a) using the shooting technique for solving the problem, and enter the calculated results in the same table.

(c)   Compare the two solutions of parts (a) and (b). Please comment on which solution technique you prefer.

(d)   Repeat part (a) for $C_{As} = 1 \times 10^{-5}$ g-mol/cm^3 and $C_{Bs} = 3 \times 10^{-5}$ g-mol/cm^3 using the solution technique that you prefer.

10.7.4   Solution (Suggestions)

This problem requires the simultaneous solution of differential Equation (10-59) with the initial condition given by Equation (10-60) and differential Equation (10-62) with the boundary condition given by Equation (10-56). Thus the boundary conditions for the two differential equations are split, with one being an initial condition and the other being a final condition.

(a) Implicit Finite Difference (IFD) Solution

This solution involves utilizing finite difference formulas for the differential equations and boundary conditions. The resulting system of algebraic equations can be solved for the variables at the various node points utilized in the problem. The 10 intervals with 11 node points for this problem are illustrated in Figure 10–6. Equation (10-59) can be rewritten using a second-order central difference formula for the first derivative [see Equation (A-6) of Appendix A] as

$$\frac{dC_{A_n}}{dz} = \frac{C_{A_{n+1}} - C_{A_{n-1}}}{2\Delta z} = \frac{(C_{A_n}/C_T)(N_{A_n}/2) - N_{A_n}}{D_e} \quad \text{for } (2 \leq n \leq 10) \quad \textbf{(10-63)}$$

The known initial condition of Equation (10-60) gives $C_{A_1} = C_{As}$.

For node 11 at the bottom of the catalyst layer, the second-order backward difference approximation [see Equation (A-7) of Appendix A] for the first derivative can be used, giving

$$\left.\frac{dC_A}{dz}\right|_{n=11} = \frac{3C_{A_{11}} - 4C_{A_{10}} + C_{A_9}}{2\Delta z} = \frac{(C_{A_{11}}/C_T)(N_{A_{11}}/2) - N_{A_{11}}}{D_e} \quad \textbf{(10-64)}$$

The boundary condition for N_A, Equation (10-54), results in the right-hand side of Equation (10-64) becoming zero; therefore, the resulting simplification yields

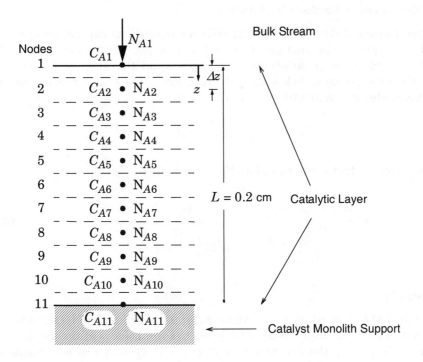

Figure 10–6 Diffusion with Reversible Chemical Reaction in a Catalytic Layer

the following algebraic equation at node 11:

$$C_{A_{11}} = (4C_{A_{10}} - C_{A_9})/3 \qquad \text{(10-65)}$$

Similarly, for the flux N_A, differential Equation (10-62) becomes

$$\frac{dN_{A_n}}{dz} = \frac{N_{A_{n+1}} - N_{A_{n-1}}}{2\Delta z} = -k\left[C_{A_n}^2 - \frac{(C_T - C_{A_n})}{K_C}\right] \quad \text{for } (2 \le n \le 10) \qquad \text{(10-66)}$$

The known final condition of Equation (10-54) gives zero flux at the bottom of the catalyst layer.

$$N_{A_{11}} = 0 \qquad \text{(10-67)}$$

For node 1 at the top of the catalyst layer, the second-order forward difference approximation for the first derivative [Equation (A-5) of Appendix A] can be used, giving

$$\left.\frac{dN_A}{dz}\right|_{n=1} = \frac{-3N_{A_1} + 4N_{A_2} - N_{A_3}}{2\Delta z} = -k\left[C_{A_1}^2 - \frac{(C_T - C_{A_1})}{K_C}\right] \qquad \text{(10-68)}$$

Effectiveness Factor Calculation

The effectiveness factor (see Problem 10.5 for definition) can be obtained by a material balance at the surface of the catalytic layer. In this case, the surface flux of A acting over an arbitrary area must equal the average reaction rate within the catalyst layer below the arbitrary area. Thus, in terms of the finite difference nodes for an arbitrary area A_c,

$$N_{A_1} A_c = (-r'_A)_{\text{AVG}} A_c L \tag{10-69}$$

The effectiveness factor can therefore be expressed as

$$\eta = \frac{(-r'_A)|_{\text{AVG}}}{(-r'_A)|_{\text{SURFACE}}} = \frac{N_{A_1}/L}{k\left[C_{A_1}^2 - \dfrac{(C_T - C_{A_1})}{K_C}\right]} \tag{10-70}$$

Results

The POLYMATH *Simultaneous Nonlinear Equation Solver* can be used to solve Equations (10-63) through (10-68) as applied to the 11 nodes of the problem plus Equation (10-70) for the effectiveness factor. The results are summarized in Table 10–8, and the calculated effectiveness factor is 0.3022.

Table 10–8 Comparison of Implicit Finite Difference and Shooting Method Solutions

z	Implicit Finite Difference Solution		Shooting Method Solution	
	C_A	N_A	C_A	N_A
0	3E–05	4.271E–06	3E–05	4.250E–06
0.02	2.549E–05	3.064E–06	2.526E–05	3.066E–06
0.04	2.165E–05	2.269E–06	2.154E–05	2.239E–06
0.06	1.887E–05	1.661E–06	1.866E–05	1.649E–06
0.08	1.657E–05	1.242E–06	1.644E–05	1.218E–06
0.1	1.493E–05	9.076E–07	1.475E–05	8.954E–07
0.12	1.362E–05	6.632E–07	1.349E–05	6.469E–07
0.14	1.273E–05	4.548E–07	1.258E–05	4.482E–07
0.16	1.209E–05	2.904E–07	1.196E–05	2.823E–07
0.18	1.174E–05	1.361E–07	1.161E–05	1.362E–07
0.2	1.162E–05	0	1.149E–05	–5.150E–10

(b) Shooting Technique Solution

The concentration and flux profiles can be determined during the numerical solution of differential Equation (10-59) with a known initial condition and differential Equation (10-62), which requires determination of an initial condition that satisfies the boundary condition of zero flux when $z = L$, as given in Equation (10-54).

Effectiveness Factor Calculation

The effectiveness factor for the catalyst layer is given by the average rate within the catalytic layer divided by the rate at the surface of the catalytic layer. Thus

$$\eta = \frac{\int_0^L \left(-k_1\left(C_A^2 - \dfrac{C_B}{K_C}\right)\right)dz}{-k_1\left(C_{As}^2 - \dfrac{C_{Bs}}{K_C}\right)L} = \frac{\int_0^L \left(C_A^2 - \dfrac{C_B}{K_C}\right)dz}{\left(C_{As}^2 - \dfrac{C_{Bs}}{K_C}\right)L} \tag{10-71}$$

which can be differentiated with respect to z, yielding

$$\frac{d\eta}{dz} = \frac{\left(C_A^2 - \dfrac{C_B}{K_C}\right)}{\left(C_{As}^2 - \dfrac{C_{Bs}}{K_C}\right)L} \tag{10-72}$$

The initial condition for Equation (10-72) is

$$\eta\big|_{z\,=\,0} = 0 \tag{10-73}$$

and the effectiveness factor is the value of η at $z = L$. Thus the inclusion of differential Equation (10-72) with the boundary condition of Equation (10-73) completes the problem formulation.

Split Boundary Value Solution

The POLYMATH *Simultaneous Ordinary Differential Equation Solver* can be used with a "shooting technique" (see Problem 6.5 for details) to solve the differential equations. With this technique, the initial condition for N_A can be determined.

Initial Condition Estimate for N_A

The shooting technique requires that the initial condition of N_A be determined, which gives a final value of nearly zero. All the techniques require an initial estimate of N_A at the surface of the catalytic layer. Often this proves to be a difficult challenge because an incorrect initial condition may lead to numerical results that give profiles that cannot be optimized easily to improve the initial estimate.

A simple material balance at the catalytic surface can provide an upper bound on this initial value when negligible diffusional resistance is assumed. This balance sets the molar mass transfer rate at the surface equal to the reaction rate in the catalyst layer that is occurring at the surface concentrations of A and B. Thus

$$N_{A_{EST}}\bigg|_{z=0} = -r_A L = kL\left[C_{As}^2 - \frac{(C_T - C_{As})}{K_C}\right] \tag{10-74}$$

Another alternative that is useful for this problem is to take the calculated value for N_A from the IFD solution in part (a) as an initial estimate for the shooting solution.

Results
The POLYMATH *Simultaneous Differential Equation Solver* can be used with a shooting technique to solve the three ordinary differential equations describing this problem. The results are also summarized in Table 10–8, and the calculated effectiveness factor is 0.3007 versus 0.3022 for the IFD solution.

(c) Comparison of Solution Methods
A comparison of the calculated concentrations and fluxes in Table 10–8 indicates good agreement between the IFD and shooting methods. The IFD solution's accuracy can be improved by more node points within the catalytic layer.

10.8 SIMULTANEOUS MULTICOMPONENT DIFFUSION OF GASES

10.8.1 Concepts Demonstrated

Application of the Stefan-Maxwell equations to describe the multicomponent molecular diffusion of gases.

10.8.2 Numerical Methods Utilized

Numerical integration of a system of simultaneous ordinary differential equations with optimization of two parameters in order to match split boundary conditions.

10.8.3 Problem Statement[*]

Gases A and B are diffusing through stagnant gas C at a temperature of 55°C and a pressure of 0.2 atmospheres. This process involves molecular diffusion between two points, where the compositions are known, as summarized in Table 10–9. The distance between the points is 10^{-3} m.

(a) Use the Stefan Maxwell equations to calculate the molar fluxes of both gases A and B from point 1 to point 2. Suggestion: An initial approximate solution can be determined by first considering the binary diffusion of only A through component C and then separately considering the binary diffusion of only B through component C.

(b) Plot the mole fractions of the gases as a function of distance from point 1 to point 2.

Table 10–9 Data for Multicomponent Diffusion (from Geankoplis[6] with permission)

Component	Point 1 Concentration kg-mol/m^3	Point 2 Concentration kg-mol/m^3	Diffusivities at 0.2 atm m^2/s
A	2.229×10^{-4}	0	$D_{AC} = 1.075 \times 10^{-4}$
B	0	2.701×10^{-3}	$D_{BC} = 1.245 \times 10^{-4}$
C	7.208×10^{-3}	4.730×10^{-3}	$D_{AB} = 1.47 \times 10^{-4}$

[*] This problem is adapted from Geankoplis[6] with permission.

Additional Information and Data

The kinetic theory of gases can be used to derive the Stefan-Maxwell equations in the z direction as (see Bird et al.[1] or Geankoplis[6])

$$\frac{dC_i}{dz} = \sum_{i=1}^{n} \frac{(x_i N_j - x_j N_i)}{D_{ij}} \tag{10-75}$$

where C_i represents the concentration of diffusing component i in kg-mol/m^3, x_i is the mole fraction of component i, N_i is the molar flux of component i in kg-mol/m$^2 \cdot$s, n is the number of components, and D_{ij} is the binary molecular diffusivity for components i and j in m^2/s.

Application of Equation (10-75) to a three-component mixture yields the equations

$$\frac{dC_A}{dz} = \frac{(x_A N_B - x_B N_A)}{D_{AB}} + \frac{(x_A N_C - x_C N_A)}{D_{AC}} \tag{10-76}$$

$$\frac{dC_B}{dz} = \frac{(x_B N_A - x_A N_B)}{D_{AB}} + \frac{(x_B N_C - x_C N_B)}{D_{BC}} \tag{10-77}$$

$$\frac{dC_C}{dz} = \frac{(x_C N_A - x_A N_C)}{D_{AC}} + \frac{(x_C N_B - x_B N_C)}{D_{BC}} \tag{10-78}$$

where the appropriate equalities for the binary molecular diffusivities have been substituted for $D_{BA} = D_{AB}$, $D_{CA} = D_{AC}$, and $D_{CB} = D_{BC}$.

Typical boundary conditions for the preceding equations are dictated by the physical or chemical process. For example, if the diffusion is to a catalyst surface where the reaction rate is very fast, then the corresponding concentration of the limiting reactant at the catalyst surface may be assumed to be zero. If the diffusion process leads to a bulk stream, then the concentrations in the bulk stream are usually assumed to be at the bulk stream concentrations and diffusing components not in the bulk stream will have zero concentrations. Often the boundary conditions involving concentrations are split between two locations.

If relationships are known between the fluxes due to reaction stoichiometry or any of the fluxes are zero (stagnant component), then these relationships can be substituted into the preceding equations or expressed separately.

10.8.4   Solution

For the three-component system of this problem, the differential equations of Equations (10-76) to (10-78) directly apply. Since component C is stagnant, then the flux of this component is zero. Thus $N_C = 0$. The problem solution requires that the two fluxes N_A and N_B must be optimized until the boundary conditions of the concentrations of Table 10–9 are satisfied. The initial conditions are the known concentrations at point 1, and the final conditions are the known concentrations at point 2.

(a) and (b) The POLYMATH *Simultaneous Differential Equation Solver* can be used to solve the differential equations as a split boundary value problem with the initial conditions at point 1. In order to converge on the values of fluxes N_A and N_B, error functions for the matching of the boundary conditions at point 2 can be defined by

$$\varepsilon(C_A) = 0 - C_A\big|_{z = 0.001} \tag{10-79}$$

$$\varepsilon(C_B) = 2.701 \times 10^{-3} - C_B\big|_{z = 0.001} \tag{10-80}$$

Note that these error functions should go to zero when convergence is obtained.

Utilizing the preceding error functions and adding definitions for the mole fractions of the three components, one can write an initial POLYMATH equation set as given in Table 10–10.

Table 10–10 POLYMATH Program (File **P10-08AB1.POL**)

Line	Equation
1	d(CA)/d(z)=(xA*NB-xB*NA)/DAB+(xA*NC-xC*NA)/DAC
2	d(CB)/d(z)=(xB*NA-xA*NB)/DAB+(xB*NC-xC*NB)/DBC
3	d(CC)/d(z)=(xC*NA-xA*NC)/DAC+(xC*NB-xB*NC)/DBC
4	NA=2.396E-5
5	NB=-3.363E-4
6	DAB=1.47E-4
7	NC=0
8	DBC=1.245E-4
9	DAC=1.075E-4
10	CT=0.2/(82.057E-3*328)
11	errA=CA-0
12	errB=CB-2.701E-3
13	xB=CB/CT
14	xA=CA/CT
15	xC=CC/CT
16	z(0)=0
17	CA(0)=0.0002229
18	CB(0)=0
19	CC(0)=0.007208
20	z(f)=0.001

Note that the initial estimate for N_A in the preceding POLYMATH equation set is obtained from an application of Fick's law for just simple binary diffusion of A in C while the other diffusional transport is neglected. Thus the initial estimate for N_A is

$$N_A = -D_{AC} \frac{\left(C_A\big|_2 - C_A\big|_1\right)}{(z\big|_2 - z\big|_1)} = -1.075 \times 10^{-4} \frac{(0 - 2.229 \times 10^{-4})}{(0.001 - 0)} = 2.396 \times 10^{-5} \quad \textbf{(10-81)}$$

Similarly, for N_B the initial estimate is

$$N_B = -D_{BC} \frac{\left(C_B\big|_2 - C_B\big|_1\right)}{(z\big|_2 - z\big|_1)} = -1.245 \times 10^{-4} \frac{(2.701 \times 10^{-3} - 0)}{(0.001 - 0)} = -3.363 \times 10^{-4} \quad \textbf{(10-82)}$$

www

The POLYMATH problem solution file for the initial solution is found in directory Chapter 10 with file named **P10-08AB1.POL**.

Optimization of N_A and N_B

A simple way to optimize these two fluxes is first to hold N_B fixed and then to converge upon an improved value of N_A by minimizing the error calculated in Equation (10-79). This iterative shooting method solution can be easily accomplished by trial and error or by the secant method, as discussed in Problem 6.5. Then the improved value of N_A can be held fixed, and an improved value of N_B can be obtained by minimizing the error calculated by Equation (10-80). Note that this simple optimization technique really involves searching along each parameter until a local minimum is obtained in an objective function, and then searching in turn along the other parameter to satisfy another objective function.

Typical progress in the solution for these local searches of this problem is summarized in Table 10–11. Note that the initial optimization holds N_B and searches for the value of N_A. Then the next step involves holding N_A and searching for an improved value of N_B. Convergence is obtained with just two searches for each flux, and the resulting values are found to be $N_A = 2.12 \times 10^{-5}$ kg-mol/m$^2 \cdot$s and $N_B = -4.14 \times 10^{-4}$ kg-mol/m$^2 \cdot$s. The resulting mole fraction profiles are nonlinear, as shown in Figure 10–7, and the final result is significantly different from the initial solution calculated from binary diffusion consideration only.

Table 10–11 Iterative Search for Fluxes N_A and N_B

Search	N_A	$\varepsilon(C_A)$	N_B	$\varepsilon(C_B)$
Start	2.396×10^{-5}	-1.692×10^{-5}	-3.363×10^{-4}	-4.170×10^{-4}
1	2.174×10^{-5}	4.224×10^{-8}	-3.363×10^{-4}	-4.196×10^{-4}
2	2.174×10^{-5}	-4.309×10^{-8}	-4.141×10^{-4}	6.325×10^{-8}
3	2.115×10^{-5}	9.811×10^{-10}	-4.141×10^{-4}	-7.510×10^{-7}
4	2.115×10^{-5}	-1.017×10^{-8}	-4.143×10^{-4}	2.827×10^{-7}

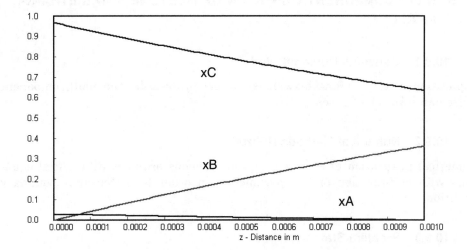

Figure 10–7 Mole Fractions Profiles for Components A, B, and C

10.9 MULTICOMPONENT DIFFUSION OF ACETONE AND METHANOL IN AIR

10.9.1   Concepts Demonstrated

Application of the Stefan-Maxwell equations to describe the multicomponent molecular diffusion of gases.

10.9.2   Numerical Methods Utilized

Numerical integration of a system of simultaneous ordinary differential equations with optimization of two parameters in order to match split boundary conditions.

10.9.3   Problem Statement

Carty and Schrodt[7] have conducted experiments involving the transport from the surface of a mixture of acetone (1) and methanol (2) through a simple diffusion tube into a stream of flowing air (3). Measurements were conducted in a Stefan tube (see Problem 10.1) at 328.5 K, where the pressure was 99.4 kPa. The gas phase composition at the liquid surface was measured to be $x_1 = 0.319$ and $x_2 = 0.528$. The length of the diffusion path was 0.238 m. The binary molecular diffusivities are estimated (see Taylor and Krishna,[2] pages 21–23) to be $D_{12} = 8.48 \times 10^{-6} \text{ m}^2/\text{s}$, $D_{13} = 13.72 \times 10^{-6} \text{ m}^2/\text{s}$, and $D_{23} = 19.91 \times 10^{-6} \text{ m}^2/\text{s}$.

(a)   Calculate the molar fluxes of both acetone and methanol from the liquid surface to the flowing air stream.

(b)   Plot the mole fractions of acetone, methanol, and air from the liquid surface to the flowing air stream.

(c)   Compare the calculated results to the data in the paper by Carty and Schrodt.[7]

(d)   Verify the calculated binary molecular diffusivities used in the problem.

10.10 MULTICOMPONENT DIFFUSION IN A POROUS LAYER COVERING A CATALYST

10.10.1 Concepts Demonstrated

Application of the Stefan-Maxwell equations to describe the multicomponent molecular diffusion of gases and binary diffusion approximations for multicomponent diffusion.

10.10.2 Numerical Methods Utilized

Numerical integration of a system of simultaneous ordinary differential equations with optimization of a single variable in order to match split boundary conditions.

10.10.3 Problem Statement

In a particular reactor for the catalytic oxidation of carbon monoxide to carbon dioxide in an mixture of N_2 and O_2, the very active catalyst is covered by a porous layer through which the reactants and products must diffuse as shown in Figure 10–8. The oxidation reaction at 1 atm and 200°C is essentially irreversible and given by

$$CO + 1/2 \, O_2 \rightarrow CO_2$$

Consider the entrance to the reactor where the composition in the bulk gas stream is 2 mol% CO, 3 mol% O_2, and 95 mol% N_2. The O_2 and CO must diffuse through the porous layer of thickness 0.001 m to the catalyst surface while the product CO_2 must diffuse out through the same layer to the bulk gas stream. The reaction is so rapid that the concentration of the limiting reactant CO is essentially zero at the catalyst surface. The effective molecular diffusivities that take into account the porosity and tortuosity of the porous layer are summarized in Table 10–12. These effective molecular diffusivities can be utilized directly in the diffusional equations.

(a) Calculate and plot the concentrations of O_2, CO, and CO_2 as a function of the distance into the porous layer to the surface of the catalyst. Use multicomponent diffusion as described by the Stefan-Maxwell equations discussed in Problem 10.8.

(b) What are the values of the fluxes of the gases of part (a) at the surface of the catalyst considering the z direction of Figure 10–8 to be positive?

(c) Work part (b) by considering only binary diffusion of CO in N_2 with the effective diffusion coefficient for CO in N_2 from Table 10–12.

(d) Comment upon the approximate solution of part (c) relative to the more exact solution of parts (a) and (b).

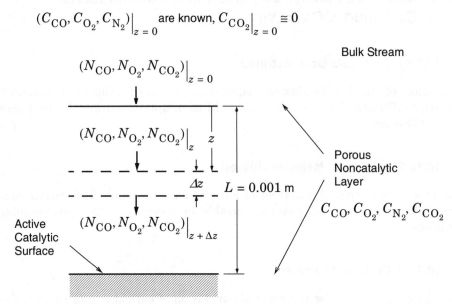

Figure 10–8 Gas Diffusion through Porous Layer Covering Active Catalyst

Table 10–12 Effective Molecular Diffusivities for Porous Layer at 1 atm and 200°C

Effective Molecular Diffusivity D_{ij} in m^2/s	N_2	O_2	CO	CO_2
N_2		7.20×10^{-6}	6.93×10^{-6}	6.20×10^{-6}
O_2	7.20×10^{-6}		7.61×10^{-6}	6.67×10^{-6}
CO	6.93×10^{-6}	7.61×10^{-6}		6.22×10^{-6}
CO_2	6.20×10^{-6}	6.67×10^{-6}	6.22×10^{-6}	

10.10.4 Solution (Suggestions)

(a) and (b) The reactions will define ratios of fluxes, which can be used in the Stefan-Maxwell equations. A positive flux should be considered to be in the z direction toward the catalyst, as shown in Figure 10–8.

(c) and (d) This problem can be approximated by CO diffusing through stagnant N_2 when the concentrations and diffusional transport of O_2 and CO_2 are neglected.

10.11 SECOND-ORDER REACTION WITH DIFFUSION IN LIQUID FILM

10.11.1 Concepts Demonstrated

Molecular diffusion of two components with a simultaneous irreversible second-order reaction in a finite liquid film.

10.11.2 Numerical Methods Utilized

Solution of simultaneous second-order ordinary differential equations using implicit finite difference techniques, and simultaneous solution of nonlinear algebraic equations and explicit algebraic equations.

10.11.3 Problem Statement

Gas absorption into a liquid film can be enhanced by chemical reaction. Consider the important case in which the reaction in the liquid film is second order and irreversible.

An example where gas A dissolves at the surface of turbulent liquid D is shown in Figure 10–9. Component B, present in the bulk liquid D, undergoes a liquid phase reaction with dissolved A to give liquid product C:

$$A + B \rightarrow C \qquad \qquad (10\text{-}83)$$

where the reaction rate is elementary and the reaction rate expression is given by $r_C = kC_A C_B$. In this system, both components A and B are very dilute solutions in liquid D. The only transport through the liquid film is by diffusion, which is influenced by the chemical reaction. One surface of the liquid film is exposed to gaseous A and thus has a known concentration of A that is designated as C_{As}. Component B has a very low vapor pressure and is not present in the gas phase, and there is no transport of B into the gas phase. The other surface of the liquid film is exposed to the bulk liquid, which is well mixed and has a negligible concentration of dissolved A. Operation is such that there is always excess B within the liquid film and bulk liquid in relation to the amount needed for reaction with A.

Steady-state material balances on a differential volume within the film that include binary diffusion approximations for both A and B in D yield

$$\frac{d^2 C_A}{dx^2} = \frac{k}{D_{AD}} C_A C_B \qquad \qquad (10\text{-}84)$$

and

$$\frac{d^2 C_B}{dx^2} = \frac{k}{D_{BD}} C_A C_B \qquad \qquad (10\text{-}85)$$

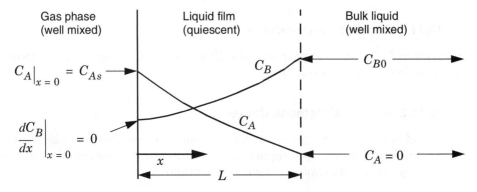

Figure 10–9 Diffusion with Concurrent Reaction of A and B in Liquid Film

where

C_A = concentration of dissolved A in kg-mol/m^3

C_B = concentration of B in kg-mol/m^3

k = reaction rate constant = 1.6×10^{-3} m^3/kg-mol · s

D_{AD} = liquid phase diffusivity of A in D = 2×10^{-10} m^2/s

D_{BD} = liquid phase diffusivity of B in D = 4×10^{-10} m^2/s

L = the total film thickness = 2×10^{-4} m

The gas phase concentration of A results in a liquid phase concentration of $C_{As} = 10$ kg-mol/m^3 at $x = 0$. There is no flux of B at $x = 0$, which means that $dC_B/dx = 0$ at the film surface. C_B is known to be 10 kg-mol/m^3, and C_A can be assumed to be zero at the interface between the bulk liquid and the liquid film, where $L = 2 \times 10^{-4}$ m.

(a) Calculate and plot the concentrations of both A and B within the liquid film as a function of x.

(b) Calculate the and plot the flux of A and B into the film in the x direction as a function of x.

(c) What is the flux of product C through the film and into the bulk liquid if C has a very low vapor pressure and is not present in the gas phase?

(d) What would be the maximum flux of A through the film if the reaction rate constant is doubled?

(e) What would be the maximum flux of A through the film if the reaction rate is negligible?

10.12 SIMULTANEOUS HEAT AND MASS TRANSFER IN CATALYST PARTICLES

10.12.1 Concepts Demonstrated

Simultaneous diffusion with nonisothermal chemical reactions in spherical catalyst particles for first-order reactions.

10.12.2 Numerical Methods Utilized

Solution of simultaneous ordinary differential equations with spit boundary conditions.

10.12.3 Problem Statement

Many porous catalyst systems involve reactions in which both diffusion and effective heat transfer must be included in a complete description of the catalytic reaction. The differential equations that describe the mass transfer in a spherical catalyst particle for an irreversible first-order reaction are given by the material balance on a differential balance within the sphere:

$$\frac{d}{dr}(N_A r^2) = -k''a C_A r^2 \quad \text{where } N_A = 0 \text{ at } r = 0 \qquad \text{(10-86)}$$

and by Fick's law for diffusion, which utilizes an effective diffusion coefficient and in which the bulk flow terms are neglected.

$$\frac{dC_A}{dr} = \frac{N_A}{-D_e} \quad \text{where } N_A = 0 \text{ at } r = 0 \text{ and } C_A = C_{As} \text{ at } r = R_s \qquad \text{(10-87)}$$

These differential equations and their boundary conditions are discussed in Problem 10.5, where the temperature is assumed constant. Note that the numerical solution also requires the algebraic equation that relates N_A to the combined quantity $N_A r^2$.

$$N_A = \frac{N_A r^2}{r^2} \qquad \text{(10-88)}$$

The effect of temperature on the reaction rate constant can be described by the Arrhenius expression with a dimensionless activation energy called the Arrhenius number. Thus at any temperature T, the rate constant is given by

$$k''|_T = k''|_{T_s} \exp\left[-\varepsilon\left(\frac{T_s}{T} - 1\right)\right] \qquad \text{(10-89)}$$

where $k''|_{T_s}$ is the first-order reaction rate constant based on the catalytic area at the particle surface temperature T_s, and ε is the dimensionless Arrhenius number. The value of ε is determined by the temperature dependency of the reaction, given by

$$\varepsilon = \frac{E}{RT_s} \qquad \text{(10-90)}$$

where E is the activation energy for the reaction and R is the gas constant. The temperature effect on the effective diffusion coefficient is usually neglected.

A similar treatment to the mass balance and application of Fick's law can be made for an energy balance on a differential volume within the pellet and application of Fourier's law for heat conduction. The resulting equations and boundary conditions are

$$\frac{d}{dr}(Q_r r^2) = -k''aC_A r^2 \Delta H_R \text{ where } Q_r = 0 \text{ at } r = 0 \qquad \text{(10-91)}$$

$$\frac{dT}{dr} = \frac{Q_r}{-k_e} \text{ where } Q_r = 0 \text{ at } r = 0 \text{ and } T = T_s \text{ at } r = R_s \qquad \text{(10-92)}$$

where ΔH_R is the heat of reaction and k_e is the effective thermal conductivity of the catalyst particle. Note that the numerical solution to these two equations also requires the algebraic equation that relates Q_r to the combined quantity $Q_r r^2$.

$$Q_r = \frac{Q_r r^2}{r^2} \qquad \text{(10-93)}$$

Thus the combined effects of mass and heat transfer within the spherical catalyst particle can be determined by the simultaneous solution of Equations (10-86) to (10-93), which involves four ODEs and three explicit algebraic equations.

Simplification of Heat Transfer Equations

The solution can be simplified for the heat transfer by considering an energy balance at any radius r, where the heat transfer must correspond to the complete reaction of all of the reactant mass that is transferred at that radius. Thus the energy flux must equal the heat of reaction multiplied by the mass flux in the opposite direction at any radius. Thus

$$Q_r\big|_r = (-\Delta H_R)(-N_A)\big|_r \qquad \text{(10-94)}$$

where the negative sign with the heat of reaction is necessary because of the thermodynamic sign convention, and the negative sign with the flux of A is because the heat flux and mass flux must be in opposite directions.

Substitution of Equations (10-87) and (10-92) into Equation (10-94) yields

$$-k_e \frac{dT}{dr} = -\Delta H_R D_e \frac{dC_A}{dr} \tag{10-95}$$

This equation can be integrated easily with the known boundary conditions to give

$$T = T_s + \frac{\Delta H_R D_e}{k_e}(C_A - C_{As}) \tag{10-96}$$

This algebraic equation can be used to replace Equations (10-91) to (10-93) in calculation of the temperature profile within the catalyst particle.

Nonisothermal Effectiveness Factor

The equivalent effectiveness factor when the catalyst particle is not isothermal can be calculated from

$$\eta = \frac{\int_0^{R_s} k''|_T \, a C_A (4\pi r^2) dr}{(k''|_{T_s} a C_{As})\left(\frac{4}{3}\pi R_s^3\right)} = \frac{3}{k''|_{T_s} a C_{As} R_s^3}\int_0^{R_s} k''|_{T_s} e^{\left[-\varepsilon\left(\frac{T_s}{T}-1\right)\right]} a C_A r^2 \, dr \tag{10-97}$$

which simplifies to

$$\eta = \frac{3}{C_{As} R_s^3}\int_0^{R_s} e^{\left[-\varepsilon\left(\frac{T_s}{T}-1\right)\right]} C_A r^2 \, dr \tag{10-98}$$

The concentration C_A is considered to be for a liquid or a gas that is not affected by temperature changes. As in Problem 10.5, the effectiveness expression of Equation (10-98) can be differentiated with respect to r, yielding

$$\frac{d\eta}{dr} = \frac{3 e^{\left[-\varepsilon\left(\frac{T_s}{T}-1\right)\right]} C_A r^2}{C_{As} R_s^3} \tag{10-99}$$

with the initial condition that $\eta = 0$ at $r = 0$. This differential equation can be solved simultaneously with the other equations for this problem in order to yield the effectiveness factor of η when the integration reaches the final value of r at R_s.

Common Dimensionless Variables

The literature in nonisothermal effectiveness factors utilizes several dimensionless variables. The Thiele modulus, which comes from the form of a combination of Equations (10-86) and (10-87), is given by

$$\phi = R_s \sqrt{\frac{k''|_{T_s} a C_{As}^{n-1}}{D_e}} \qquad (10\text{-}100)$$

which for a first-order reaction with $n = 1$ simplifies to

$$\phi = R_s \sqrt{\frac{k''|_{T_s} a}{D_e}} \qquad (10\text{-}101)$$

The heat transfer is characterized by a dimensionless parameter, which is typically designated by β and defined by

$$\beta = \frac{C_{As}(-\Delta H_R)D_e}{k_e T_s} \qquad (10\text{-}102)$$

A typical numerical solution to the nonisothermal effectiveness factor problem is shown in Figure 10–10, which is a function of ε, ϕ, and β and indicates several interesting features for first-order reactions. For example, some conditions for exothermic reactions where β is positive lead to effectiveness factors that can be greater than unity. This is due to the temperature rise internal to the particle, which accelerates the reaction in spite of the reduction of the reactant concentration due to diffusion into the particle. Another feature is that there are multiple values for the effectiveness factors at certain values of the Thiele modulus.

Consider a first-order reaction of reactant A in a spherical porous catalyst particle for which $C_{As} = 0.01$ kg-mol/m3, $T_s = 400$ K, $D_e = 10^{-6}$ m^2/s, $R_s = 0.01$ m, and $\Delta H_R = -8 \times 10^7$ kJ/kg-mol.
(a) Calculate and plot the concentration C_A and the temperature T as a function of radius r for the case where $\varepsilon = 30$, $\phi = 1$, and $\beta = 0.2$.
(b) Calculate the nonisothermal effectiveness factor for part (a), and compare your result with that found in Figure 10–10.
(c) Select another case for $\varepsilon = 30$ and different values of ϕ and β, where the effectiveness factor η is expected to be less than unity. Verify the point on Figure 10–10 by calculating η.
(d) Figure 10–10 suggests multiple steady states for positive β's when $\phi < 1$. Select fixed values of ϕ and β for $\varepsilon = 30$ from Figure 10–10 where this should occur, and calculate the upper and lower effectiveness factors.
(e) Explain the results of part (d).

10.12.4 Solution (Suggestions)

This problem is similar to Problem 10.5, but it is complicated by need for the energy balance. Figure 10–10 gives some indication of the gradients within the

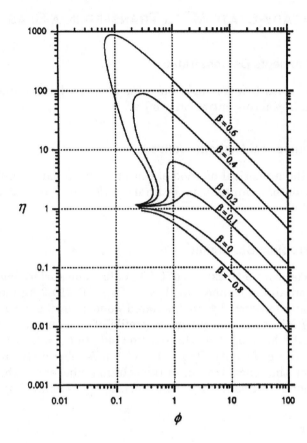

Figure 10–10 Nonisothermal Effectiveness Factor for ε = 30 (Reprinted from Fogler,[4] p. 620, with permission)

particle. Increasing values of β lead to large temperature gradients, while higher values of φ indicate greater reaction rates. As the reaction rate constant is increased, the reaction becomes more localized nearer the particle surface so that the interior concentrations are nearly zero and the interior temperature profile is nearly constant. The numerical results should indicate these trends.

Multiple steady states can be demonstrated numerically by utilizing different starting points for the solution of the differential equations. This is most conveniently accomplished by utilizing a simple trial-and-error procedure to converge upon the desired solution that matches the boundary condition if shooting techniques are used. For solutions involving the implicit finite difference approach, the initial estimates of the variables at the various node points usually determine which steady state is reached.

10.13 UNSTEADY-STATE MASS TRANSFER IN A SLAB

10.13.1 Concepts Demonstrated

Unsteady-state mass transfer in a one-dimensional slab having only one face exposed and an initial concentration profile.

10.13.2 Numerical Methods Utilized

Application of the numerical method of lines to solve a partial differential equation, and solution of simultaneous ordinary differential equations and explicit algebraic equations.

10.13.3 Problem Statement[*]

A slab of material with a thickness of 0.004 m has one surface suddenly exposed to a solution containing component A with $C_{A0} = 6 \times 10^{-3}$ kg-mol/m^3 while the other surface is supported by an insulated solid allowing no mass transport. There is an initial linear concentration profile of component A within the slab from $C_A = 1 \times 10^{-3}$ kg-mol/m^3 at the solution side to $C_A = 2 \times 10^{-3}$ kg-mol/m^3 at the solid side. The diffusivity $D_{AB} = 1 \times 10^{-9}$ m^2/s. The distribution coefficient relating between the concentration in the solution adjacent to the slab C_{ALi} and the concentration in the solid slab at the surface C_{Ai} is defined by

$$K = \frac{C_{ALi}}{C_{Ai}} \tag{10-103}$$

where $K = 1.5$. The convective mass transfer coefficient at the slab surface can be considered as infinite.

The unsteady-state diffusion of component A within the slab is described by the partial differential equation

$$\frac{\partial C_A}{\partial t} = D_{AB}\frac{\partial^2 C_A}{\partial x^2} \tag{10-104}$$

The initial condition of the concentration profile for C_A is known to be linear at $t = 0$. Since the differential equation is second order in C_A, two boundary conditions are needed. Utilization of the distribution coefficient at the slab surface gives

$$C_{Ai}\Big|_{x=0} = \frac{C_{A0}}{K} \tag{10-105}$$

[*] Adapted from Geankoplis,[3] pp. 471–473, with permission.

and the no diffusional flux condition at the insulated slab boundary gives

$$\frac{\partial C_A}{\partial x}\bigg|_{x = 0.004} = 0 \qquad\qquad\qquad \textbf{(10-106)}$$

(a) Calculate the concentrations within the slab after 2500 s. Utilize the numerical method of lines with an interval between nodes of 0.0005 m.

(b) Compare the results obtained with those reported by Geankoplis,[3] p. 473, and summarized in Table 10–15.

(c) Plot the concentrations versus time to 20000 s at $x = 0.001, 0.002, 0.003,$ and 0.004 m.

(d) Repeat part (a) with an interval between nodes of 0.00025. Compare results with those of part (a).

(e) Repeat parts (a) and (c) for the case where mass transfer is present at the slab surface. The external mass transfer coefficient is $k_c = 1.0 \times 10^{-6}$ m/s.

The Numerical Method of Lines

The method of lines (MOL) is a general technique for the solution of partial differential equations that has been introduced in Problem 6.8. This method utilizes ordinary differential equations for the time derivative and finite differences on the spatial derivatives. The finite difference elements for this problem are shown in Figure 10–11, where the interior of the slab has been divided into $N = 8$ intervals involving $N + 1 = 9$ nodes.

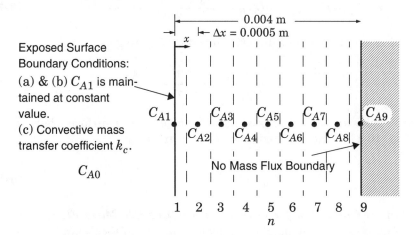

Figure 10–11 Unsteady-State Mass Transfer in a One-Dimensional Slab

Equation (10-104) can be written using a central difference formula for the second derivative and replacing the partial time derivatives with ordinary derivatives as given in Equation (10-7).

$$\frac{dC_{An}}{dt} = \frac{D_{AB}}{(\Delta x)^2}(C_{An+1} - 2C_{An} + C_{An-1}) \quad \text{for } (2 \leq n \leq 8) \tag{10-107}$$

Boundary Condition for Exposed Surface

The general surface boundary condition is obtained from a mass balance at the interface, which equates the mass transfer to the surface via the mass transfer coefficient to the mass transfer away from the surface due to diffusion within the slab. Thus at any time for mass transfer normal to the slab surface in the x direction,

$$k_c(C_{A0} - KC_{A1}) = -D_{AB}\frac{\partial C_A}{\partial x}\bigg|_{x=0} \tag{10-108}$$

where k_c is the external mass transfer coefficient in m/s and the partition coefficient K is used to have the liquid phase concentration driving force.

The derivative on the right side of Equation (10-108) can be written in finite difference form using the second-order three-point forward difference expression at node 1 [see Equation (A-5) of Appendix A].

$$\frac{\partial C_A}{\partial x}\bigg|_{x=0} = \frac{(-C_{A3} + 4C_{A2} - 3C_{A1})}{2\Delta x} \tag{10-109}$$

Thus substitution of Equation (10-109) into Equation (10-108) yields

$$k_c(C_{A0} - KC_{A1}) = -D_{AB}\frac{(-C_{A3} + 4C_{A2} - 3C_{A1})}{2\Delta x} \tag{10-110}$$

The preceding equation can be directly solved for C_{A1} to give

$$C_{A1} = \frac{2k_c C_{A0}\Delta x - D_{AB}C_{A3} + 4D_{AB}C_{A2}}{3D_{AB} + 2k_c K\Delta x} \tag{10-111}$$

which is the general result. For good mass transfer to the surface where $k_c \to \infty$ in Equation (10-111), the expression for C_{A1} is given by

$$C_{A1} = \frac{C_{A0}}{K} \tag{10-112}$$

Boundary Condition for Insulated Surface (No Mass Flux)

The mass flux is zero at the insulated surface; thus from Fick's law

$$\frac{\partial C_A}{\partial x}\bigg|_{x=0.004} = 0 \tag{10-113}$$

Utilizing the second-order approximation for the three-point forward difference for the preceding derivative [Equation (A-5) of Appendix A] yields

$$\frac{\partial C_{A9}}{\partial x} = \frac{3C_{A9} - 4C_{A8} + C_{A7}}{2\Delta x} = 0 \tag{10-114}$$

which can be solved for C_{A9} to yield

$$C_{A9} = \frac{4C_{A8} - C_{A7}}{3} \tag{10-115}$$

Initial Concentration Profile

The initial profile is known to be linear, so the initial concentrations at the various nodes can be calculated as summarized in Table 10–13.

Table 10–13 Initial Concentration Profile in Slab

x in m	C_A	node n
0	1.0×10^{-3}	1
0.0005	1.125×10^{-3}	2
0.001	1.25×10^{-3}	3
0.0015	1.375×10^{-3}	4
0.002	1.5×10^{-3}	5
0.0025	1.625×10^{-3}	6
0.003	1.75×10^{-3}	7
0.0035	1.825×10^{-3}	8
0.004	2.0×10^{-3}	9

10.13.4 Solution (Partial)

(a), (b), and (c) The problem is solved by the numerical solution of Equations (10-107), (10-112), and (10-115), which results in seven simultaneous ordinary differential equations and two explicit algebraic equations for the nine concentration nodes. This set of equations can be entered into the POLYMATH *Simultaneous Differential Equation Solver*. Note that the equations for nodes 1 and 9 need to use an "if ... then ... else ..." statement in POLYMATH to provide the desired initial values of the concentrations of A. Also, the POLYMATH full-screen editor can be used to duplicate equations as an aid for entering the finite difference equations.

The resulting equation set is given in Table 10–14.

Table 10–14 POLYMATH Program (File **P10-13AB.POL**)

Line	Equation
1	d(CA2)/d(t)=DAB*(CA3-2*CA2+CA1)/deltax^2
2	d(CA3)/d(t)=DAB*(CA4-2*CA3+CA2)/deltax^2
3	d(CA4)/d(t)=DAB*(CA5-2*CA4+CA3)/deltax^2
4	d(CA5)/d(t)=DAB*(CA6-2*CA5+CA4)/deltax^2
5	d(CA6)/d(t)=DAB*(CA7-2*CA6+CA5)/deltax^2
6	d(CA7)/d(t)=DAB*(CA8-2*CA7+CA6)/deltax^2
7	d(CA8)/d(t)=DAB*(CA9-2*CA8+CA7)/deltax^2
8	DAB=1.0E-9
9	deltax=0.0005
10	CA9=if(t==0)then(2.0E-3)else((4*CA8-CA7)/3)
11	CA0=6.0E-3
12	K=1.5
13	CA1=if(t==0)then(1.0E-3)else(CA0/K)
14	t(0)=0
15	CA2(0)=0.001125
16	CA3(0)=0.00125
17	CA4(0)=0.001375
18	CA5(0)=0.0015
19	CA6(0)=0.001625
20	CA7(0)=0.00175
21	CA8(0)=0.001825
22	t(f)=2500

The concentration variables at t = 2500 are summarized in Table 10–15 where a comparison with the approximate hand calculations by Geankoplis[3] shows reasonable agreement.

Table 10–15 Results for Unsteady-state Mass Transfer in One-Dimensional Slab at t = 2500 s

Distance from Slab Surface in m	Geankoplis[3] $\Delta x = 0.001$ m		Method of Lines (a) $\Delta x = 0.0005$ m	
	n	C_A in kg-mol/m^3	n	C_A in kg-mol/m^3
0	1	0.004	1	0.004
0.001	2	0.003188	3	0.003169
0.002	3	0.002500	5	0.002509
0.003	4	0.002095	7	0.002108
0.004	5	0.001906	9	0.001977

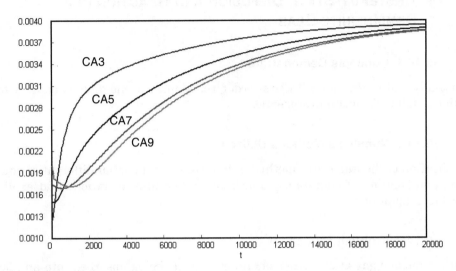

Figure 10–12 Calculated Concentration Profiles for C_A at Selected Node Points

The calculated concentration profiles for C_A at nodes 3, 5, 7, and 9 to $t = 20000$ s are presented in Figure 10–12, where the dynamics of the interior points show the effects of the initial concentration profile.

The POLYMATH problem solution files for parts (a), (b), and (c) are found in directory Chapter 10 with files named **P10-13AB.POL** and **P10-13C.POL**.

10.14 UNSTEADY-STATE DIFFUSION AND REACTION IN A SEMI-INFINITE SLAB

10.14.1 Concepts Demonstrated

Unsteady-state diffusion with first-order reaction in a one-dimensional semi-infinite slab with constant properties.

10.14.2 Numerical Methods Utilized

Application of the numerical method of lines to solve a partial differential equation, and solution of simultaneous ordinary differential equations and explicit algebraic equations.

10.14.3 Problem Statement[*]

Carbon dioxide gas at one atmosphere pressure is being absorbed into an alkaline solution containing a catalyst. The dissolved CO_2 designated as reactant A undergoes diffusion and an irreversible first-order reaction within the solution that is described by the following partial differential equation:

$$\frac{\partial C_A}{\partial t} = D_{AB}\frac{\partial^2 C_A}{\partial x^2} - k'C_A \tag{10-116}$$

where C_A is the concentration of dissolved CO_2 in kg-mol/m^3, t is the time in s, D_{AB} is the diffusivity of CO_2 within the alkaline solution B in m^2/s, x is the distance from the top of the solution in m, and k' is the first-order reaction rate constant in s^{-1}. The partial pressure of CO_2 is $p_{A0} = 1.0132\times10^5$ Pa, and the solubility of CO_2 is $S = 2.961\times10^{-7}$ kg-mol/Pa.

The initial condition for Equation (10-16) is given by

$$C_A = 0 \text{ at initial time } t = 0 \text{ and for all } x \tag{10-117}$$

and the boundary conditions are

$$C_A = C_{As} \quad \text{for } t > 0 \text{ and } x = 0 \tag{10-118}$$

and

$$C_A = 0 \quad \text{for } t > 0 \text{ and } x = \infty \tag{10-119}$$

Geankoplis[3] provides the following data for this system:

$C_{As} = p_{A0}S = (1.0132\times10^5 \text{ Pa})(2.961\times10^{-7} \text{ kg-mol/Pa}) = 0.03 \text{ kg-mol/m}^3$

$D_{AB} = 1.5\times10^{-9}$ m^2/s and $k' = 35$ s^{-1}

[*] Adapted from Geankoplis,[3] pp. 460–461, with permission.

(a) Calculate the concentration of dissolved A within the solution after 0.01 s by utilizing the numerical method of lines with 11 nodes (10 intervals) and an interval between nodes of 1.0×10^{-6} m. Consider the concentration C_A at the deepest node to be zero.

(b) Compute the total amount of A absorbed to a time of 0.01 s and compare your results with the value of 1.458×10^{-7} kg-mol/m^2 as calculated from an analytical solution by Geankoplis,[3] p. 461.

(c) Plot the concentration of A versus time to steady state at nodes 2, 3, 4, and 5, and tabulate the concentrations at $t = 0.01$ s.

(d) Repeat parts (a) through (b) with 21 nodes (20 intervals) and an interval between nodes of 5.0×10^{-7} m. Compare the concentrations at equivalent node locations with results of part (c) at $t = 0.01$ s.

(e) Repeat parts (a) and (c) for the case where the total pressure is doubled but the partial pressure of A remains the same. The additional component in the gas phase is not soluble in the solution. The external mass transfer coefficient is $k_G = 1.0 \times 10^{-10}$ kg-mol/s $\cdot$ m$^2 \cdot$ Pa. What is the percent reduction in amount of A absorbed to 0.01s that is due to the external mass transfer resulting from the pressure increase?

10.14.4 Solution (Partial)

(a) The numerical method of lines is introduced in Problem 6.8 and in other problems, including Problem 10.13, for a finite slab that involves only unsteady-state diffusion. A similar treatment here can be utilized for 10 finite difference elements at the slab surface by writing Equation (10-116) as

$$\frac{dC_{An}}{dt} = \frac{D_{AB}}{(\Delta x)^2}(C_{An+1} - 2C_{An} + C_{An-1}) - k'C_{An} \quad \text{for } (2 \le n \le 10) \quad \textbf{(10-120)}$$

where the second-order central difference approximation of the second derivative is used from Equation (A-9) of Appendix A.

The initial and boundary Equations (10-117) and (10-118) can be written as

$$C_{A1} = 0 \quad \text{when } t = 0 \quad \textbf{(10-121)}$$

and

$$C_{A1} = p_{CO_2}S_{CO_2} = (1.0132 \times 10^5)(2.961 \times 10^{-7}) = 0.03 \text{ kg-mol/m}^3 \quad \textbf{(10-122)}$$

for $t > 0$.

The boundary condition of Equation (10-119) under the assumption stated in part (a) becomes

$$C_{A11} = 0 \text{ for } t \geq 0 \tag{10-123}$$

The finite difference equations for the 11 nodes as given in Equations (10-120) to (10-123) can be solved by the POLYMATH *Simultaneous Differential Equation Solver*. Equations (10-121) and (10-122) can be entered using the "if ... then ... else ..." capability, as demonstrated in Table 10–16.

Table 10–16 POLYMATH Program (File **P10-14A.POL**)

Line	Equation
1	d(CA2)/d(t)=DAB*(CA3-2*CA2+CA1)/deltax^2-kprime*CA2
2	d(CA3)/d(t)=DAB*(CA4-2*CA3+CA2)/deltax^2-kprime*CA3
3	d(CA4)/d(t)=DAB*(CA5-2*CA4+CA3)/deltax^2-kprime*CA4
4	d(CA5)/d(t)=DAB*(CA6-2*CA5+CA4)/deltax^2-kprime*CA5
5	d(CA6)/d(t)=DAB*(CA7-2*CA6+CA5)/deltax^2-kprime*CA6
6	d(CA7)/d(t)=DAB*(CA8-2*CA7+CA6)/deltax^2-kprime*CA7
7	d(CA8)/d(t)=DAB*(CA9-2*CA8+CA7)/deltax^2-kprime*CA8
8	d(CA9)/d(t)=DAB*(CA10-2*CA9+CA8)/deltax^2-kprime*CA9
9	d(CA10)/d(t)=DAB*(CA11-2*CA10+CA9)/deltax^2-kprime*CA10
10	DAB=1.5E-9
11	CA1=if(t==0)then(0)else(0.03)
12	deltax=1.E-6
13	kprime=35
14	CA11=0
15	t(0)=0
16	CA2(0)=0
17	CA3(0)=0
18	CA4(0)=0
19	CA5(0)=0
20	CA6(0)=0
21	CA7(0)=0
22	CA8(0)=0
23	CA9(0)=0
24	CA10(0)=0
25	t(f)=0.01

Calculations show that the "if ... then ... else ..." statement for C_{A1} in line 11 of the equation set is really not necessary and that C_{A1} can just be set equal to 0.03 for all time t.

The POLYMATH problem solution file for part (a) is found in directory Chapter 10 with file named **P10-14A.POL**.

www

(b) An unsteady-state material balance at the surface of the slab yields

$$\frac{dQ}{dt} = -D_{AB}\frac{dC_A}{dx}\bigg|_{x=0} \tag{10-124}$$

where Q is the total amount of A transferred to the solution in kg-mol/m^2. The initial condition is that $Q = 0$ at $t = 0$, and the integration to any time t yields the value of Q over that time period. Equation (10-124) can be written using a second-order forward finite difference approximation as [see Equation (A-5) of Appendix A]

$$\frac{dQ}{dt} = -D_{AB}\frac{(-3C_{A1} + 4C_{A2} - C_{A3})}{2\Delta x} \tag{10-125}$$

Integration of Equations (10-125) can be accomplished along with the numerical method of lines solution for the concentrations yielding the corresponding value of Q.

(e) In the case of external mass transfer, the flux of A to the surface at any time t can be calculated by using the mass transfer coefficient as

$$N_A = k_G\left(P_{A_0} - \frac{C_{As}}{S}\right) \tag{10-126}$$

The same flux of A is also given by applying Fick's law at the surface of the slab at any time t.

$$N_A = -D_{AB}\frac{dC_A}{dx} \tag{10-127}$$

Setting the fluxes equal in Equations (10-126) and (10-127) gives the general expression for the effect of external mass transfer at the slab surface.

$$k_G\left(P_{A_0} - \frac{C_{As}}{S}\right) = -D_{AB}\frac{dC_A}{dx} \tag{10-128}$$

Use of the preceding equation in finite difference notation, which incorporates the second-order forward finite difference formula for the derivative dC_A/dx yields

$$k_G\left(P_{A_0} - \frac{C_{A1}}{S}\right) = -D_{AB}\frac{-3C_{A1} + 4C_{A2} - C_{A3}}{2\Delta x} \tag{10-129}$$

and Equation (10-129) can be solved for C_{A1} to give

$$C_{A1} = \frac{2k_G P_{A0}\Delta x - D_{AB}C_{A3} + 4D_{AB}C_{A2}}{3D_{AB} + (2k_G\Delta x)/S} \tag{10-130}$$

Thus, the finite difference equations for parts (a) and (c) only need to be modified using Equation (10-130) to account for the external mass transfer.

10.15 Diffusion and Reaction in a Falling Laminar Liquid Film

10.15.1 Concepts Demonstrated

Unsteady-state mass transfer with gas absorption, liquid-phase diffusion, and first-order reaction in a falling Newtonian fluid of finite thickness.

10.15.2 Numerical Methods Utilized

Application of the numerical method of lines to solve a partial differential equation which can be expressed as a system of simultaneous ordinary differential equations.

10.15.3 Problem Statement

Consider the absorption of CO_2 gas into a falling liquid film of alkaline solution in which there is a first-order irreversible reaction. A similar process without flow is discussed in Problem 10.14. The resulting concentration of the dissolved CO_2 in the film is quite small so that the viscosity of the liquid is not affected, and the mass transport in the liquid by bulk flow is negligible. The steady-state laminar flow of a Newtonian fluid down a vertical wall results in a velocity distribution, which is given by

$$v_z = \frac{\rho g \delta^2}{2\mu}\left[1-\left(\frac{x}{\delta}\right)^2\right] = v_{z_{max}}\left[1-\left(\frac{x}{\delta}\right)^2\right] \qquad (10\text{-}131)$$

as has been discussed in Problem 8.3.

A steady-state material balance on a differential volume within the liquid film yields the partial differential equation given by

$$v_z\frac{\partial C_A}{\partial z} = D_{AB}\frac{\partial^2 C_A}{\partial x^2} - k'C_A \qquad (10\text{-}132)$$

where v_z is the velocity in m/s, C_A is the concentration of dissolved CO_2 in kg-mol/m^3, D_{AB} is the diffusivity of dissolved CO_2 in the alkaline solution with units of m^2/s, and k' is a first-order reaction rate constant for the neutralization reaction in s^{-1}. The numerical values of all variables are the same as in Problem 10.14 except for k'. The film thickness is $\delta = 3\times10^{-4}$ m, $v_{z_{max}} = 0.6$ m/s, and $v_{z_{avg}} = (2/3)v_{z_{max}} = 0.4$ m/s.

The boundary condition for Equation (10-132) in the z direction is that C_A is zero at the point where the film begins to flow down the wall. Thus

$$C_A\big|_{z=0} = 0 \qquad (10\text{-}133)$$

The first boundary condition in the x direction is that C_A is known at the surface of the film, and this can be expressed as

$$C_A\big|_{x=0, z>0} = C_{As} = 0.03 \qquad \text{(10-134)}$$

which implies that the external mass transfer coefficient is very large. The second boundary condition in the x direction is that there is no mass transfer at the wall. Thus

$$\frac{\partial C_A}{\partial x}\bigg|_{x=\delta, z\geq 0} = 0 \qquad \text{(10-135)}$$

(a) Calculate the concentration of dissolved A at each node point within the liquid when there is no reaction and at $z = 1$ m. Utilize the numerical method of lines with 11 nodes (10 intervals), as shown in Figure 10–13.

(b) Extend part (a) by calculating the average flux of A absorbed by the film in kg-mol/s to $z = 1$ m.

(c) Plot the concentration of A versus z at nodes 3, 5, 7, and 9 for part (a).

(d) Verify the results of part (a) by calculating the molar rate of A absorbed at the film surface and comparing this with the calculated molar rate of A exiting in the liquid film at $z = 1$ m. Consider the film to be 1 m wide.

(e) Repeat parts (a) and (c) for the case where a weak alkaline solution causes an irreversible first-order reaction of A with a rate constant of $k' = 1 \text{ s}^{-1}$.

(f) What is the percentage increase in absorption of A from the gas phase because of the reaction in part (e) relative to that of part (a) that had no reaction?

10.15.4 Solution (Partial)

(a) The numerical method of lines is introduced in Problem 6.8, and it is applied in Problem 10.13 to unsteady-state diffusion in a finite slab with no reaction. However, this current problem is at steady state, with C_A being a function of depth within the film, designated as x, and the distance from the top of the falling film, designated as z. The finite difference elements for this problem are shown in Figure 10–13, where the interior of the slab has been divided into $N = 10$ intervals involving $N + 1 = 11$ nodes.

The method of lines allows ordinary differential equations to describe the variation of C_A with the z direction and can utilize finite elements to describe the variation in the x direction. This treatment gives a working equation set from

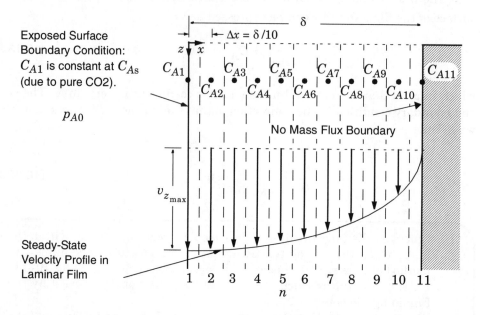

Exposed Surface
Boundary Condition:
C_{A1} is constant at C_{As}
(due to pure CO2).

p_{A0}

$v_{z_{max}}$

Steady-State
Velocity Profile in
Laminar Film

Figure 10–13 Mass Transfer with Reaction within a Falling Laminar Film

Equation (10-132) as

$$\frac{dC_{An}}{dz} = \left(\frac{D_{AB}}{(\Delta x)^2}(C_{An+1} - 2C_{An} + C_{An-1}) - k'C_{An}\right)/v_{z_n} \quad \text{for } (2 \le n \le 10) \quad \textbf{(10-136)}$$

where the second-order central difference approximation of Equation (A-9) is used for the second derivative.

The velocity v_{z_n} in Equation (10-136) varies only with x, and this can be expressed by writing Equation (10-131) as

$$v_{z_n} = v_{z_{max}}\left[1 - \left(\frac{(n-1)\Delta x}{\delta}\right)^2\right] \quad \text{for } (2 \le n \le 10) \quad \textbf{(10-137)}$$

Boundary Conditions
The initial condition of Equation (10-133) applies to the C_A in each of the internal finite elements. Thus

$$C_{An} = 0 \text{ at } z = 0 \text{ for } (2 \le n \le 10) \quad \textbf{(10-138)}$$

The boundary condition given by Equation (10-134) applies to the first finite element, giving

$$C_{A1} = 0.03 \text{ for } z \ge 0 \quad \textbf{(10-139)}$$

Equation (10-135) involves the derivative of C_A at the wall, which can be obtained using the second-order backward finite difference approximation of Equation (A-7) for the derivative as

$$\frac{\partial C_{A11}}{\partial x} = \frac{3C_{A11} - 4C_{A10} + C_{A9}}{2\Delta x} = 0 \tag{10-140}$$

The preceding equation can be solved for C_{A11} to yield

$$C_{A11} = \frac{4C_{A10} - C_{A9}}{3} \tag{10-141}$$

Sometimes numerical noise may enter into the preceding equation to yield negative values. This can be handled by logic to keep C_{A11} at zero whenever a negative value is calculated.

Numerical Solution

The general finite difference expression for Equation (10-132) with the velocity expression from Equation (10-137) can be combined and written as

$$\frac{dC_{An}}{dz} = \left[\frac{D_{AB}}{(\Delta x)^2}(C_{An+1} - 2C_{An} + C_{An-1}) - k'C_{An}\right] \Big/ \left\{v_{z_{max}}\left[1 - \left(\frac{(n-1)\Delta x}{\delta}\right)^2\right]\right\}$$

$$\text{for } (2 \leq n \leq 10) \tag{10-142}$$

The initial conditions are all zero from the boundary condition of Equation (10-138). These nine ordinary differential equations plus Equations (10-139) and (10-141) from the remaining boundary conditions allow the C_A's at the 11 nodes to be calculated as a function of z. Note that more accurate results could be obtained by utilizing more node points.

The POLYMATH *Simultaneous Differential Equation Solver* or any other ODE package can be used to solve this set of equations. The equation set for POLYMATH is given in Table 10–17 where the capability of the full-screen editor to duplicate an equation was used during problem entry of the repetitive differential equations so that only the node values needed to be changed.

Table 10–17 POLYMATH Program (File **P10-15A.POL**)

Line	Equation
1	d(CA2)/d(z)=(DAB*(CA3-2*CA2+CA1)/deltax^2-kprime*CA2)/(vmax*(1-((2-1)*deltax/delta)^2))
2	d(CA4)/d(z)=(DAB*(CA5-2*CA4+CA3)/deltax^2-kprime*CA4)/(vmax*(1-((4-1)*deltax/delta)^2))
3	d(CA5)/d(z)=(DAB*(CA6-2*CA5+CA4)/deltax^2-kprime*CA5)/(vmax*(1-((5-1)*deltax/delta)^2))
4	d(CA3)/d(z)=(DAB*(CA4-2*CA3+CA2)/deltax^2-kprime*CA3)/(vmax*(1-((3-1)*deltax/delta)^2))
5	d(CA6)/d(z)=(DAB*(CA7-2*CA6+CA5)/deltax^2-kprime*CA6)/(vmax*(1-((6-1)*deltax/delta)^2))
6	d(CA7)/d(z)=(DAB*(CA8-2*CA7+CA6)/deltax^2-kprime*CA7)/(vmax*(1-((7-1)*deltax/delta)^2))
7	d(CA8)/d(z)=(DAB*(CA9-2*CA8+CA7)/deltax^2-kprime*CA8)/(vmax*(1-((8-1)*deltax/delta)^2))
8	d(CA9)/d(z)=(DAB*(CA10-2*CA9+CA8)/deltax^2-kprime*CA9)/(vmax*(1-((9-1)*deltax/delta)^2))
9	d(CA10)/d(z)=(DAB*(CA11-2*CA10+CA9)/deltax^2-kprime*CA10)/(vmax*(1-((10-1)*deltax/delta)^2))

Table 10–17 (Continued) POLYMATH Program (File **P10-15A.POL**)

Line	Equation
10	DAB=1.5E-9
11	kprime=0
12	vmax=0.6
13	delta=3.E-4
14	CA1=0.03
15	CA11=if(4*CA10<CA9)then(0)else((4*CA10-CA9)/3)
16	deltax=0.1*delta
17	vavg=(2/3)*vmax
18	z(0)=0
19	CA2(0)=0
20	CA4(0)=0
21	CA5(0)=0
22	CA3(0)=0
23	CA6(0)=0
24	CA7(0)=0
25	CA8(0)=0
26	CA9(0)=0
27	CA10(0)=0
28	z(f)=1

www

The POLYMATH problem solution file for part (a) is found in directory Chapter 10 with file named **P10-15A.POL**.

(b) The average flux of A to a liquid film of height H is given by

$$N_{A_{avg}} = \frac{\int_0^H \left(-D_{AB} \frac{dC_A}{dx} \Big|_{x=0, z} \right) dz}{H} \tag{10-143}$$

where $N_{A_{avg}}$ is the average flux of A transferred to the liquid film in kg-mol/$m^2 \cdot$s and H is the film height in m. Equation (10-143) can be differentiated to yield

$$\frac{dN_{A_{avg}}}{dz} = \frac{\left(-D_{AB} \frac{dC_A}{dx} \Big|_{x=0, z} \right)}{H} \tag{10-144}$$

with an initial condition that $N_{A_{avg}} = 0$ at $z = 0$. Integration of this equation to any distance z yields the value of $N_{A_{avg}}$ over the film height.

Equation (10-144) can be written using a second-order forward finite difference approximation as

$$\frac{dN_{A_{avg}}}{dz} = -\frac{D_{AB}}{H}\frac{(-3C_{A1} + 4C_{A2} - C_{A3})}{2\Delta x} \tag{10-145}$$

The integration of the preceding equation for $H = 1$ m simultaneously with the equation set from part (a) to a final value of $z = 1$ m allows the determination of the requested $N_{A_{avg}}$.

(d) An overall steady-state material balance on A within the film when there is no reaction requires that the A that is transferred at the film surface must equal the A that flows out with the film. For a film of height H in m and width W in m, the input is given by

$$M_A = N_{A_{avg}} H W \tag{10-146}$$

where M_A is in kg-mol/s.

The output of A that exits the film at height H can be calculated from

$$M_A = W \int_0^{\delta} v_z C_A \, dx \tag{10-147}$$

in which v_z varies with z according to Equation (10-131) and C_A is the concentration profile determined from the numerical solution in part (a) at height H.

In order to evaluate Equation (10-147), the numerical values of C_A at the 11 node locations from the solution of part (a) can be used. The integral can be evaluated by fitting the product of $v_z C_A$ versus x with a cubic spline or polynomial and evaluating the integral with the POLYMATH *Curve Fitting and Regression Program*. The comparison of the calculations from Equations (10-146) and (10-147) should be made with $H = W = 1$ m.

REFERENCES

1. Bird, R. B., Stewart, W. E., and Lightfoot, E. N. *Transport Phenomena*, New York: Wiley, 1960.
2. Taylor, R., and Krishna, R. *Multicomponent Mass Transport*, New York: Wiley, 1993.
3. Geankoplis, C. J. *Transport Processes and Unit Operations*, 3rd ed., Englewood Cliffs, NJ: Prentice Hall, 1993.
4. Fogler, H. S. *Elements of Chemical Reaction Engineering*, 2nd ed., Englewood Cliffs, NJ: Prentice Hall, 1992.
5. Aris, R., *Chem. Eng. Sci.*, *6*, 265 (1957).
6. Geankoplis, C. J. *Mass Transport Phenomena*, Minneapolis: C. J. Geankoplis, 1972. (Available from The Ohio State University Bookstore, Columbus, OH)
7. Carty, R., and Schrodt, J. T., "Concentration Profiles in Ternary Gaseous Diffusion," *Ind. Eng. Chem. Fundam.*, *14*, 276–278 (1975).

CHAPTER **11**

Chemical Reaction Engineering[*]

11.1 PLUG-FLOW REACTOR WITH VOLUME CHANGE DURING REACTION

11.1.1 Concepts Demonstrated

Calculation of conversion in a gas-phase, isothermal, plug-flow reactor at constant pressure for a reaction with a change in number of moles.

11.1.2 Numerical Methods Utilized

Solution of simultaneous ordinary differential equations.

11.1.3 Problem Statement

The irreversible decomposition of the di-*tert*-butyl peroxide is to be carried out in an isothermal plug-flow reactor in which there is no pressure drop. Symbolically, this reaction can be written $A \rightarrow B + 2C$. The feed consists of di-*tert*-butyl peroxide and inert nitrogen. The reactor volume is 200 dm^3, and the entering volumetric flow rate is maintained constant at 10 dm^3/min. The reaction rate constant k for this first-order reaction is 0.08 min^{-1} which is based on reactant A.

(a) Determine and plot the conversion as a function of reactor volume for a feed stream of pure A at a concentration of 1.0 g-mol/dm^3. Also make a similar conversion plot when the feed is only 5% A with the balance being an inert component.

(b) Repeat (a) for a reaction $3A \rightarrow B$ with the same rate constant that is based on reactant A. All other variables remain the same.

(c) Summarize the results of (a) and (b) and discuss the effect of concentration level and reaction stoichiometry on this first-order reaction.

[*] The notation and equations of this chapter follow those of Fogler, H. S., *Elements of Chemical Reaction Engineering,* 4th ed., Upper Saddle River, NJ: Prentice Hall, 2006.

445

11.1.4   Solution (Partial)

(a) The material balance in moles for the plug-flow reactor is

$$\frac{dX}{dV} = \frac{-r_A}{F_{A0}} \qquad (11\text{-}1)$$

where the rate law is given by

$$-r_A = kC_A \qquad (11\text{-}2)$$

Stoichiometric considerations allow the concentration of reactant to be given by

$$C_A = C_{A0}\frac{(1-X)}{(1+\varepsilon X)}\frac{P}{P_0}\left(\frac{T_0}{T}\right) \qquad (11\text{-}3)$$

where

$$\varepsilon = y_{A0}\delta = y_{A0}(3-1) = 2y_{A0} \qquad (11\text{-}4)$$

and

$$F_{A0} = C_{A0}v_0 \qquad (11\text{-}5)$$

Since the reaction is isothermal and there is no pressure drop, the pressure ratio and temperature ratio are unity. Equation (11-3) can be rewritten as

$$C_A = C_{A0}\frac{(1-X)}{(1+\varepsilon X)} \qquad (11\text{-}6)$$

Equation (11-1) can be simplified by Equations (11-2) and (11-4) through (11-6) to

$$\frac{dX}{dV} = \frac{k(1-X)}{v_0(1+\varepsilon X)} \qquad (11\text{-}7)$$

where the initial condition is $X = 0$ at $V = 0$. The final value is $V = 200$. The value of ε is calculated from Equation (11-4) to be $\varepsilon = 2$ for the pure reactant case and $\varepsilon = 0.1$ for the 5% reactant case.

The numerical integration of the differential equation given by Equation (11-7) for the pure reactant case can be accomplished by the POLYMATH *Simultaneous Differential Equation Solver* using the equation set given in Table 11–1.

Table 11–1 POLYMATH Program (File **P11-01A1.POL**)

Line	Equation
1	d(X)/d(V)=k*(1-X)/(v0*(1+epsilon*X))
2	k=.08
3	v0=10

Line	Equation
4	epsilon=2
5	V(0)=0
6	X(0)=0
7	V(f)=200

The plot of the conversion versus reactor volume can be obtained easily as shown in Figure 11–1.

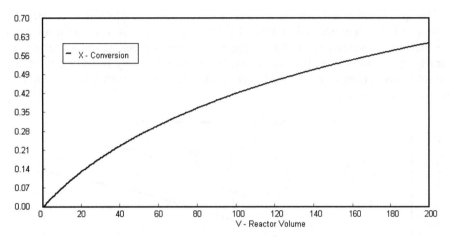

Figure 11–1 Conversion versus Reactor Volume for Pure Reactant

The solution for the dilute reactant case is obtained by changing only $\varepsilon = 0.1$ in the equation set.

The POLYMATH problem solution files for both cases of part (a) are found in directory Chapter 11 with files named **P11-01A1.POL** and **P11-01A2.POL**.

Graphical Comparison of Results

Often it is desirable to plot solutions for different cases for comparisons, such as in this problem solution. This can be accomplished easily by using indices to denote each separate case and solving all cases simultaneously. For part (a) of this problem, a POLYMATH equation set for comparison of both reactant feed cases using index 1 for the pure feed and index 2 for the 5% feed case is given in Table 11–2.

Table 11–2 POLYMATH Program (File **P11-01A3.POL**)

Line	Equation
1	d(X1)/d(V)=k*(1-X1)/(v0*(1+epsilon1*X1))
2	d(X2)/d(V)=k*(1-X2)/(v0*(1+epsilon2*X2))
3	k=.08
4	v0=10

Table 11–2 (Continued) POLYMATH Program (File **P11-01A3.POL**)

Line	Equation
5	epsilon2=0.1
6	epsilon1=2
7	V(0)=0
8	X1(0)=0
9	X2(0)=0
10	V(f)=200

The graphical result for conversion in both cases is plotted in Figure 11–2. The effect of the concentration level on the conversion is very evident for the pure reactant as compared to the 5% reactant. The increase in moles for the pure reactant greatly increases the volumetric flow rate, thereby reducing the time in the reactor when compared to the 5% reactant. This results in a lower conversion for the pure reactant case.

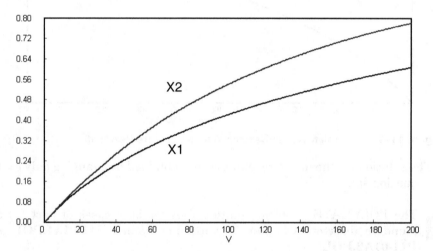

Figure 11–2 Comparison of Conversion for Pure Feed (X1) and 5% Feed (X2) (File **P11-01A3.POL**)

 The POLYMATH problem solution file for the comparison in part (a) is found in directory Chapter 11 with file named **P11-01A3.POL**.

Tabulated Results

There are instances when the actual values are required during problem output. This is illustrated for part (a), where the actual conversions are tabulated by requesting the POLYMATH *Simultaneous Differential Equation Solver* to generate tabular results. A example of the POLYMATH tabular results for 10 increments of V is given in Table 11–3 for the two cases of part (a).

Table 11–3 Tabular Output for Conversions

V	X1	X2
0	0	0
20	0.13153	0.14684
40	0.22783	0.27058
60	0.30434	0.37530
80	0.36778	0.46421
100	0.42183	0.53992
120	0.46873	0.60453
140	0.50999	0.65978
160	0.54666	0.70710
180	0.57951	0.74768
200	0.60915	0.78253

(b) For this reaction the same equations can be used as in part (a), where the change in the reaction gives

$$\varepsilon = y_{A0}\delta = y_{A0}\left(\frac{1}{3} - 1\right) = -\frac{2}{3}y_{A0} \tag{11-8}$$

11.2 VARIATION OF CONVERSION WITH REACTION ORDER IN A PLUG-FLOW REACTOR

11.2.1   Concepts Demonstrated

The effect of reaction order on conversion in a constant-volume, isothermal, plug-flow reactor.

11.2.2   Numerical Methods Utilized

Solution of ordinary differential equations.

11.2.3   Problem Statement

Consider the effect of reaction order on conversion in a plug-flow reactor whose volume is 1.5 dm^3. The irreversible liquid phase reaction $A \rightarrow B$ is taking place in which the feed concentration is C_{A0} = 1.0 g-mol/dm^3 and the volumetric flow rate is v_0 = 0.9 dm^3/min. The reaction rate constants for the various reaction orders are fixed at 1.1 for this comparison with units that are consistent with the aforementioned concentration and flow rate.

(a)   Plot the conversion in a plug-flow reactor as a function of reactor volume for a zero-, first-, second-, and third-order reaction to a reactor volume of 1.5.dm^3.

(b)   Repeat part (a) for feed concentrations of C_{A0} at 0.5 and 2.0 g-mol/dm^3.

11.2.4   Solution (Partial)

For the plug-flow reactor, a differential mole balance gives the following differential equation for the conversion of reactant A

$$\frac{dX}{dV} = \frac{-r_A}{F_{A0}} \tag{11-9}$$

where the rate law is

$$-r_A = kC_A^\alpha \tag{11-10}$$

and α = 0, 1, 2, or 3 according to the order of the reaction. The usual initial condition for Equation (11-9) is that there is no conversion at the reactor inlet that is expressed as $X = 0$ when $V = 0$.

Stoichiometric considerations at constant temperature and pressure allow the concentration of reactant A to be expressed using conversion as

$$C_A = C_{A0}(1 - X) \tag{11-11}$$

Special attention is necessary for the reaction of zero order because the rate proceeds at a constant value until the conversion reaches 1.0 and then the rate becomes zero.

(a) This problem can be conveniently solved with the POLYMATH *Simultaneous Differential Equation Solver*. The special logic required for the zero order reaction can utilize the "if ... then ... else ..." capability within POLYMATH. Also, in the solution with POLYMATH, it is convenient to consider all of the reaction orders in a single solution to allow for convenient plotting of the results. This can be accomplished by defining different conversion variables for each reaction order, such as x0 for the conversion from zero order, x1 for conversion for first order, etc. The POLYMATH equation set for part (a) of this problem is provided in Table 11–4.

Table 11–4 POLYMATH Program (File **P11-02A.POL**)

Line	Equation
1	d(x0)/d(V)=if(x0<=1.0)then(k/(v0*CA00))else(0.)
2	d(x1)/d(V)=k*CA1/(v0*CA10)
3	d(x2)/d(V)=k*CA2^2/(v0*CA20)
4	d(x3)/d(V)=k*CA3^3/(v0*CA30)
5	k=1.1
6	v0=.9
7	CA00=1.
8	CA10=1.
9	CA20=1.
10	CA30=1.
11	CA0=CA00*(1-x0)
12	CA1=CA10*(1-x1)
13	CA2=CA20*(1-x2)
14	CA3=CA30*(1-x3)
15	V(0)=0
16	x0(0)=0
17	x1(0)=0
18	x2(0)=0
19	x3(0)=0
20	V(f)=1.5

The graphical results for this problem are given in Figure 11–3, where the effect of reaction order on conversion is clearly shown.

The POLYMATH problem solution file for part (a) is found in directory Chapter 11 with file named **P11-02A.POL**.

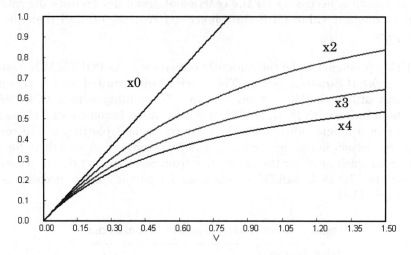

Figure 11–3 Conversion versus Volume for Various Reaction Orders
(File **P11-02A.POL**)

11.3 GAS PHASE REACTION IN A PACKED BED REACTOR WITH PRESSURE DROP

11.3.1   Concepts Demonstrated

Calculation of conversion and pressure drop in an isothermal, gas phase, packed bed reactor (PBR) with a change in moles during reaction.

11.3.2   Numerical Methods Utilized

Solution of simultaneous ordinary differential equations and explicit algebraic equations.

11.3.3   Problem Statement

The catalytic gas phase reaction $A \overset{k}{\to} B$ is to be carried out in a packed bed reactor under isothermal operation. The reactant is pure A with an inlet concentration of 1 g-mol/dm^3, the entering pressure is 25 atm, and the entering volumetric flow rate is 1 dm^3/min. The first-order reaction rate constant k, which is based on reactant A, is 1 dm^3/kg-cat·min.

(a)   Study the effect of the pressure drop on the conversion for a first-order reaction, and plot the conversion X and relative pressure y as a function of the weight of the packing W for three different values of the pressure drop parameter α. Consider that $0.05 \text{ kg}^{-1} \leq \alpha \leq 0.2 \text{ kg}^{-1}$). Plot both X and y versus W from zero to $W_{\max} = 2$ kg.

(b)   Repeat part (a) for a first-order reaction in which there is a change in the number of moles during the reaction $(\delta > 0)$, $A \to 3B$.

(c)   Repeat part (a) for a first-order reaction in which $(\delta < 0)$, $A \to \frac{1}{3}B$.

(d)   Summarize the results of parts (a), (b), and (c).

11.3.4   Solution (Partial)

(a) The mole balance equation for this reactor is given by

$$\frac{dX}{dW} = \frac{-r_A'}{F_{A0}} \qquad (11\text{-}12)$$

where the first-order rate law is

$$-r_A' = kC_A \qquad (11\text{-}13)$$

and the stoichiometry yields

$$C_A = C_{A0}\frac{(1-X)}{(1+\varepsilon X)}\frac{P}{P_0} \tag{11-14}$$

Since there is no change in the number of moles, $\varepsilon = 0$. Defining $y = P/P_0$ and substituting into Equation (11-14) yields

$$C_A = C_{A0}(1-X)y \tag{11-15}$$

where the pressure drop in a packed bed reactor is given by Fogler[1] as

$$\frac{dy}{dW} = \frac{-\alpha}{2}\left(\frac{1+\varepsilon X}{y}\right) \tag{11-16}$$

Suggestions The technique utilized for comparisons in Problem 11.2, in which different conversion variables were used for each case, might be convenient to apply in this problem as well. In this problem, different relative pressures will also need to be used.

(b) Equations (11-12) through (11-16) are again applicable here, but the value of ε changes, as given by $\varepsilon = y_{A0}\delta = y_{A0}(3-1) = 2y_{A0} = 2$.

(c) The solution is the same as in part (a) except that the value of ε is calculated from

$$\varepsilon = y_{A0}\delta = y_{A0}\left(\frac{1}{3}-1\right) = -\frac{2y_{A0}}{3} = -\frac{2}{3} \tag{11-17}$$

11.4 CATALYTIC REACTOR WITH MEMBRANE SEPARATION

11.4.1 Concepts Demonstrated

Calculation of flow rate and concentrations of the reactants and products in an isothermal catalytic reactor with product removal by a membrane where there is pressure drop and mass transfer dependency on local velocity.

11.4.2 Numerical Methods Utilized

Solution of simultaneous ordinary differential equations and explicit algebraic equations.

11.4.3 Problem Statement

The dehydrogenation of a compound (elementary kinetics) is taking place in a selective membrane reactor under isothermal conditions.

$$A \rightleftarrows B + \frac{1}{2}C \tag{11-18}$$

The reaction is reversible, and K_C is the equilibrium constant. The membrane reactor in which the preceding reaction is occurring is shown in Figure 11–4.

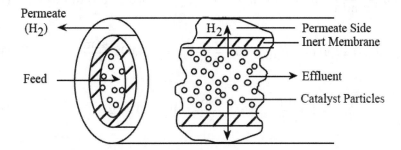

Figure 11–4 Catalytic Reactor with Product Removal by a Membrane (from Fogler,[1] with permission)

The advantage in using a membrane reactor is that by having one of the products selectively pass through the membrane, the reaction is driven toward completion. In this case hydrogen diffuses out through the sides of the membrane, thereby allowing the reaction to proceed further to the right. Isothermal conditions can be assumed.

The rate of the mass transfer across the membrane depends not only on the resistance offered by the membrane, but also on any boundary layers on each side of the membrane. As the flow rate past the membrane surface increases, the boundary layer thickness decreases, as does the resistance. Consequently, the mass transfer coefficient increases. The manner in which the k_C increases with flow rate depends upon the flow geometry. One common correlation relating the mass transfer coefficient with velocity is

$$k_C(@v) = k_C(@v_0)\left(\frac{v}{v_0}\right)^{1/2} \tag{11-19}$$

where $k_C(@v_0)$ is the mass transfer coefficient at volumetric flow rate v_0, and $k_C(@v)$ is the corresponding coefficient at v. However, if transport through the membrane is the limiting step, then the mass transfer coefficient will be independent of velocity.

In this problem the mass transfer dependence on velocity is given by

$$k_C = k_{C0}\left(\frac{v}{v_0}\right)^{1/2} \tag{11-20}$$

and the membrane transport of C is given by $k_C a C_C$ when permeate concentration of C is low.

The following parameter values apply:

$$k = 0.5\frac{dm^3}{kg \cdot min} \quad v_0 = 50\ dm^3/min \quad K_C = 0.5(kg \cdot mol/dm^3)^{1/2} \tag{11-21}$$

$$P_0 = 10\ atm \quad F_{A0} = 10\frac{kg \cdot mol}{min} \quad \alpha = 0.002\frac{atm}{kg} \tag{11-22}$$

$$k_{C0} = 0.1\frac{dm}{min} \quad a = 2\frac{dm^2}{kg} \quad W = 200\ kg \tag{11-23}$$

Compare calculated output values of F_B, the molar flow rate of B, for the following:
(a)   Base case
(b)   Base case with no membrane transport
(c)   Base case with no pressure drop
(d)   Base case with no membrane transport and no pressure drop

11.4.4   Solution (Equations)

The applicable equations for this case are as follows:

Mole balances

$$\frac{dF_A}{dW} = r'_A$$

$$\frac{dF_B}{dW} = -r'_A \tag{11-24}$$

$$\frac{dF_C}{dW} = -\frac{1}{2}r'_A - k_C a C_C$$

Rate law

$$r'_A = -k\left(C_A - \frac{C_B C_C^{1/2}}{K_C}\right) \tag{11-25}$$

Stoichiometry

$$C_A = \frac{F_A}{v} \qquad C_B = \frac{F_B}{v} \qquad C_C = \frac{F_C}{v} \tag{11-26}$$

$$v = v_0\left(\frac{F_A + F_B + F_C}{F_{A0}}\right)\frac{P_0}{P} = v_0\left(\frac{F_A + F_B + F_C}{F_{A0}}\right)\frac{1}{y} \tag{11-27}$$

where

$$y = \frac{P}{P_0} \tag{11-28}$$

Pressure drop

$$\frac{dy}{dW} = \frac{-\alpha\left(\dfrac{F_A + F_B + F_C}{F_{A0}}\right)}{2y} \tag{11-29}$$

Mass transfer coefficient

$$k_C = k_{C0}\left(\frac{v}{v_0}\right)^{1/2} \tag{11-30}$$

11.5 SEMIBATCH REACTOR WITH REVERSIBLE LIQUID PHASE REACTION

11.5.1 Concepts Demonstrated

Calculation of conversion in an isothermal liquid phase reaction carried out in a semibatch reactor under both equilibrium and rate-controlling assumptions.

11.5.2 Numerical Methods Utilized

Solution of simultaneous ordinary differential equations and explicit algebraic equations.

11.5.3 Problem Statement

Pure butanol is to be fed into a semibatch reactor containing pure ethyl acetate to produce butyl acetate and ethanol. The reaction

$$CH_3COOC_2H_5 + C_4H_9OH \rightleftarrows CH_3COOC_4H_9 + C_2H_5OH \qquad \text{(11-31)}$$

which can be expressed as

$$A + B \rightleftarrows C + D \qquad \text{(11-32)}$$

is elementary and reversible. The reaction is carried out isothermally at 300 K. At this temperature the equilibrium constant based on concentrations is 1.08 and the reaction rate constant is 9×10^{-5} dm^3/g-mol. Initially there are 200 dm^3 of ethyl acetate in the reactor, and butanol is fed at a rate of 0.05 dm^3/s for a period of 4000 seconds from the start of reactor operation. At the end of the butanol introduction, the reactor is operated as a batch reactor. The initial concentration of ethyl acetate in the reactor is 7.72 g-mol/dm^3 and the feed butanol concentration is 10.93 g-mol/dm.^{3}

(a) Calculate and plot the concentrations of A, B, C, and D within the reactor for the first 5,000 seconds of reactor operation.

(b) Simulate reactor operation in which reaction equilibrium is always attained by increasing the reaction rate constant by a factor of 100 and repeating the calculations and plots requested in part (a). Note that this is a difficult numerical integration.

(c) Compare the conversion of ethyl acetate under the conditions of part (a) with the equilibrium conversion of part (b) during the first 5,000 seconds of reactor operation.

(d) If the reactor down time between successive semibatch runs is 2,000 seconds, calculate the reactor operation time that will maximize the rate of butyl acetate production.

11.5.4   Solution (Partial)

The mole balance, rate law, and stoichiometry equations applicable to the semi-batch reactor are as follows:

Mole balances

$$\frac{dN_A}{dt} = r_A V \qquad \frac{dN_B}{dt} = r_A V \tag{11-33}$$

$$\frac{dN_C}{dt} = -r_A V \qquad \frac{dN_D}{dt} = -r_A V \tag{11-34}$$

Rate law

$$-r_A = k\left(C_A C_B - \frac{C_C C_D}{K_e}\right) \tag{11-35}$$

Stoichiometry

$$C_A = \frac{N_A}{V} \qquad C_B = \frac{N_B}{V} \tag{11-36}$$

$$C_C = \frac{N_C}{V} \qquad C_D = \frac{N_D}{V} \tag{11-37}$$

Overall material balance

$$\frac{dV}{dt} = v_0 \tag{11-38}$$

Definition of conversion

$$x_A = \frac{N_{A0} - N_A}{N_{A0}} \tag{11-39}$$

Definition of production rate of butyl acetate

$$P = \frac{N_C}{(t_p + 2000)} \tag{11-40}$$

At equilibrium the net rate is equal to zero or $r_A = 0$. A convenient way to achieve this with the problem described with differential equations is to give the rate constant a large value, such as suggested in part (b).

(a) The equation set for part (a) is shown in Table 11–5 for use with the POLYMATH *Simultaneous Differential Equation Solver*. Note that the POLYMATH "if ... then ... else ... " capability is utilized to introduce the volumetric feed rate of butanol, v0, to the reactor during the first 4,000 s of operation.

Table 11–5 POLYMATH Program (File **P11-05A.POL**)

Line	Equation
1	d(V)/d(t)=v0
2	d(NA)/d(t)=rA*V
3	d(NB)/d(t)=rA*V+v0*CB0
4	d(NC)/d(t)=-rA*V
5	d(ND)/d(t)=-rA*V
6	v0=if(t<=4000)then(.05)else(0.)
7	CB0=10.93
8	k=9.e-5
9	CA=NA/V
10	CB=NB/V
11	CC=NC/V
12	CD=ND/V
13	Ke=1.08
14	NA0=200*7.72
15	P=NC/(t+2000)
16	xA=(NA0-NA)/NA0
17	rA=-k*(CA*CB-CC*CD/Ke)
18	t(0)=0
19	V(0)=200
20	NA(0)=1544
21	NB(0)=0
22	NC(0)=0
23	ND(0)=0
24	t(f)=5000

The POLYMATH problem solution file for part (a) is found in directory Chapter 11 with file named **P11-05A.POL**.

(c) A comparison of the rate-based conversion and the equilibrium-based conversion is shown in Figure 11–5. Clearly, the equilibrium conversion is higher than the rate-based conversion at all times, as expected. This graph can be generated by solving both the rate-based case of part (a) and the equilibrium-based case of part (b) in the same POLYMATH equation set. This useful technique for graphical comparisons has been applied in Problem 11.1.

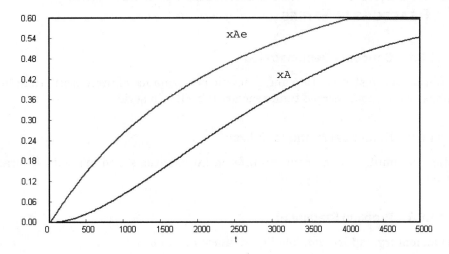

Figure 11–5 Comparison of Rate-Based and Equilibrium-Based Conversions

11.6 OPERATION OF THREE CONTINUOUS STIRRED TANK REACTORS IN SERIES

11.6.1 Concepts Demonstrated

Calculation of the steady-state and dynamic performance of three isothermal liquid-phase continuous stirred tank reactors (CSTRs) in series.

11.6.2 Numerical Methods Utilized

Solution of simultaneous ordinary differential equations and explicit algebraic equations.

11.6.3 Problem Statement

The elementary and irreversible liquid phase reaction

$$A + B \rightarrow C \tag{11-41}$$

is to be carried out in a series of three CSTRs as shown in Figure 11–6.   Species A and B are fed in separate streams to the first CSTR, with the volumetric flow of each stream controlled at 6 dm³/min. The volume of each CSTR is 200 dm³, and each reactor is initially filled with inert solvent. The initial concentrations of the reactants are $C_{A0} = C_{B0} = 2.0$ g-mol/dm³, and the reaction rate coefficient is $k = 0.5$ dm³/g-mol·min.

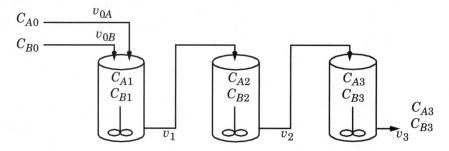

Figure 11–6 Train of Three Continuous Stirred Reactors (CSTRs)

(a)   Calculate the steady-state concentrations of all reacting components exiting the third reactor. Suggestion: Set up this problem solution as ordinary differential equations that describe the unsteady-state reactor operation, and integrate these equations to a final time when concentrations do not change (steady state).

(b)   Determine the time necessary to reach steady state (i.e., when C_A exiting the third reactor is 99% of the steady-state value).

(c)   Plot the concentration of A exiting each tank during start up to the steady-state time determined in part (b).

(d)   Consider reactor operation when the feed for species B is split equally between each reactor. Repeat parts (a), (b), and (c).

(e)   Compare the results of (a) with that of a plug-flow reactor with the same total volume of 600 dm^3 and the same reactor feed.

11.6.4 Solution

(a) – (c) Material balances can be made on each reactor i. For a liquid phase reaction, the volume change with reaction can be neglected. Thus the unsteady-state balances yield the following ordinary differential equations:

$$\frac{dC_{A1}}{dt} = (v_{0A}C_{A0} - v_1 C_{A1} - kC_{A1}C_{B1}V_1)/V_1 \qquad \text{(11-42)}$$

$$\frac{dC_{B1}}{dt} = (v_{0B}C_{B0} - v_1 C_{B1} - kC_{A1}C_{B1}V_1)/V_1 \qquad \text{(11-43)}$$

$$\frac{dC_{A2}}{dt} = (v_1 C_{A1} - v_2 C_{A2} - kC_{A2}C_{B2}V_2)/V_2 \qquad \text{(11-44)}$$

$$\frac{dC_{B2}}{dt} = (v_1 C_{B1} - v_2 C_{B2} - kC_{A2}C_{B2}V_2)/V_2 \qquad \text{(11-45)}$$

$$\frac{dC_{A3}}{dt} = (v_2 C_{A2} - v_3 C_{A3} - kC_{A3}C_{B3}V_3)/V_3 \qquad \text{(11-46)}$$

$$\frac{dC_{B3}}{dt} = (v_2 C_{B2} - v_3 C_{B3} - kC_{A3}C_{B3}V_3)/V_3 \qquad \text{(11-47)}$$

where

$$C_{A0} = C_{B0} = 2.0; \quad k = 0.5; \quad V_1 = V_2 = V_3 = 200$$
$$v_{0A} = v_{0B} = 6; \quad v_1 = v_2 = v_3 = 12 \qquad \text{(11-48)}$$

A convenient method for determining steady-state operation is to integrate the differential equations to large time, where the derivatives go to zero. Thus, this corresponds to steady-state operation. An alternate solution would be to solve the simultaneous nonlinear equations that result from setting the time derivatives to zero. Once steady state is determined, then the dynamic response can be determined from the numerical solution.

(d) Material balances for each reactor must be revised to include the fresh feed stream of component B. Thus equations (11-45) and (11-47) become

$$\frac{dC_{B2}}{dt} = (v_{0B}C_{B0} + v_1 C_{B1} - v_2 C_{B2} - kC_{A2}C_{B2}V_2)/V_2$$

$$\frac{dC_{B3}}{dt} = (v_{0B}C_{B0} + v_2 C_{B2} - v_3 C_{B3} - kC_{A3}C_{B3}V_3)/V_3$$

(11-49)

The volumetric flow rates must be altered to $v_{0B} = 2$, $v_1 = 8$, $v_2 = 10$, and $v_3 = 12$.

(e) The material balances for each reactant in a plug-flow reactor yield the following ordinary differential equations, where $v_0 = 12$ dm^3/min. For the combined reactor feed, the initial conditions are given by $C_{A0} = C_{B0} = 1.0$ g-mol/dm^3.

$$\frac{dC_A}{dV} = -\frac{kC_A C_B}{v_0}$$

(11-50)

$$\frac{dC_B}{dV} = -\frac{kC_A C_B}{v_0}$$

(11-51)

The preceding differential equations must be integrated from $V = 0$ to $V = 600$ dm^3.

11.7 DIFFERENTIAL METHOD OF RATE DATA ANALYSIS IN A BATCH REACTOR

11.7.1 Concepts Demonstrated

Determination of reaction order and rate constant by applying the differential method of rate data analysis to batch reactor data.

11.7.2 Numerical Methods Utilized

Fitting a polynomial to experimental data, differentiation of tabular data, and linear and nonlinear regression of algebraic expressions with data.

11.7.3 Problem Statement

Data on the liquid phase bromination of xylene at 17°C have been reported by Hill.[2] This reaction has been studied by the introduction of iodine (a catalyst) and small quantities of reactant bromine into a batch reactor containing reactant xylene in considerable excess. The concentrations of reactant xylene and the iodine catalyst are approximately constant during the reaction. A material balance on the batch reactor yields

$$\frac{dC_{Br2}}{dt} = -k(C_{Br2})^n \qquad \text{(11-52)}$$

where C_{Br2} is the concentration of bromine in g-mol/dm,[3] k is a pseudo rate constant that depends on the iodine and xylene concentrations, and n is the reaction order.

Equation (11-52) can easily be linearized by taking the ln of each side of the equation, yielding

$$\ln\left(-\frac{dC_{Br2}}{dt}\right) = \ln k + n\ln(C_{Br2}) \qquad \text{(11-53)}$$

This expression can be used in linear regression to determine k and n.

Nonlinear regression can also be used directly on Equation (11-52) in order to determine k and n.

(a) Find a polynomial that represents the data of C_{Br2} versus t given in Table 11–6.

(b) Prepare a table of $d(C_{Br2})/dt$ versus C_{Br2} at the experimental data points.

(c) Determine k and n from the linearized form of Equation (11-53).

(d) Use nonlinear regression to estimate k and n from the nonlinear form of Equation (11-52).

(e) Compare the results of parts (c) and (d). Discuss any differences.

Table 11–6 Measured Bromine Concentration versus Time (File **P11-07.POL**)

Time, t min	Bromine Concentration g-mol/dm^3	Time, t min	Bromine Concentration g-mol/dm^3
0	0.3335	19.60	0.1429
2.25	0.2965	27.00	0.1160
4.50	0.2660	30.00	0.1053
6.33	0.2450	38.00	0.0830
8.00	0.2255	41.00	0.0767
10.25	0.2050	45.00	0.0705
12.00	0.1910	47.00	0.0678
13.50	0.1794	57.00	0.0553
15.60	0.1632	63.00	0.0482
17.85	0.1500		

11.7.4   Solution (Suggestions)

The regression of experimental data is discussed in Chapter 3.

www

The POLYMATH problem data file is found in directory Chapter 11 with file named **P11-07.POL**.

11.8 INTEGRAL METHOD OF RATE DATA ANALYSIS IN A BATCH REACTOR

11.8.1 Concepts Demonstrated

Determination of a reaction rate constant by applying the integral method of rate data analysis to batch reactor data.

11.8.2 Numerical Methods Utilized

Linear and nonlinear regression of algebraic expressions with data.

11.8.3 Problem Statement

The gas phase decomposition of dimethyl ether was studied by Hinshelwood and Askey[3] in a constant volume batch reactor at 552°C.

$$CH_3OCH_3 \rightarrow CH_4 + CO + H_2$$

Experiments were conducted by measuring the total pressure as a function of reaction time. Typical results are given in Table 11–7.

A material balance on the batch reactor for a first-order reaction gives

$$\ln\left(\frac{3P_0 - P}{2P_0}\right) = -kt \qquad (11\text{-}54)$$

where P_0 is the initial pressure, P is the measured pressure, and k is the rate coefficient.

(a) Determine the first-order rate constant and confidence intervals from the data given in Table 11–7 by a direct regression of Equation (11-54).
(b) Solve Equation (11-54) for P and use nonlinear regression to determine the first-order rate constant and the 95% confidence intervals.

Table 11–7 Total Pressure Variation in Decomposition of Dimethyl Ether

Time, t s	Pressure mm Hg	Time, t s	Pressure mm Hg
0	420	182	891
57	584	219	954
85	662	261	1013
114	743	299	1054
145	815		

The POLYMATH problem data file is found in directory Chapter 11 with file named **P11-08.POL**.

11.9 INTEGRAL METHOD OF RATE DATA ANALYSIS—BIMOLECULAR REACTION

11.9.1   Concepts Demonstrated

Determination of reaction order and rate constant by application of the integral method of rate data analysis to batch reactor data.

11.9.2   Numerical Methods Utilized

Linear regression with transformed data.

11.9.3   Problem Statement

The liquid phase reaction between ethylene dibromide and potassium iodide in 99% methanol was investigated by R. T. Dillon[4] at 60°C.

$$C_2H_4Br_2 + 3KI \rightarrow C_2H_4 + 2KBr + KI_3$$

$$A + 3B \rightarrow C + 2D + E$$

The concentration of dibromide was measured as a function of time in a batch reactor. The data are organized in Table 11–8 in terms of dibromide concentration C_A and calculated conversion X_A as a function of time t. The initial concentrations were $C_{A0} = 0.02864$ kg-mol/m^3 and $C_{B0} = 0.1531$ kg-mol/m^3.

Convenient analysis of these data to determine the reaction order and corresponding rate constant can be based on conversion as discussed by Hill.[2] The material balances on the batch reactor utilizing conversion yield the following:

(a) 0th order

$$C_{A0}X_A = kt \tag{11-55}$$

(b) 1st order with respect to A, 0th order with respect to B

$$\ln[1/(1-X_A)] = kt \tag{11-56}$$

(c) 1st order with respect to A and B

$$\frac{1}{(\theta_B - 3)C_{A0}}\ln\left(\frac{1 - 3X_A/\theta_B}{1-X_A}\right) = kt \tag{11-57}$$

where t is the time in s and k is the reaction rate coefficient. The initial concentration ratio of reactants is given by

$$\theta_B = \frac{C_{B0}}{C_{A0}} \tag{11-58}$$

(a) Determine the order of the reaction and the appropriate reaction rate constant from Table 11–8 by fitting Equations (11-55) through (11-57).
(b) Derive Equations (11-55) through (11-57) by making material balances on a batch reactor and applying the reaction stoichiometry.

Table 11–8 Reaction Rate for the Reaction Between Ethylene Dibromide and Potassium Iodide
(File **P11-09.POL**)

Time, t s	Conversion, X_A
0	0
29.7	0.2863
40.5	0.3630
47.7	0.4099
55.8	0.4572
62.1	0.4890
72.9	0.5396
83.7	0.5795

www

The POLYMATH problem data file is found in directory Chapter 11 with file named **P11-09.POL**.

11.10 INITIAL RATE METHOD OF DATA ANALYSIS

11.10.1 Concepts Demonstrated

Determination of the orders of a reaction and the reaction rate constant using initial reaction rate data from a batch reactor.

11.10.2 Numerical Methods Utilized

Nonlinear regression of data to obtain model parameters.

11.10.3 Problem Statement

The initial rate data, given in Table 11–9 as $(-dP_A/dt)$, have been determined for the gas phase reaction $A + 2B \rightarrow C$. It can be assumed that the rate expression is of the form

$$-r_A = k(P_A)^\alpha (P_B)^\beta \tag{11-59}$$

where k is the reaction rate coefficient, α is the reaction order with respect to A, and β is the reaction order with respect to B.

(a) Use nonlinear regression to find the numerical values for α, β, and k.
(b) Set the orders of reaction to integer or simple fractional values on the basis of part (a) and then determine the corresponding value of k and the corresponding 95% confidence intervals.

Table 11–9 Initial Reaction Rate versus Initial Pressure Data (File **P11-10.POL**)

Run	P_A torr	P_B torr	$-r_A$ torr/s
1	6	20	0.420
2	8	20	0.647
3	10	20	0.895
4	12	20	1.188
5	16	20	1.811
6	10	10	0.639
7	10	20	0.895
8	10	40	1.265
9	10	60	1.550
10	10	100	2.021

The POLYMATH problem data file is found in directory Chapter 11 with file named **P11-10.POL**.

11.11 HALF-LIFE METHOD FOR RATE DATA ANALYSIS

11.11.1 Concepts Demonstrated

Determination of reaction order, rate coefficient, and activation energy by utilizing the half-life method.

11.11.2 Numerical Methods Utilized

Transformation of data and multiple linear regression.

11.11.3 Problem Statement

Hinshelwood and Burk[5] have studied the thermal decomposition of nitrous oxide in a constant volume batch reactor where the reaction is

$$2\,N_2O \rightarrow 2\,N_2 + O_2$$

or

$$2A \rightarrow 2B + C$$

Representative half-life data are given in Table 11–10 for isothermal conditions (T = 1055 K) and in Table 11–11 for nonisothermal conditions.

Table 11–10 Half-Life Data at 1055 K

No.	Half-life $t_{1/2}$ s	Temperature T K	Concentration C_{A0} g-mol/dm$^3 \times 10^3$
1	1048	1055	1.6334
2	919	1055	1.8616
3	704	1055	2.4315
4	537	1055	3.1533
5	474	1055	3.6092
6	409	1055	4.1031
7	382	1055	4.4830
8	340	1055	4.9769

Table 11–11 Half Life Data at Various Temperatures

No	Half-life $t_{1/2}$ s	Temperature T K	Concentration C_{A0} g-mol/dm$^3 \times 10^3$
1	1240	1060	1.2478
2	1352	975	2.5899
3	510	970	3.0250
4	918	1035	4.0495
5	455	1035	4.2599
6	318	1050	5.3060

Fogler[1] gives the integrated design equation for a batch reactor with the rate given by $-r_A = kC_A^\alpha$ and expressed in terms of the half-life as

$$t_{1/2} = \frac{2^{\alpha-1}-1}{k(\alpha-1)}\left(\frac{1}{C_{A0}^{\alpha-1}}\right) \tag{11-60}$$

The preceding equation can be linearized to a form suitable for linear regression by taking the natural log of both sides of the equation, giving

$$\ln t_{1/2} = \ln\frac{2^{\alpha-1}-1}{k(\alpha-1)} + (1-\alpha)\ln C_{A0} \tag{11-61}$$

When the temperature of the rate data varies, the Arrhenius expression is used to represent the effect of temperature on the rate constant:

$$k = A\exp[-E/(RT)] \tag{11-62}$$

where typically T is the absolute temperature, R is the gas constant (8.314 kJ/g-mol·K), E is the activation energy (units of kJ/g-mol), and A is the frequency factor with units of the rate constant.

Introduction of the Arrhenius expression into Equation (11-60) yields

$$t_{1/2} = \frac{2^{\alpha-1}-1}{\{A\exp[-E/(RT)]\}(\alpha-1)}\left(\frac{1}{C_{A0}^{\alpha-1}}\right) \tag{11-63}$$

which can be used to correlate nonisothermal half-life data.

The corresponding linearized expression for Equation (11-63) is given by

$$\ln t_{1/2} = \ln\left(\frac{2^{\alpha-1}-1}{\alpha-1}\right) - \ln A + \frac{E}{RT} + (1-\alpha)\ln C_{A0} \tag{11-64}$$

(a) Assuming that the rate expression for this reaction is of the form $-r_A = kC_A^\alpha$, apply linear regression to Equation (11-61) to determine the reaction rate constant and the order of the reaction at 1055 K using the data from Table 11–10.

(b) Repeat part (a) by applying nonlinear regression to Equation (11-60) and using the converged parameter values of part (a) as initial estimates for the nonlinear regression.

(c) Utilize linear regression on Equation (11-64) with the combined data from Tables 11–10 and 11–11 to estimate the order of the reaction, the corresponding frequency factor, and the activation energy of the rate coefficient in kJ/g-mol.

(d) Repeat part (c) by applying nonlinear regression to Equation (11-63) and using the converged parameter values of part (c) as initial estimates for the nonlinear regression.

11.11.4 Solution (Suggestions)

The linear and nonlinear regression required for this problem can be accomplished with the POLYMATH *Polynomial, Multiple Linear, and Nonlinear Regression Program*. The linear and nonlinear regression of data has been discussed extensively in Chapter 3. The most detailed discussions are found in Problems 3.3 and 3.8.

The POLYMATH problem data files for parts (a & b) and (c & d) are found in directory Chapter 11 with respective files named **P11-11AB.POL** and **P11-11CD.POL**.

11.12 METHOD OF EXCESS FOR RATE DATA ANALYSIS IN A BATCH REACTOR

11.12.1 Concepts Demonstrated

Determination of reaction order and kinetic reaction rate constant using the method of excess in a batch reactor.

11.12.2 Numerical Methods Utilized

Fitting a polynomial to experimental data, differentiation of tabular data, and linear or nonlinear regression.

11.12.3 Problem Statement

The reaction of acetic acid and cyclohexanol in dioxane solution is catalyzed by sulfuric acid. This reaction was studied in a well-stirred batch reactor at 40°C. The reaction can be described by $A + B \rightarrow C + D$. For initial equal concentrations of $C_{A0} = C_{B0} = 2$ g-mol/dm^3, the concentration of A changed with time, as shown in Table 11–12. Another experiment was performed in which $C_{A0} = 1$ g-mol/dm^3 and $C_{B0} = 8$ g-mol/dm^3. The results of this run are summarized in Table 11–13.

For a constant volume batch reactor, the material balance on reactant A gives the time derivative that is equal to the reaction rate and rate expression

$$\frac{dC_A}{dt} = -kC_A^\alpha C_B^\beta \tag{11-65}$$

where t is the time in min, k is the reaction rate constant, α is the reaction order with respect to C_A, and β is the reaction order with respect to C_B.

(a) Use the data in Table 11–12 to calculate the reaction rates given by the derivative dC_A/dt at the given experimental times. From these rates, calculate the reaction rate coefficient and the total reaction order: $\alpha + \beta$.

(b) Utilize the data in Table 11–13 to calculate the reaction rates at the given experimental times. By assuming that C_B does not change significantly during this experiment, determine the approximate reaction order α with respect to component A.

(c) Repeat part (b) by calculating the C_B at the various times utilizing the reaction stoichiometry.

(d) Employ nonlinear regression on the combined reaction rates determined in parts (a) and (b) to determine accurate values for k, α and β. (Adjust α and β to integer values or simple fractions before reporting final results for k.)

Table 11–12 Concentration of Acetic Acid versus Time for C_{A0} = 2 g-mol/dm^3 and C_{B0} = 2 g-mol/dm^3 (File **P11-12A.POL**)

No.	Time t min	Concentration C_A g-mol/dm$^3 \times 10^3$
1	0	2.000
2	120	1.705
3	150	1.647
4	180	1.595
5	210	1.546
6	240	1.501
7	270	1.460
8	300	1.421

Table 11–13 Concentration of Acetic Acid versus Time for C_{A0} = 1 g-mol/dm^3 and C_{B0} = 8 g-mol/dm^3 (File **P11-12B.POL**)

No.	Time t min	Concentration C_A g-mol/dm$^3 \times 10^3$
1	0	1.000
2	30	0.959
3	45	0.939
4	75	0.903
5	120	0.854
6	150	0.824
7	210	0.771
8	255	0.734

The POLYMATH problem data files for parts (a) and (b) are found in directory Chapter 11 with respective files named **P11-12A.POL** and **P11-12B.POL**.

11.13 RATE DATA ANALYSIS FOR A CSTR

11.13.1 Concepts Demonstrated

Determination of the order of reaction and corresponding reaction rate constant and for an irreversible gas-phase reaction carried out in a CSTR.

11.13.2 Numerical Methods Utilized

Regression of a data set to obtain parameters of an algebraic expression.

11.13.3 Problem Statement

A homogeneous irreversible gas phase reaction whose stoichiometry can be represented by $A \rightarrow B + 2C$ is carried out in a 1 dm^3 CSTR at 300°C and 0.9125 atm. The data for conversion X_A versus the feed flow rate v_0 at reactor conditions where the reactor feed consists of pure reactant A are summarized in Table 11–14.

A material balance on reactant A for this CSTR yields

$$v_0 = \frac{V k C_{A0}^n (1 + X_A)^n}{C_{A0} X_A (1 + 2X_A)^n} = \frac{V k C_{A0}^{n-1} (1 + X_A)^n}{X_A (1 + 2X_A)^n} \tag{11-66}$$

where v_0 is the volumetric flow rate in dm^3/s, V is the reactor volume in dm^3, k is the rate constant, C_{A0} = 0.1942 g-mol/dm^3, and n is the order of the reaction. Note that the factor of 2 in the denominator of Equation (11-66) accounts for the volumetric change during reaction.

Determine the reaction order with respect to A (an integer value) and calculate the corresponding value of the reaction rate coefficient.

Table 11–14 Conversion Data from a CSTR[a] (File **P11-13.POL**)

v_0 dm^3/s	X_A	v_0 dm^3/s	X_A
250	0.45	5	0.8587
100	0.5562	2.5	0.8838
50	0.6434	1	0.9125
25	0.7073	0.5	0.95
10	0.7874		

[a]These data were generated using the results of McCracken and Dickson.[6]

The POLYMATH problem data file is found in directory Chapter 11 with file named **P11-13.POL**.

11.14 DIFFERENTIAL RATE DATA ANALYSIS FOR A PLUG-FLOW REACTOR

11.14.1 Concepts Demonstrated

Differential analysis of the reaction order and corresponding reaction rate constant for a liquid phase reaction carried out in a plug-flow reactor.

11.14.2 Numerical Methods Utilized

Fitting and differentiation of tabular data, and multiple linear or nonlinear regression of data.

11.14.3 Problem Statement[*]

Oxygen is transported to living tissues by being bound to hemoglobin in arterial blood and transported to individual cells through tissue capillaries. The initial phases of deoxygenation of hemoglobin at the capillary-cell interface are assumed to be irreversible, where the deoxygenation is represented by

$$HbO_2 \rightarrow Hb + O_2$$

or

$$A \xrightarrow{k} B + C$$

The kinetics of this deoxygenation have been studied in a 0.158-cm-diameter plug-flow reactor. A solution of HbO_2 with a concentration of 1×10^{-5} g-mol/cm^3 was fed to the reactor at a rate of 19.60 cm^3/s. Oxygen concentrations were measured by electrodes placed at 2.5-cm intervals along the length of the tube. The resulting data are summarized in Table 11–15 for a single run at a constant feed rate of the HbO_2 solution.

The material balance for hemoglobin in this plug-flow reactor is given by the differential equation

$$\frac{dX_A}{dV} = \frac{-r_A}{F_{A0}} \tag{11-67}$$

where X_A is the conversion of HbO_2, V is the reactor volume in cm^3, r_A is the reaction rate expression in g-mol/cm^3, and F_{A0} is the feed rate of A to the reactor in g-mol/s.

In the differential method of rate data analysis, the differential equation form of the material balance (reactor design equation) is fitted to the experimental data. In this case, direct use of Equation (11-67) requires the derivative information from the conversion versus reactor volume measurements.

---

[*] Adapted from Fogler, H. S. *Elements of Chemical Reaction Engineering*, 1st ed., Englewood Cliffs, NJ: Prentice Hall, 1986.

(a) Fit a polynomial to the data in Table 11–15 and calculate the value for dX_A/dV at the various electrode positions.

(b) Assuming that the rate expression for this reaction is of the form $-r_A = kC_A^{\alpha}$, use nonlinear regression directly on Equation (11-57) to determine the reaction order α as an integer or simple fraction. Find the corresponding value of the reaction rate constant k.

(c) Verify that the reaction order obtained in part (b) is correct by analytically integrating Equation (11-67) and then showing that a linear plot can be obtained with the original data. Some transformations of the original data may be necessary.

Table 11–15 Deoxygenation of Hemoglobin as Function of Distance from Reactor Inlet
(File **P11-14.POL**)

Electrode Position z in cm	Conversion of HbO$_2$ X_A
0.0	0.000
2.5	0.0096
5.0	0.0192
7.5	0.0286
10.0	0.0380
12.5	0.0472
15.0	0.0564
17.5	0.0655

www

The POLYMATH problem data file is found in directory Chapter 11 with file named **P11-14.POL**.

11.15 INTEGRAL RATE DATA ANALYSIS FOR A PLUG-FLOW REACTOR

11.15.1 Concepts Demonstrated

Integral analysis of the reaction order and corresponding reaction rate constant for a gas-phase reaction carried out in a plug-flow reactor.

11.15.2 Numerical Methods Utilized

Fitting and differentiation of tabular data, multiple linear or nonlinear regression of data.

11.15.3 Problem Statement

Acetaldehyde undergoes decomposition at elevated temperatures.

$$CH_3CHO \rightarrow CH_4 + CO$$

or

$$A \rightarrow C + D$$

In an experiment, pure acetaldehyde vapor was passed through a reaction tube maintained by a surrounding furnace at 510°C. The reaction tube had an inside diameter of 2.5 cm and a length of 50 cm, and the pressure was atmospheric. Decomposition rates were obtained for various flow rates are summarized in Table 11–16.

Application of the integral analysis of reactor data involves integration of the reactor design equation (material balance) for a particular reaction rate expression. For a general first-order gas phase reaction with the stoichiometry of the acetaldehyde decomposition, the reaction rate expression is given by

$$r_A = -kC_{A0}\frac{(1-X_A)}{(1+X_A)} \tag{11-68}$$

where C_{A0} is the feed concentration entering the reactor.

The integral form of the design equation for a first-order reaction is given by Fogler as

$$V = F_{A0}\int_0^{X_A} \frac{dX_A}{(-r_A)} = \frac{F_{A0}}{kC_{A0}}\int_0^{X_A} \frac{(1+X_A)}{(1-X_A)}dX_A \tag{11-69}$$

which can be analytically integrated to

$$V = \frac{F_{A0}}{kC_{A0}}\left\{2\ln\left[\frac{1}{(1-X_A)}\right] - X_A\right\} \tag{11-70}$$

A convenient rearrangement of Equation (11-70) yields the relationship between the molar feed rate and the conversion from which the rate constant can be determined by a linear regression.

$$F_{A0} = \frac{kC_{A0}V}{\left\{ 2 \ln\left[\frac{1}{(1-X_A)}\right] - X_A \right\}} \tag{11-71}$$

A similar treatment for a second-order reaction gives

$$V = \frac{F_{A0}}{kC_{A0}^2}\left[4 \ln(1-X_A) + X_A + \frac{4X_A}{(1-X_A)} \right] \tag{11-72}$$

which can be rearranged for regression as

$$F_{A0} = \frac{kC_{A0}^2 V}{\left[4 \ln(1-X_A) + X_A + \frac{4X_A}{(1-X_A)} \right]} \tag{11-73}$$

(a)   Utilize linear regression on the data of Table 11–16 to estimate a first-order rate constant from Equation (11-71) and second-order rate constant from Equation (11-73). Please use units of h, cm^3, and g-mol.

(b)   Select the reaction order and corresponding rate constant that best represents the data. Justify your selection.

Table 11–16 Acetaldehyde Decomposition Data (File **P11-15.POL**)

No.	Flow Rate F_{A0} g-mol/h	Conversion X_A	No.	Flow Rate F_{A0} g-mol/h	Conversion X_A
1	9.09	0.524	5	4.55	0.652
2	2.05	0.775	6	3.27	0.705
3	0.909	0.871	7	2.68	0.738
4	1.73	0.797	8	0.682	0.894

The POLYMATH problem data file is found in directory Chapter 11 with file named **P11-15.POL**.

11.16 DETERMINATION OF RATE EXPRESSIONS FOR A CATALYTIC REACTION

11.16.1 Concepts Demonstrated

Evaluation of catalytic reaction rate expressions with experimental data.

11.16.2 Numerical Methods Utilized

Nonlinear regression of complex expressions and multiple linear regression, including linearization of complex expressions with transformation of data prior to linear regression.

11.16.3 Problem Statement

Cutlip and Peters[7] investigated the catalytic heterogeneous dehydration of tertiary butyl alcohol (A) to isobutylene (B) and water (W). The reaction was studied at atmospheric pressure and various temperatures. The observed reaction rates at a temperature of 533.1 K for various partial pressures of the reactant and the products are given in Table 11–17.

Consideration of a number of different reaction mechanisms with different rate-controlling steps resulted in 24 expressions for reaction rate as function of the partial pressures. After initial screening, the following four rate expressions needed to be evaluated:

$$\text{Model 1} \qquad r' = \frac{kK_AP_A}{1 + K_AP_A + K_WP_W} \qquad\qquad (11\text{-}74)$$

$$\text{Model 2} \qquad r' = \frac{kK_AP_A}{1 + K_AP_A + K_WP_W + K_BP_B} \qquad\qquad (11\text{-}75)$$

$$\text{Model 3} \qquad r' = \frac{kK_AP_A}{(1 + K_AP_A + K_WP_W)^2} \qquad\qquad (11\text{-}76)$$

$$\text{Model 4} \qquad r' = \frac{kK_AP_A}{(1 + K_AP_A + K_WP_W + K_BP_B)^2} \qquad\qquad (11\text{-}77)$$

where
r' = reaction rate in g-mol/h·g
k = rate constant in g-mol/h·g
P_A, P_W, and P_B = partial pressures of alcohol, water, and butylene, respectively, in atm
K_A, K_W, and K_B = adsorption constants in atm^{-1}

Table 11–17 Reaction Rates as Function of Partial Pressure at 533.1 K (File **P11-16.POL**)

No.	Average Partial Pressures (atm)			Reaction Rate g-mol/h·g
	Alcohol	Water	Butylene	
1	0.7913	0.0177	0.0172	0.005047
2	0.6349	0.0159	0.0156	0.004409
3	0.4788	0.0149	0.0146	0.003857
4	0.3339	0.0157	0.0163	0.003048
5	0.6362	0.0146	0.1736	0.004464
6	0.4864	0.0128	0.3252	0.003671
7	0.3302	0.0135	0.4819	0.002716
8	0.651	0.1629	0.0104	0.004271
9	0.474	0.3374	0.0122	0.003244
10	0.3167	0.4982	0.01	0.002348
11	0.3506	0.0121	0.314	0.002841
12	0.3973	0.2705	0.0121	0.002903
13	0.3661	0.186	0.0083	0.002995
14	0.3219	0.0117	0.1819	0.002801
15	0.4737	0.0135	0.1821	0.003622
16	0.4857	0.1687	0.0089	0.003523

Cutlip and Peters[7] also considered an empirical power-law rate expression to correlate their data.

Model 5 $r' = k(P_A)^a (P_W)^w (P_B)^b$ **(11-78)**

where a, w, and b are unknown exponents.

(a) Select one of the five models for fitting to the rate data of Table 11–17. Use multiple linear regression on a linearized form of the rate expression to determine the model parameters.

(b) Use direct nonlinear regression on the selected model with the same rate data, where the initial estimates of the model parameters are the parameter values from part (a). Note that this procedure provides reasonable initial estimates for the nonlinear regression from the linear regression results.

(c) Repeat parts (a) and (b) for the additional models, and determine the variances of the various nonlinear regressions. Select the "best" model as the one that has the lowest variance.

(d) Compare your results in part (c) with those of Cutlip and Peters.[7]

11.16.4 Solution (Suggestions and Partial Results)

(a) The POLYMATH *Polynomial, Multiple Linear, and Nonlinear Regression Program* has options for both multiple linear regression and nonlinear regression. Linear and nonlinear regressions have been discussed in Problems 3.3, 3.5, 3.6, 3.10, 3.11, and 3.12.

In order to apply multiple linear regression to these rate equation, the various expressions must be linearized and the data appropriately transformed. Consider Model 1. This model can be linearized by taking the reciprocal of both sides of the Equation (11-74):

$$\frac{1}{r'} = \frac{1}{k} + \frac{1}{kK_AP_A} + \frac{K_WP_W}{kK_AP_A} \tag{11-79}$$

The preceding equation can be expressed in the general form of a linear regression

$$y = a_0 + a_1x_1 + a_2x_2 \tag{11-80}$$

by introducing the transformation functions

$$x_1 = 1/P_A \qquad x_2 = P_W/P_A \tag{11-81}$$

where

$$a_0 = \frac{1}{k} \qquad a_1 = \frac{1}{kK_A} \qquad a_2 = \frac{K_W}{kK_A} \tag{11-82}$$

The multiple linear regression for Model 1 yields the values for a_0, a_1, and a_2 from which simple calculations utilizing Equation (11-82) give $k = 0.01102$ g-mol/h·g, $K_A = 1.0575$ atm^{-1} and $K_W = 0.49706$ atm^{-1}.

Model 3 A linearized form of Model 3 can be obtained by first taking the reciprocal of Equation (11-76) followed by taking the square root, yielding

$$\frac{1}{r'^{1/2}} = \frac{1}{(kK_A)^{1/2}P_A^{1/2}} + \frac{K_AP_A}{(kK_A)^{1/2}P_A^{1/2}} + \frac{K_WP_W}{(kK_A)^{1/2}P_A^{1/2}} \tag{11-83}$$

The appropriate transformation functions for Equation (11-83) are

$$\begin{aligned} y &= 1/(r'^\wedge 0.5) \\ x_1 &= 1/(P_A^\wedge 0.5) \\ x_2 &= P_A^\wedge 0.5 \\ x_3 &= P_W/(P_A^\wedge 0.5) \end{aligned} \tag{11-84}$$

where the multiple linear function is

$$y = a_1x_1 + a_2x_2 + a_3x_3 \tag{11-85}$$

Note that there is no constant a_0 on the right side of Equation (11-85); thus the POLYMATH option "to regress with $a_0 = 0$" should be selected when solving this problem.

(c) Statistical principles for the selection of the best model are discussed in Chapter 3 and by Constantinides.[8]

 The POLYMATH problem data file for this problem, including the transformations for part (a), is found in directory **Chapter 11** with file named
P11-16A.POL.

11.17 PACKED BED REACTOR DESIGN FOR A GAS PHASE CATALYTIC REACTION

11.17.1 Concepts Demonstrated

Calculation of conversion in an isothermal packed bed catalytic reactor for different catalytic rate expressions with pressure drop.

11.17.2 Numerical Methods Utilized

Solution of simultaneous ordinary differential equations.

11.17.3 Problem Statement

The irreversible gas phase catalytic reaction

$$A + B \rightarrow C + D \tag{11-86}$$

is to be carried out in a packed bed reactor with four different catalysts. For each catalyst, the rate expression has a different form.

$$-r'_{A1} = \frac{kC_A C_B}{1 + K_A C_A} \tag{11-87}$$

$$-r'_{A2} = \frac{kC_A C_B}{1 + K_A C_A + K_C C_C} \tag{11-88}$$

$$-r'_{A3} = \frac{kC_A C_B}{(1 + K_A C_A + K_B C_B)^2} \tag{11-89}$$

$$-r'_{A4} = \frac{kC_A C_B}{(1 + K_A C_A + K_B C_B + K_C C_C)^2} \tag{11-90}$$

The molar feed flow rate of A is $F_{A0} = 1.5$ g-mol/min, and the initial concentrations of the reactants are $C_{A0} = C_{B0} = 1.0$ g-mol/dm^3 at the reactor inlet. There is a total of $W = 2$ kg of each catalyst used in the reactor. The reaction rate constant and the various catalytic rate parameters are given by

$$k = 10 \ \text{dm}^6/\text{kg} \cdot \text{min}$$

$$K_A = 1 \ \text{dm}^3/\text{g-mol}$$

$$K_B = 2 \ \text{dm}^3/\text{g-mol}$$

$$K_C = 20 \ \text{dm}^3/\text{g-mol}$$

(a)   Calculate and plot the conversion versus catalyst weight for each of the catalytic rate expressions when the reactor operation is at constant pressure. Summarize the expected outlet conversions.

(b)   Repeat part (a) when the pressure ratio within the reactor is given by

$$\frac{dy}{dW} = \frac{-\alpha}{2y} \qquad \textbf{(11-91)}$$

where $y = P/P_0$ and α is a constant ($\alpha = 0.4$).

11.17.4 Solution (Partial with Suggestions)

(a) The equations applicable for part (a) are as follows:

Mole balance

$$\frac{dX}{dW} = \frac{-r_A'}{F_{A0}} \qquad \textbf{(11-92)}$$

Stoichiometry

$$C_A = C_B = C_{A0}(1 - X)$$
$$C_C = C_D = C_{A0}X \qquad \textbf{(11-93)}$$

It is also convenient to solve the four cases at the same time in order to make a single plot comparing the four conversions. This strategy is discussed in Problem 11.2.

The four cases for part (a) can be integrated with the POLYMATH *Simultaneous Differential Equation Solver* utilizing the equation set of Table 11–18.

Table 11–18 POLYMATH Program (File **P11-17A.POL**)

Line	Equation
1	d(XA1)/d(W)=(k/FA0)*(CA0*(1-XA1))^2/(1+KA*CA0*(1-XA1))
2	d(XA2)/d(W)=(k/FA0)*(CA0*(1-XA2))^2/(1+KA*CA0*(1-XA2)+KC*CA0*XA2)
3	d(XA3)/d(W)=(k/FA0)*(CA0*(1-XA3))^2/(1+KA*CA0*(1-XA3)+KB*CA0*(1-XA3))^2
4	d(XA4)/d(W)=(k/FA0)*(CA0*(1-XA4))^2/(1+KA*CA0*(1-XA4)+KB*CA0*(1-XA4)+KC*CA0*XA4)^2
5	k=10
6	FA0=1.5
7	CA0=1
8	KA=1
9	KC=20
10	KB=2
11	W(0)=0
12	XA1(0)=0
13	XA2(0)=0
14	XA3(0)=0
15	XA4(0)=0
16	W(f)=2

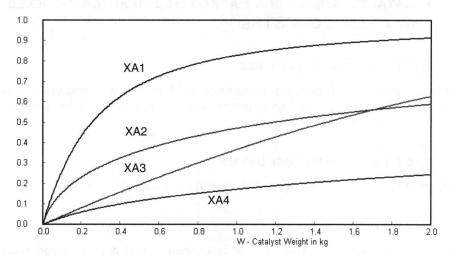

Figure 11–7 Calculated Conversions versus Catalyst Weight for Four Different
Catalysts

A plot of the conversions versus the weight of catalyst W is given in Figure
11–7.

 The POLYMATH problem solution file for part (a) is found in directory
Chapter 11 with file named **P11-17A.POL**.

(b) When the pressure varies within the reactor, the stoichiometric equations relating the concentrations to conversion need to be altered to include the
change of pressure by

$$C_A = C_B = C_{A0}(1-X)y$$
$$C_C = C_D = C_{A0}Xy$$

(11-94)

The differential equation for pressure drop given by Equation (11-91) must be
solved simultaneously with the differential equations for conversion.

11.18 CATALYST DECAY IN A PACKED BED REACTOR MODELED BY A SERIES OF CSTRS

11.18.1 Concepts Demonstrated

Determination of the change of reactant and product concentration and catalyst decay with time in a packed bed reactor that is approximated by a series of CSTRs with and without pressure drop.

11.18.2 Numerical Methods Utilized

Solution of simultaneous ordinary differential equations.

11.18.3 Problem Statement

A gas phase catalytic reaction $A \overset{k}{\rightarrow} B$ is carried out in a packed bed reactor where the catalyst activity is decaying. The reaction with deactivation follows the rate expression given by

$$-r_A = akC_A \tag{11-95}$$

where a is the catalyst activity that follows either the deactivation kinetics

$$\frac{da}{dt} = -k_{d1}a \tag{11-96}$$

or

$$\frac{da}{dt} = -k_{d2}aC_B \tag{11-97}$$

The packed bed reactor can be approximated by three CSTRs in series, as shown in Figure 11–8.

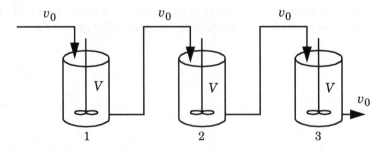

Figure 11–8 Reactor Approximation by a Train of Three CSTRs

The volumetric flow rate to each reactor is v_0. The reactor feed is pure A at concentration C_{A0} to the first reactor. The pressure drop can be neglected. At time zero there is only inert gas in all of the reactors.

The following parameter values apply:

$$k_{d1} = 0.01 \text{ min}^{-1} \qquad k_{d2} = 1.0 \frac{\text{dm}^3}{\text{g-mol} \cdot \text{min}}$$

$$k = 0.9 \frac{\text{dm}^3}{\text{dm}^3 (\text{of catalyst}) \text{min}}$$

$$C_{A0} = 0.01 \frac{\text{g-mol}}{\text{dm}^3} \qquad V = 10 \text{ dm}^3 \qquad \upsilon_0 = 5 \frac{\text{dm}^3}{\text{min}}$$

(a) Plot the concentration of A in each of the three reactors as a function of time to 60 minutes using the activity function given by Equation (11-96). Create a separate plot for the activities in all three reactors.

(b) Repeat part (a) for the activity function as given in Equation (11-97).

(c) Compare the outlet concentration of A for parts (a) and (b) at 60 minutes deactivation to that from a plug-flow packed bed reactor with no deactivation (total volume of 30 dm³) and the three CSTR reactors in series model with no deactivation.

11.18.4 Partial Solution

The respective material balances on components A and B yield the following differential equations for the first CSTR, where the subscript 1 indicates the concentrations in the first reactor:

$$\frac{dC_{A1}}{dt} = \frac{(C_{A0} - C_{A1})\upsilon_0}{V} + r_{A1}$$

$$\frac{dC_{B1}}{dt} = \frac{-C_{B1}\upsilon_0}{V} - r_{A1}$$

(11-98)

For the second and third CSTR, where $i = 2$ and 3, the balances yield

$$\frac{dC_{Ai}}{dt} = \frac{(C_{A(i-1)} - C_{A(i)})\upsilon_0}{V} + r_{Ai}$$

$$\frac{dC_{Bi}}{dt} = \frac{(C_{B(i-1)} - C_{B(i)})\upsilon_0}{V} + r_{Ai}$$

(11-99)

These equations, together with Equations (11-95) and (11-96) or (11-97), provide the equations that need to be solved simultaneously in this problem.

(a) For this part, only the material balances involving A are needed in addition to Equations (11-95) and (11-96). This set of equations of Table 11–19 can be entered into the POLYMATH *Simultaneous Differential Equation Solver*.

Table 11–19 POLYMATH Program (File **P11-18A.POL**)

Line	Equation
1	d(a1)/d(t)=-.01*a1
2	d(a2)/d(t)=-.01*a2
3	d(a3)/d(t)=-.01*a3
4	d(CA2)/d(t)=(CA1-CA2)*5/10-a2*.9*CA2
5	d(CA1)/d(t)=(.01-CA1)*5/10-a1*.9*CA1
6	d(CA3)/d(t)=(CA2-CA3)*5/10-a3*.9*CA3
7	t(0)=0
8	a1(0)=1
9	a2(0)=1
10	a3(0)=1
11	CA2(0)=0
12	CA1(0)=0
13	CA3(0)=0
14	t(f)=60

The resulting graph of the concentration of A in each of the three reactors is given in Figure 11–9.

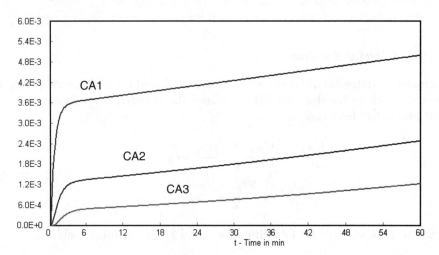

Figure 11–9 Concentration of A in Each Reactor

 The POLYMATH problem solution file for part (a) is found in directory Chapter 11 with file named **P11-18A.POL**

(b) The material balances involving both components A and B are needed along with Equations (11-95) and (11-97). Note that the activities in each reactor will be different because of the dependency of the activity relationship of Equation (11-97) on the concentration of B.

491 DESIGN FOR CATALYST DEACTIVATION IN A STRAIGHT-THROUGH REACTOR   491

11.19 DESIGN FOR CATALYST DEACTIVATION IN A STRAIGHT-THROUGH REACTOR

11.19.1 Concepts Demonstrated

The effect of various types of catalyst deactivation on a catalytic gas phase reaction in a straight-through transport reactor where the catalyst bed is moving through the reactor.

11.19.2 Numerical Methods Utilized

Solution of simultaneous ordinary differential equations.

11.19.3 Problem Statement[*]

The gas phase cracking of a Salina light gas oil can be represented by a lumped parameter kinetic model that incorporates the general reaction

$$\text{Gas Oil (g)} \rightarrow \text{Products (g)} + \text{coke (s)}$$

which can be expressed as the gas phase reaction

$$A \rightarrow B \tag{11-100}$$

A typical application is to carry out this reaction in a straight-through transport reactor. This reactor, which contains a catalyst that decays as a result of coking, is shown in Figure 11–10. The catalyst particles are assumed to move with the mean gas velocity given by $u = 8.0$ m/s in this case. The reaction is to be carried out at 750°F under constant temperature and pressure. The entering concentration of A is 0.2 kg-mol/m^3. The reactor length is 6 m. For this problem, the volume change with reaction, pressure drop, and temperature variation may be neglected.

The catalytic activity, denoted by a, is usually defined as the ratio of the reaction rate for a catalyst subjected to deactivation for time t to the reaction rate for fresh catalyst. For a moving bed of catalyst traveling through the reactor with a plug-flow velocity u, the time that the catalyst has been in the reactor when it reaches reactor height z is given by

$$t = \frac{z}{u} \tag{11-101}$$

Three types of catalyst deactivation are to be examined in this problem. The first is catalyst coking, which has the form

$$a_1 = \frac{1}{1 + A't^{1/2}} \tag{11-102}$$

---

[*] This problem adapted from Fogler,[1] with permission.

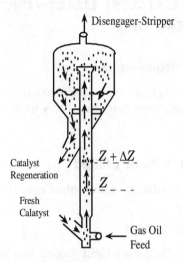

Figure 11-10 Straight-Through Transport Reactor (from Fogler,[1] with permission)

where A' is the coking parameter.

The second type of rate law is deactivation by sintering:

$$\frac{da_2}{dt} = -k_{d2}a_2^{\,2} \qquad \text{(11-103)}$$

The third type is deactivation by poisoning:

$$\frac{da_3}{dt} = -k_{d3}a_3C_B \qquad \text{(11-104)}$$

The catalytic reaction rate per unit volume of catalyst bed for this problem is simply the particular activity multiplied by the catalytic reaction rate expression

$$-r_A = \frac{akC_A}{1 + K_AC_A} \qquad \text{(11-105)}$$

where $k = 30$ s^{-1} and $K_A = 5.0$ m^3/kg-mol.

> (a)   Plot the conversion of A and the catalyst activity versus the reactor length z for the case of no catalyst deactivation.
> (b)   Repeat part (a) for the three types of catalyst deactivation, where $A' = 12$ s$^{-1/2}$, $k_{d2} = 17.5$, s^{-1} and $k_{d3} = 140$ dm^3/mol·s.

11.19.4 Solution

The following equations are needed for the base case where A_C is the cross-sectional area of the reactor:

Mole Balance on Differential Volume of Reactor Bed (expressed in terms of conversion x_A)

$$v_0 C_{A0} \frac{dx_A}{dz} = -r_A A_C \quad \text{I.C. } x_A = 0 \text{ at } z = 0 \tag{11-106}$$

Rate Law (modified by the catalytic activity)

$$-r_A = \frac{akC_A}{1 + K_A C_A} \tag{11-107}$$

Stoichiometry (for gas phase reaction with $\varepsilon = 0$)

$$v = v_0$$

$$u = \frac{v_0}{A_C} \tag{11-108}$$

$$C_A = C_{A0}(1 - x_A)$$

$$C_B = C_{A0} x_A$$

(a) No Deactivation The base case with no deactivation is described by the preceding equations, where the activity is constant at unity. Equation (11-106) can be rearranged with u from Equation Set (11-108) to yield

$$\frac{dx_A}{dz} = \frac{akC_A}{(1 + K_A C_A)u} \quad \text{I.C. } x_A = 0 \text{ at } z = 0 \tag{11-109}$$

The POLYMATH *Simultaneous Differential Equation Solver* can be used to solve Equation (11-109) with the equation set given in Table 11–20.

Table 11–20 POLYMATH Program (File **P11-19A.POL**)

Line	Equation
1	d(xA)/d(z)=a*30*(1-xA)/((1+5*.2*(1-xA))*8)
2	a=1
3	z(0)=0
4	xA(0)=0
5	z(f)=6

 The POLYMATH problem solution file for part (a) is found in directory Chapter 11 with file named **P11-19A.POL**.

(b) Deactivation

Coking   The introduction of Equation (11-101) into Equation (11-102) yields the activity as a function of the reactor length z:

$$a_1 = \frac{1}{1 + \left(\dfrac{A'}{u^{1/2}}\right) z^{1/2}} \qquad \textbf{(11-110)}$$

where the subscript 1 indicates deactivation by coking.

Sintering   The derivative of Equation (11-101)

$$dt = \frac{dz}{u} \qquad \textbf{(11-111)}$$

can be used to rewrite Equation (11-103) can be written in terms of the reactor length z as

$$\frac{da_2}{dz} = \frac{-k_{d2} a_2^{\,2}}{u} \qquad \textbf{(11-112)}$$

where the subscript 2 is used for the deactivation by sintering.

Poisoning   Equation (11-104) can also be expressed in terms of z by using

$$\frac{da_3}{dz} = \frac{-k_{d3} a_3 C_B}{u} \qquad \textbf{(11-113)}$$

where the subscript 3 is used for the deactivation by poisoning.

POLYMATH Solution   It is convenient to enter the equation sets for all four cases of catalytic activity that have been considered in this problem simultaneously by using subscript notation as shown in Table 11–21.

Table 11–21 POLYMATH Program (File **P11-19B.POL**)

Line	Equation
1	d(xA)/d(z)=a*30*(1-xA)/((1+5*.2*(1-xA))*8)
2	a=1
3	d(xA1)/d(z)=a1*30*(1-xA1)/((1+5*.2*(1-xA1))*8)
4	a1=1/(1+(12/(8^.5))*z^(.5))
5	d(xA2)/d(z)=a2*30*(1-xA2)/((1+5*.2*(1-xA2))*8)
6	d(a2)/d(z)=-17.5*a2*a2/8
7	d(xA3)/d(z)=a3*30*(1-xA3)/((1+5*.2*(1-xA3))*8)
8	d(a3)/d(z)=-140.*a3*(.2*xA3)/8
9	xA(0)=0
10	xA1(0)=0
11	xA2(0)=0
12	a2(0)=1
13	xA3(0)=0
14	a3(0)=1
15	z(0)=0
16	z(f)=6

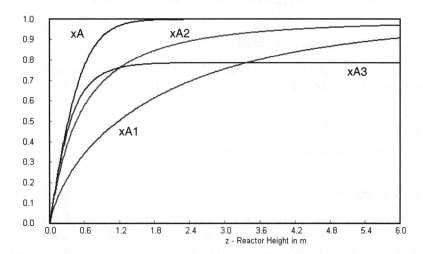

Figure 11–11 Conversion versus Reactor Height

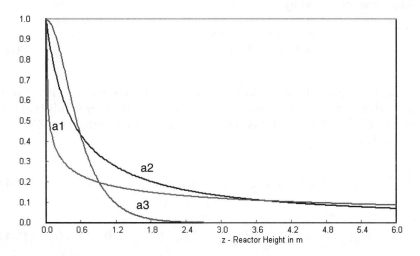

Figure 11–12 Activity Factors versus Reactor Height

The various conversions versus reactor height are given in Figure 11–11, and the profiles for the activities are given in Figure 11–12.

The POLYMATH problem solution file for part (b) is found in directory Chapter 11 with file named **P11-19B.POL**.

11.20 ENZYMATIC REACTIONS IN A BATCH REACTOR

11.20.1 Concepts Demonstrated

Enzymatic reaction kinetics, batch reactor design equations for enzymatic reactions, pseudo-steady-state hypothesis and the validity of its application.

11.20.2 Numerical Methods Utilized

Solution of simultaneous ordinary differential equations.

11.20.3 Problem Statement

One of the simplest models of enzyme catalysis is the second-order reversible binding of a substrate S (reactant) to an enzyme catalyst E to give an enzyme-substrate complex $E \cdot S$, which then decomposes by a first-order reaction to give product P and regenerates the enzyme catalyst E. This model of enzyme catalysis is referred to as Michaelis-Menten kinetics when the pseudo-steady-state hypothesis is utilized. (See Fogler[1] for more details.) The elementary steps in the reaction sequence are given by

$$S + E \underset{k_2}{\overset{k_1}{\rightleftarrows}} (ES) \overset{k_3}{\rightarrow} E + P \tag{11-114}$$

Material balances can be made on a constant volume batch reactor in the preceding reactions, where the initial enzyme concentration and substrate concentration are C_{E_0} and C_{S_0}, respectively. No product concentration C_P is initially present. Thus

$$\frac{dC_S}{dt} = -k_1 C_S C_E + k_2 C_{E \cdot S} \quad \text{I.C.} \quad C_S = C_{S_0} \text{ at } t = 0 \tag{11-115}$$

$$\frac{dC_{E \cdot S}}{dt} = k_1 C_S C_E - k_2 C_{E \cdot S} - k_3 C_{E \cdot S} \quad \text{I.C.} \quad C_{E \cdot S} = 0 \text{ at } t = 0 \tag{11-116}$$

$$\frac{dC_P}{dt} = k_3 C_{E \cdot S} \quad \text{I.C.} \quad C_P = 0 \text{ at } t = 0 \tag{11-117}$$

$$C_{E_0} = C_E + C_{E \cdot S} \tag{11-118}$$

The preceding set of equations describing the interactions in a batch reactor cannot be solved analytically. The pseudo-steady-state hypothesis (also known as the quasi-steady-state assumption) is normally used by assuming that the enzyme-substrate complex is an active intermediate present at very low concentration and the time-derivative of this concentration is zero:

$$\frac{dC_{E \cdot S}}{dt} = 0 \tag{11-119}$$

The preceding equation is a normal assumption of the pseudo-steady-state hypothesis that leads to the Michaelis-Menton rate expression for a batch reactor.

(a) Show that the pseudo-steady-state hypothesis results in the following batch reactor design expression involving the Michaelis-Menten rate expression:

$$\frac{dC_S}{dt} = \frac{-k_3 C_{E_0} C_S}{K + C_S} = -\frac{dC_P}{dt} \qquad \text{where} \qquad K = \frac{k_2 + k_3}{k_1} \tag{11-120}$$

(b) Analytically integrate Equation (11-120) for a batch reactor to show that

$$k_3 C_{E_0} t = C_{S_0} - C_S + K \ln\left(\frac{C_{S_0}}{C_S}\right) \tag{11-121}$$

(c) Demonstrate the validity of the pseudo steady-state hypothesis from Equation (11-120) to predict the concentration transients as calculated by the rigorous solution of the equation set (11-115) to (11-118). The following parameter values apply: $C_{S_0} = 1.0$ g-mol/dm^3, $C_{E_0} = 10^{-3}$ g-mol/dm^3, $t_{\text{final}} = 48.0$ h, $k_1 = 2.0 \times 10^{3}$ dm^3/g-mol·h, $k_2 = 3.0 \times 10^5$ h^{-1}, and $k_3 = 1.0 \times 10^4$ h^{-1}.

11.20.4 Solution (Suggestions)

(c) A comparison may be made by carrying out the rigorous solution (using Equations (11-115) to (11-118)) and the pseudo-steady-state solution (using Equation (11-120)) in the same numerical solution utilizing the POLYMATH *Simultaneous Differential Equation Solver*. The solution may require the use of a stiff integration algorithm.

11.21 ISOTHERMAL BATCH REACTOR DESIGN FOR MULTIPLE REACTIONS

11.21.1 Concepts Demonstrated

Calculation of concentration profiles in a constant volume batch reactor at constant temperature in which a number of simultaneous elementary reactions are occurring.

11.21.2 Numerical Methods Utilized

Numerical integration of a system of simultaneous ordinary differential equations subject to known initial conditions.

11.21.3 Problem Statement[*]

A seven-step mechanism can be used to describe the atmospheric reactions involving formaldehyde and nitrogen oxides at 1 atm and $T = 298$ °C, as presented by Seinfeld.[9] These photochemical reactions lead to the formation of ozone.

Step 1	$NO_2 + h\nu \rightarrow NO + O$	$k_1 = 0.533$ min^{-1}
Step 2	$O + O_2 + M \rightarrow O_3 + M$	$k_2 = 2.21 \times 10^{-5}$ ppm^{-2} min^{-1}
Step 3	$O_3 + NO \rightarrow NO_2 + O_2$	$k_3 = 26.7$ ppm^{-1} min^{-1}
Step 4a	$HCHO + h\nu \rightarrow 2HO_2\cdot + CO$	$k_{4a} = 1.6 \times 10^{-3}$ min^{-1}
Step 4b	$HCHO + h\nu \rightarrow H_2 + CO$	$k_{4b} = 2.11 \times 10^{-3}$ min^{-1}
Step 5	$HCHO + OH\cdot \rightarrow HO_2\cdot + CO + H_2O$	$k_5 = 1.62 \times 10^4$ ppm^{-1} min^{-1}
Step 6	$HO_2\cdot + NO \rightarrow NO_2 + OH\cdot$	$k_6 = 1.22 \times 10^4$ ppm^{-1} min^{-1}
Step 7	$OH\cdot + NO_2 \rightarrow HNO_3$	$k_7 = 1.62 \times 10^4$ ppm^{-1} min^{-1}

The photolysis reactions of Steps 1, 4a, and 4b are written for the case where the incoming solar radiation is assumed to be constant at the mid-day value of k_{iM}, thus leading to the given rate constants. A more detailed model would be to have the rate constants for these steps vary with the time of day

---

[*] This problem was adapted from an exercise developed by Joseph J. Helble.

according to

$$k_i = k_{iM} \sin\left[\frac{2\pi(t-6)}{24}\right] \quad \text{for} \quad \sin\left[\frac{2\pi(t-6)}{24}\right] > 0$$

$$k_i = 0 \quad \text{for all other values of } t$$

(11-122)

where t is the time in hours on a 24 hour/day basis and the solar radiation begins at 6 a.m. and ends at 6 p.m.

All of the reactions in the mechanism are considered to be elementary as written. For example, the rate of Step 2 is given by

$$r_2 = k_2 C_O C_{O_2} C_M$$

(11-123)

where the concentrations are in ppm. C_M is the concentration of third bodies, which is considered to be at a constant value of 10^6 ppm, and C_{O_2}, the concentration of oxygen, is constant at 0.21×10^6 ppm.

The simplest application of chemical reaction engineering to urban smog simulation involves considering the atmosphere below an inversion layer to be a batch reactor with no input or output terms. The resulting material balance for each of the species in the mechanism can be made and expressed as ordinary differential equations. Thus for ozone

$$\frac{dC_{O_3}}{dt} = k_2 C_O C_{O_2} C_M - k_3 C_{O_3} C_{NO}$$

(11-124)

Similar material balances can be made for the other reacting components, including the free radicals. Typical initial conditions for all free radicals are that the concentrations are essentially zero.

Consider a simplified model for photochemical smog that is represented by the seven-step mechanism for formaldehyde.

(a) Verify the concentration profiles for O_3, NO_2, NO, and HCHO, which are presented in Figure 11–13 for the given initial conditions and for the constant values of photolysis reaction rate constants of Steps 1, 4a, and 4b. Note that a similar figure is given by Seinfeld.[9]

(b) Repeat part (a) for a single 24-hour day starting at midnight with the same initial conditions but with the three photolysis reactions varying according to the general relationship of Equation (11-122).

(c) Stricter NO_x controls have been suggested as a means of reducing ozone. Comment on the wisdom of this approach after making several additional solutions with different initial concentrations of NO.

11.21.4 Solution (Suggestions)

(a) The various material balances on the batch reactor yield the following ordinary differential equations for the reacting species:

$$\frac{dC_{NO_2}}{dt} = -k_1 C_{NO_2} + k_3 C_{O_3} C_{NO} + k_6 C_{HO_2} . C_{NO} - k_7 C_{OH} . C_{NO_2} \qquad \text{(11-125)}$$

$$\frac{dC_{O_3}}{dt} = k_2 C_O C_{O_2} C_M - k_3 C_{O_3} C_{NO} \qquad \text{(11-126)}$$

$$\frac{dC_{NO}}{dt} = k_1 C_{NO_2} - k_3 C_{O_3} C_{NO} - k_6 C_{HO_2} . C_{NO} \qquad \text{(11-127)}$$

$$\frac{dC_{HO_2}.}{dt} = 2k_{4a} C_{HCHO} + k_5 C_{HCHO} C_{OH} . - k_6 C_{HO_2} . C_{NO} \qquad \text{(11-128)}$$

$$\frac{dC_{OH}.}{dt} = -k_5 C_{HCHO} C_{OH} . + k_6 C_{HO_2} C_{NO} - k_7 C_{OH} . C_{NO_2} \qquad \text{(11-129)}$$

$$\frac{dC_O}{dt} = k_1 C_{NO_2} - k_2 C_O C_{O_2} C_M \qquad \text{(11-130)}$$

$$\frac{dC_{HCHO}}{dt} = -k_{4a} C_{HCHO} - k_{4b} C_{HCHO} - k_5 C_{HCHO} C_{OH} . \qquad \text{(11-131)}$$

The POLYMATH *Simultaneous Differential Equation Solver* can be used to solve the preceding differential equations. Results are presented in Figure 11–13 for the given initial conditions. All initial conditions that are not specified with particular values are zero. The equation set may require the use of a "stiff" integration algorithm.

(b) The variation of light intensity with time of day can be handled with the "if ... then ... else ... " capability available within POLYMATH. Implementation of the variation of k_1 with the time of day can be accomplished by

```
k1=if(sign(sin(2*3.1416/24*(t-6)))<=0)then(0)else
    ((0.533*60)*sin(2*3.1416/24*(t-6)))
```

where the intrinsic sign function returns either +1, 0, or –1, depending upon the value of the function argument being either positive, zero, or negative, respectively. The net result is that the rate constant only has values between 6 a.m. to 6 p.m. and is zero at other times. This same type of statement can be used for the other photolysis reactions that are sunlight dependent.

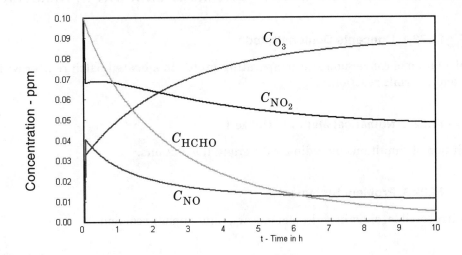

Figure 11–13 Formaldehyde Photooxidation in a Mixture of NO, NO$_2$, and air.
Initial Conditions:

C_{NO_2} = 0.1 ppm, C_{NO} = 0.01 ppm

C_{HCHO} = 0.1 ppm, C_{O_3} = 0.0 ppm

11.22 MATERIAL AND ENERGY BALANCES ON A BATCH REACTOR

11.22.1 Concepts Demonstrated

Calculation of conversion and temperature profile in a heated batch reactor with an endothermic reaction.

11.22.2 Numerical Methods Utilized

Solution of simultaneous ordinary differential equations.

11.22.3 Problem Statement

An irreversible, endothermic, elementary, liquid phase reaction

$$A + B \xrightarrow{k} C \tag{11-132}$$

is to be carried out in an agitated batch reactor that is heated by a steam jacket on the reactor exterior. Initially the concentrations are $C_{A0} = 2.5$ g-mol/dm^3, $C_{B0} = 5$ g-mol/dm^3, and $C_{C0} = 0$. The reactor volume is $V = 1200$ dm^3, and the steam in the surrounding jacket is kept at $T_j = 150$ °C. Additional data are given in Table 11–22.

(a) Plot the conversion of A, x_A, and the reactor temperature during the first 60 minutes of reactor operation for an initial reactor temperature of $T_0 = 30$°C.

(b) Calculate the minimal initial heating time to assure that $x_A = 0.99$ after 60 minutes of reactor operation for an initial reactor temperature of $T_0 = 30$°C.

Table 11–22 Additional Data for Heated Batch Reactor

Activation Energy	$E = 83.6$ kJ/g-mol
Rate Coefficient	$k = 0.001$ dm^3/g-mol·min at 27°C
Heat of Reaction at 273.13 K	$\Delta H_R = 27.85$ kJ/g-mol
Mean Heat Capacities	$\tilde{C}_{PA} = 14$ J/g-mol·K
	$\tilde{C}_{PB} = 28$ J/g-mol·K
	$\tilde{C}_{PC} = 42$ J/g-mol·K
Heat Transfer Area	$A = 5$ m^2
Heat Transfer Coefficient	$U = 3.76$ kJ/min·m^2·K

11.22.4 Solution (Suggestions)

The equations representing this system are as follows:

Mole balance (Design equation expressed in terms of conversion x_A)

$$\frac{dx_A}{dt} = \frac{-r_A}{C_{A0}} \qquad (11\text{-}133)$$

Energy balance[1]

$$\frac{dT}{dt} = \frac{UA(T_j - T) + r_A V \Delta H_R}{N_{A0}(C_{PA} + \theta_B C_{PB})} \qquad (11\text{-}134)$$

Rate law

$$r_A = -kC_{A0}^{\,2}(1 - x_A)(\theta_B - x_A) \qquad (11\text{-}135)$$

$$k = 0.001 \, \exp\left[\frac{-E}{R}\left(\frac{1}{T} - \frac{1}{300}\right)\right] \qquad (11\text{-}136)$$

where N_{A0} is the initial molar amount of A that is calculated from $N_{A0} = VC_{A0}$, x_A is the conversion of A, θ_B is the initial molar ratio given by C_{B0}/C_{A0}, and T is the absolute temperature.

These equations, together with the numerical data provided, can be entered into the POLYMATH *Simultaneous Differential Equation Solver*, and a numerical solution can be generated for the time interval indicated. Please be sure to use consistent units in the working equations, particularly for the various temperatures.

11.23 OPERATION OF A COOLED EXOTHERMIC CSTR

11.23.1 Concepts Demonstrated

Material and energy balances on a CSTR with an exothermic reaction and cooling jacket.

11.23.2 Numerical Methods Utilized

Solution of a system of simultaneous nonlinear algebraic equations, and conversion of the system of equations into one equation to examine multiple steady-state solutions.

11.23.3 Problem Statement[*]

An irreversible exothermic reaction $A \xrightarrow{k} B$ is carried out in a perfectly mixed CSTR, as shown in Figure 11–14. The reaction is first order in reactant A and

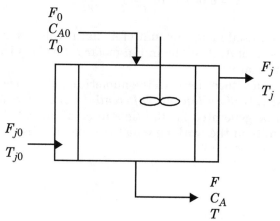

Figure 11–14 Cooled Exothermic CSTR

has a heat of reaction given by λ, which is based on reactant A. Negligible heat losses and constant densities can be assumed. A well-mixed cooling jacket surrounds the reactor to remove the heat of reaction. Cooling water is added to the jacket at a rate of F_j and at an inlet temperature of T_{j0}. The volume V of the contents of the reactor and the volume V_j of water in the jacket are both constant. The reaction rate constant changes as function of the temperature according to the equation

$$k = \alpha \exp(-E/RT) \tag{11-137}$$

The feed flow rate F_0 and the cooling water flow rate F_j are constant. The jacket water is assumed to be completely mixed. Heat transferred from the reac-

[*] This problem is adapted from Luyben,[10] with permission.

tor to the jacket can be calculated from

$$Q = UA(T - T_j) \qquad \qquad \text{(11-138)}$$

where Q is the heat transfer rate, U is the overall heat transfer coefficient, and A is the heat transfer area. Detailed data for the process from Luyben[10] are shown in the Table 11–23.

Table 11–23 CSTR Parameter Values[a]

F_0	40 ft^3/h	U	150 btu/h $\cdot$ ft$^2 \cdot$ °R
F	40 ft^3/h	A	250 ft^2
C_{A0}	0.55 lb-mol/ft^3	T_{j0}	530 °R
V	48 ft^3	T_0	530 °R
F_j	49.9 ft^3/h	λ	−30,000 btu/lb-mol
C_P	0.75 btu/lb$_m \cdot$ °R	C_j	1.0 btu/lb$_m \cdot$ °R
α	7.08×10^{10} h^{-1}	E	30,000 btu/lb-mol
ρ	50 lb$_m$/ft^3	ρ_j	62.3 lb$_m$/ft^3
R	1.9872 btu/lb-mol $\cdot$ °R	V_j	12 ft^3

[a]Data are from Luyben,[10] with permission.

(a) Formulate the material and energy balances that apply to the CSTR and the cooling jacket.

(b) Calculate the steady-state values of C_A, T_j, and T for the operating conditions of Table 11–23.

(c) Identify all possible steady-state operating conditions, as this system may exhibit multiple steady states.

(d) Solve the unsteady-state material and energy balances to identify whether any of the possible multiple steady states are unstable.

11.23.4 Solution (Partial)

(a) There are three balance equations that can be written for the reactor and the cooling jacket. These include the material balance on the reactor, the energy balance on the reactor, and the energy balance on the cooling jacket.

Mole balance on CSTR for reactant A

$$F_0 C_{A0} - F C_A - V k C_A = 0 \tag{11-139}$$

Energy balance on the reactor

$$\rho C_P (F_0 T_0 - FT) - \lambda V k C_A - UA(T - T_j) = 0 \tag{11-140}$$

Energy balance on the cooling jacket

$$\rho_j C_j F_j (T_{j0} - T_j) + UA(T - T_j) = 0 \tag{11-141}$$

(b) Introduction of the numerical parameter values into Equations (11-139), (11-140), (11-141), and (11-137) and entry of the system of equations into the POLYMATH *Nonlinear Algebraic Equation Solver* are presented in Table 11–24.

Table 11–24 POLYMATH Program (File **P11-23B.POL**)

Line	Equation
1	f(CA)=40.*(0.55-CA)-48.*k*CA
2	f(T)=50.*0.75*40.*(530.-T)+30000.*48.*k*CA-150.*250.*(T-Tj)
3	f(Tj)=62.3*1.0*49.9*(530.-Tj)+150.*250.*(T-Tj)
4	k=7.08e10*exp(-30000./(1.9872*T))
5	CA(0)=0.55
6	T(0)=530
7	Tj(0)=530

A reasonable initial assumption is that there is no reaction; therefore, $C_{Ao} = 0.55$, $T_0 = 530$, and $T_{j0} = 530$. The solution obtained with these initial estimates is summarized in Table 11–25.

Table 11–25 Steady-State Operating Conditions for CSTR

Variable	Solution Value	$f()$
C_A	0.52139	−4.441e−16
T	537.855	1.757e−9
T_j	537.253	−1.841e−9
k	0.0457263	

The POLYMATH data file for part (b) is found in directory Chapter 11 with file named **P11-23B.POL**.

(c) Several different steady states may be possible with exothermic reactions in a CSTR. One possible method to determine these different steady states is to solve the system of nonlinear equations with different initial estimates of the final solution. While this approach is not very sophisticated, it can be of benefit in very complex systems of equations.

Another approach is to convert the system of equations into a single implicit and several explicit or auxiliary equations. (Incidentally, this is a good way to show that a particular system does not have a solution at all.) In this particular case, the material balance of Equation (11-139) can be solved for C_A.

$$C_A = \frac{F_0 C_{A0}}{(F + Vk)} \tag{11-142}$$

Also, the energy balance of Equation (11-141) on the cooling jacket can be solved for T_j.

$$T_j = \frac{\rho_j C_j F_j T_{j0} + UAT}{(\rho_j C_j F_j + UA)} \tag{11-143}$$

Thus the problem has been converted to a single nonlinear equation given by Equation (11-140) and three explicit equations given by Equations (11-137), (11-142), and (11-143). The POLYMATH solution is given in Table 11–26.

Table 11–26 POLYMATH Program (File **P11-23C.POL**)

Line	Equation
1	f(T)=50.*0.75*40.*(530.-T)+30000.*48.*k*CA-150.*250.*(T-Tj)
2	k=7.08e10*exp(-30000./(1.9872*T))
3	Tj=(62.3*1.0*49.9*530.+150.*250.*T)/(62.3*1.0*49.9+150.*250.)
4	CA=40.*0.55/(40.+48.*k)
5	T(min)=500
6	T(max)=700

When $f(T)$ is plotted versus T in the range $500 \le T \le 700$, three solutions can be clearly identified, as shown in Figure 11–15. The first one is at low temperature as this is the solution that was initially identified. The second is at an intermediate temperature, and the third is at a high temperature. The three resulting solutions are summarized in Table 11–27.

Table 11–27 Multiple Steady-State Solutions for CSTR

	Solution Number		
	1	2	3
T	537.86	590.35	671.28
C_A	0.5214	0.3302	0.03542
T_j	537.25	585.73	660.46

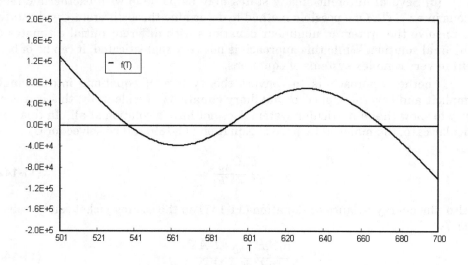

Figure 11–15 Graphical Indication of Multiple Steady-State Solutions

The POLYMATH solution file for part (c) is found in directory Chapter 11 with file named **P11-23C.POL**.

11.24 EXOTHERMIC REVERSIBLE GAS PHASE REACTION IN A PACKED BED REACTOR

11.24.1 Concepts Demonstrated

Design of a gas-phase catalytic reactor with pressure drop for an exothermic reversible gas-phase reaction.

11.24.2 Numerical Methods Utilized

Integration of simultaneous ordinary differential equations with known initial conditions.

11.24.3 Problem Statement

The elementary gas phase reaction $2A \rightleftharpoons C$ is carried out in a packed bed reactor. There is a heat exchanger surrounding the reactor, and there is a pressure drop along the length of the reactor as shown in Figure 11–16.

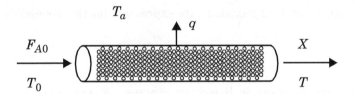

Figure 11–16 Packed Bed Catalytic Reactor

The various parameters values for this reactor design problem are summarized in Table 11–28.

Table 11–28 Parameter Values for Packed Bed Reactor

$C_{PA} = 40.0$ J/g-mol·K	$R = 8.314$ J/g-mol·K
$C_{PC} = 80.0$ J/g-mol·K	$F_{A0} = 5.0$ g-mol/min
$\Delta H_R = -40{,}000$ J/g-mol	$U_a = 0.8$ J/kg·min·K
$E_A = 41{,}800$ J/g-mol·K	$T_a = 500$ K
$k = 0.5$ dm^6/kg·min·mol @ 450 K	$\alpha = 0.015$ kg^{-1}
$K_C = 25{,}000$ dm^3/g-mol @ 450 K	$P_0 = 10$ atm
$C_{A0} = 0.271$ g-mol/dm^3	$y_{A0} = 1.0$ (Pure A feed)
$T_0 = 450$ K	

(a)   Plot the conversion (X), reduced pressure (y), and temperature ($T \times 10^{-3}$) along the reactor from $W = 0$ kg up to $W = 20$ kg.

(b)   Around 16 kg of catalyst you will observe a "knee" in the conversion profile. Explain why this knee occurs and what parameters affect the knee.

(c)   Plot the concentration profiles for reactant A and product C from $W = 0$ kg up to $W = 20$ kg.

Additional Information

The notation used here and the following equations and relationships for this particular problem are adapted from the textbook by Fogler.[1] The problem is to be worked assuming plug flow with no radial gradients of concentrations and temperature at any location within the catalyst bed. The reactor design will use the conversion of A designated by X and the temperature T which are both functions of location within the catalyst bed specified by the catalyst weight W.

The general reactor design expression for a catalytic reaction in terms of conversion is a mole balance on reactant A given by

$$F_{A0}\frac{dX}{dW} = -r'_A \tag{11-144}$$

The simple catalytic reaction rate expression for this reversible reaction is

$$-r'_A = k\left[C_A^2 - \frac{C_C}{K_C}\right] \tag{11-145}$$

where the rate constant is based on reactant A and follows the Arrhenius expression

$$k = (k@T = 450 \text{ K})\exp\frac{E_A}{R}\left[\frac{1}{450} - \frac{1}{T}\right] \tag{11-146}$$

and the equilibrium constant variation with temperature can be determined from van't Hoff's equation with $\Delta \tilde{C}_P = 0$:

$$K_C = (K_C@T = 450 \text{ K})\exp\frac{\Delta H_R}{R}\left[\frac{1}{450} - \frac{1}{T}\right] \tag{11-147}$$

The stoichiometry for $2A \rightleftarrows C$ and the stoichiometric table for a gas allow the concentrations to be expressed as a function of conversion and temperature while allowing for volumetric changes due to decrease in moles during the reaction. Therefore,

$$C_A = C_{A0}\left(\frac{1-X}{1+\varepsilon X}\right)\frac{P}{P_0}\frac{T_0}{T} = C_{A0}\left(\frac{1-X}{1-0.5X}\right)y\frac{T_0}{T} \tag{11-148}$$

and

$$y = \frac{P}{P_0}$$

$$C_C = \left(\frac{0.5 C_{A0} X}{1 - 0.5 X}\right) y \frac{T_0}{T}$$

(11-149)

The pressure drop can be expressed as a differential equation (see Fogler[1] for details):

$$\frac{d\left(\frac{P}{P_0}\right)}{dW} = \frac{-\alpha(1 + \varepsilon X)}{2} \frac{P_0}{P} \frac{T}{T_0}$$

(11-150)

or

$$\frac{dy}{dW} = \frac{-\alpha(1 - 0.5 X)}{2y} \frac{T}{T_0}$$

(11-151)

The general energy balance may be written as

$$\frac{dT}{dW} = \frac{U_a(T_a - T) + r'_A (\Delta H_R)}{F_{A0}\left(\sum \theta_i C_{Pi} + X \Delta \tilde{C}_P\right)}$$

(11-152)

which for only reactant A in the reactor feed simplifies to

$$\frac{dT}{dW} = \frac{U_a(T_a - T) + r'_A (\Delta H_R)}{F_{A0}(C_{PA})}$$

(11-153)

www

The POLYMATH solution file is found in directory Chapter 11 with file named **P11-24ABC.POL.**

This problem is also solved with Excel, Maple, MathCAD, MATLAB, Mathematica, and POLYMATH as Problem 9 in the Set of Ten Problems discussed on the book web site.

11.25 TEMPERATURE EFFECTS WITH EXOTHERMIC REACTIONS

11.25.1 Concepts Demonstrated

Reactor design considerations for a reversible exothermic reaction carried out in various ideal reactors.

11.25.2 Numerical Methods Utilized

Solutions of simultaneous nonlinear algebraic equations and simultaneous ordinary differential equations.

11.25.3 Problem Statement

The reversible exothermic and elementary liquid phase reaction

$$A \leftrightarrow R \tag{11-154}$$

is studied in a variety of reactors some of which are adiabatic. The rate expression and rate constants are

$$r_A = -k_1 C_A + k_2 C_R$$

$$k_1 = 5.2 \times 10^7 e^{-\frac{12,000}{RT}} \text{ min}^{-1} \quad \text{and} \quad k_2 = 2.8 \times 10^{18} e^{-\frac{30,000}{RT}} \text{ min}^{-1} \tag{11-155}$$

where $R = 1.9872$ cal/g-mol $\cdot$ °K and T is in °K.

For the base case, the reactor V is 12 dm^3 and the volumetric flow rate $v_0 = 24$ dm^3/min. The feed concentration of A is $C_{A0} = 1$ g-mol/dm^3. Consider the heat capacity of the reacting mixture based on a g-mol of A to be defined as

$$C_P = \sum_{i=1}^{n} \Theta_i \tilde{C}_{pi} \tag{11-156}$$

which is constant. The change in heat capacity during reaction is zero

$$\Delta \tilde{C}_p = 0 \tag{11-157}$$

and the ratio of the heat of reaction to the heat capacity is given by

$$\Delta H_{R,T_R}/C_p = -200 \, °C \tag{11-158}$$

Details on the preceding equations are found in Fogler.[1]

(a) Calculate and plot the conversion and reactor temperature versus volume for an adiabatic plug-flow reactor in which the feed temperature is 60°C.

(b) Determine what feed temperature would maximize the output conversion in part (a).

(c) If the reactor of part (a) is operated isothermally at 60°C, what would be the resulting conversion?

(d) If the plug-flow reactor of part (a) were operated at the most optimal temperature profile (where T in the reactor cannot exceed a temperature of 105°C), what would be the resulting conversion?

(e) The plug-flow reactor of part (a) is to be replaced by a series of three adiabatic CSTRs, with each having one third the volume of the original reactor. What feed temperatures should be used for each CSTR to give the optimal output conversion? What is the overall optimal conversion?

(f) Compare the temperature and conversion profiles down the plug-flow reactor of part (a) to a space time of 0.5 minutes for a feed temperature of 60°C with different rates of external cooling to an ambient temperature of 20°C. Consider Ua/C_p to be 0.1, 0.5, and 1.0 g-mol/dm^3·min, where C_p is the constant heat capacity of the mixture per g-mol of A. Which heat transfer rate gives the best conversion and why?

(g) For the adiabatic plug-flow reactor of part (a), vary the heat of reaction to one half and to one quarter of the base case value. Plot the temperatures and conversions as a function of reactor space time to 1.0 min. Why would a reduction of the heat of reaction result in higher conversions?

11.26 DIFFUSION WITH MULTIPLE REACTIONS IN POROUS CATALYST PARTICLES

11.26.1 Concepts Demonstrated

Selectivity in multiple reactions occurring with a solid catalyst comprised of two components with different activities and effective diffusivities.

11.26.2 Numerical Methods Utilized

Simultaneous ordinary differential equations with split boundary conditions.

11.26.3 Problem Statement

The catalytic cracking of gas-oil, A, to gasoline, B, is accompanied by the production of light gases, coke, and coke precursors, C. A schematic of this reaction system is indicated in Figure 11–17, where both reactions involving reactant A are second order and the reaction involving reactant B is first order.

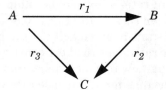

Figure 11–17 Reactions Involved in a Lumped Parameter Model for Catlayst Cracking

The industrial catalysts for this catalytic process are typically mixtures of 5.0 weight % zeolite Z in a silica-alumina matrix M. The zeolite catalysis is more catalytically active and is more selective to the desired gasoline product than the matrix, but the zeolite has a much lower effective diffusivities. Table 11–29 presents typical data for these catalytic reaction systems as adapted from Martin et al.[11]

In a particular application, the catalyst particle can be approximated by a slab of total thickness L (one surface exposed as in Problem 6.5) where the kinetic and transport properties can be considered as the weighted average of the zeolite designated Z and the matrix designated M as given by

$$D_{Ax} = uD_{AZ} + (1-u)D_{AM} \quad \text{and} \quad D_{Bx} = uD_{BZ} + (1-u)D_{BM} \qquad \textbf{(11-159)}$$

$$k_{ix} = uk_{iZ} + (1-u)k_{iM} \quad \text{for } i = 1, 2, \text{ and } 3 \qquad \textbf{(11-160)}$$

where u is the local mass fraction of zeolite. Note that the total mass fraction of zeolite is always 5% and cannot exceed 20% at any location in the catalyst particle. Thus

$$\frac{1}{L}\int_0^L u \, dx = 0.05 \quad \text{where} \quad u \le 0.20 \qquad \textbf{(11-161)}$$

The zero flux boundary conditions at the particle center are given by

$$\frac{dC_A}{dx} = \frac{dC_B}{dx} = 0 \qquad \text{at} \qquad x = 0 \qquad \textbf{(11-162)}$$

and the boundary conditions for the gas concentrations at the particle surface are given by

$$C_A = 1.6 \times 10^{-5} \text{gmol/cm}^3 \quad \text{and} \quad C_B = C_C = 0 \quad \text{at} \quad x = L \qquad \textbf{(11-163)}$$

The selectivity for B is the mass flux of B in the x direction divided by the mass flux of A in the opposite direction. Thus

$$S_B = \frac{-D_{Bx}(dC_B/dx)}{D_{Ax}(dC_A/dx)} \qquad \text{evaluated at} \qquad x = L \qquad \textbf{(11-164)}$$

The appropriate differential equations for the diffusion with reaction are

$$D_{Ax}\frac{d^2 C_A}{dx^2} = k_{3x}C_A^2 + k_{1x}C_A^2 \qquad \textbf{(11-165)}$$

$$D_{Bx}\frac{d^2 C_B}{dx^2} = k_{2x}C_B - k_{1x}C_A^2 \qquad \textbf{(11-166)}$$

Table 11–29 Parameter Values (adapted from Martin et al.[11])

$k_{1Z} = 10^8$ cm^3/g-mol·s	$k_{1M} = 10^7$ cm^3/g-mol·s
$k_{2Z} = 8 \times 10^2$ s^{-1}	$k_{2M} = 10^2$ s^{-1}
$k_{3Z} = 3 \times 10^6$ cm^3/g-mol·s	$k_{3M} = 8 \times 10^6$ cm^3/g-mol·s
$D_{AZ} = 10^{-6}$ cm^2/s	$D_{AM} = 10^{-3}$ cm^2/s
$D_{BZ} = 10^{-5}$ cm^2/s	$D_{BM} = 10^{-2}$ cm^2/s
$L = 0.002$ cm	

(a) Derive and verify Equations (11-164), (11-165), and (11-166).
(b) Summarize all of the equations needed to determine the selectivity of B for the case where the zeolite catalyst is evenly distributed throughout the catalyst particle.
(c) Repeat part (b) for the case where the zeolite is located in a 20% layer at the surface of the catalyst particle.
(d) Calculate the selectivity for part (b) using the values of Table 11–29.
(e) Calculate the selectivity for part (c) using the values of Table 11–29.
(f) Discuss the benefits of placing the zeolite layer at the surface of the catalyst particle.

11.27 Nitrification of Biomass in a Fluidized Bed Reactor

11.27.1 Concepts Demonstrated

Steady-state operation of a fluidized sand bed reactor for nitrification of biomass incorporating a simple gas absorption unit to provide needed dissolved oxygen from air.

11.27.2 Numerical Methods Utilized

Solution of a system of nonlinear equations and optimization of some problem variables.

11.27.3 Problem Statement (adapted from Tanaka et al.[12] and Dunn et al.[13])

An important process for the removal of nitrogen-containing organic components from waste streams involves the conversion of ammonium ions to nitrate ions by microbial catalysts. The overall reaction

$$NH_4^+ + 2O_2 \rightarrow NO_3^- + H_2O + 2H^+ \qquad \text{(11-167)}$$

is considered to occur by two separate biochemical reactions

$$NH_4^+ + \frac{3}{2}O_2 \rightarrow NO_2^- + H_2O + 2H^+ \quad \text{with rate } r_1 \qquad \text{(11-168)}$$

$$NO_2^- + \frac{1}{2}O_2 \rightarrow NO_3^- \quad \text{with rate } r_2 \qquad \text{(11-169)}$$

where dissolved oxygen is required.

A fluidized biofilm sand bed reactor can be used for nitrification as is indicated in Figure 11–18. Continuous operation involves an oxygen absorption unit that provides dissolved oxygen from air. The fluidized bed can be modeled as four equal-volume CSTR's in series with stage indicated by subscript n. The biokinetic reaction rates of the nitrogen components in mg/L·h for stage n can be approximated by

$$r_{1,n} = \frac{\mu_1 S_{1,n} O_n}{(K_1 + S_{1,n})(K_{O1} + O_n)} \qquad \text{(11-170)}$$

$$r_{2,n} = \frac{\mu_2 S_{2,n} O_n}{(K_2 + S_{2,n})(K_{O2} + O_n)} \qquad \text{(11-171)}$$

where subscript i on $S_{i,j}$ with a value of 1 represents NH_4^+ concentration, 2 represents NO_2^-, concentration and 3 represents NO_3^- concentration. Gaseous O_2 is designated O_{In} at the absorber inlet. The inlet dissolved O_2 concentration the absorber is O. All concentrations are in g/L. The recycle ratio is α. The saturation

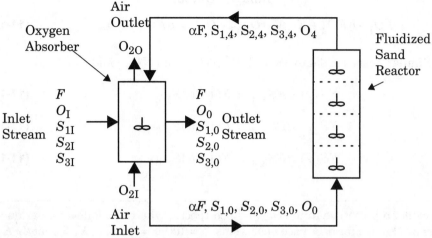

Figure 11–18 Process Flow Diagram for Nitrification of Biomass

concentration of dissolved O_2 in the absorber is O_{SAT} and the dissolved O_2 mass transfer coefficient based on the absorber volume is K_{Lv} in h/L.

Balances on Stages of the Fluidized Sand Reactor

A steady-state material balance on NH_4^+ in each of the four stages of the fluidized sand reactor with stage volume V yields

$$\text{Input + Generation = Output}$$

$$\alpha F S_{1, n-1} - r_{1, n} V = \alpha F S_{1, n} \tag{11-172}$$

where μ_{net} is the net growth rate for the cells in g/dm^3 and V is in L.

Similar stage balances on NO_2^-, NO_3^-, and O yield

$$\alpha F S_{2, n-1} + r_{1, n} V - r_{2, n} V = \alpha F S_{2, n} \tag{11-173}$$

$$\alpha F S_{3, n-1} + r_{2, n} V = \alpha F S_{3, n} \tag{11-174}$$

$$\alpha F O_{n-1} - Y_1 r_{1, n} V - Y_2 r_{2, n} V = \alpha F O_n \tag{11-175}$$

where Y_1 is the yield coefficient for O_2 in reaction 1, and Y_2 is the yield coefficient for O_2 in reaction 2.

Balances on the Oxygen Absorber

A simple single-stage dissolved oxygen balance on the absorber is a good approximation for this unit where the saturation dissolved oxygen concentration from air is O_{SAT}, and an overall mass transfer coefficient $K_L v$ based on liquid volume V_A is adequate. Thus

$$\text{Input} = \text{Output}$$

$$FO_I + \alpha FO_4 + K_{Lv}V_A(O_{SAT} - O_0) = \alpha FO_0 + FO_0 \tag{11-176}$$

Similar absorber balances on NH_4^+, NO_2^-, and NO_3^- yield

$$FS_{1I} + \alpha FS_{1,4} = F(1 + \alpha)S_{1,0} \tag{11-177}$$

$$FS_{2I} + \alpha FS_{2,4} = F(1 + \alpha)S_{2,0} \tag{11-178}$$

$$FS_{3I} + \alpha FS_{3,4} = F(1 + \alpha)S_{3,0} \tag{11-179}$$

Consider a prototype fluidized sand bed reactor where the following parameters are known for base case operations: $\alpha = 100$, F = 0.5 L/h, $K_1 = 5$ mg/L, $K_2 = 3$ mg/L, $K_{Lv} = 40$ $(L \cdot h)^{-1}$, $K_{O1} = 0.25$ mg/L, $K_{O2} = 0.5$ mg/L, $\mu_1 = 30$ mg/L·h, $\mu_2 = 20$ mg/L·h, $S_{1I} = 300$ mg/L, $O_I = S_{2I} = S_{3I} = 0$, $O_{SAT} = 8$ mg/L, V = 0.5 L, $V_A = 1.0$ L, $Y_1 = 3.5$ mg O_2/mg NH_4^+, $Y_2 = 1.1$ mg O_2/mg NO_2^-.

(a) Calculate the outlet concentrations of all the reacting species for the conditions of base case operation.

(b) Plot the effect of the mass transfer coefficient on the exit concentration of NO_3^- over the range from 5 to 80 $(L \cdot h)^{-1}$. Explain these results.

(c) What flow rate is required for the base conditions with NH_4^+ inlet concentration at $S_{1I} = 100$ mg/L that will reduce the NH_4^+ outlet concentration to 5 mg/L?

(d) Plot the steady-state outlet concentrations of NH_4^+, NO_2^-, and NO_3^- for the base case with NH_4^+ inlet concentration at $S_{1I} = 100$ mg/L for the flow rate range to the reactor from 0.05 L/h to 0.5 L/h.

11.28 STERILIZATION KINETICS AND EXTINCTION PROBABILITIES IN BATCH FERMENTERS

11.28.1 Concepts Demonstrated

Kinetics of the death rates of microorganisms due to temperature and probability of extinction and unsuccessful sterilization of microorganisms in a homogeneous population.

11.28.2 Numerical Methods Utilized

Solution of an ordinary differential equation.

11.28.3 Problem Statement

The reaction kinetics of the death rates of microorganisms by heating is correlated by a first-order death rate according to the following differential equation that applies to a constant volume batch reactor

$$\frac{dN}{dt} = -kN \qquad (11\text{-}180)$$

where N is the number of microorganisms at time t. k_d is the rate constant which may be expressed as an Arrhenius relationship with temperature

$$k_d = \alpha \exp\left(-\frac{E_{0d}}{RT}\right) \qquad (11\text{-}181)$$

where α is the frequency factor in time^{-1}, E_{0d} is the activation energy in kJ/g-mol, R is the gas constant, and T is the absolute temperature in K.

When the sterilization process can be considered to be at a constant temperature, Equation (11-180) can be integrated to give

$$\ln\frac{N_t}{N_0} = -k_d t \qquad (11\text{-}182)$$

where N_0 is the original number of microorganisms at $t = 0$, and N_t is the number remaining at time t. This approach is valid as N_t is reduced to about 10 to 100 as discussed by Schuler and Kargi.[14]

As the temperature changes during the sterilization of liquids such as in a fermenter, for example, the differential Equation (11-180) must be solved along with Equation (11-181) as the temperature profile is varied with time, as for example during the heating and cooling of the sterilization process in a fermenter.

The probability of extinction of microorganisms in a homogeneous population is given by

$$P_0(t) = [1 - \exp(-k_d t)]^{N_0} \tag{11-183}$$

and the corresponding probability of a unsuccessful sterilization is

$$P_U = 1 - P_0(t) = 1 - [1 - \exp(-k_d t)]^{N_0} \tag{11-184}$$

The sterilization process is often performed on media that contains vitamins and other important growth factors that also can be inactivated by elevated temperature. Thus it may be important to achieve sterilization with minimal inactivation of the growth factors. The reaction kinetics for the inactivation of many growth factors are approximately first-order and can also be described by Equations (11-180) to (11-182).

(a) A liquid medium is to be sterilized at 121°C to obtain an unsatisfactory sterilization probability of 0.001. The initial concentration of spores is 10^6/L. Estimate the time in minutes required for a 5 L laboratory fermenter and a 5,000 L production fermenter. Do not consider the influence of heating and cooling to the sterilization temperature. The kinetic parameters are $\alpha = 2.\times10^{36}$ min^{-1} and $E_{0d} = 271$ kJ/g-mol.

(b) Calculate the time necessary to reduce the number of spores to 0.1% of their original value in each reactor for the conditions of part (a).

(c) The treatment described in part (a) also partially inactivates a vitamin in the medium that has first-order kinetic parameters of $\alpha = 1.\times10^{10}$ min^{-1} and $E_{0d} = 90$ kJ/g-mol. The original concentration of the vitamin is 40 mg/L. For a set sterilization time of 10 minutes, what is the final concentration of the vitamin? Do not consider the influence of heating and cooling to the sterilization temperature.

(d) A sterilization run has been made in the 5,000 L fermenter. The temperature-time curve is well represented by the following polynomial from t = 0 to t = 120 min. The temperature is in °C. TC = - 0.8817*t + 0.122*t^2 − 0.0018328*t^3 + 0.00000744*t^4 + 35.78. Calculate and plot the number of spores and the concentration of the vitamin in the large fermenter during this sterilization run over this time interval from t = 0 to t = 120 min.

REFERENCES

1. Fogler, H. S. *Elements of Chemical Reaction Engineering*, 2nd ed., Englewood Cliffs, NJ: Prentice Hall, 1992

2. Hill, C. G. *An Introduction to Chemical Engineering Kinetics & Reactor Design*, New York: Wiley, 1977.

3. Hinshelwood, C. N., and Askey, P. J. *Proc. Roy. Soc.*, *A115*, 215 (1927).

4. Dillon, R. T. *J. Am. Chem. Soc.*, *54*, 952 (1932).

5. Hinshelwood, C. N., and Burke *Proc. Roy. Soc. (London)*, *A106*, 284 (1924).

6. McCracken and Dickson, *Ind. Eng. Chem. Proc. Des. and Dev.*, *6*, 286 (1967).

7. Cutlip, M. B., and Peters, M. S., "Heterogeneous Catalysis over an Organic Semiconducting Polymer Made from Polyacrylonitrile," *Chem. Eng. Progr. Symp. Ser.*, *64* (89), 1–11 (1968).

8. Constantinides, A. *Applied Numerical Methods with Personal Computers*, New York: McGraw-Hill, 1987.

9. Seinfeld, J. H., *Atmospheric Chemistry and Physics of Air Pollution*, New York: Wiley, 1986.

10. Luyben, W. L. *Process Modeling Simulation and Control for Chemical Engineers*, 2nd ed., New York: McGraw-Hill, 1990.

11. Martin, G. R., White, C. W., and Dadyburjor, D. B., "Design of Zeolite/Silica-Alumina Catalysts for Triangular Cracking Reactions", *Journal of Catalysis*, *106*, 116–124 (1987).

12. Tanaka, H., Uzman, S., and Dunn, I. J. "Kinetics of Nitrification Using a Fluidized Sand Bed Reactor, Biotechnology and Bioengineering, *23*, No. 8, pp. 1683-1702 (1981).

13. Dunn, I. J., Heinzle, E., Ingham, J. and Prenosil, J. E. *Biological Reaction Engineering*, 2nd ed, Weinheim: Wiley-VCH, 2003.

14. Schuler, M. L. and F. Kargi, *Bioprocess Engineering*, 2nd ed., Upper Saddle River, NJ: Prentice Hall, 2002.

CHAPTER 12

Phase Equilibria and Distillation

12.1 THREE STAGE FLASH EVAPORATOR FOR RECOVERING HEXANE FROM OCTANE

12.1.1 Concepts Demonstrated

Bubble and dew point temperature calculations for a binary mixture. Isothermal and adiabatic flash calculations. Vapor and liquid enthalpy estimations for pure components and ideal mixtures.

12.1.2 Numerical Methods Utilized

Solution of systems of nonlinear algebraic equations.

12.1.3 Problem Statement

Seader and Henley[1] (p. 171) investigated the process shown in Figure 12–1 where n-hexane (designated as component 1) is separated from n-octane by a series of three flash units at 1 atm. The feed to the first stage is an equimolar bubble point liquid at a flow rate of 100 kg-mol/h. The flash evaporator stages operate isothermally at the given temperatures specified in Figure 12–1. The condensers at the exits from stages 1 and 2 condense the vapor to bubble point liquid.

(a) Calculate the unknown temperatures, flow rates, compositions, and heat duties indicated by question marks in Figure 12–1.

(b) Recalculate the process variables for each stage when the feed flow rate is changed by plus or minus 10% and the heat duties of the coolers and heaters remain unchanged from the values calculated in part (a) and given in Table 12–1. Thus the stage temperatures will change and the condensers can be considered to be adiabatic. Condenser temperatures are not required.

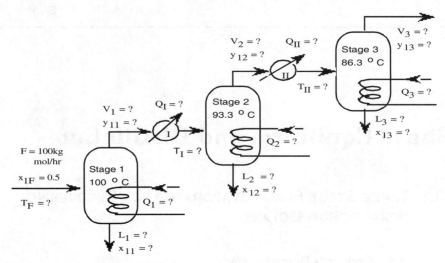

Figure 12–1 Three-Stage Flash Evaporation Process

Additional Information and Data

The vapor liquid equilibrium ratios can be calculated assuming ideal solution, thus $k_j = P_j / P$. For vapor pressure calculations, the Antoine equation can be used with the constants shown in Table 12–1. The molar enthalpies of the pure compounds can be calculated using the equations also given in Table 12–1.

Table 12–1   Vapor Pressure and Molar Enthalpy Correlations for *n*-Hexane and *n*-Octane

	Hexane	Octane
Vapor Pressure (mmHg)[a]	log(P1) = 6.87776 – 1171.53 / (224.366 + T)	log(P2) = 6.92374 – 1355.126 / (209.517 + T)
Liquid Molar Enthalpy (cal/g-mol)[a]	h1 = 51.72 * T	h2 = 66.07 * T
Vapor Molar Enthalpy (cal/g-mol)[a]	H1 = 7678 + 31.83*T + 0.0903 * T^2 / 2	H2 = 10444.7 + 41.836*T + 0.1218 * T^2 / 2

[a]T—Temperature in $°C$

12.1.4   Solution (Partial)

(a) The equations associated with the feed, stage 1 flash, and the first condenser, as entered into the POLYMATH *Nonlinear Equations Solver* program, are shown in Table 12–2. Note that the row numbers have been added only to help with the explanation as they are not part of the POLYMATH input.

The bubble point temperature of feed (TF) is calculated in line 3 to line 6. The enthalpy of the feed is approximated at the bubble point using the ideal liquid mixture mixing rule in line 7. The fraction evaporated (alpha1), liquid composition (x11 and x21), vapor (V1) and liquid (L1) flow rates and vapor composition (y11 and y21) are calculated in lines 10 through 19. The enthalpies

Table 12–2 Equations Associated with the Feed, the First-Stage Flash, and the First Condenser (File **P12-01A.POL**)

Line	Equation
1	F1 = 100 #Total feed rate to stage 1 in kg-mol/h
2	# Feed enthalpy calculation
3	f(TF) = 0.5 * k1F + 0.5 * k2F - 1 #Bubble point calculation of feed stream
4	TF(0) = 100 #Initial estimate of bubble point temperature for feed stream
5	k1F = 10 ^ (6.87776 - 1171.53 / (224.366 + TF)) / 760 #k factor calculation for component 1
6	k2F = 10 ^ (6.92374 - 1355.126 / (209.517 + TF)) / 760 #k factor calculation for component 2
7	hIF = F1 * (0.5 * 51.72 * TF + 0.5 * 66.07 * TF) #Enthalpy calculation at bubble point
8	# Stage 1 calculations
9	T1 = 100 #Temperature of stage 1
10	f(alpha1) = x11 * (1 - k11) + x21 * (1 - k21) #Calculation of alpha1 - fraction evaporated in stage 1
11	alpha1(0) = 0.5 #Initial estimate for alpha1
12	x11 = 0.5 / (1 + alpha1 * (k11 - 1)) #Mole fraction of component 1 in liquid leaving stage 1
13	x21 = 0.5 / (1 + alpha1 * (k21 - 1)) #Mole fraction of component 2 lin liquid leaving stage 1
14	k11 = 10 ^ (6.87776 - 1171.53 / (224.366 + T1)) / 760 #k factor calculation for component 1
15	k21 = 10 ^ (6.92374 - 1355.126 / (209.517 + T1)) / 760 #k factor calculation for component 1
16	L1 = (1 - alpha1) * F1 #Liquid flow rate exiting stage 1
17	V1 = alpha1 * F1 #Vapor flow rate exiting stage 1
18	y21 = k21 * x21 #Mole fraction of component 1 in vapor leaving stage 1
19	y11 = k11 * x11 #Mole fraction of component 2 in vapor leaving stage 1
20	HV1 = V1 * (y11 * (7678 + T1 * (31.83 + 0.0903 * T1 / 2)) + y21 * (10444.7 + T1 * (41.836 + 0.1218 * T1 / 2))) #Enthalpy of vapor leaving stage 1
21	hl1 = L1 * (x11 * 51.72 * T1 + x21 * 66.07 * T1) #Enthalpy of liquid leaving stage 1
22	Q1 = HV1 + hl1 - hIF #Energy balance on stage 1
23	# Cooler I heat duty calculations
24	f(TI) = y11 * k1I + y21 * k2I - 1 #Bubble point calculation for condenser I
25	TI(0) = 100 #Initial estimate of bubble point calculation for condenser I
26	k1I = 10 ^ (6.87776 - 1171.53 / (224.366 + TI)) / 760 #k factor calculation for component 1
27	k2I = 10 ^ (6.92374 - 1355.126 / (209.517 + TI)) / 760 #k factor calculation for component 2
28	hlI = V1 * (y11 * 51.72 * TI + y21 * 66.07 * TI) #Enthalpy of liquid leaving condenser I
29	QI = hlI - HV1 #Energy balance on condenser I

of the vapor (HV1) and liquid (hl1) streams are calculated in lines 20 and 21. An energy balance on the first stage flash yields the heat added to this drum (Q1, see line 22). The exit temperature from condenser I (TI, bubble point temperature of liquid of the same composition as the stage 1 vapor), and the rate of heat removal from this condenser (QI) are calculated in lines 24 through 29.

A summary of the solution to the set of equations in Table 12–2 is shown in Figure 12–2. This includes all the variable values associated with the feed (TF), the first stage (V1, y11, L1, x11, and Q1), and the first condenser (TI and QI).

Solution of the complete process requires the addition of the flash calculations for stages 2 and 3. This can be accomplished by repeating the equations of lines 9 through 22 in Table 12–2 using the appropriate variable names. The equations for condenser II can include a modified form of the equations of lines 24 through 29 in Table 12–2. The solution of this extended set of equations provides the values for the heat duties given in Table 12–3.

Variable	Value	f(x)	Initial Guess
1 alpha1	0.5793148	-1.11E-16	0.5
2 TF	86.78427	8.127E-14	100.
3 TI	79.62836	1.98E-10	100.

Variable	Value
1 F1	100.
2 hl1	2.614E+05
3 hlF	5.111E+05
4 hlI	2.608E+05
5 HV1	7.317E+05
6 k11	2.427689
7 k1F	1.705251
8 k1I	1.390458
9 k21	0.4620944
10 k2F	0.2947494

Variable	Value
11 k2I	0.2271257
12 L1	42.06852
13 Q1	4.82E+05
14 QI	-4.709E+05
15 T1	100.
16 V1	57.93148
17 x11	0.2736605
18 x21	0.7263395
19 y11	0.6643626
20 y21	0.3356374

Figure 12–2 Numerical Results Obtained by Solution of the Equation Set Shown in Table 12–2 (File **P12-01A.POL**)

Table 12–3 Heat Duties of the Heaters and the Coolers

Unit	Variable	Heat Duty (kcal/h)
Stage 1	Q1	4.82E+05
Stage 2	Q2	3.321E+05
Stage 3	Q3	2.349E+05
Cooler I	QI	-4.709E+05
Cooler II	QII	-3.282E+05

The POLYMATH problem solution file is found in directory Chapter 12 with file named **P12-01A1.POL**.

(b) Since the heat duties are known, the dew point calculations for the two condensers at the exits from stages 1 and 2 are not needed. The temperatures of the three stages can be determined from adiabatic energy balances. Thus the equation set of Table 12–2 must be modified.

The explicit energy balance on the first stage and condenser (shown in line 22 of Table 12–2) can be replaced by the constant value of Q1 given in Table 12–3 and an implicit energy balance is used to calculate the temperature T1.

$$f(T1) = hlF + Q1 - (HV1 + hl1) \tag{12-1}$$

The rate of heat removal by condenser I (QI = -4.709E+05) can be directly used instead of the bubble point temperature calculation and energy balance on the condenser (shown in lines 24 through 29 of Table 12–2). Similar changes need be carried out on stages 2 and 3 and condenser 2. The exit temperature calculations for the condensers (not required for this problem) can be carried out separately by determining first the phase condition of the exit stream and then determining the temperature from enthalpy balances on the condensers.

12.2 NON-IDEAL VAPOR-LIQUID AND LIQUID-LIQUID EQUILIBRIUM

12.2.1 Concepts Demonstrated

Calculation of bubble point and dew point temperatures. Isothermal flash calculations involving two- and three-phase regions.

12.2.2 Numerical Methods Utilized

Solution of systems of nonlinear algebraic equations and selection of initial estimates for system variables.

12.2.3 Problem Statement

Isobutanol (20 mol%) and water (80 mol%) is a mixture for which two liquid phases are known to exist at the bubble point temperature and at atmospheric pressure. A simple flash separator is to be designed.
(a) Calculate the mixture bubble point temperature (T_b) and dew point temperature (T_d) at atmospheric pressure $(P = 760$ mm Hg$)$.
(b) Calculate the fraction evaporated at various temperatures between the bubble point and the dew point temperatures (at atmospheric pressure) and plot the boiling temperature versus the fraction evaporated (α) on a mole basis (separation curve).

Additional Information and Data

The vapor liquid equilibrium ratios can be calculated assuming that the modified Raoult's law apply, thus $k_{ij} = \gamma_{ij} P_i / P$ where γ_{ij} is the activity coefficient of component i in liquid phase number j, P_i is the vapor pressure of component i, and P is the total pressure. For vapor pressure and activity coefficient calculations, the correlation equations given in Table 12–4 may be used.

Table 12–4 Vapor Pressure and Activity Coefficient Correlations for the Isobutanol-Water mixture (from Henley and Rosen[2])

	Isobutanol	Water
Vapor Pressure (mmHg)[a]	$\log(P1) = (7.62231 - 1417.9 / (191.15 + T))$	$\log(P2) = (8.10765 - 1750.29 / (235 + T))$
Activity Coefficient[b]	$\log\gamma_{1,j} = \dfrac{1.7 x_{2,j}^2}{\left(\dfrac{1.7}{0.7} x_{1,j} + x_{2,j}\right)^2}$	$\log\gamma_{2,j} = \dfrac{0.7 x_{1,j}^2}{\left(x_{1,j} + \dfrac{0.7}{1.7} x_{2,j}\right)^2}$

[a]T—Temperature in $^\circ C$
[b]$x_{1,j}$ is mole fraction of the 1st component in liquid phase number j

12.2.4 Solution

The behavior of this system will be explained with reference to the separation curve shown in Figure 12–3. As a mixture of isobutanol and water is heated at a constant pressure of 760 mmHg, two liquid phases coexist. One of these liquid phases is composed principally of water and the other one is composed principally of isobutanol. When the bubble point temperature is reached at 88.54 °C, the two liquid phases are in equilibrium with the vapor phase. When the temperature is raised above the bubble point, only a single liquid phase and the vapor phase are present until the dew point temperature is reached. Note that these phenomena are a consequence of the phase rule.

Separate equations need to be used for modeling the three-phase bubble point calculation, the two-phase dew point, and the flash calculations. The various models will be explained with reference to the three-phase flash unit depicted schematically in Figure 12–4. The variables shown in this figure are: F – feed flow rate (mole/hr), z_i – mole fraction of the i-th component in the feed, V – vapor flow rate (mole/hr), y_i – mole fraction of the i-th component in the vapor, L_1, L_2 – liquid flow rates (mole/hr) in the various liquid phases and

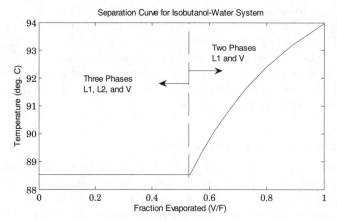

Figure 12–3 Separation Curve for the Isobutanol and Water System at 760 mmHg

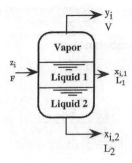

Figure 12–4 Three-Phase Flash Unit

$x_{i,j}$ – mole fraction of the i-th component in the j-th liquid phase. The temperature and pressure in the flash drum are constant at T and P, respectively. The total number of compounds in the system is n_c. The following model equations are applicable to the three-phase flash.

Overall Material Balance $\qquad F = L_1 + L_2 + V \qquad$ (one equation)**(12-2)**

Material Balance for Component i $\; F z_i = L_1 x_{i,1} + L_2 x_{i,2} + V y_i \;$ (n_c equations) **(12-3)**

Phase Equilibrium $\qquad\qquad y_i = k_{i,1} x_{i,1} \qquad\qquad$ (n_c equations) **(12-4)**

$\qquad\qquad\qquad\qquad\qquad\quad y_i = k_{i,2} x_{i,2} \qquad\qquad$ (n_c equations) **(12-5)**

Sum of Vapor Mole Fractions $\quad \displaystyle\sum_{i=1}^{n_c} y_i = 1 \qquad\qquad$ (one equation)**(12-6)**

Sum of Liquid Mole Fractions $\quad \displaystyle\sum_{i=1}^{n_c} x_{i,1} = 1 \qquad\qquad$ (one equation)**(12-7)**

Sum of Liquid Mole Fractions $\quad \displaystyle\sum_{i=1}^{n_c} x_{i,2} = 1 \qquad\qquad$ (one equation)**(12-8)**

The total number of unknowns is $3n_c + 3$ ($3n_c$ unknowns $x_{i,1}, x_{i,2}$, and y_i and three single variables V, L_1, and L_2). The total number of equations is $3n_c + 4$, but Equation (12-2) is linearly dependent on Equations (12-3) to (12-8). Thus, the total number of independent equations is $3n_c + 3$. The number of independent equations is equal to the number of unknowns, thus there is a unique solution to the system of equations.

For effective problem solution, the equations should be rewritten and organized differently for the two- and three-phase regions, namely for calculation of the bubble point temperature, dew point temperature, and the two-phase flash.

(a1) Calculation of the Bubble Point Temperature and the Composition of the Two Liquid Phases

At the bubble point $V = 0$. Dividing Equation (12-3) by F yields

$$z_i = x_{i,1}\left[\beta + (1-\beta)\frac{k_{i,1}}{k_{i,2}}\right] \qquad (12\text{-}9)$$

where $\beta = L_1/F$. Equations (12-4) and (12-5) are combined to yield the composition in the 2nd liquid phase.

$$x_{i,2} = (k_{i,1} x_{i,1})/k_{i,2} \qquad (12\text{-}10)$$

The mole fraction summation equations are rewritten

$$\sum_i x_{i,1} - \sum_i y_i = 0 \tag{12-11}$$

$$\sum_i x_{i,1} - \sum_i x_{i,2} = 0 \tag{12-12}$$

In this case there are $2n_c + 2$ unknowns: T, β, $x_{1,1}$, $x_{2,1}$, $x_{1,2}$, and $x_{2,2}$. Because of the dependency of the activity coefficients on the liquid phase compositions, Equations (12-9) to (12-12) (a total of $2n_c + 2$ equations) must be solved as implicit equations.

The POLYMATH implementation of this system of equations is shown in Table 12–5. The implicit equations in lines 2 and 3 are based on Equations (12-9), the equations in lines 4 and 5 are based on Equations (12-10), the equation in line 6 is based on Equation (12-11), and that in line 7 is based on Equation (12-12). Note that POLYMATH requires assigning one of the unknowns for each implicit equation in order to keep track of the defined and undefined variables.

Table 12–5 POLYMATH Program for Three-phase Bubble Point Calculations (File **P12-02A1.POL**)

Line	Equation
1	#Problem 12.2 - Three-phase Bubble Point
2	f(x11) = x11 - 0.2 / (beta + (1 - beta) * k11 / k12) # Mole fraction of comp. 1 in liquid phase 1
3	f(x21) = x21 - 0.8 / (beta + (1 - beta) * k21 / k22) # Mole fraction of comp. 2 in liquid phase 1
4	f(x12) = x12 - x11 * k11 / k12 # Mole fraction of comp. 1 in liquid phase 2
5	f(x22) = x22 - x21 * k21 / k22 # Mole fraction of comp. 2 in liquid phase 2
6	f(T) = x11 * (1 - k11) + x21 * (1 - k21) # Bubble point temperature (deg. C)
7	f(beta) = (x11 - x12) + (x21 - x22) # Liquid phase split ratio [L1/(L1+L2)]
8	# The explicit equations
9	p1 = 10 ^ (7.62231 - 1417.9 / (191.15 + T)) # Vapor pressure of 1st component (mmHg)
10	p2 = 10 ^ (8.10765 - 1750.29 / (235 + T)) # Vapor pressure of 2nd component (mmHg)
11	A = 1.7 # Van Laar equations constant A
12	B = 0.7 # Van Laar equations constant B
13	gamma11 = 10 ^ (A * x21 * x21 / ((A * x11 / B + x21) ^ 2)) # Activity coefficient of comp. 1 in liquid phase 1
14	gamma21 = 10 ^ (B * x11 * x11 / ((x11 + B * x21 / A) ^ 2)) # Activity coefficient of of comp. 2 in liquid phase 1
15	gamma12 = 10 ^ (A * x22 * x22 / ((A * x12 / B + x22) ^ 2)) # Activity coefficient of of comp. 1 in liquid phase 2
16	gamma22 = 10 ^ (B * x12 * x12 / ((x12 + B * x22 / A) ^ 2)) # Activity coefficient of of comp. 2 in liquid phase 2
17	k11 = gamma11 * p1 / 760 # Vapor liquid equilibrium ratio of comp. 1 in liquid phase 1
18	k21 = gamma21 * p2 / 760 # Vapor liquid equilibrium ratio of comp. 2 in liquid phase 1
19	k12 = gamma12 * p1 / 760 # Vapor liquid equilibrium ratio of comp. 1 in liquid phase 2
20	k22 = gamma22 * p2 / 760 # Vapor liquid equilibrium ratio of comp. 2 in liquid phase 2
21	y1 = k11 * x11 # Mole fraction of comp. 1 in the vapor phase
22	y2 = k21 * x21 # Mole fraction of comp. 2 in the vapor phase
23	# Initial estimates for nonlinear equations variables
24	x11(0) = 0

Line	Equation
25	x21(0) = 1
26	x12(0) = 1
27	x22(0) = 0
28	T(0) = 100
29	beta(0) = 0.8

The explicit equations include calculations of vapor pressures of the pure compounds, activity coefficients, vapor liquid equilibrium ratios (using the equations shown in Table 12–4), and vapor composition (using Equation (12-4)).

The equations of non-ideal phase equilibrium are very non-linear, and there are often difficulties in achieving a numerical solution. It is very important to provide reasonable initial estimates (in this case the assumption is that each one of the two liquid phases is contains only one of the compounds). If the implicit equations include terms with some unknowns in the denominator, the problem solution may require multiplication of the nonlinear expressions by the denominator term in order to avoid possible division by zero.

The problem solution obtained with POLYMATH is shown in Figure 12–5. The first liquid phase contains mainly water and very small amount of isobutanol. In the second liquid phase, the two compounds are divided in the approximate ratios of 2/3 (isobutanol) and 1/3 (water). The concentration of the isobutanol in the vapor phase is higher than in the feed, as can be expected. Note the extremely large activity coefficient value of the isobutanol in the 1st liquid phase (γ_{11}) where its mole fraction is very small.

(a2) Calculation of the Dew Point Temperature and the Composition of the Liquid Phase

At the dew point $L_1 = L_2 = 0$, $V = F$, $y_i = z_i$, and $x_i \equiv x_{i,1}$. There are $n_c + 1$ unknowns: T, x_1, and x_2. Equations (12-4) and (12-7) (n_c+1 equations) must be

Calculated values of NLE variables

	Variable	Value	f(x)	Initial Guess
1	beta	0.7329991	0	0.8
2	T	88.53783	2.418E-13	100.
3	x11	0.0226982	-3.469E-15	0
4	x12	0.6867476	4.681E-13	1.
5	x21	0.9773018	1.221E-15	1.
6	x22	0.3132524	-2.387E-15	0

	Variable	Value
1	A	1.7
2	B	0.7
3	gamma11	33.36649
4	gamma12	1.102821
5	gamma21	1.004606
6	gamma22	3.134222
7	k11	15.67568
8	k12	0.5181085
9	k21	0.6591518
10	k22	2.056457
11	p1	357.0503
12	p2	498.6588
13	y1	0.3558097
14	y2	0.6441903

Figure 12–5 POLYMATH Results for the Three Phase Bubble-Point Calculations (File **P12-02A1.POL**)

solved implicitly to obtain the dew point temperature and the liquid phase composition.

The POLYMATH implementation of the model is shown in Table 12–6 and the results are presented in Figure 12–6. These results indicate that at the dew point there is very small amount of isobutanol in the liquid phase but still the dew point temperature (93.97°C) is considerably lower than the boiling point temperature of the water (100°C) at atmospheric pressure. This is due to the extremely high activity coefficient value for isobutanol.

 The POLYMATH problem solution files for part (a) are found in directory Chapter 12 with files named **P12-02A1.POL** and **P12-02A1.POL**.

www

Table 12–6 POLYMATH Equations for Dew-Point Temperature Calculations (File **P12-02A2.POL**)

Line	Equation
1	# Problem 12.2 - Dew Point
2	f(x1) = x1 - y1 / k1 # Mole fraction of comp. 1 in the liquid phase
3	f(x2) = x2 - y2 / k2 # Mole fraction of comp. 2 in the liquid phase
4	f(T) = x1 + x2 - 1 # Dew point temp. Deg C.
5	# The explicit equations
6	p1 = 10 ^ (7.62231 - 1417.9 / (191.15 + T)) # Vapor pressure of 1st component (mmHg)
7	p2 = 10 ^ (8.10765 - 1750.29 / (235 + T)) # Vapor pressure of 2nd component (mmHg)
8	A = 1.7 # Van Laar equations constant A
9	B = 0.7 # Van Laar equations constant B
10	gamma1 = 10 ^ (A * x2 * x2 / ((A * x1 / B + x2) ^ 2)) # Activity coefficient of comp. 1 in the liquid phase
11	gamma2 = 10 ^ (B * x1 * x1 / ((x1 + B * x2 / A) ^ 2)) # Activity coefficient of comp. 2 in the liquid phase
12	k1 = gamma1 * p1 / 760 # Vapor liquid equlibrium ratio of comp. 1
13	k2 = gamma2 * p2 / 760 # Vapor liquid equlibrium ratio of comp. 2
14	y1 = 0.2 # Mole fraction of comp. 1 in the vapor phase
15	y2 = 0.8 # Mole fraction of comp. 2 in the vapor phase
16	# Initial Guess for nonlinear equations variables
17	x1(0) = 0.2
18	x2(0) = 0.8
19	T(0) = 100

Calculated values of NLE variables

	Variable	Value	f(x)	Initial Guess
1	T	93.96707	-1.904E-13	100.
2	x1	0.0078755	4.767E-13	0.2
3	x2	0.9921245	-3.476E-10	0.8

	Variable	Value
1	A	1.7
2	B	0.7
3	gamma1	43.28155
4	gamma2	1.000577
5	k1	25.39535
6	k2	0.8063504
7	p1	445.9282
8	p2	612.4731
9	y1	0.2
10	y2	0.8

Figure 12–6 POLYMATH Results for the Dew-Point Calculations (File **P12-02A2.POL**)

(b) Isothermal Flash Calculations in the Two Phase Region

The two-phase region is between the temperatures T_d = 93.97 °C and T_b = 88.54 °C. To calculate the fraction evaporated and the liquid and vapor compositions at various temperature values in this region, Equation (12-3) should be rewritten

$$x_i = \frac{(1 - z_i)}{1 + \alpha(k_i - 1)} \qquad \text{(12-13)}$$

where α is the fraction evaporated ($\alpha = V/F$).

For two-phase flash calculations, the POLYMATH input shown in Table 12–7 can be used except that the flash temperature should be specified and the implicit equations should be replaced by Equations (12-13) and (12-11). The resultant POLYMATH set of equations is shown in Figure 12–7.

Table 12–7 POLYMATH Equations for Two-Phase Flash Calculations (File **P12-02B1.POL**)

Line	Equation
1	# Example 12.2 - Two-phase Flash
2	f(x1) = x1 - z1 / (1 + alpha * (k1 - 1)) # Mole fraction of comp. 1 in the liquid phase
3	f(x2) = x2 - z2 / (1 + alpha * (k2 - 1)) # Mole fraction of comp. 2 in the liquid phase
4	f(alpha) = x1 + x2 - (y1 + y2) # Fraction evaporated
5	# The explicit equations
6	T = 88.538 # Temperature deg. C
7	z1 = 0.2 # Mole fraction of comp. 1 in the feed
8	z2 = 0.8 # Mole fraction of comp. 2 in the feed
9	p1 = 10 ^ (7.62231 - 1417.9 / (191.15 + T)) # Vapor pressure of 1st component (mmHg)
10	p2 = 10 ^ (8.10765 - 1750.29 / (235 + T)) # Vapor pressure of 2nd component (mmHg)
11	A = 1.7 # Van Laar equations constant A
12	B = 0.7 # Van Laar equations constant B
13	gamma1 = 10 ^ (A * x2 * x2 / ((A * x1 / B + x2) ^ 2)) # Activity coefficient of comp. 1 in the liquid phase
14	gamma2 = 10 ^ (B * x1 * x1 / ((x1 + B * x2 / A) ^ 2)) # Activity coefficient of comp. 2 in the liquid phase
15	k1 = gamma1 * p1 / 760 # Vapor liquid equlibrium ratio of comp. 1
16	k2 = gamma2 * p2 / 760 # Vapor liquid equlibrium ratio of comp. 2
17	y1 = k1 * x1 # Mole fraction of comp. 1 in the vapor phase
18	y2 = k2 * x2 # Mole fraction of comp. 2 in the vapor phase
19	# Initial Guess for nonlinear equations variables
20	x1(0) = 0
21	x2(0) = 1
22	alpha(0) = 0.5

The results obtained at the bubble point temperature are shown in Figure 12–7. Fifty-three percent of the feed evaporates at the bubble point temperature. The gas phase at this point includes about 95% of the more volatile isobutanol.

Calculated values of NLE variables

	Variable	Value	f(x)	Initial Guess
1	alpha	0.5322678	2.365E-14	0.5
2	x1	0.0226975	-2.082E-16	0
3	x2	0.9773025	6.661E-16	1.

	Variable	Value
1	A	1.7
2	B	0.7
3	gamma1	33.3669
4	gamma2	1.004605
5	k1	15.67598
6	k2	0.659156
7	p1	357.0528
8	p2	498.6621
9	T	88.538
10	y1	0.3558052
11	y2	0.6441948
12	z1	0.2
13	z2	0.8

Figure 12–7 POLYMATH Results for the Two-Phase Flash Calculations at 88.538°C (File **P12-02B1.POL**)

At the bubble point temperature, both three-phase and two-phase solutions are valid. According to the phase rule, in a binary mixture, the temperature, and the composition of the different phases won't change in the three-phase region (for equilibrium conditions) as α, the amount evaporated, changes.

The plotting of the separation curve requires several reruns of the POLY-MATH solution at various temperature values between the dew point and the bubble point. A separate plot must be constructed using the POLYMATH *Regression and Data Analysis* program. An Excel or MATLAB solution can automate the solution and automate the plotting. MATLAB was used to plot the separation curve of Figure 12–4.

The POLYMATH problem solution files for part (b) are found in directory Chapter 12 with files named **P12-02B1.POL, P12-02B2.POL,** and **P12-02B3.PLG**.

12.3 CALCULATION OF WILSON EQUATION COEFFICIENTS FROM AZEOTROPIC DATA

12.3.1   Concepts Demonstrated

Calculation of the Wilson equation coefficients from binary azeotropic data. Calculation of bubble point temperature and vapor and liquid phase compositions of a non-ideal mixture.

12.3.2   Numerical Methods Utilized

Solution of systems of nonlinear algebraic equations.

12.3.3   Problem Statement

The Wilson[3] equations are used to correlate the activity coefficients of strongly non-ideal, but miscible systems. The Wilson equations for activity coefficients in binary system are

$$\ln\gamma_1 = -\ln(\theta_1) + \frac{x_2(G_{12}\theta_2 - G_{21}\theta_1)}{\theta_1\theta_2}$$

$$\ln\gamma_2 = -\ln(\theta_2) - \frac{x_1(G_{12}\theta_2 - G_{21}\theta_1)}{\theta_1\theta_2}$$

(12-14)

where $\theta_1 = x_1 + x_2 G_{12}$ and $\theta_2 = x_2 + x_1 G_{21}$. G_{12} and G_{21} are adjustable parameters which must be determined from experimental data, and x_1 and x_2 are mole fractions of components 1 and 2 respectively.

Vapor liquid equilibrium ratios can be calculated by assuming that the modified Raoult's law apply; thus $k_i = \gamma_i P_i / P$, where γ_i is the activity coefficient of component i in the liquid phase, P_i is the vapor pressure of component i, and P is the total pressure.

Vapor pressures at various temperatures can be obtained from the Antoine equation

$$\log P_i = A_i - \frac{B_i}{C_i + T}$$

(12-15)

where T is the temperature in °C, P_i is the vapor pressure of component i in mmHg, and A_i, B_i and C_i are the Antoine equation constants given in Table 12–9.

Azeotropes represent a condition where vapor and liquid phases have identical composition, and the mixture has a constant boiling temperature at a given pressure. At the azeotrope, all the vapor liquid equilibrium coefficients are given by $k_i = \gamma_i P_i / P = 1$. Thus at the temperature of the azeotrope, the activity coefficients are given by $\gamma_i = P / P_i$. This allows the coefficients of the Wilson equations to be calculated from one set of azeotropic data.

(a)   Select one of the binary systems in Table 12–8 and calculate the Wilson equation coefficients for that system using the vapor pressures calculated by the Antoine equation with the constants given in Table 12–9.

(b)   Use the Wilson equation constants determined in part (a) to calculate the bubble point temperature at $x_1=0$, $x_1=1$ and at the azeotropic point. Compare the azeotropic point (temperature and vapor composition) with the data given in Table 12–8. Also compare the predicted bubble points temperatures (boiling points) of the pure components with the known boiling point temperatures.

(c)   Calculate and plot the bubble point temperature curve for the selected system by using the Wilson equation coefficients determined in part (a).

Table 12–8 Binary Azeotropes Containing Ethyl Alcohol[4]

Ethyl Alcohol	B.P @ 760 mmHg		% by Weight	
(B.P. 78.3°)	Other Component	Azeotrope	Alcohol	Other Component
Methyl Acetate	57.0	56.9	3	97
Methyl Propionate	79.7	72.0	33	67
Ethyl Acetate	77.1	71.8	30.8	69.2
Ethyl Propionate	99.2	78	75	25
n-Propyl Formate	80.8	71.8	38	62
n-Propyl Acetate	101.6	78.2	85	15
iso-Propyl Acetate	91.0	76.8	57	43
Benzene	80.2	68.2	32.4	67.6
Toluene	110.8	76.7	68	32
n-Pentane	36.2	34.3	5	95
n-Hexane	68.9	58.7	21	79
n-Heptane	98.5	70.9	49	51
n-Octane	125.6	77.0	78	22
Carbon Tetrachloride	76.8	65.1	15.8	84.2

Table 12–9 Antoine Equation Constants for Vapor Pressure of Various Hydrocarbons[4]

Name	Range, $°C$	A	B	C
Ethyl Alcohol (Ethanol)		8.04494	1554.3	222.65
Methyl Acetate		7.20211	1232.83	228.0

Name	Range, $°C$	A	B	C
Methyl Propionate	-2.5 to 257	7.12841	1257.14	216.4
Ethyl Propionate	-20 to 160	7.07293	1298.30	210.7
n-Propyl Acetate	0 to 170	7.06665	1304.10	210.0
n-Propyl Formate		7.04006	1235.00	216.1
Benzene		6.90565	1211.033	220.79
Toluene		6.95464	1344.80	219.482
n-Pentane		6.85221	1064.63	232.00
n-Hexane		6.87776	1171.530	1171.530
n-Heptane		6.90240	1268.115	216.900
n-Octane		6.92374	1355.126	209.517
Carbon Tetrachloride		6.9339	1242.43	230.0

12.3.4 Solution: System Ethanol (1) and *n*-Hexane (2)

(a) The POLYMATH program for determination of the Wilson equations coefficients for the binary system, ethanol (No. 1, MW = 46.07) and *n*-hexane (No. 2, MW = 86.18), is given in Table 12–10. The POLYMATH *Simultaneous Algebraic Equation Solver* is used to calculate the two Wilson coefficients. The temperature of the azeotrope from Table 12–8 is 58.7°C. The corresponding weight percent of ethanol is 21% and that of *n*-Hexane is 79%.

Table 12–10 POLYMATH Model for the Calculation of the Wilson Equation Coefficients

Line	Equation
1	# Problem 12-3 - Wilson Coefficients from Ethyl Alcohol - n-Hexane Binary Azeotrope
2	T = 58.7
3	pw1 = 21
4	x1 = (pw1/46.07) / (pw1 / 46.07 + (100 - pw1) / 86.18)
5	x2 = 1 - x1
6	P1 = 10 ^ (8.04494 - 1554.3 / (222.65 + T))
7	P2 = 10 ^ (6.87776 - 1171.53 / (224.366 + T))
8	gamma2 = 760 / P2
9	gamma1 = 760 / P1
10	f(G21) = t1 * t2 * (ln(gamma2) + ln(t2)) + x1 * (G12 * t2 - G21 * t1)
11	G21(0) = 0.1
12	f(G12) = t1 * t2 * (ln(gamma1) + ln(t1)) - x2 * (G12 * t2 - G21 * t1)
13	G12(0) = 0.1
14	t1 = x1 + x2 * G12
15	t2 = x2 + x1 * G21

The calculation of the Wilson coefficients requires mole fractions, so the weight percents must be converted to mole fractions (see lines 3 through 5 in Table 12–10). The Antoine equation is then used for calculating the vapor pressures of the two compounds at the azeotrope's temperature (in lines 6 and 7). The relationship $\gamma_i = P/P_i$ is used to calculate the activity coefficients (lines 8 and 9). Finally, the mole fraction and the activity coefficient values are introduced into the nonlinear Equations (12-14) in lines 10 through 15. Note that Equations (12-14) have been made less sensitive to the initial estimates. Division by the unknowns G_{12} and G_{21} has been eliminated by multiplying both sides of these equations by $\theta_1\theta_2$ to avoid possible division by zero errors.

These equations can be solved simultaneously for the constants G_{12} and G_{21}. This is the form in which the implicit equations are presented (in lines 10 and 12). The resulting POLYMATH solution for this case is presented in Figure 12–8.

Calculated values of NLE variables

	Variable	Value	f(x)	Initial Guess
1	G12	0.0785889	9.671E-11	0.1
2	G21	0.3017536	2.93E-11	0.1

	Variable	Value
1	gamma1	2.2925
2	gamma2	1.386023
3	P1	331.5158
4	P2	548.3315
5	pw1	21.
6	T	58.7
7	t1	0.3846004
8	t2	0.7681041
9	x1	0.3321118
10	x2	0.6678882

Figure 12–8 POLYMATH Results for the Wilson Equation Coefficients

www
The POLYMATH problem solution files are found in directory Chapter 12 with files named **P12-03A.POL**

(b) For miscible systems, such as the systems shown in Table 12–8, that form one liquid phase at the bubble point temperature (T_b), a nonlinear equation can be written for the sum of the gas phase mole fractions equaling unity as (after introducing $y_i = k_ix_i$)

$$f(T_b) = 1 - \sum_{i=1}^{n_c} k_ix_i \tag{12-16}$$

For a particular liquid composition this equation can be solved for the bubble point temperature.

The POLYMATH program for ethanol (1) and n-hexane (2) system at the composition of the azeotrope where x1 = 0.332 (see Table 12–8) as they are introduced into the POLYMATH *Simultaneous Algebraic Equation Solver* are shown in Table 12–11.

Table 12–11 POLYMATH Model for the Calculation of the Bubble Point Temperature and Vapor and Liquid Compositions

Line	Equation
1	# Problem 12-3 - POLYMATH Program for the Calculation of the Bubble Point Temperature and Vapor and Liquid Compositions
2	x1 = 0.332
3	x2 = 1 - x1
4	G12 = 0.0785889
5	G21 = 0.3017536
6	t1 = x1 + x2 * G12
7	t2 = x2 + x1 * G21
8	gamma2 = exp(-ln(t2) - (x1 * (G12 * t2 - G21 * t1)) / (t1 * t2))
9	gamma1 = exp(-ln(t1) + (x2 * (G12 * t2 - G21 * t1)) / (t1 * t2))
10	f(Tb) = 1 - k1 * x1 - k2 * x2
11	Tb(min) = 50
12	Tb(max) = 70
13	P1 = 10 ^ (8.04494 - 1554.3 / (222.65 + Tb))
14	P2 = 10 ^ (6.87776 - 1171.53 / (224.366 + Tb))
15	k1 = gamma1 * P1 / 760
16	k2 = gamma2 * P2 / 760
17	y1 = k1 * x1
18	y2 = k2 * x2

The Wilson equations parameters are specified in lines 4 and 5 and the Wilson equations are used to calculate the activity coefficients for the specified liquid composition in line 4 through 9. The implicit equation for bubble point calculation is shown in line 10 while the interval where the root can be found is specified in lines 11 and 12. Note that in a mixture which creates an azeotrope the boiling point can be lower than the minimum boiling points of the pure compounds or higher than the maximum boiling point of the pure compounds. This should be taken into consideration when defining the interval where the solution is expected to be located. The vapor liquid equilibrium coefficients and the mole fractions of the two compounds in the vapor phase are calculated in lines 13 through 18.

The POLYMATH solution, presented in Figure 12–9, shows that the calculated bubble point temperature at the azeotrope as 58.7°C which is the same as the azeotrope temperature shown in Table 12–8. Furthermore, the values of k_1 and k_2 are very close to 1 (unity), and the vapor composition is nearly equal to the liquid composition as y1 = 0.3321 for x1 = 0.332. These comparisons verify the results as these are additional characteristic properties of an azeotrope.

For $x_1 = 0$ (pure *n*-hexane), the calculated bubble point temperature is 68.74°C and for $x_2 = 0$ (pure ethanol), the temperature is 78.33°C. These results both very close the pure component boiling points of 68.9°C and 78.3°C respectively as given in Table 12–8.

The POLYMATH problem solution files are found in directory Chapter 12 with files named **P12-03B.POL**

Calculated values of NLE variables

	Variable	Value	f(x)	Initial Guess
1	Tb	58.7	-2.676E-14	60. (50. < Tb < 70.)

	Variable	Value
1	G12	0.0785889
2	G21	0.3017536
3	gamma1	2.293211
4	gamma2	1.385809
5	k1	1.00031
6	k2	0.9998459
7	P1	331.5158
8	P2	548.3316
9	t1	0.3844974
10	t2	0.7681822
11	x1	0.332
12	x2	0.668
13	y1	0.3321029
14	y2	0.6678971

Figure 12–9 POLYMATH Solution for the Calculation of the Bubble Point
Temperature and Vapor and Liquid Compositions

(c) There are several ways to calculate and plot the bubble point as a function of the composition for the ethanol–n-hexane system. Using POLYMATH, one can solve for the bubble point at various individual values and then make a graph with the POLYMATH *Polynomial, Multiple Linear, and Nonlinear Regression Program*. Other options involve solutions in Excel or MATLAB. The bubble point temperature curve generated by a MATLAB program is shown in Figure 12–10. Note that there is indeed a minimum temperature azeotrope for this mixture, at temperature of 58.7°C at x1 = 0.332 as also given in Table 12–8.

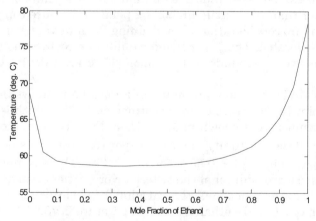

Figure 12–10 Bubble Point Temperature Curve for the Ethanol–n-Hexane System

The MATLAB problem solution files are found in directory Chapter 12 with files named **Prob_12_3.m**, **bubbleW.m** and **Wilsonf.m**.

12.4 VAN LAAR EQUATIONS COEFFICIENTS FROM AZEOTROPIC DATA

12.4.1 Concepts Demonstrated

Estimation of the coefficients of the Van Laar equations from binary azeotropic data. Calculation of bubble point temperatures and corresponding vapor and liquid phase compositions for a non-ideal mixture.

12.4.2 Numerical Methods Utilized

Solution of systems of nonlinear algebraic equations.

12.4.3 Problem Statement

The Van Laar equations for a binary system are often utilized to correlate activity coefficients

$$\log\gamma_1 = \frac{Ax_2^2}{\left(\frac{A}{B}x_1 + x_2\right)^2}$$

$$\log\gamma_2 = \frac{Bx_1^2}{\left(x_1 + \frac{B}{A}x_2\right)^2}$$

(12-17)

where A and B are adjustable parameters that must be determined from experimental data and x_1 and x_2 are mole fractions of components 1 and 2 respectively.

(a) Select one of the binary systems in Table 12–8 and calculate the coefficients for the Van Laar equations for that system using the vapor pressures calculated by the Antoine equation with the constants given in Table 12–9.

(b) Use the Van Laar equation constants determined in part (a) to calculate the bubble point temperature at $x_1=0$, $x_1=1$ and at the azeotropic point. Compare the calculated azeotropic point (temperature and vapor composition) with the data given in Table 12–8. Also compare the predicted bubble point temperatures (boiling points) of the pure components with the known boiling point temperatures.

(c) Calculate and plot the bubble point temperature curve for the selected system by using the Van Laar equation coefficients determined in part

12.4.4 Solution (Suggestion)

The same procedure that presented in Problem 12.3 can be followed except that the Wilson equations are replaced by the Van Laar equations.

12.5 NON-IDEAL VLE FROM AZEOTROPIC DATA USING THE VAN LAAR EQUATIONS

12.5.1 Concepts Demonstrated

Estimation of the Van Laar equation coefficients from binary azeotropic data. Calculation of bubble-point temperatures and corresponding vapor and liquid phase compositions for a non-ideal mixture. Comparison of calculated and experimental phase equilibrium data.

12.5.2 Numerical Methods Utilized

Solution of systems of nonlinear algebraic equations.

12.5.3 Problem Statement

Experimental vapor-liquid equilibrium data[5] at one atmosphere for the system chloroform (1)–methanol (2) are shown in Table 12–12.
(a) Use the information given in Table 12–12 regarding the azeotropic point of this system and the Antoine equation coefficients shown in Table 12–13 to calculate the Van Laar equations coefficients for the chloroform–methanol system.
(b) Calculate the bubble point temperatures (T_b) and the vapor compositions (y_1) for the values of x_1 shown in Table 12–12.
(c) Compare the experimental and calculated T_b and y_1 values and comment on the precision of the activity coefficient estimation procedure.

The vapor pressure in the pertinent temperature range can be calculated from the Antoine equation (12-15) using the coefficients found in Table 12–13.

Table 12–12 Vapor-Liquid Equilibrium Data, at 1 atm, for the System Chloroform (1) and Methanol (2).

x_1	y_1	$T_b(°C)$
0.0400	0.1020	63.0
0.0950	0.2150	60.9
0.1960	0.3780	57.8
0.2870	0.4720	55.9
0.4250	0.5640	54.3
0.6500	0.6500	53.5(azeotrope)
0.7970	0.7010	53.9
0.9040	0.7680	55.2
0.9700	0.87540	57.9

Table 12–13 Antoine Equation Coefficients for Chloroform and Methanol[4]

	A	B	C
Chloroform	6.90328	1163.03	227.4
Methanol	7.87863	1473.11	230.0

12.5.4 Solution Suggestions

(a) The same solution logic presented in Section 12.3.4(a) can be followed except that the Wilson equations are replaced by the Van Laar equations and there is no need to convert weight percent into mole fraction.

(b) The method for calculating the bubble point temperature is demonstrated in detail in Section 12.3.4(b). The Wilson equations must be replaced by the Van Laar equations.

(c) The experimental and calculated values from the respective equations can be exported either to Excel or MATLAB where the comparisons can be conveniently done. The comparison of experimental and calculated bubble point temperature values, as obtained by a MATLAB program is shown in Figure 12–11.

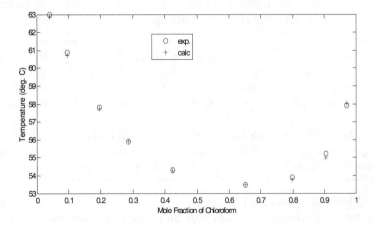

Figure 12–11 Bubble Point Temperature Curves, Experimental and Calculated, for the Chloroform–Methanol System

12.6 FENSKE-UNDERWOOD-GILLILAND CORRELATIONS FOR SEPARATION TOWERS

12.6.1   Concepts Demonstrated

Short-cut distillation calculations. The Fenske correlation to estimate minimum number of trays. The Underwood correlation to estimate the minimum reflux ratio. The Gilliland correlation for approximating the actual number of trays required to achieve a specified separation.

12.6.2   Numerical Methods Utilized

Solution of systems of nonlinear algebraic equations.

12.6.3   Problem Statement[*]

A distillation column is required for separating butanes from lighter compounds. The feed with the composition shown in Table 12–14 enters the distillation column as liquid at its bubble point. The pressure in the column is P = 7 bar. It is required that 95% of the i-butane in the feed be recovered in the bottoms, and the bottoms stream must contain no more than 0.1% propane. A sharp separation may be assumed for ethane and n-butane.

(a)   Calculate the minimum number of trays (N_{min}) required to achieve the desired separation at total reflux using the Fenske equation.
(b)   Estimate the minimum reflux (R_{min}) required to achieve the desired separation with infinite number of trays using the Underwood equations.
(c)   Approximate the number of theoretical stages required if the actual reflux ratio $R = 1.5 * R_{min}$ using the Gilliland correlation. Find the optimal feed stage location using the Kirkbride correlation.

Additional Information and Data

The vapor liquid equilibrium ratios can be calculated assuming ideal solution, thus $k_j = P_j/P$. Needed vapor pressures can be calculated from the Antoine equation given in Equation (12-15) with the coefficients shown in Table 12–14. Note that the coefficients shown in Table 12–14 are consistent with temperature units in K and the pressure units in bars.

12.6.4   Solution

The first step of the solution involves mole balances for calculating the distillate composition (x_{Dj}) and bottoms composition (x_{Bj}). The POLYMATH equations for

[*] Seader, J. D., and Henley, E. J., *Separation Process Principles*, 2nd Ed., Wiley: New York, 2006, pp. 344-360.

Table 12–14 Feed Composition and Antoine Equation Coefficients[6] for the Butane Separation Tower

No.	Component	Feed Composition (Mole Fractions)	A	B	C
1	ethane	0.15	3.93835	659.739	-16.719
2	propane (light key)	0.18	4.53678	1149.36	24.906
3	i-butane (heavy key)	0.18	4.3281	1132.108	0.918
4	n-butane	0.49	4.35576	1175.581	-2.071

carrying out the mole balance are shown in Table 12–15.

Table 12–15 POLYMATH Program for Distillation Tower Material Balance Equations

Line	Equation	Line	Equation
1	# Feed Composition	18	# Distillate composition
2	xF1 = 0.15	19	xD1 = D1 / tD
3	xF2 = 0.18	20	xD2 = D2 / tD
4	xF3 = 0.18	21	xD3 = D3 / tD
5	xF4 = 0.49	22	xD4 = D4 / tD
6	# Distillate flow rates	23	# Bottoms composition
7	D1 = xF1	24	xB1 = B1 / tB
8	D2 = xF2 - B2	25	xB2 = B2 / tB
9	D3 = 0.05 * xF3	26	xB3 = B3 / tB
10	D4 = 0	27	xB4 = B4 / tB
11	tD = D1 + D2 + D3 + D4		
12	# Bottoms flow rates		
13	B1 = 0		
14	B2 = 0.001 * (B3 + B4) / 0.999		
15	B3 = 0.95 * xF3		
16	B4 = xF4		
17	tB = B1 + B2 + B3 + B4		

Note that propane (component number 2) is assigned as the "light key" (LK) because it is the most volatile component that appears in the bottoms. i-Butane (component No. 3) is assigned as "heavy key" (HK) because it is the least volatile component that appears in the distillate.

The basis for the calculations is a feed flow rate of 1 lb-mol/h. The first 5 lines provide the feed compositions and flow rates. Lines 6 through 11 calculate the distillate flow rates with the specification that 95% of the i-butane in the feed be recovered in the bottoms (or 5% is in the distillate as specified in Line 9). Lines 12 through 17 calculate the bottoms flow rates using simple materials balances and the specification that the bottoms stream must contain no more than 0.1% propane (see Line 14). The resulting mole fractions of the distillate and bottoms are calculated in Lines 18 to 27.

12.6.5   Solution

(a) The Fenske[7] equation is used to calculate the minimum number of trays required for separation at total reflux

$$N_{min} = \frac{\log\left[\left(\frac{x_{LK}}{x_{HK}}\right)_D \left(\frac{x_{HK}}{x_{LK}}\right)_B\right]}{\log\alpha_{(LK/HK)\text{avg}}} \tag{12-18}$$

where the subscripts D and B refer to distillate and bottoms respectively and

N_{min} = minimum number of trays, including re-boiler
and partial condenser
x_{LK} = mole fraction of the light key component
x_{HK} = mole fraction of the heavy key component
α_j = relative volatility of component j defined as $\alpha_j = k_j / k_{HK}$
$\alpha_{(LK/HK)\text{avg}}$ = geometric average volatility of the light key component

The calculation of $\alpha_{(LK/HK)\text{avg}}$ involves

1. Calculation of the temperature at the top of the column (dew point temperature T_D of the distillate if total condenser is used) and the temperature at the bottom of the column (bubble point temperature T_B of the bottoms)
2. Calculation of the relative volatility of the light key $\alpha_{(LK/HK)}$ at T_D ($=\alpha_{TD}$) and at T_B ($=\alpha_{TB}$)
3. Calculation of the geometric average volatility from $\alpha_{(LK/HK)\text{avg}} = \sqrt{\alpha_{TD}\alpha_{TB}}$

The pertinent mole fraction values and the average volatility are introduced into Equation (12-18) to calculate N_{min}. The additional POLYMATH program statements for carrying out these calculations is shown in Table 12–16.

Table 12–16 POLYMATH Program Statements for Calculation of the Temperatures at the Top and at the Bottom of the Tower, and Calculation of N_{min}

Line	Equation
28	P = 7 # pressure bar
29	# Dew point temperature at the top of the column
30	f(TD) = xD1 / k1D + xD2 / k2D + xD3 / k3D - 1
31	TD(0) = 270
32	k1D = 10 ^ (3.93835 - 659.739 / (TD - 16.719)) / P
33	k2D = 10 ^ (4.53678 - 1149.36 / (TD + 24.906)) / P
34	k3D = 10 ^ (4.3281 - 1132.108 / (TD + 0.918)) / P
35	# Bubble point temperature at the bottom of the column
36	f(TB) = xB2 * k2B + xB3 * k3B + xB4 * k4B - 1
37	TB(0) = 300
38	k2B = 10 ^ (4.53678 - 1149.36 / (TB + 24.906)) / P
39	k3B = 10 ^ (4.3281 - 1132.108 / (TB + 0.918)) / P
40	k4B = 10 ^ (4.35576 - 1175.581 / (TB - 2.071)) / P
41	# Fenske Correlation
42	Nmin = log((xD2 / xD3) * (xB3 / xB2)) / log(sqrt((k2D / k3D) * (k2B / k3B)))

(b) The Underwood[8] equations are used to calculate the minimum reflux required to achieve the desired separation with an infinite number of trays. The use of these equations involves two steps. First Equation (12-19) must be solved for the parameter θ.

$$1 - q = \sum_{j=1}^{n_c} \frac{\alpha_j x_{jF}}{(\alpha_{jF} - \theta)} \tag{12-19}$$

where

$\quad q \quad$ = feed thermal condition ($q = 1$ for saturated liquid, $q = 0$ for saturated vapor).

$\quad x_{jF}$ = mole fraction of component j in the feed

$\quad \alpha_{jF}$ = relative volatility of component j in the feed

$\quad n_c \quad$ = total number of components

Note that Equation (12-19) is an implicit equation which has several roots. The desired root is in the interval $\alpha_{LKF} > \theta > 1$. When solving this equation, it is important to select an initial estimate inside this interval, otherwise most solution techniques will converge to an undesired root.

After solving Equation (12-19) for θ, the minimum reflux ratio is calculated from the equation

$$R_{min} + 1 = \sum_{j=1}^{n_c} \frac{\alpha_{jD} x_{jD}}{(\alpha_{jD} - \theta)} \tag{12-20}$$

where

$\quad \alpha_{jD}$ = relative volatility of component j in the distillate

$\quad x_{jD}$ = mole fraction of component j in the distillate

In order to determine the relative volatilities of the various compounds at the feed conditions (saturated liquid), the bubble point temperature of the feed (T_F) must be calculated. The set of POLYMATH equations for calculating T_F, θ, and R_{min} are shown in Table 12–17. Note than in the specified conditions $q = 1$.

Table 12–17 Calculation of Feed Temperature and Underwood Parameters θ and R_{min}.

Line	Equation
43	# Bubble point temperature of the feed
44	f(TF) = xF1 * k1F + xF2 * k2F + xF3 * k3F + xF4 * k4F - 1
45	TF(0) = 300
46	k1F = 10 ^ (3.93835 - 659.739 / (TF - 16.719)) / P
47	k2F = 10 ^ (4.53678 - 1149.36 / (TF + 24.906)) / P
48	k3F = 10 ^ (4.3281 - 1132.108 / (TF + 0.918)) / P
49	k4F = 10 ^ (4.35576 - 1175.581 / (TF - 2.071)) / P

Table 12–17 (Continued) Calculation of Feed Temperatureand Underwood Parameters θ and R_{min}.

Line	Equation
50	# Underwood Equations
51	f(theta) = xF1 * (k1F / k3F) / (k1F / k3F - theta) + xF2 * (k2F / k3F) / (k2F / k3F - theta) + xF3 / (1 - theta) + xF4 * (k4F / k3F) / (k4F / k3F - theta)
52	theta(0) = 1.5
53	Rmin = xD1 * (k1D / k3D) / (k1D / k3D - theta) + xD2 * (k2D / k3D) / (k2D / k3D - theta) + xD3 / (1 - theta) - 1

(c) Gilliland's[9] correlation is an experimental curve which gives the connection between X and Y where these variables are defined as

$$X = \frac{R - R_{min}}{R_{min} + 1} \qquad Y = \frac{N - N_{min}}{N_{min} + 1} \tag{12-21}$$

where R is the actual reflux ratio and N is the actual number of trays. Several correlations were fitted to Gilliland's experimental curve. A simple relationship[9] is given by

$$Y = 0.75(1 - X^{0.5658}) \tag{12-22}$$

Equations (12-21) and (12-22) enable the calculation of the actual number of trays when the reflux ratio is specified. The number of trays above and below the feed tray can be calculated using the Kirkbride equation

$$\log\frac{m}{p} = 0.206\log\left[\frac{B}{D}\left(\frac{x_{HK}}{x_{LK}}\right)_F\left(\frac{(x_{LK})_B}{(x_{HK})_D}\right)^2\right] \tag{12-23}$$

where

m = number of trays above the feed tray

p = number of trays below the feed tray

B = molar flow rate of the bottoms

D = molar flow rate of the distillate

Introduction of the numerical values into Equation (12-23) yields the ratio (m/p). This ratio and the total number of stages (N) can be used to calculate the number of trays below and above the feed tray.

The POLYMATH coding that implements Equations (12-21) through (12-23) is provided in Table 12–18.

Table 12–18 Calculation of the Actual Number of Theoretical Stages and Feed Tray Location

Line	Equation
54	# Number of theoretical trays
55	X = (1.5 - 1) * Rmin / (Rmin + 1)
56	Y = 0.75 * (1 - X ^ 0.5658)

Line	Equation
57	N = Y * (Nmin + 1) + Nmin
58	# Number of stages above and below the feed stage
59	m_on_p = 10 ^ (0.206 * log((tB / tD) * (xF3 / xF2) * (xB2 / xD3) ^ 2))
60	p = N / (1 + m_on_p)
61	m = N - p
62	# Specify the reflux ratio R
63	R=1.5*Rmin

Calculated values of NLE variables

	Variable	Value	f(x)	Initial Guess
1	TB	333.1795	6.661E-14	300.
2	TD	274.0267	-1.11E-16	270.
3	TF	285.404	2.22E-16	300.
4	theta	1.673146	-1.11E-16	1.5

	Variable	Value			Variable	Value			Variable	Value
1	B1	0		15	k3D	0.2319624		29	X	0.1960766
2	B2	0.0006617		16	k3F	0.3380909		30	xB1	0
3	B3	0.171		17	k4B	0.9124767		31	xB2	0.001
4	B4	0.49		18	k4F	0.2299043		32	xB3	0.2584402
5	D1	0.15		19	m	2.944521		33	xB4	0.7405598
6	D2	0.1793383		20	m_on_p	0.297132		34	xD1	0.4433432
7	D3	0.009		21	N	12.85433		35	xD2	0.5300562
8	D4	0		22	Nmin	8.543802		36	xD3	0.0266006
9	k1D	3.382548		23	p	9.909806		37	xD4	0
10	k1F	4.343251		24	P	7.		38	xF1	0.15
11	k2B	3.033514		25	R	0.9677271		39	xF2	0.18
12	k2D	0.7027538		26	Rmin	0.6451514		40	xF3	0.18
13	k2F	0.9722379		27	tB	0.6616617		41	xF4	0.49
14	k3B	1.242929		28	tD	0.3383383		42	Y	0.4516571

Figure 12–12   Results for the Multi-component Separation Calculations (File **P12-06.POL**)

POLYMATH results for the complete set of equations with $R = 1.5R_{min}$, presented in Tables 12–15 through 12–18, are shown in Figure 12–12. The column temperatures are 274 K at the top, 333.2 K at the bottom, and the temperature of the feed is 285.4 K. The minimum number of stages $N_{min} = 8.54$, minimum reflux ratio $R_{min} = 0.645$, the actual number of theoretical stages is 13, and the number of stages below (and including) the feed tray is 10.

The POLYMATH problem solution file is found in directory Chapter 12 with file named **P12-06.POL**.

12.7 FENSKE-UNDERWOOD-GILLILAND CORRELATIONS IN DEPROPANIZER DESIGN

12.7.1 Concepts Demonstrated

Short-cut distillation calculations. Use of the Fenske correlation to estimate minimum number of trays, the Underwood correlation to estimate the minimum reflux ratio, and the Gilliland correlation for calculating actual number of trays required to achieve a specified separation.

12.7.2 Numerical Methods Utilized

Solution of systems of nonlinear algebraic equations.

12.7.3 Problem Statement

A distillation column is to be designed by using short-cut methods for use as a depropanizer. Feed with the composition shown in Table 12–19 enters the distillation column as saturated vapor. The pressure in the column is $P = 250$ psig. The flow rate of n-butane in the distillate should not exceed 0.3 lb-mol/h and the flow rate of propane in the bottoms must not exceed 0.4 lb-mol/h. Assume sharp separation for methane, ethane, n-pentane and n-hexane. The vapor liquid equilibrium ratios can be calculated assuming ideal solution, thus $k_j = P_j/P$.

Table 12–19 Feed Flow Rates and Antoine Equation Coefficients for the Depropanizer[a]

No.	Component	Feed Flow Rate lb-mol/h	A	B	C
1	methane	26	6.6438	395.74	266.681
2	ethane	9	6.82915	663.72	256.681
3	propane (light key)	25	6.80338	804	247.04
4	n-butane (heavy key)	17	6.80776	935.77	238.789
5	n-pentane	11	6.85296	1064.84	232.012
6	n-hexane	12	6.87601	1171.17	224.408

[a]Note that these Antoine equation coefficients require temperature in °C and yield the vapor pressure in mmHg.

(a) Calculate the minimum number of trays (N_{min}) required to achieve the desired separation at total reflux using the Fenske equation.

(b) Estimate the minimum reflux (R_{min}) required to achieve the desired separation with infinite number of trays, using the Underwood equations.

(c) Approximate the number of theoretical stages required if the actual reflux ratio $R = 1.5R_{min}$ using the Gilliland correlation. Find the optimal feed stage location using the Kirkbride correlation.

(d) Repeat (a), (b) and (c) for various pressures between $P = 200$ psig and $P = 400$ psig. Is there an optimum pressure with regard to the number of theoretical stages required?

12.8 RIGOROUS DISTILLATION CALCULATIONS FOR A SIMPLE SEPARATION TOWER

12.8.1 Concepts Demonstrated

Rigorous distillation calculations using the material balance, energy balance, and phase equilibrium equations (the MESH equations) for each stage.

12.8.2 Numerical Methods Utilized

Solution of systems of nonlinear algebraic equations.

12.8.3 Problem Statement

A simple distillation column with three theoretical stages is used to separate n-butane and n-pentane. A feed stream that consists of 0.23 lb-mol/h butane and 0.77 lb-mol/h pentane enters the column as liquid at its bubble point on tray 2. The column operating pressure is $P = 120$ psia. A total condenser is used, and the amount of heat added to the reboiler is 10000 Btu/h. Distillate is removed at the rate of $D = 0.25$ lb-mol/h.

(a) Carry out complete material and energy balances to obtain the stage and condenser temperatures; the liquid and vapor flow rates and compositions inside the column; and the flow rates, compositions, and temperatures of the distillate and the bottoms.

(b) Change the feed composition, column pressure, and distillate flow rate, and investigate how those changes effect the recovery of the n-pentane in the bottoms using part (a) as a base case.

Additional Information and Data

The vapor liquid equilibrium ratios can be calculated assuming ideal solution, thus $k_j = P_j/P$. The Antoine equations for estimating the vapor pressures and the equations for calculating the molar enthalpies of the pure compounds are shown in Table 12–20. Ideal mixtures should be assumed for calculating mixture enthalpies. Initial estimates for stage temperatures and vapor flow rates are provided in Table 12–21.

Table 12–20 Vapor Pressure and Molar Enthalpy Correlations for n-Butane and n-Pentane[a]

	n-Butane (1)	n-Pentane (2)
Vapor Pressure (mmHg)	$\log(P1) = 6.80776 - 935.77 / (T + 238.789)$	$\log(P2) = 6.85296 - 1064.0 / (232.012 + T)$
Liquid Molar Enthalpy (BTU/lb-mol)	$h1 = 29.6T + 0.04\ T^2$	$h2 = 38.5T + 0.025T^2$
Vapor Molar Enthalpy (BTU/lb-mol)	$H1 = 8003 + 43.8T - 0.04\ T^2$	$H2 = 12004 + 31.7T + 0.007\ T^2$

[a]Note that T is the temperature in °C for vapor pressure and in °F for enthalpies.

Table 12–21 Initial Estimates for Butane-Pentane Tower

Tray No (j)	T_j(°F)	V_j (lb-mol/h)
1	145	1.1
2	190	1.0
3	210	1.1

12.8.4   Solution

The material and energy balance equations for a single stage (the MESH equations[*]) can be formulated with reference to the jth ideal stage that is depicted in Figure 12–13.

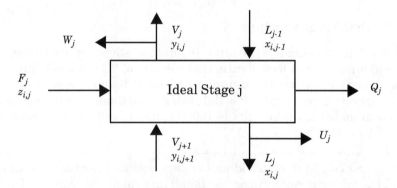

Figure 12–13 General Equilibrium Stage

The equations associated with the distillation column and the individual equilibrium stages utilize the variables summarized in Table 12–22.

Table 12–22 Column and Equilibrium Stage Variables

n	total number of stages	F_j	feed flow rate to the j-th stage (lb-mol/h)
j	stage number	$k_{i,j}$	equilibrium ratio for comp. i on stage j
i	component number	H_{Fj}	enthalpy of the j-th feed stream (BTU/lb-mol)
n_c	total number of components	h_j	enthalpy of the liquid stream leaving stage j (Btu/lb-mol)
$x_{i,j}$	liquid mole fraction for i-th component leaving j-th stage	L_j	flow rate of the liquid from the j-th stage to the j+1 stage (lb-mol/h)
$y_{i,j}$	vapor mole fraction for i-th component leaving j-th stage	W_j	flow rate of the liquid side stream from the j-th stage (lb-mol/h)
$z_{i,j}$	mole fraction of i-th component entering j-th stage	V_j	flow rate of the vapor from the j-th stage to the j-1 stage (lb-mol/hr)
T_j	temperature on the j-th stage (°F)	U_j	flow rate of the vapor side stream from the j-th stage (lb-mol/h)
B	bottoms product (lb-mol/hr)	H_j	enthalpy of vapor stream leaving stage j (Btu/lb-mol)
D	distillate product (lb-mol/hr)	h_j	enthalpy of liquid stream leaving stage j (Btu/lb-mol)
		Q_j	heat transfer from tray j (Btu/h)

[*] Seader and Henley[1], p. 366

The MESH equations for one equilibrium stage are

1. Mole balances on component j (M-equations, n_c equations for each stage)

$$L_{j-1}x_{i,j-1} + F_j z_{i,j} + V_{j+1}y_{i,j+1} = (L_j + U_j)x_{i,j} + (V_j + W_j)y_{i,j} \qquad \textbf{(12-24)}$$

2. Vapor-liquid equilibrium (E-equations, n_c equations for each stage)

$$y_{i,j} = k_{i,j}x_{i,j} \qquad \textbf{(12-25)}$$

3. Summation of mole fractions (S-equations)

$$\sum_{i=1}^{n_c} y_{i,j} = 1 \qquad \sum_{i=1}^{n_c} x_{i,j} = 1 \qquad \textbf{(12-26)}$$

4. Enthalpy balance (H-equations)

$$L_{j-1}h_{j-1} + F_j h_{Fj} + V_{j+1}H_{j+1} = (L_j + U_j)h_j + (V_j + W_j)H_j \qquad \textbf{(12-27)}$$

One of the S-equations can be replaced by the total mole balance equation

$$L_{j-1} + F_j + V_{j+1} = L_j + U_j + V_j + W_j \qquad \textbf{(12-28)}$$

(a) The distillation column specified in the "Problem Definition" section is depicted in Figure 12–14. Note that the stages are numbered from the top to the bottom. Since the condenser is assumed to condense all the vapor that enters it (a total condenser), it is not considered as an equilibrium stage, and the number 0 (zero) is used to identify all the associated variables.

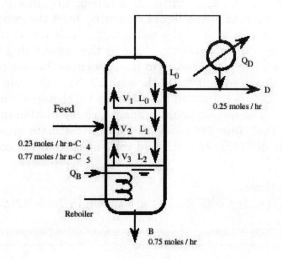

Figure 12–14 Simple n-Butane, n-Pentane Separation Tower

Problem Specification and Condenser Calculations

The equations associated with the problem specification and the condenser, as entered into the POLYMATH *Nonlinear Equations Solver* program, are shown in Table 12–23.

Table 12–23 Distillation Column Specifications and Equations Associated with the Condenser

Line	Equation
1	# Specifications
2	F = 1 # Feed flow rate
3	z1 = 0.23 # Mole fraction of component 1 in the feed
4	z2 = 1 - z1 # Mole fraction of component 2 in the feed
5	B = 0.75 # Bottoms flow rate
6	D = 0.25 # Distillate flow rate
7	P = 760 * 120 / 14.7 # Columns pressure (mmHg)
8	Q = 10000 # Reboiler's heat duty
9	
10	# Condenser flow rate, temperature and enthalpy
11	L0 = V1 - D # Liquid returning from the condenser
12	x10 = k11 * x11
13	x20 = k21 * x21
14	f(t0) = k10 * x10 + k20 * x20 - 1 # Bubble point temperature
15	t0(0) = 200
16	k10 = 10 ^ (6.80776 - 935.77 / ((t0 - 32) * 5 / 9 + 238.789)) / P
17	k20 = 10 ^ (6.85296 - 1064.84 / ((t0 - 32) * 5 / 9 + 232.012)) / P
18	h0 = t0 * (29.6 + 0.04 * t0) * x10 + t0 * (38.5 + 0.025 * t0) * x20 # Liquid from condenser
19	

The general specification are entered in lines 2-8. In line 11, the total material balance on the upper part of the column is used to define an explicit equation to calculate the flow rate of the liquid returning from the condenser to the column. The liquid composition in the condenser is calculated in lines 12 and 13 assuming that it is of the same composition as the vapor rising from the 1st stage (a total condenser). The temperature in the condenser (bubble point temperature of the liquid) is calculated in lines 14 through 17. Note that the calculation of the vapor liquid equilibrium ratios (on lines 16 and 17) requires temperature conversion from °F to °C. The bubble point temperature calculation involves solution of an implicit equation (line 14) using the initial estimate specified on line 15. Finally the molar enthalpy of the liquid returning from the condenser is calculated in line 18.

Stage 1 Calculations

The equations associated with Stage 1 are shown in Table 12–24.

Table 12–24 Equations Associated with Stage 1 of the Distillation Column

Line	Equation
20	# Stage 1 flow rates, temperatures, enthalpies and composition
21	L1 = V2 - D # Liquid flow rate from stage 1

Line	Equation
22	f(x11) = -((V1 - L0) * k11 + L1) * x11 + V2 * k12 * x12 # Component 1 stage 1 Mole Balance
23	x11(0) = 0.65
24	f(x21) = -((V1 - L0) * k21 + L1) * x21 + V2 * k22 * x22 # Component 2 stage 1 Mole Balance
25	x21(0) = 0.35
26	f(t1) = k11 * x11 + k21 * x21 - 1 # Stage 1 Bubble Point Temp.
27	t1(0) = 145
28	f(V1) = -V1 * hv1 + V2 * hv2 - L1 * hl1 + L0 * h0 # Stage 1 Enthalpy Balance
29	V1(0) = 1.1
30	# Stage 1 phase equilibrium ratios and enthalpies
31	k11 = 10 ^ (6.80776 - 935.77 / ((t1 - 32) * 5 / 9 + 238.789)) / P
32	k21 = 10 ^ (6.85296 - 1064.84 / ((t1 - 32) * 5 / 9 + 232.012)) / P
33	hl1 = t1 * (29.6 + 0.04 * t1) * x11 + t1 * (38.5 + 0.025 * t1) * x21 # Stage 1 liquid
34	hv1 = (8003 + t1 * (43.8 - 0.04 * t1)) * k11 * x11 + (12004 + t1 * (31.7 + 0.007 * t1)) * k21 * x21 # Stage 1 vapor
35	

The liquid flow from the stage 1 is calculated from an explicit equation which is obtained from a mole balance on the upper part of the column (in line 21). The mole balance equations on component j are revised by substituting Equation (12-25) into Equation (12-24) and setting $F_j = U_j = W_j = 0$. The resultant implicit equations are defines in lines 22 and 24 while initial estimates for x_{11} and x_{21} are provided in lines 23 and 25, respectively. It is assumed that the liquid on this stage is at its bubble point temperature, and this temperature is calculated in the implicit equation (based on Equation (12-26)) presented in line 26 with initial estimate on line 27. The enthalpy balance equation (Equation (12-27)) is written in an implicit form to determine the vapor flow rate from stage 1 (see lines 28 and 29). The vapor liquid equilibrium ratios and the molar enthalpies of the liquid and vapor are calculated using the equations shown on lines 31 through 34.

Stage 2 and 3 Calculations

The equations associated with the 2nd and 3rd stages are very similar to the equations of the 1st stage. Note that stage 3 is the column reboiler. The feed stream temperature and enthalpy must also be calculated. This can be accomplished using a set of equations similar to the set used for the condenser. The POLYMATH coding is presented in Table 12–25.

Table 12–25 POLYMATH Coding for Column Feed and Remaining Stages

Line	Equation
36	# Feed temperature and enthalpy
37	f(tf) = k1f * z1 + k2f * z2 - 1 # Feed (bubble point) temperature
38	tf(0) = 200
39	k1f = 10 ^ (6.80776 - 935.77 / ((tf - 32) * 5 / 9 + 238.789)) / P
40	k2f = 10 ^ (6.85296 - 1064.84 / ((tf - 32) * 5 / 9 + 232.012)) / P
41	hf = tf * (29.6 + 0.04 * tf) * z1 + tf * (38.5 + 0.025 * tf) * z2 # Enthalpy of feed

Table 12–25 (Continued) POLYMATH Coding for Column Feed and Remaining Stages

Line	Equation
42	
43	# Stage 2 flow rates, temperatures, enthalpies and composition
44	L2 = V3 + F - D # Stage 2
45	f(x12) = L1 * x11 - (V2 * k12 + L2) * x12 + V3 * k13 * x13 + z1 * F # Component 1 stage 2
46	x12(0) = 0.43
47	f(x22) = L1 * x21 - (V2 * k22 + L2) * x22 + V3 * k23 * x23 + z2 * F # Component 2 stage 2
48	x22(0) = 0.57
49	f(t2) = k12 * x12 + k22 * x22 - 1 # Stage 2
50	t2(0) = 190
51	f(V2) = -V2 * hv2 + V3 * hv3 + hf + L1 * hl1 - L2 * hl2 # Stage 2
52	V2(0) = 1
53	# Stage 2 phase equilibrium ratios and enthalpies
54	k12 = 10 ^ (6.80776 - 935.77 / ((t2 - 32) * 5 / 9 + 238.789)) / P
55	k22 = 10 ^ (6.85296 - 1064.84 / ((t2 - 32) * 5 / 9 + 232.012)) / P
56	hl2 = t2 * (29.6 + 0.04 * t2) * x12 + t2 * (38.5 + 0.025 * t2) * x22 # Stage 2 liquid
57	hv2 = (8003 + t2 * (43.8 - 0.04 * t2)) * k12 * x12 + (12004 + t2 * (31.7 + 0.007 * t2)) * k22 * x22 # Stage 2 vapor
58	
59	# Stage 3 flow rates, temperatures, enthalpies and composition
60	L3 = B # Stage 3
61	f(x13) = L2 * x12 - (V3 * k13 + B) * x13 # Component 1 stage 3
62	x13(0) = 0.33
63	f(x23) = L2 * x22 - (V3 * k23 + B) * x23 # Component 2 stage 3
64	x23(0) = 0.76
65	f(t3) = k13 * x13 + k23 * x23 - 1 # Stage 3
66	t3(0) = 210
67	f(V3) = -V3 * hv3 + Q + L2 * hl2 - L3 * hl3 # Stage 3
68	V3(0) = 1.1
69	# Stage 3 phase equilibrium ratios and enthalpies
70	k13 = 10 ^ (6.80776 - 935.77 / ((t3 - 32) * 5 / 9 + 238.789)) / P
71	k23 = 10 ^ (6.85296 - 1064.84 / ((t3 - 32) * 5 / 9 + 232.012)) / P
72	hl3 = t3 * (29.6 + 0.04 * t3) * x13 + t3 * (38.5 + 0.025 * t3) * x23 # Stage 3 liquid
73	hv3 = (8003 + t3 * (43.8 - 0.04 * t3)) * k13 * x13 + (12004 + t3 * (31.7 + 0.007 * t3)) * k23 * x23 # Stage 3 vapor

Calculated Results (Partial Solution)

(a) Table 12–26 shows selected results for the n-butane, n-pentane separation column. It can be seen that the temperature in the condenser is the lowest (188.28°F) and increases monotonically toward the bottoms of the tower, reaching 226.55°F at the bottoms. As expected, the concentration of the n-butane is the highest at the top ($x_1 = 0.5526$) and the concentration of n-pentane is the highest at the bottom of the column ($x_2 = 0.8775$).

Table 12–26 Selected Results for the n-Butane, n-Pentane Separation Column

	Condenser	Stage 1	Feed	Stage 2	Stage 3
T (°F)	188.28	206.43	215.49	218.55	226.55
V (lb-mol/hr)	-	1.0537	-	1.0388	1.0417
L (lb-mol/hr)	0.8037	0.7888	1	1.792	0.75
x_1	0.5526	0.3272	0.23	0.1991	0.1225
x_2	0.4474	0.6728	0.77	0.8009	0.8775
h (BTU/lb-mol)	7503.1	8620.8	9176.3	9363.4	9852.7
H (BTU/lb-mol)	-	1.69E+04	-	1.79E+04	1.86E+04

(b) The recovery of the n-pentane (η) can be calculated from

$$\eta = \frac{F_2 z_{2,2}}{B x_{2,3}} \tag{12-29}$$

In order to carry out parametric runs it is preferable to export the equations to MATLAB and modify the main program as needed. Use the MATLAB file **Prob_12_8.m** and MATLAB function **conles.p** to solve the system of equations in MATLAB.

The POLYMATH and MATLAB problem solution files are found in directory Chapter 12 with files named **P12-08.POL**, **Prob_12_8.m,** and www **conles.p**.

12.9 RIGOROUS DISTILLATION CALCULATIONS FOR HEXANE-OCTANE SEPARATION TOWER

12.9.1 Concepts Demonstrated

Rigorous distillation calculations using the material balance, energy balance, and phase equilibrium equations (the MESH equations) for each stage.

12.9.2 Numerical Methods Utilized

Solution of systems of nonlinear algebraic equations.

12.9.3 Problem Statement

A simple distillation column with three theoretical stages is used to separate n-hexane and n-octane. The feed to tray 2 is equimolar bubble point liquid at a flow rate of 100 kg-mol/h. The column operates at one atmosphere. The heat rate input to the reboiler is 10^6 kcal/hr. The vapor entering the condenser is condensed to a bubble point liquid. Distillate is removed at the rate of 30 kg-mol / h.

(a) Carry out complete material and energy balances to obtain the stage and condenser temperatures; the liquid and vapor flow-rates and compositions inside the column; and the distillate and bottoms flow rates, compositions, and temperatures.

(b) Investigate how changes in the feed composition, column pressure, and distillate flow rate affect the recovery of the n-hexane in the distillate using part (a) as a base case.

Additional Information and Data

The vapor liquid equilibrium ratios can be calculated assuming ideal solution, thus $k_j = P_j/P$. For vapor pressure calculation the Antoine equations shown in Table 12–1 can be used. The equations for calculating the molar enthalpies of the pure compounds are given in Table 12–1. Assume ideal mixtures for calculating mixture enthalpies. For stage temperatures and vapor flow rates, the initial estimates shown in Table 12–27 may be used.

Table 12–27 Initial Estimates for Tray Temperatures and Vapor Flow Rates

Tray No.	$T_j\,(^\circ C)$	$V_j\,$(kg-mol/h)
1	86	110
2	94	100
3	100	110

12.9.4 Solution (Suggestion)

The method of solution is explained in detail in Problem 12.8.

12.10 BATCH DISTILLATION OF A WATER-ETHANOL MIXTURE

12.10.1 Concepts Demonstrated

Binary batch distillation of a non-ideal system.

12.10.2 Numerical Methods Utilized

Solution of a differential-algebraic system of equations.

12.10.3 Problem Statement

Figure 12–15 shows a single-stage batch distillation apparatus that is used for separation of alcohol from water. A liquid mixture of 40 mol% water ($x_1 = 0.4$) and 60 mol% ethanol is charged initially to the still pot. The amount of the initial charge is $L = 100$ kg-moles. The distillation is carried out at 1 atm total pressure and is continues until the water mole fraction reaches 0.80 ($x_1 = 0.8$).

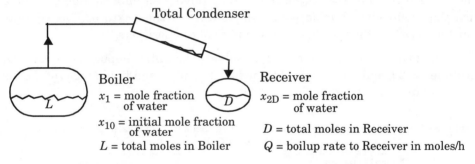

Figure 12–15 Single Stage Batch Still

(a) Find the temperature in the batch still when the boiling starts and at the end point of the distillation.

(b) Calculate the kg-moles of liquid remaining in the still when the mole fraction of the water reaches $x_1 = 0.8$. Carry out the calculations both rigorously (considering the change of the vapor-liquid equilibrium ratios with the temperature and composition) and by assuming constant relative volatility of the two compounds. Compare the two results.

(c) Repeat parts (a) and (b) when a vacuum distillation is used at a working pressure of 0.2 atm.

Additional Information and Data

The vapor liquid equilibrium ratios can be calculated assuming that the modified Raoult's law apply, thus $k_i = \gamma_i P_i / P$ where γ_i is the activity coefficient of component i, P_i is the vapor pressure of component i, and P is the total pressure. Table 12–28 provides the Antoine equations and constants for estimating the needed vapor pressures. The activity coefficients can be calculated using the equations given in Table 12–28.

Table 12–28 Vapor Pressure and Molal Enthalpy Correlations for Water-Ethanol Mixtures

	Water (1)	Ethanol (2)
Vapor Pressure (mmHg)[a]	log(P1) = 7.96681 - 1668.21 / (T + 228)	log(P2) = 8.04494 - 1554.3 / (T + 222.65)
Activity Coefficient[b]	log(γ1) = x22* (0.3781 + 2 * x1 * (0.6848 - 0.3781))	log(γ2) = x12 * (0.6848 + 2 * x2 * (0.3781 - 0.6848))

[a]Dean[4] with T-Temperature in °C
[b]Holmes and Van Winkle[10]

12.10.4 Solution (Partial)

(a) The initial and final temperatures, the vapor liquid equilibrium ratios and relative volatilities can be calculated by solving an implicit algebraic equation that defines the bubble point temperature of the mixture. This equation must be solved at the specified initial and final compositions.

The POLYMATH equation set for solving the initial bubble point is shown in Table 12–29, and the results obtained at the initial and final stages of the distillation are summarized in Table 12–30.

Table 12–29 POLYMATH Equations for Bubble Point Temperature of the Water-Ethanol Mixture (File **P12-10A1.POL**)

Line	Equation
1	f(T) = 1 - k1 * x1 - k2 * x2
2	T(min) = 70
3	T(max) = 100
4	x1 = 0.4
5	x2 = 1 - x1
6	gamma1 = 10 ^ (x2 * x2 * (0.3781 + 2 * x1 * (0.6848 - 0.3781)))
7	gamma2 = 10 ^ (x1 * x1 * (0.6848 + 2 * x2 * (0.3781 - 0.6848)))
8	P1 = 10 ^ (7.96681 - 1668.21 / (T + 228))
9	P2 = 10 ^ (8.04494 - 1554.3 / (T + 222.65))
10	k2 = gamma2 * P2 / 760
11	k1 = gamma1 * P1 / 760
12	alpha12 = k1 / k2

Table 12–30 Initial and Final Conditions in the Batch Still

	x_1	x_2	$T°C$	k_1	k_2	α_{12}
Initial	0.4	0.6	79.17	0.7577	1.162	0.652
Final	0.8	0.2	82.85	0.5676	2.729	0.208

The POLYMATH problem solution files are found in directory Chapter 12 with files named **P12-10A1.POL** and **P12-10A2.POL.**

(b) The moles remaining in the boiler can be determined from the differential equation given by Equation (12-30) (Seader and Henley[1], p. 467). The final composition is obtained when the differential equation is integrated from $x_1(0) = 0.4$ up to $x_1(f) = 0.8$.

$$\frac{dL}{dx_1} = \frac{L}{y_1 - x_1} \tag{12-30}$$

Analytical solution to Equation (12-30) can be derived if either the vapor liquid equilibrium ratios or the relative volatility α_{12} are constant. For constant relative volatility, the integration of Equation (12-30) from $x_1 = x_{10}$ and $L = L_0$ to to x_1 and L, yields

$$\ln\left(\frac{L}{L_0}\right) = \frac{1}{\alpha_{12} - 1} \ln\left[\frac{x_1(1 - x_{10})}{x_{10}(1 - x_1)}\right] + \ln\left(\frac{1 - x_{10}}{1 - x_1}\right) \tag{12-31}$$

Introduction of the numerical values $x_{10} = 0.4$, $L_0 = 100$, $x_1 = 0.8$, and the average value of the relative volatility $\alpha_{12} = (0.652+0.208)/2 = 0.43$ into Equation (12-31) yields the amount remaining of L = 12.94 kg-mol. However, this change in the relative volatility may be too large (see Table 12–30) to justify the assumption of constant volatility so it is necessary to verify this result.

A rigorous solution of Equation (12-30) requires integration while the temperature in the boiler follows the bubble point curve. This can be achieved using the *controlled integration technique* which is described in detail in Problem 6.7.

The equation input to the POLYMATH *Ordinary Differential Equation Solver* is presented in Table 12–31. Note that the differential equation in line 1 is used like a proportional controller with a large proportional gain K_c that adjusts the boiler temperature to maintain the vapor-liquid equilibrium as the distillation process proceeds.

Table 12–31 POLYMATH Equation Set for the Controlled Integration Solution of the Binary Batch Distillation Problem (File **P12-10B.POL**)

Line	Equation
1	d(T)/d(x1) = Kc * err
2	T(0) = 79.1685
3	d(L)/d(x1) = L / (x1 * (k1 - 1))
4	L(0) = 100
5	err = 1 - k1 * x1 - k2 * x2
6	Kc = 500000
7	x1(0) = 0.4
8	x1(f) = 0.8
9	x2 = 1 - x1
10	gamma1 = 10 ^ (x2 * x2 * (0.3781 + 2 * x1 * (0.6848 - 0.3781)))

Line	Equation
11	gamma2 = 10 ^ (x1 * x1 * (0.6848 + 2 * x2 * (0.3781 - 0.6848)))
12	P1 = 10 ^ (7.96681 - 1668.21 / (T + 228))
13	P2 = 10 ^ (8.04494 - 1554.3 / (T + 222.65))
14	k2 = gamma2 * P2 / 760
15	k1 = gamma1 * P1 / 760

The amount of the liquid remaining obtained by the rigorous solution is L = 13.02 kg-mol which is almost identical to the analytical solution of L = 12.94 kg-mol.

The POLYMATH problem solution file is found in directory Chapter 12 with file named **P12-10B.POL.**

www

(c) The use of other operating pressures may yield significant variation between the analytical and rigorous solutions.

12.11 DYNAMICS OF BATCH DISTILLATION OF FERMENTER BROTH

12.11.1 Concepts Demonstrated

Dynamics of the binary batch distillation of a non-ideal system.

12.11.2 Numerical Methods Utilized

Solution of a differential-algebraic system of equations.

12.11.3 Problem Statement

The dynamics of the ethanol-water batch distillation discussed in problem 12.10 are to be investigated. The boilup rate at which the liquid L in the boiler can be assumed to be constant at 5 kg-mol/h. The liquid mixture for distillation is from a fermentation process and contains an ethanol mole fraction of 0.025. The initial charge for the batch distillation is $L = 75$ kg-moles. The distillation is carried out at 1 atm total pressure and is continued until 95% of the ethanol is recovered in the distillate in the receiver. The condenser totally condenses all the entering vapor.

(a) Determine the expected distillation time in hours for the ethanol recovery. The data of Table 12–28 should be used. The time starts at zero when the boiler liquid has just reached the boiling point.

(b) Plot the expected time dependency of the boiler temperature, the mole fraction of ethanol in the boiler liquid, the mole fraction of ethanol in the boiler vapor, and the mole fraction of ethanol in the receiver condensate liquid.

12.11.4 Solution (Suggestions)

The differential equations for the molar contents of the boiler and the receiver depend upon the boilup rate Q (kg-mol/h) as the content in the condenser and vapor volumes can be neglected.

$$\frac{dL}{dt} = -15Q \qquad (12\text{-}32)$$

$$\frac{dD}{dt} = 15Q \qquad (12\text{-}33)$$

Both the boiler and receiver can be considered to be completely mixed at all times.

REFERENCES

1. Seader, J. D., Henley, E. J. *Separation Process Principles*, 2nd Ed., Wiley: New York, 2006.
2. Henley, E. J., Rosen, E.M. *Material and Energy Balance Computation*, Wiley: New York, 1969.
3. Wilson, G. M. *J. Amer. Chem. Soc.*, *86*, 127-130 (1964).
4. Dean. A. (Ed.), *Lange's Handbook of Chemistry*, McGraw Hill: New York, 1973.
5. *J. Chem. Eng. Data.*, *7*, 367 (1962).
6. NIST Chemistry WebBook (http://webbook.nist.gov/chemistry/).
7. Fenske, M.R. *Ind. Eng. Chem. 24*, 482, (1932).
8. Underwood, A.J.V. *J. Inst. Petrol.*, *31*, 111 (1945).
9. Gilliland, E.R. *Ind. Eng. Chem.*, *32*, 1101-1106 (1940).Aduljee, H. *Hydrocarbon. Process.*, p. 120, Sept. (1975).
10. Holmes, M. J., and Van Winkle, M. "Prediction of Ternary Vapor-Liquid Equilibrium from Binary Data," *Ind. Eng. Chem.*, *62* (1), 21 (1970).
11. Shacham, M., Brauner, N., and Pozin, M. "Application of Feedback Control Principles for Solving Differential-Algebraic Systems of Equations in Process Control Education," *Computers Chem. Engng.*, *20*, Suppl. pp S1329-S1334 (1996).

Process Dynamics and Control

13.1 MODELING THE DYNAMICS OF FIRST- AND SECOND-ORDER SYSTEMS

13.1.1 Concepts Demonstrated

Dynamics of step response for general linear first-order and linear second-order systems. Comparisons of analytical solutions with numerical solutions. Identification and characterization of typical system responses.

13.1.2 Numerical Methods Utilized

Solution of first-order and second-order ordinary differential equations.

13.1.3 Problem Statement[*]

Mathematical descriptions of many processes can be accomplished by models that utilize ordinary differential equations. Control theory often treats these processes as either first-order or second-order systems. Laplace transforms and other methods can be used to obtain solutions to these systems. Most control theory utilizes transfer functions so that algebraic solutions with Laplace transforms are enabled for the analysis of many control systems has been described by many authors including Bequette[1] and Stephanopoulos.[2] This transfer function approach is supported by MATLAB.

First-Order Systems

A first-order linear system is one in which the output $y(t)$ is described by a first-order ordinary differential equation

$$a_1\frac{dy}{dt} + a_0 y = b_0\, u(t) \tag{13-1}$$

[*] Materials for this problem were adapted from Bequette[1] and Stephanopoulos.[2]

where $u(t)$ is the input variable also known as the forcing function. If $u(t)$ has a constant value represented by Δu and $a_0 \neq 0$, then Equation (13-1) can be expressed in a equation form used in process control by

$$\tau \frac{dy}{dt} + y = k\Delta u \tag{13-2}$$

where the time constant is given by $\tau = a_1/a_0$ and the gain is given by $k = b_0/a_0$.

When the model is expressed in terms of deviation variables (the initial conditions are all zero at $t = 0$) and a step change Δu, the analytical solution is easily determined to be

$$y(t) = k\Delta u(1 - e^{-t/\tau}) \tag{13-3}$$

The analytical solution given in Equation (13-3) can be rearranged into the dimensionless graph given in Figure 13–1.

Second-Order Systems

A second-order linear system is one in which the output $y(t)$ is described by a second-order ordinary differential equation (Bequette[1])

$$a_2\frac{d^2y}{dt^2} + a_1\frac{dy}{dt} + a_0y = b_0u(t) \tag{13-4}$$

where $u(t)$ is known as the forcing function or input variable. This equation is usually written in terms of deviation variables so the initial conditions are all zero at $t = 0$. If $a_0 \neq 0$, then Equation (13-4) can be expressed in a equation form

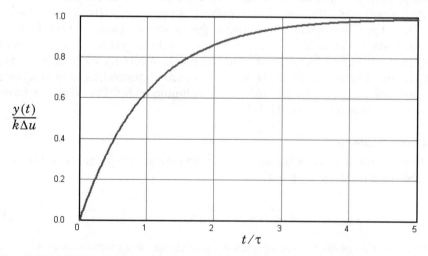

Figure 13–1 Dimensionless Response of Linear First-Order System to a Step Function at t = 0.

used in process control with the forcing function $u(t)$ being a constant given by Δu. Thus Equation (13-4) becomes

$$\tau^2 \frac{d^2 y}{dt^2} + 2\zeta\tau\frac{dy}{dt} + y = k\Delta u \qquad \text{(13-5)}$$

where the time constant is $\tau = a_2/a_0$ (units of time), the damping factor is $\zeta = a_1/(\sqrt{a_0 a_2})$ (dimensionless), and the gain is given by $k = b_0/a_0$ (units of output/input).

The response of the second-order system to a constant step input of Δu at $t = 0$ is strongly dependent upon the damping factor ζ.

Case 1 where $\zeta > 1$–Overdamped System Response
The solution of this case utilizing Laplace transforms yield a time-domain solution

$$\frac{y(t)}{k\Delta u} = 1 + \frac{\tau_1 e^{-t/\tau_1} - \tau_2 e^{-t/\tau_2}}{\tau_2 - \tau_1} \qquad \text{(13-6)}$$

where

$$\tau_1 = \frac{\tau}{\zeta - \sqrt{\zeta^2 - 1}} \qquad \text{(13-7)}$$

$$\tau_2 = \frac{\tau}{\zeta + \sqrt{\zeta^2 - 1}} \qquad \text{(13-8)}$$

This solution is plotted in Figure 13–2 for various values of the damping factor ζ.

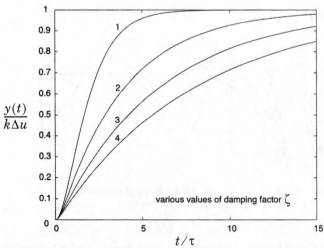

Figure 13–2 Step Function Response of Linear Second-Order Systems for $\zeta \geq 1$ from Bequette[1]

Case 2 where $\zeta = 1$–Critically Damped System Response

The time-domain solution to the critically damped case is given by

$$\frac{y(t)}{k\Delta u} = 1 - \left(1 + \frac{t}{\tau}\right)e^{-t/\tau} \tag{13-9}$$

that is also shown in Figure 13–2 for $\zeta = 1$.

Case 3 where $\zeta < 1$–Underdamped System Response

The analytical solution to the underdamped response as found by Laplace transforms is provided by

$$\frac{y(t)}{k\Delta u} = 1 - \frac{1}{\sqrt{1-\zeta^2}}e^{-\zeta t/\tau}\sin(\alpha t + \phi) \tag{13-10}$$

where

$$\alpha = \frac{\sqrt{1-\zeta^2}}{\tau} \tag{13-11}$$

$$\phi = \tan^{-1}\left(\frac{\sqrt{1-\zeta^2}}{\tau}\right) \tag{13-12}$$

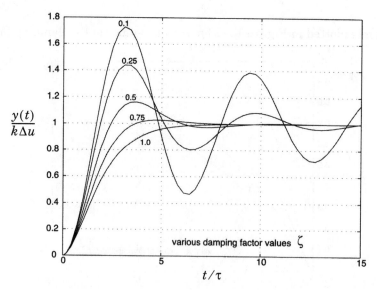

Figure 13–3 Step Function Response of Linear Second-Order Systems for $\zeta \le 1$ from Bequette[1]

(a) Determine an arbitrary solution point, $y(t)/(k\Delta u)$ and t/τ, on Figure 13–1 for a first-order system in which $\tau = 3$, $k = 3$, and $\Delta u = 2$. Verify the solution for your selected point by numerically solving the differential equation (13-2). Compare numerical solution values for $y(t)/(k\Delta u)$ with the analytical solution values calculated from Equation (13-3) over the range of t/τ used in Figure 13–1.

(b) Use the numerical solution of part (a) to determine the dimensionless output response for a first-order system when $t/\tau = 1$. Calculate the values of t when the output response reaches 95% and 99% of its final value.

(c) Determine an arbitrary solution point, $y(t)/(k\Delta u)$ and t/τ, at a particular value of ζ for a second-order system shown in Figure 13–2. The parameter values are $\tau = 5$, $k = 4$, and $\Delta u = 10$. Numerically solve differential equation (13-2) and compare the calculated dimensionless response with the analytical solution response as determined from Equations (13-6) though (13-8). Generate the solution over the range of t/τ used in Figure 13–2.

(d) Repeat part (c) for $\zeta = 1$ and compare the calculated dimensionless response with the analytical solution response as determined from Equation (13-10). Generate the solution over the range of t/τ used in Figure 13–2.

(e) Repeat part (c) for a selected value of $\zeta < 1$ shown in Figure 13–3 and compare the calculated dimensionless response with the analytical solution response as determined from Equations (13-10) though (13-12). Generate the solution over the range of t/τ used in Figure 13–3.

(f) What do you conclude about the accuracy of numerical solutions to the control differential equations that were solved in this problem?

13.1.4 Solution (Partial)

(a) Consider an arbitrary point on Figure 13–1 to be at $t/\tau = 2$. From the figure, the value of $y(t)/(k\Delta u)$ is approximately 0.87. The analytical solution is given by Equation (13-3). The numerical solution is found by solving Equation (13-1) with the initial condition that $y = 0$ when $t = 0$ to a final value of $t = 6$. The POLYMATH coding for both the analytical and numerical solutions at the selected arbitrary point is given in Table 13–1, and the results are summarized in Figure 13–4. Note that the solutions must be compared at the final value of $t = 6$ where the analytical and numerical solutions both yield the same result of 0.8646647 for the dimensionless response.

Table 13–1 Analytical and Numerical Solutions for Linear First-Order ODE (**File P13-01A.POL**)

Line	Equation
1	#General First-Order Linear ODE with Step Change at t = 0
2	tOVERtau=2 #Arbitrary selection: t/tau = 2
3	tau=3 #Given value of tau
4	k=3 #Given value of gain
5	du=2 #Given value of forcing function (disturbance)
6	d(y) / d(t) = k*du/tau-y/tau #Differential equation
7	yOVERkduANAL=(1-exp(-tOVERtau)) #Calculated value of dimensionless response
8	yANAL=yOVERkduANAL*k*du #Calculated value of y(t)
9	y(0) = 0 #Initial value of y(t)
10	t(0) = 0 #initial value of t
11	t(f) = 6 # Final value of t calcul.ated from t(f) = tOVERtau*tau = 2*3 = 6

Calculated values of DEQ variables

	Variable	Initial value	Minimal value	Maximal value	Final value
1	du	2.	2.	2.	2.
2	k	3.	3.	3.	3.
3	t	0	0	6.	6.
4	tau	3.	3.	3.	3.
5	tOVERtau	2.	2.	2.	2.
6	y	0	0	5.187988	5.187988
7	yANAL	5.187988	5.187988	5.187988	5.187988
8	yOVERkduANAL	0.8646647	0.8646647	0.8646647	0.8646647

Figure 13–4 Results for a First-Order System (**File P13-01A.POL**)

(b) The value of the independent variable when a certain condition is reached can be easily determined by using the 'if … then … else …" statement in POLYMATH.

d(t95)/d(t) = if (y<0.95*k*du) then(1) else (0)

This statement yields the value of the desired time t95 (time to reach 95% of the total response) by changing this value directly with the independent variable t until the condition is met, and then holding this value by setting the time derivative to be zero. This is demonstrated in file **P13-01B.POL**.

(c) The standard form of the second-order system written as Equation (13-5) can be rearranged and then numerically solved by conversion to two simultaneous first-order ordinary differential equations given by

$$\frac{dy}{dt} = dydt \quad \text{I. C. } y = 0 \text{ when } t = 0 \tag{13-13}$$

$$\frac{d(dydt)}{dt} = -\frac{2\zeta}{\tau}(dydt) - \frac{y}{\tau^2} + \frac{kdu}{\tau^2} \quad \text{I. C. } dydt = 0 \text{ when } t = 0 \tag{13-14}$$

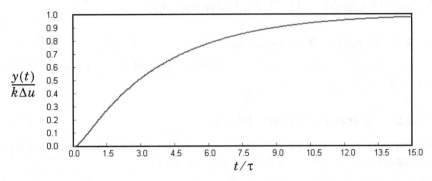

Figure 13–5 Numerical Solution to Second-Order Linear Ordinary Differential Equation for the Given Parameters (File **P13-01C2.POL**)

The second-order numerical and analytical solutions are generated by the POLY-MATH coding presented in Table 13–2, and the graphical display of the numerical results is given in Figure 13–5. At an arbitrary value of $t/\tau = 5$, both the analytical and numerical solutions agree in the calculation of $y(t)/(k\Delta u)$ to be 0.7178288 as found with file **P13-01C2.POL**.

Table 13–2 Analytical and Numerical Solutions for Linear Second-Order ODE (File **P13-01C1.POL**)

Line	Equation
1	#General Linear Second-Order ODE with Step Change at t = 0
2	#Case 1 - Overdamped Response when zeta>1
3	zeta=2 #Arbitrary selection
4	tau=5
5	k=4
6	du=10
7	t(0) = 0
8	t(f) = 75
9	d(y) / d(t) = dydt
10	y(0) = 0
11	d(dydt)/d(t)=-2*zeta*dydt/tau-y/tau^2+k*du/tau^2
12	dydt(0)=0
13	tOVERtau=t/tau
14	yOVERkdu=y/(k*du)
15	yOVERkduANAL=1+(tau1*exp(-t/tau1)-tau2*exp(-t/tau2))/(tau2-tau1)
16	tau1=tau/(zeta-sqrt(zeta^2-1))
17	tau2=tau/(zeta+sqrt(zeta^2-1))

The POLYMATH problem solution files are found in directory Chapter 13 with files named **P13-01A.POL**, **P13-01B.POL**, **P13-01C1.POL**, and **P13-01C2.POL**.

13.2 DYNAMICS OF A U-TUBE MANOMETER

13.2.1   Concepts Demonstrated

Analysis of the step response of a second-order system using analytical and numerical solutions.

13.2.2   Numerical Methods Utilized

Conversion of a second-order differential equation to two first-order equations. Solution of systems of ordinary differential equations.

13.2.3   Problem Statement

A conventional U-tube manometer is shown in Figure 13–6. When subjected to a time dependent pressure differential, the level of liquid in the legs changes with time according to the equation (assuming laminar flow in the tubes)

$$\frac{L}{g}\frac{d^2 h}{dt^2} + \frac{8\mu L}{\rho g R^2}\frac{dh}{dt} + h = \frac{\Delta P}{\rho g} \tag{13-15}$$

where L is the total length of the liquid in the tube, h is the height differential and ΔP is the pressure differential between the two legs. R is the radius of the manometer tube, μ is the viscosity, and ρ is the density of the manometer fluid.

(a) Determine and plot the response of the manometer height h using both analytical and numerical solutions to a step change

$$\frac{\Delta P}{\rho g} = \begin{cases} 0 & t \leq 0 \\ 50 & t > 0 \end{cases}$$

when $L = 150$ cm, $\mu = 0.01$ poise, $\rho = 1.0$ g/cm^3, $g = 980.7$ cm/s^2, $R = 0.25$ cm, and $h_0 = 0$.

(b) Repeat part (a) for $R = 0.1$ cm.

(c) Comment on the accuracy of the numerical solution in comparison to the analytical solution for parts (a) and (b).

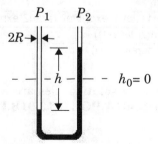

Figure 13–6 U-Tube Manometer

13.2.4 Solution (Partial)

(a) Equation (13-15) for the manometer is a second-order linear ordinary differential equation that can be placed in the same form as Equation (13-5) where

$$\tau^2 \frac{d^2 y}{dt^2} + 2\zeta\tau\frac{dy}{dt} + y = k\Delta u \tag{13-16}$$

when $\tau = \sqrt{\frac{L}{g}}$, $\zeta = \frac{4\mu}{\rho R^2}\sqrt{\frac{L}{g}}$, and $k\Delta u = \frac{\Delta P}{\rho g}$.

The numerical value of ζ is < 1 so that the problem is underdamped and thus the analytical solution is given by Equation (13-10). The general second-order differential equation can be solved numerically as discussed in Problem 13.1. The analytical and numerical solutions are identical, and the $y(t)$ representing the manometer height h is plotted versus time in Figure 13–7.

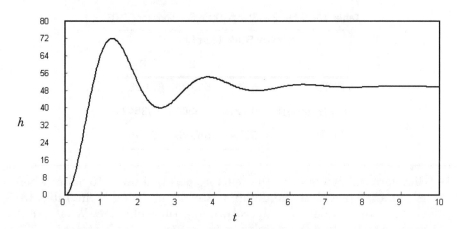

Figure 13–7 Manometer Height (cm) as Function of Time (s) (File **P13-02A.POL**)

 The POLYMATH problem solution file for part (a) is found in directory Chapter 13 with file name **P13-02A.POL**.

13.3 DYNAMICS AND STABILITY OF AN EXOTHERMIC CSTR

13.3.1 Concepts Demonstrated

Investigation of multiple steady states and the stability of an exothermic CSTR by dynamic simulation.

13.3.2 Numerical Methods Utilized

Solution of a system of ordinary differential equations.

13.3.3 Problem Statement

The irreversible exothermic reaction $A \xrightarrow{k} B$ is carried out in a jacketed CSTR as described in Problem 11.23. This reactor is shown to have three possible steady-state operating conditions as summarized in Table 13–3.

Table 13–3 Multiple Steady-State Solutions for CSTR

	Steady State Number.		
	1	**2**	**3**
T °R	537.86	590.35	671.28
C_A lb-mol/ft^3	0.5214	0.3302	0.03542
T_j °R	537.25	585.73	660.46

(a) Simulate the dynamics of the reactor operation up to 10 hours when it is initially started at the different steady states given in Table 13–3. Observe the change of temperature and concentration. What can you conclude regarding the stability of the different steady states?

(b) Simulate the reactor dynamics at initial reactor conditions for temperatures that are 2°R higher and also 2°R lower than for steady state number 2 given in Table 13–3. Initial C_A and T_j are unchanged. How do these results affect your stability conclusions from part (a)?

(c) Safety considerations require modeling an abrupt loss of cooling water flow rate to the reactor jacket as the reactor feed continues. Predict the reactor temperature dynamics from operation at steady state number 1 of Table 13–3.

(d) Repeat part (c) but cut off the reacting stream flow rate entering and exiting the reactor. Thus the reactor becomes a batch reactor with no cooling feed rate to the jacket. Would this be a good operational procedure when the cooling jacket pump completely fails?

(e) Repeat parts (c) and (d) for operation starting at steady state number 3 of Table 13–3.

13.3.4   Solution (Partial)

(a) The dynamic balances for this problem include the material balance on the reactor, the energy balance on the reactor, and the energy balance on the cooling jacket.

Dynamic mole balance on CSTR for reactant A

$$\frac{d(VC_A)}{dt} = F_0 C_{A0} - FC_A - VkC_A \tag{13-17}$$

Dynamic energy balance on the reactor

$$\frac{d(\rho VC_P T)}{dt} = \rho C_P (F_0 T_0 - FT) - \lambda VkC_A - UA(T - T_j) \tag{13-18}$$

Dynamic energy balance on the cooling jacket

$$\frac{d(\rho_j V_j C_j T_j)}{dt} = \rho_j C_j F_j (T_{j0} - T_j) + UA(T - T_j) \tag{13-19}$$

The balance equations can be solved simultaneously for the parameters given in Table 11–23 of Problem 11.23.

A POLYMATH solution file starting at the solution 1 steady state gives the results summarized in Figure 13–8. Note that these results indicate a very stable steady-state and also confirm the steady-state calculation carried out in Problem 11.23 that involved solving a nonlinear equation to obtain solution 1.

Calculated values of DEQ variables

	Variable	Initial value	Minimal value	Maximal value	Final value
3	CA	0.5214	0.5213904	0.5214	0.5213905
16	t	0	0	10.	10.
17	T	537.86	537.8548	537.86	537.8548
19	Tj	537.25	537.25	537.2559	537.2534

Figure 13–8 Transient Initiated from Solution 1 Steady State (File **P13-03A1.POL**)

The POLYMATH problem solution file for part (a) is found in directory Chapter 13 with file name **P13-03A1.POL**.

www

13.4 FITTING A FIRST-ORDER PLUS DEAD-TIME MODEL TO PROCESS DATA

13.4.1   Concepts Demonstrated

Fitting process data resulting from a step change in a process input variable to a first-order plus dead-time model.

13.4.2   Numerical Methods Utilized

Nonlinear regression of data.

13.4.3   Problem Statement

The most common plant test that yields data for control system design involves a simple step change in the controller output and the response of the measured process variable is recorded. This step change is typically made during plant operation, so it must be large enough to allow the output to be analyzed for control parameters. However, it must be small enough so as not to disturb the overall plant process to produce "off specification" product during the step-change test.

First-Order Plus Dead-Time Model (FOPDT Model)[*]
The expected deviation response, $y(t) = y - y_S$ where y_S is the steady state value of the measured process variable y, to a simple step change in the controller output (or manipulated variable), Δu, is given by

$$y(t) = 0 \quad \text{for} \ \ t < \theta_P \tag{13-20}$$

$$y(t) = K_P \Delta u \left\{ 1 - e^{[(-t + \theta_P)/\tau_P]} \right\} \quad \text{for} \ \ t \geq \theta_P \tag{13-21}$$

where K_P is the process gain, τ_P is the process time constant, and θ_P is the dead time. This response always is measured from the process at steady state.

The parameters that can be determined from the step-change response include K_P, τ_P, θ_P, and possibly the steady-state value of the measured variable given by y_S. Thus for a measured variable, temperature T in this example, the FOPDT response to the step test is described by

$$T = T_s + K_P \Delta u \left\{ 1 - e^{[(-t + \theta_P)/\tau_P]} \right\} \tag{13-22}$$

When the step-change is t_d units of time later than initial time for data collection at $t = 0$, the step-response testing can be written as

$$T = T_S \quad \text{for} \ \ 0 \leq t < (t_d + \theta_P) \tag{13-23}$$

$$T = T_s + K_P \Delta u \left\{ 1 - e^{[(-t + t_d + \theta_P)/\tau_P]} \right\} \quad \text{for} \ \ t \geq (t_d + \theta_P) \tag{13-24}$$

---

[*] Notation used here is that of Cooper[3]

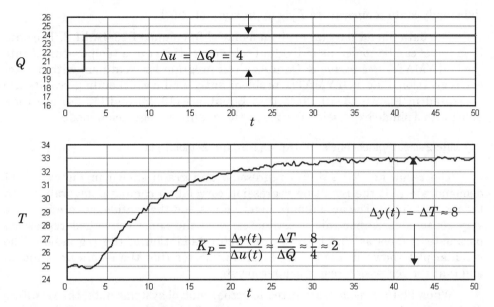

Figure 13–9 Step-Change Response Data

A typical response of measured process variable given by $y(t)$, temperature T in this case, to a step change in the controller output $u(t) = \Delta u$, volumetric flow rate Q in this case, at $t_d = 2$ minutes is shown in Figure 13–9.

(a)   Utilize the data displayed in Figure 13–9 and available as a data file to determine parameters of a first-order plus dead-time model given by Equation (13-24).

(b)   Repeat part (a) with the provided data file where the controller output is stepped –6 units at $t = 6$. Data file is **P13-04B.POL** for POLYMATH.

13.4.4   Solution (Partial)

(a) **First-Order Plus Dead Time-Model** Nonlinear regression can be used to determine up to four parameters of Equation (13-24). In this example, t_d is set to be 2 min the change in the controller output during the step change is set at $\Delta u(t) = 4$ ft^3/min.

Initial estimates for the various parameters are required for the use of non-linear regression to determine the parameters of the FOPDT equation. The dead time θ_P is approximately 1 minute from examination of the data given in Figure 13–9. The total measured process variable change is about 8°F and occurs over a time interval of approximately 40 minutes. The process gain K_P is estimated to be 2°F/(ft^3/min) as shown in Figure 13–9. The process time constant τ_P can be approximated by using the relationship that 63.8% change in the response occurs in one time constant. This is about 9 minutes from Figure 13–9. The initial steady-state temperature T_s is approximately 25°F.

The data file for part (a) is given as Excel file named P13-4A.XLS. The data from Excel can be copied as individual columns of time t and temperature T into the POLYMATH *Regression & Data Analysis* program for treatment with nonlinear regression. The POLYMATH problem entry and corresponding results are presented in Figure 13–10. Note that Equations (13-23) and (13-24) are combined into a single expression for use as the nonlinear regression model.

T=if(t<(2+thetaP))then(Ts)else(Ts+Kp*4*(1-exp((-t+2+thetaP)/tauP)))

The resulting POLYMATH regression and graph shown in Figure 13–11 demonstrate that the resulting regression correlation compares favorably with the response data. The actual parameter values and the confidence intervals are very reasonable. Other initial estimates yield the same converged regression parameter values as this is always a useful validation of the regression. The residual plot shown in Figure 13–11 also indicates that the errors are random and that the regression result is appropriate.

When FOPDT models are fitted to actual control systems data, the resulting model will only approximate the measured variable response as most processes are nonlinear. Collection of data in the vicinity of the controller set point will allow the FOPDT model to be a very useful tool in the design of control systems.

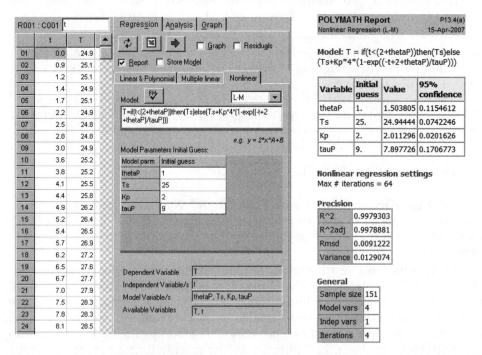

Figure 13–10 POLYMATH Nonlinear Regression of First-Order plus Dead-Time Model (File **P13-04A.POL**)

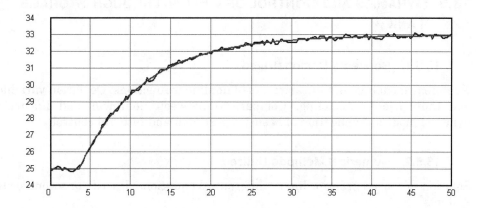

Figure 13–11 POLYMATH Nonlinear Regression Result for First-Order plus Dead-Time Model (File **P13-04A.POL**)

$T - T_{\text{FOPDT}}$

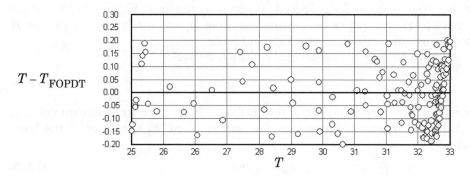

Figure 13–12 POLYMATH Nonlinear Regression Result for First-Order plus Dead-Time Model (File **P13-04A.POL**)

The program and data files are provided in directory Chapter 13 and named **P13-04A.POL** and **P13-04B.POL.**

13.5 DYNAMICS AND CONTROL OF A FLOW-THROUGH STORAGE TANK

13.5.1 Concepts Demonstrated

Nonlinear modeling and linearization of nonlinear functions. Deviation variables and their use in modeling. Comparison between nonlinear and linearized dynamic models. Application of basic proportional and integral control.

13.5.2 Numerical Methods Utilized

Solution of second-order ordinary differential equations describing second-order systems.

13.5.3 Problem Statement

A well-mixed tank has a single feed stream as is shown in Figure 13–13. The single exit stream can either be a simple draining pipe or a valve whose volumetric flow rate can be controlled with a PI controller.

A mass balance on the contents of the tank yields

$$A_c \frac{dh}{dt} = F_1 - F_2 \qquad \text{(13-25)}$$

where F_1 is the input volumetric flow rate and F_2 is the output stream volumetric flow rate. When the outlet pipe flow is turbulent, F_2 is related to the liquid level h by

$$F_2 = c\sqrt{h} \qquad \text{(13-26)}$$

where c can be considered to be constant. Thus Equation (13-25) becomes a nonlinear model given by

$$A_c \frac{dh}{dt} = F_1 - c\sqrt{h} \qquad \text{(13-27)}$$

which is not a first-order linear ordinary differential equation in the form of Equation (13-42). However, the numerical integration of Equation (13-27) can be easily accomplished with POLYMATH.

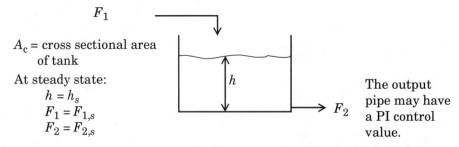

F_1

A_c = cross sectional area
    of tank

At steady state:
  $h = h_s$
  $F_1 = F_{1,s}$
  $F_2 = F_{2,s}$

h

F_2

The output
pipe may have
a PI control
value.

Figure 13–13 Flow-through Storage Tank with Constant Cross-sectional Area

Linearization of a Nonlinear Function

Nonlinear functions can converted into linear functions by use of a Taylor series expansion. The function of the $\sqrt{h}$ in Equation (13-27) in a region close to the steady state where $h = h_s$ can be approximated by

$$f(h) = f(h_s) + \frac{\partial}{\partial h}f(h)\Big|_{h_0} (h - h_s) + \frac{1}{2!}\frac{\partial^2}{\partial h^2}f(h)\Big|_{h_0} (h - h_s)^2 + \text{higher order terms} \quad \textbf{(13-28)}$$

The second and higher ordered terms can be neglected so that the function $f(h) = \sqrt{h}$ can be linearized to

$$\sqrt{h} = \sqrt{h_s} + \frac{(h - h_s)}{2\sqrt{h_s}} \quad \textbf{(13-29)}$$

Thus Equation (13-27) can be rewritten as a linear first-order ordinary differential equation with the linearization for the $\sqrt{h}$ as

$$A_c\frac{dh_L}{dt} + \frac{c}{2\sqrt{h_s}}h_L = F_1 - \frac{c}{2}\sqrt{h_s} \quad \textbf{(13-30)}$$

Deviation Variables

Often in control problem formulation, the variables and equations are transformed into deviation variables. Note that the general solutions to the first- and second-order ordinary differential equations are usually given for deviation variables (see Problem 13.6).

   The use of deviation variables in Equation (13-25) can be introduced by first writing this equation at the steady state where $h = h_s$, $F_1 = F_{1,s}$, and $F_2 = F_{2,s}$.

$$A_c\frac{dh_s}{dt} = F_{1,s} - F_{2,s} \quad \textbf{(13-31)}$$

Equation (13-31) can be subtracted from Equation (13-25) and written as

$$A_c\frac{d}{dt}(h - h_s) = (F_1 - F_{1,s}) - (F_2 - F_{2,s}) \quad \textbf{(13-32)}$$

Thus defining the deviation variables by

$$h' = h - h_s \quad ; \quad F_1' = F_1 - F_{1,s} \quad ; \quad F_2' = F_2 - F_{2,s} \quad \textbf{(13-33)}$$

yields Equation (13-30) written as a linear first-order ordinary differential equation.

$$A_c\frac{d}{dt}h' = F_1' - F_2' \quad \textbf{(13-34)}$$

Basic Proportional and Integral (PI) Control

When the outlet volumetric flow rate from the storage tank is under PI control, the basic model for the system is given by Equation (13-34) with the controller equation providing the outlet flow rate from the tank. Here K_c is the proportional gain and τ_I is the integral time constant for the controller.

$$F_2' = K_c h' + \frac{K_c}{\tau_I} \int_0^t h' dt \tag{13-35}$$

With PI control action for F_2 and the use of deviation variables, Equation (13-34) can be written as

$$A_c \frac{dh'}{dt} = F_1' - \left(K_c h' + \frac{K_c}{\tau_I} \int_0^t h' dt \right) \text{ I. C. } h = h_s \text{ at } t = 0 \tag{13-36}$$

Numerical Solution of Control Equations

The numerical solution of Equation (13-36) can be achieved by defining a variable for the integral term which represents the deviation between the actual height in the tank and the desired height (set point height). Thus

$$\frac{d(\text{errsum})}{dt} = h' \text{ I. C. errsum} = 0 \text{ at } t = 0 \tag{13-37}$$

and Equation (13-36) becomes

$$A_c \frac{dh'}{dt} = F_1' - \left(K_c h' + \frac{K_c}{\tau_I} \text{errsum} \right) \text{ I. C. } h' = 0 \text{ at } t = 0 \tag{13-38}$$

Thus differential equations (13-37) and (13-38) must be solved simultaneously.

Analytical Solution of Control Equations

The analytical solution of Equation (13-36) can be obtained by first taking the time derivative of Equation (13-38) to yield

$$A_c \frac{d^2 h'}{dt^2} + K_c \frac{dh'}{dt} + \frac{K_c}{\tau_I} h' = \frac{dF_1'}{dt} \tag{13-39}$$

For a step change in F_1 at $t > 0$, $\frac{dF_1'}{dt} = 0$ and Equation (13-39) can be rearranged.

$$\frac{A_c \tau_I}{K_c} \frac{d^2 h'}{dt^2} + \tau_I \frac{dh'}{dt} + h' = 0 \tag{13-40}$$

This is the same general form of the second-order solution that is given in Problem 13.1 where

$$\tau = \sqrt{\frac{A_c \tau_I}{K_c}} \quad \text{and} \quad \zeta = \frac{1}{2}\sqrt{\frac{K_c \tau_I}{A_c}} \tag{13-41}$$

Therefore the dynamic responses of Equation (13-40) are expected to be highly dependent upon ζ as shown in Figures 13–2 and 13–3. Thus the application of the PI control to this first-order process yields a second-order system with a variety of overdamped to underdamped responses that are possible.

(a)   Consider the storage tank with only the exit pipe (no control valve) when $A_c = 10$ m^2, $h_s = 5$ m, and $F_{1,s} = F_{2,s} = 3$ m^3/min. Demonstrate the effect of linearization by introducing a step change in feed flow rate at $t = 0$ to the storage tank to 50% of the original steady-state feed rate. What is the new steady state value of h? Plot the calculated changes in both of the calculated tank heights, h from Equation (13-27) and h_L from Equation (13-30), to the time when h decreases to 105% of the new steady-state value. What is the % error in the linearized model calculation of h_L at $t = 10, 20, 30$ and 40 minutes and at the new steady state?

(b)   Implement PI control on the outlet valve for the storage tank in part (a) when $K_c = 2$ and $\tau_I = 0.6$. Plot the liquid level height versus time.

(c)   Repeat part (b) when the integral control action is removed. What is the deviation variable value h' (or offset) at the new steady state?

(d)   Repeat part (b) for various combinations of K_c and τ_I that will lead to overdamped, critically damped, and underdamped responses.

(e)   Investigate the response for h when $K_c = 2$, $\tau_I = 0.6$, and the feed rate F_1 is at 3 m^3/min for 2 minutes, increased to 3.5 m^3/min for 5 minutes, and then returned to 3.0 m^3/min. The steady-state conditions are as given in part (a).

13.5.4   Solution (Partial Solution and Suggestions)

(a) **Effect of Model Linearization** The steady-state tank height after the step change to $F_1 = 0.50*3 = 1.5$ m^3/min can be calculated by setting the time derivative in Equation (13-25) to zero and solving for the corresponding h to yield 1.25 m.

The actual calculation of h from Equation (13-27) and the calculation of the linearized value of h_L from Equation (13-30) can be accomplished with the POLYMATH program given in Table 13–4. Both numerical solutions are carried out simultaneously. The resulting graph of h from Equation (13-27) and h_L from Equation (13-30) is shown in Figure 13–14. Note that the linearized solution is initially accurate, but it become highly inaccurate as the system reaches the new steady state. The solution to the linearized model could also have been generated by the analytical solution given in Problem 13.1.

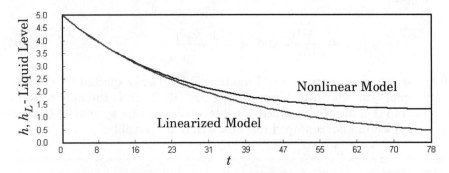

Figure 13–14 Calculated Liquid Level Heights for Nonlinear versus Linearized Model
(File **P13-05A.POL**)

Table 13–4 POLYMATH Coding for Model Comparison (File **P13-05A.POL**)

Line	Equation
1	d(h) / d(t) = (F1-c*sqrt(h))/Ac #Accurate model differential equation
2	h(0) = 5 #Initial condition for liquid level h
3	d(hl)/d(t)=(F1-c/2*sqrt(hs)-c/2*hl/sqrt(hs))/Ac #Linearized model differential equation
4	hl(0) = 5 #Initial condition for liquid level hl
5	Ac=10 #Cross sectional area of tank
6	hs=5 #Initial steady state liquid level
7	c=3/sqrt(hs) #Constant for calculation of outlet volumetric flow rate using liquid level height
8	F1=3*0.5 #Calculation of new input flow rate at 50% of the original (step function at t = 0
9	t(0) = 0 #Initial condition on time
10	t(f) = 78 #Final value of time (must be greated than tLIM)
11	hsNEW=(1.5/c)^2 #Calculation of new steady-state tank height after step function in F1
12	d(tLIM)/d(t)=if(h>1.05*hsNEW)then(1)else(0) #Variable tLIM follows the time t until desired time condition is satisfied
13	tLIM(0)=0 #Initial condition on tLIM

(b) PI Control of Liquid Level The model for the control action is given
by Equations (13-37) and (13-38). The POLYMATH coding for these equations is
given in Table 13–5.

Table 13–5 POLYMATH Program for PI Control (File **P13-05B.POL**)

Line	Equation
1	d(errsum) / d(t) = hp # The symbol for the prime ' is "p"
2	errsum(0) = 0
3	d(hp) / d(t) = (F1p-(Kc*hp+Kc*errsum/tauI))/Ac
4	t(0) = 0
5	t(f) = 40

Line	Equation
6	hp(0) = 0
7	Ac=10
8	hs=5
9	F1s=3
10	F2s=3
11	Kc= 2.
12	taul=0.6
13	F1p=F1-F1s
14	F1=0.50*F1s
15	F2p=Kc*hp+Kc*errsum/taul
16	F2=F2p+F2s
17	h=hp+hs

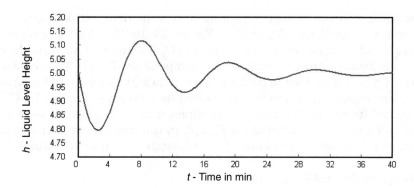

Figure 13–15 Dynamic Response of Liquid Level Height h under PI Control (File **P13-05B.POL**)

The predicted liquid-level height shown in Figure 13–15 shows a typical underdamped response of a second-order system.

(c) The integral action can be eliminated by dropping the terms containing the "errsum" variable in the related equations.

The POLYMATH problem solution files are found in directory Chapter 13 with files named **P13-05A.POL** and **P13-05B.POL**.

13.6 DYNAMICS AND CONTROL OF A STIRRED TANK HEATER

13.6.1   Concepts Demonstrated

Closed-loop dynamics of a first-order process including dead time and the Pade approximation of time delay.

13.6.2   Numerical Methods Utilized

Solution of ordinary differential equations, generation of step functions, simulation of a proportional integral controller.

13.6.3   Problem Statement

A continuous process system consisting of a well-stirred tank, heater, and PI temperature controller is depicted in Figure 13–16. The feed stream of liquid with density of ρ (kg/m^3) and heat capacity of C_p (kJ/kg·°C) flows into the heated tank at a constant rate of W (kg/min) and temperature T_i (°C). The volume of the tank is V (m^3). It is desired to heat this stream to a higher set point temperature T_r (°C). The outlet temperature is measured by a thermocouple as T_m (°C), and the required heater input q (kJ/min) is adjusted by a PI temperature controller. The control objective is to maintain $T_0 = T_r$ in the presence of a change in inlet temperature T_i which differs from the steady-state design temperature of T_{is}.

Modeling and Control Equations
An energy balance on the stirred tank yields

$$\frac{dT}{dt} = \frac{WC_p(T_i - T) + q}{\rho V C_p} \tag{13-42}$$

with initial condition $T = T_r$ at $t = 0$ which corresponds to steady-state operation

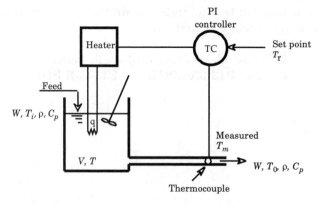

Figure 13–16 Well Mixed Tank with Heater and Temperature Controller

at the set point temperature T_r.

The thermocouple for temperature sensing in the outlet stream is described by a first-order system plus the dead time τ_d which is the time for the output flow to reach the measurement point. The dead-time expression is given by

$$T_0(t) = T(t - \tau_d) \tag{13-43}$$

The effect of dead time may be calculated for this situation by the Padé approximation which is a first-order differential equation for the measured temperature.

$$\frac{dT_0}{dt} = \left[T - T_0 - \left(\frac{\tau_d}{2} \right) \left(\frac{dT}{dt} \right) \right] \frac{2}{\tau_d} \quad \text{I. C. } T_0 = T_r \text{ at } t = 0 \text{ (steady state)} \tag{13-44}$$

The above equation is used to generate the temperature input to the thermocouple T_0.

The thermocouple shielding and electronics are modeled by a first-order system for the input temperature T_0 given by

$$\frac{dT_m}{dt} = \frac{T_0 - T_m}{\tau_m} \quad \text{I. C. } T_m = T_r \text{ at } t = 0 \text{ (steady state)} \tag{13-45}$$

where the thermocouple time constant τ_m is known.

The energy input to the tank, q, as manipulated by the proportional/integral (PI) controller can be described by

$$q = q_s + K_c(T_r - T_m) + \frac{K_c}{\tau_I} \int_0^t (T_r - T_m)dt \tag{13-46}$$

where K_c is the proportional gain of the controller, and τ_I is the integral time constant or reset time. The q_s in the above equation is the energy input required at steady state for the design conditions as calculated by

$$q_s = WC_p(T_r - T_{is}) \tag{13-47}$$

The integral in Equation (13-46) can be conveniently calculated by defining a new variable as

$$\frac{d}{dt}(\text{errsum}) = T_r - T_m \quad \text{I. C. errsum} = 0 \text{ at } t = 0 \text{ (steady state)} \tag{13-48}$$

Thus Equation (13-46) becomes

$$q = q_s + K_c(T_r - T_m) + \frac{K_c}{\tau_I}(\text{errsum}) \tag{13-49}$$

Let us consider some of the interesting aspects of this system as it responds to a variety of parameter and operational changes. The numerical values of the system and control parameters in Table 13–6 will be used for baseline steady state operation.

Table 13–6 Baseline System and Control Parameters

$\rho V C_p = 4000 \text{ kJ/°C}$	$W C_p = 500 \text{ kJ/min·°C}$
$T_i = 60 \text{ °C}$	$T_r = 80 \text{ °C}$
$\tau_d = 1 \text{ min}$	$\tau_m = 5 \text{ min}$
$K_c = 50 \text{ kJ/min·°C}$	$\tau_I = 2 \text{ min}$

(a)   Demonstrate the open loop performance (set $K_c = 0$) of this system when the system is initially operating at design steady state at a temperature of 80°C, and the inlet temperature T_i is suddenly changed to 40°C at time $t = 10$ min. Plot the temperatures T, T_0, and T_m to steady state, and verify that Padé approximation for 1 minute of dead time given in Equation (13-44) is working properly.

(b)   Demonstrate the closed-loop performance of the system for the conditions of part (a) and the baseline parameters from Table 13–6. Plot temperatures T, T_0, and T_m to steady state.

(c)   Repeat part (b) with $K_c = 500 \text{ kJ/min·°C}$.

(d)   Repeat part (c) for proportional only control action by setting the term $K_c/\tau_I = 0$.

(e)   Implement limits on q (as per Equation (13-49)) so that the maximum is 2.6 times the baseline steady-state value and the minimum is zero. Demonstrate the system response from baseline steady state for a proportional only controller when the set point is changed from 80°C to 90°C at $t = 10$ min. with $K_c = 5000 \text{ kJ/min·°C}$. Plot q and q_{lim} (the bounded value of q) versus time to steady state to demonstrate the limits. Also plot the temperatures T, T_0, and T_m to steady state to indicate controller performance.

13.6.4   Solution

This problem requires the solution of Equations (13-42) and (13-44) through (13-49) which can be accomplished with the POLYMATH *Simultaneous Differential Equation Solver*. The step change in the inlet temperature can be introduced at $t = 10$ by using the POLYMATH "if ... then ... else ..." statement to provide the logic for a variable to change at a particular value of t. The generation of a step change at $t = 10$, for example, is accomplished by the following POLYMATH program statement

```
step=if (t<10) then (0) else (1)
```

(a) Open Loop Performance The step down of 20°C in the inlet temperature at $t = 10$ is implemented in line 12 of the POLYMATH equation set shown in Table 13–7 for the case where $K_c = 0$, which gives the open-loop response.

Table 13–7 Equation Set for Open Loop Performance (File **P13-06A.POL**)

Line	Equation
1	d(T)/d(t) = (WC*(Ti-T)+q)/rhoVCp
2	d(T0)/d(t) = (T-T0-(taud/2)*dTdt)*2/taud
3	d(Tm)/d(t) = (T0-Tm)/taum #
4	d(errsum)/d(t) = Tr-Tm #
5	WC = 500 #
6	rhoVCp = 4000
7	taud = 1
8	taum = 5
9	Tr = 80
10	Kc = 0
11	taul = 2
12	step = if (t<10) then (0) else (1)
13	Ti = 60+step*(-20)
14	q = 10000+Kc*(Tr-Tm)+Kc/taul*errsum
15	dTdt = (WC*(Ti-T)+q)/rhoVCp
16	t(0)=0
17	T(0)=80
18	T0(0)=80
19	Tm(0)=80
20	errsum(0)=0
21	t(f)=60

A plot of the temperatures T, T_0, and T_m as generated by POLYMATH is given in Figure 13–17, which also verifies the steady-state operation for $t < 10$ min as there is no change in any of the temperature values.

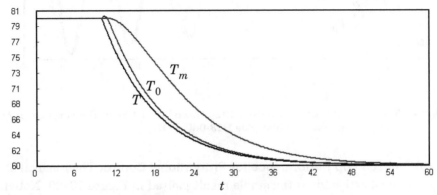

Figure 13–17 Open Loop Response to Step Down in Inlet Feed Temperature at $t = 10$ ı (File **P13-06A.POL**)

(b) Closed-Loop Performance The closed-loop performance of the PI controller requires the change of K_c from zero in part (a) to the baseline proportional gain of 50. This simple change results in the temperature transients shown in Figure 13–18.

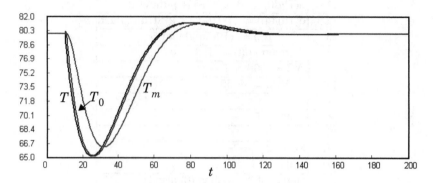

Figure 13–18 Closed-Loop Response to Step Down in Inlet Feed Temperature at t = 10 (File **P13-06B.POL**)

(c) Closed-Loop Performance for K_C = 500 The increase of a factor of 10 in the proportional gain from the baseline case gives the unstable result plotted in Figure 13–19. This is clearly an undesirable result.

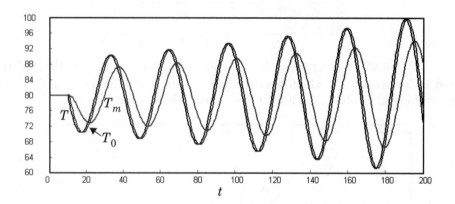

Figure 13–19 Closed-Loop Response to Step Down in Inlet Feed Temperature at $t = 10$ min for $K_c = 500$ (File **P13-06C.POL**)

(d) Closed Loop Performance for Proportional Control The removal of the integral control action gives the stable result plotted in Figure 13–20. Note that there is an offset from the set point when the system returns to steady-state operation. This offset is always the case for proportional control, and the use of integral control allows the offset to be eliminated.

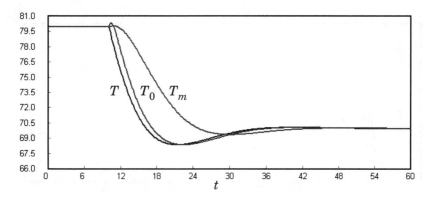

Figure 13–20 Closed Loop Response for Proportional Control
(File **P13-06D.POL**)

(e) Closed-Loop Performance with Limits on q There are many times in control when limits must be established. In this example, the limits on q can be achieved by a POLYMATH "if ... then ... else ..." statement which is indicated on the complete POLYMATH equations set given in Table 13–8.

Table 13–8 POLYMATH Equation Set for Closed Loop Performance with Limits on q (File **P13-06E.POL**)

Line	Equation
1	d(T)/d(t) = (WC*(Ti-T)+qlim)/rhoVCp
2	d(T0)/d(t) = (T-T0-(taud/2)*dTdt)*2/taud
3	d(Tm)/d(t) = (T0-Tm)/taum
4	d(errsum)/d(t) = Tr-Tm
5	WC = 500
6	Ti = 60
7	rhoVCp = 4000
8	taud = 1
9	taum = 5
10	Kc = 5000
11	taul = 2
12	step = if (t<10) then (0) else (1)
13	Tr = 80+step*(10)
14	q = 10000+Kc*(Tr-Tm)
15	qlim = if(q<0)then(0)else(if(q>=2.6*10000)then(2.6*10000)else (q))
16	dTdt = (WC*(Ti-T)+qlim)/rhoVCp
17	t(0)=0
18	T(0)=80
19	T0(0)=80
20	Tm(0)=80
21	errsum(0)=0
22	t(f)=200

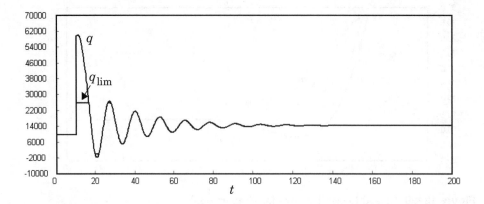

Figure 13–21 Closed-Loop Response for Proportional Control with Limits on q.

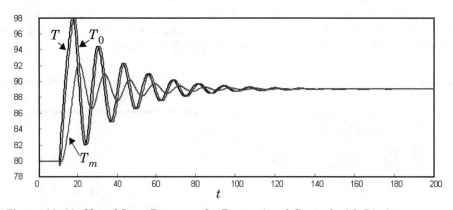

Figure 13–22 Closed-Loop Response for Proportional Control with Limits on q.

The values of q and q_{lim} plotted in Figure 13–21 indicate that this proportional controller has wide oscillations before settling to a steady state, and the limits imposed on q_{lim} are evident. The corresponding plots of the system temperatures are presented in Figure 13–22.

The POLYMATH problem solution files are found in directory Chapter 13 with files named **P13-06A.POL, P13-06B.POL, P13-06C.POL, P13-06D.POL**, and **P13-06E.POL**. This problem is also solved with Excel, Maple, MathCAD, MATLAB, Mathematica, and POLYMATH as Problem 10 in the Set of Ten Problems discussed on the book web site.

13.7 CONTROLLER TUNING USING INTERNAL MODEL CONTROL (IMC) CORRELATIONS

13.7.1   Concepts Demonstrated

Utilization of first-order plus dead-time (FOPDT) parameters to determine the Internal Model Control (IMC) controller settings. Simulation of control action for various controller settings.

13.7.2   Numerical Methods Utilized

Solution of a second-order ordinary differential equation representing a PI controller.

13.7.3   Problem Statement

The well-stirred tank heater and PI temperature controller are identical to the system discussed in Problem 13.6. The controller is to be tuned with the IMC recommended tuning correlations given in Table 13–9 for a process where K_P is the process gain, τ_P is the process time constant, and θ_P is the process dead time.

Table 13–9 IMC Controller Tuning Correlations from Cooper[3]

Desired Tuning	Controller Response Time Constant Formula
Aggressive	τ_C is the larger of $0.1\tau_P$ or $0.8\theta_P$
Moderate	τ_C is the larger of $1.0\tau_P$ or $8.0\theta_P$
Conservative	τ_C is the larger of $10.0\tau_P$ or $80.0\theta_P$
Controller Proportional Gain and Integral Time Constant Formulas	
$K_C = \dfrac{1}{K_P}\dfrac{\tau_P}{(\tau_d + \tau_C)}$ $\qquad\qquad \tau_I = \tau_P$	

Practical controller tuning can be achieved by using a first-order plus dead-time model on most chemical processes. The FOPDT (First-Order Plus Dead-Time) parameters can be determined by open-loop response testing on the process based on a small step change in the process variable as has been discussed in Problem 13.4.

The tuning parameters for the controller can be calculated from Table 13–9 using the FOPDT parameters and then used in the PI control. Thus the heat input q for the process is given by

$$q = q_s + K_c(T_r - T_m) + \frac{K_c}{\tau_I}\int_0^t (T_r - T_m)dt \qquad\qquad \textbf{(13-50)}$$

where the variables are defined in Problem 13.6.

The stirred tank heater shown in Figure 13–16 of Problem 13.6 is operating at the steady-state baseline conditions and controller parameters given in Table 13–6. A first-order plus dead-time (FOPDT) model was fitted for the measured temperature T_m as the energy input q was varied from its steady-state value. The FOPDT parameters were found to be $K_P = 0.002°C/(kJ/min)$, $\tau_P = 10.7$ min, and $\tau_d = 3.87$ min.

(a) Determine the controller tuning parameters when the IMC aggressive control is to be used based on the FOPDT parameters. Predict the tank temperature T for 60 minutes when a feed disturbance starts after five minutes at steady-state temperature T_i of 60°C, is stepped up to 65°C for 5 minutes and then stepped down to 50°C for 5 minutes, and then is held at 60°C. Also plot the resulting q for 60 minutes during the feed temperature disturbance.

(b) Repeat part (a) for the IMC moderate tuning parameters.

(c) Repeat part (a) for the IMC conservative tuning parameters.

(d) What qualitative differences are observed between the three levels of controller tuning?

13.7.4   Solution (Partial with Suggestions)

(a) The controller tuning parameters can easily be calculated from Table 13–9. Thus for aggressive control in this problem

$$\tau_C = \text{the larger of } 0.1\tau_P \text{ or } 0.8\theta_P = 0.1(10.7) \text{ or } 0.8(3.87) = 3.10$$

$$K_C = \frac{1}{K_P(\theta_P + \tau_C)}\frac{\tau_P}{} = \frac{1}{0.002}\frac{10.7}{(3.87 + 3.10)} = 768$$

$$\tau_I = \tau_P = 10.7$$

As in Problem 13.6, the Equations (13-42) and (13-44) through (13-49) that describe the control of the heated tank can be solved with the POLYMATH *Simultaneous Differential Equation Solver*. The step change in the inlet temperature can be introduced by nested "if ... then ... else ..." logic in line 12 of the POLYMATH program as shown in Table 13–10.

Ti=if(t<5)then(60)else(if(t<10)then(65)else(if(t<15)then(50)else(60)))

Table 13–10 POLYMATH Program Using IMC Tuning Parameters for Aggressive Control (File **P13-07A.POL**)

Line	Equation
1	d(T)/d(t) = (WC*(Ti-T)+q)/rhoVCp
2	d(T0)/d(t) = (T-T0-(taud/2)*dTdt)*2/taud
3	d(Tm)/d(t) = (T0-Tm)/taum
4	d(errsum)/d(t) = Tr-Tm
5	WC = 500

Line	Equation
6	rhoVCp = 4000
7	taud = 1
8	taum = 5
9	Tr = 80
10	Kc = 768
11	taul = 10.7
12	Ti = if(t<5)then(60)else(if(t<10)then(65)else(if(t<15)then(50)else(60)))
13	q = 10000+Kc*(Tr-Tm)+Kc/taul*errsum
14	dTdt = (WC*(Ti-T)+q)/rhoVCp
15	t(0)=0
16	T(0)=80
17	T0(0)=80
18	Tm(0)=80
19	errsum(0)=0
20	t(f)=60

(b) The moderate controller tuning parameters can easily be calculated from Table 13–9. Thus

$$\tau_C = \text{ the larger of } 1.0\tau_P \text{ or } 8.0\tau_d = 1.0(10.7) \text{ or } 8.0(3.87) = 31.0$$

$$K_C = \frac{1}{K_P(\tau_d + \tau_C)}\frac{\tau_P}{} = \frac{1}{0.002}\frac{10.7}{(3.87 + 31.0)} = 153$$

$$\tau_I = \tau_P = 10.7$$

The POLYMATH problem solution file for part (a) is found in directory Chapter 13 with file named **P13-07A.POL**

13.8 FIRST ORDER PLUS DEAD TIME MODELS FOR STIRRED TANK HEATER

13.8.1   Concepts Demonstrated

Fitting measured process variable data resulting from a step change in a controlled variable to obtain first-order plus dead-time model parameters for process control.

13.8.2   Numerical Methods Utilized

Nonlinear regression of data.

13.8.3   Problem Statement

(a) Determine the FOPDT parameters for stirred tank heater using the model of the control system developed in part (b) of Problem 13.6 by performing an open-loop test by stepping up the value of q by 20% of its baseline value after 10 minutes at steady state.

(b) Repeat part (a) for a step down in the value of q by 20%.

(c) How would you use this information in the design of a control system with moderate control parameters?

(d) Repeat part (a) by stepping the set point temperature of the controller to a new temperature that will also increase q by 20% at the new steady state. Use the controller setting for baseline operation under PI control as given in Problem 13.6. Does the controller action significantly alter the values of the FOPDT parameters of that were determined in part (a)?

13.8.4   Solution (Suggestions)

(a) and (b) The POLYMATH program of the control system for the stirred tank heater is given in Table 13–7 can easily be modified to incorporate the step up in set point temperature T_r while maintaining the inlet temperature T_i at the steady state value of 60°C. Make sure that the controller remains turned off by keeping $K_C = 0$ in order to determine the open loop response.

13.9 CLOSED-LOOP CONTROLLER TUNING–THE ZIEGLER-NICHOLS METHOD

13.9.1 Concepts Demonstrated

The determination of the ultimate gain and ultimate period of a closed-loop P or PI controller by the Ziegler-Nichols method and calculation of the tuning parameters.

13.9.2 Numerical Methods Utilized

Solution of systems of differential equations. Determination of amplitudes and periods for stable limit cycles.

13.9.3 Problem Statement

The Ziegler-Nichols method provides the historical foundation for oscillation-based tuning that is widely in industry (see Problem 13.10). This method utilizes the ultimate gain of the control system that can be established with P-only control. The proportional gain K_C is increased until the controlled process variable output switches from a stable oscillation to an unstable oscillation. This yields the ultimate or critical gain K_{Cu}. The time between the peaks of the oscillations is the ultimate or critical period P_u. The ultimate gain and period can be used to calculate the suggested controller tuning parameters according to Table 13–11 based on work by Ziegler-Nichols[4] and Tyreus-Luyben.[5]

Table 13–11 Tuning Parameters from Ultimate Gain and Ultimate Period

Tuning Parameters	K_C	τ_I
Ziegler-Nichols[4]	$0.45K_{Cu}$	$P_u/1.2$
Tyreus-Luyben[5]	$K_{Cu}/3.2$	$2.2P_u$

(a)   Determine the ultimate gain K_{Cu} and the ultimate period P_u for the stirred tank heater process described in Problem 13.6.

(b)   Calculate the Ziegler-Nichols tuning parameters of K_c and τ_I for a PI controller and determine the expected response of the measured temperature T_m in stirred tank heater at steady state to a step change down in the inlet temperature T_i to 50°C with these settings.

(c)   Repeat part (b) for the Tyreus-Luyben tuning parameters.

(d)   Compare the control actions of the Ziegler-Nichols and the Tyreus-Luyben tuning parameters

13.9.4   Solution (Partial)

(a) The POLYMATH code given in Table 13–7 for part (a) of Problem 13.6 (File **P13-06A.POL**) can be modified to determine the value of K_C that provides a stable oscillation with larger values yielding unstable oscillations. Table 13–12 indicates an arbitrary modification in line 9 to introduce a small pertubation of 0.1°C in the set point that is implemented between 10 and 20 minutes to check for stability of the control system. This pertubation is implemented with the logic of line 12. Also the controller is temporarily made proportional only by multiplication of the integral control code in line 14 by zero.

Table 13–12 POLYMATH Code for Determination of Ultimate Gain (File **P13-09A.POL**)

Line	Equation
1	d(T)/d(t) = (WC*(Ti-T)+q)/rhoVCp
2	d(T0)/d(t) = (T-T0-(taud/2)*dTdt)*2/taud
3	d(Tm)/d(t) = (T0-Tm)/taum
4	d(errsum)/d(t) = Tr-Tm
5	WC = 500
6	rhoVCp = 4000
7	taud = 1
8	taum = 5
9	**Tr = 80+0.1*step**
10	Kc = 7030
11	taul = 2
12	**step = if (t<10) then (0) else (if(t<20)then(1) else(0))**
13	Ti = 60
14	**q = 10000+Kc*(Tr-Tm)+Kc/taul*errsum*0**
15	dTdt = (WC*(Ti-T)+q)/rhoVCp
16	t(0)=0
17	T(0)=80
18	T0(0)=80
19	Tm(0)=80
20	errsum(0)=0
21	t(f)=100

The POLYMATH solution with the *Ordinary Differential Equation Solver* should demonstrate steady-state operation for the first 10 minutes. Graphical results from different trial values of K_C can be examined by trial and error until the ultimate gain K_{Cu} can be determined. This procedure indicates that there is apparently a stable limit cycle for K_C between 7000 and 7100 kJ/min·°C. Closer examination finds that the ultimate or critical gain is approximately 7030 kJ/min·°C and the ultimate or critical period is about 11.1 minutes as shown in Figure 13–23.

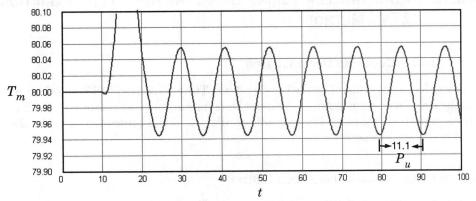

Figure 13–23 Stable Limit Cycle for Determination of Ultimate Gain K_{Cu} and Ultimate Period P_u (File **P13-09A1.POL**)

The POLYMATH problem solution file for part (a) is found in directory Chapter 13 with file named **P13-09A.POL**.

13.10 PI CONTROLLER TUNING USING THE AUTO TUNE VARIATION "ATV" METHOD

13.10.1 Concepts Demonstrated

Tuning feedback controllers using the Auto Tune Variation "ATV" method with application of various tuning settings. Determination of ultimate gain and ultimate period with a relay-type (on-off) oscillation of the controller variable above and below the set point.

13.10.2 Numerical Methods Utilized

Solution of systems of differential equations. Simulation of "relay" type perturbations.

13.10.3 Problem Statement

Many industrial controllers utilize an autotuning capability to determine the tuning parameters (Bequette[1], pp. 354–357). When this ATV tuning method is utilized (Astrom and Hagglund[6]), the controller output is placed in a temporary operational mode in which the control variable is continuously alternated above and then below the current steady-state value to determine the ultimate gain and ultimate period. This control action is based on the current error between the controller set point and the measured process variable. The period of the established limit cycle of the measured process variable is the ultimate period P_u. The ultimate gain is given by

$$K_{Cu} = \frac{4R}{a\pi} \qquad \text{(13-51)}$$

where a is amplitude of the measured process variable and R is the magnitude of the control variable perturbation introduced by the relay.

(a) Determine the ultimate frequency and ultimate gain for the baseline heating tank process described in Problem 13.6 using the ATV tuning method.

(b) Set K_c equal to the ultimate gain determined in part (a) using a proportional only controller and observe whether the theoretically expected limit cycles are obtained for the conditions of part (a).

(c) Use the Ziegler-Nichols tuning parameters calculated from Table 13–11 to determine the response of the measured temperature T_m to a 50% step down in both the feed and exit flow rates W at $t = 20$ minutes.

(d) Repeat part (c) using the Tyreus-Luyben tuning parameters from Table 13–11.

13.10.4 Solution (Partial)

(a) The general POLYMATH code for the control of the stirred tank heater is given in Table 13–7 for Problem 13.6. This code (File **P13-06A.POL**) can easily be modified to alternate the control variable above and below the steady-state value as is indicated in lines 12-14 of Table 13–13. Note that the control output variable q is perturbed at the level of 40% of the original value.

Table 13–13 Modified POLYMATH Code for Determination of Ultimate Gain and Ultimate Period with ATD Method (File **P13-10A.POL**)

Line	Equation
1	d(T)/d(t) = (WC*(Ti-T)+q)/rhoVCp
2	d(T0)/d(t) = (T-T0-(taud/2)*dTdt)*2/taud
3	d(Tm)/d(t) = (T0-Tm)/taum
4	d(errsum)/d(t) = Tr-Tm
5	WC = 500
6	rhoVCp = 4000
7	taud = 1
8	taum = 5
9	Tr = 80
10	Ti = 60
11	Kc = 0
12	**error=Tr-Tm**
13	**relay=if(error<0)then(-1)else(1)**
14	**q = 10000+relay*4000**
15	dTdt = (WC*(Ti-T)+q)/rhoVCp
16	t(0)=0
17	T(0)=80
18	T0(0)=80
19	Tm(0)=80
20	errsum(0)=0
21	t(f)=100

The action on the controller manipulated or output variable is shown in Figure 13–24. Close examination indicates that four to six cycles are necessary for the cycle to be completely established.

The response of the measured variable T_m in Figure 13–25 shows the desired oscillations where sufficient time (100 minutes) is used for the cycle to be completely established. The amplitude a is approximately 0.78°C, and the ultimate period P_u is about 11.6 minutes. The ultimate gain is calculated from Equation (13-51) with R = 4000 kJ/min to be K_{Cu} = 6530 kJ/min·°C which compares favorably with K_{Cu} = 7030 kJ/min·°C determined in Problem 13.9. Additional details on the solution are given by Brauner and others.[7]

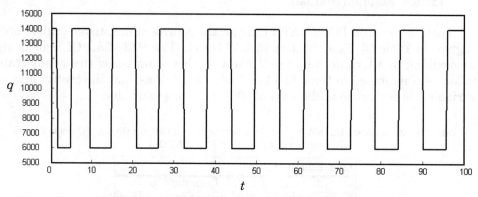

Figure 13–24 Variation of the Manipulated Variable q during ATV Tuning
(File **P13-10A.POL**)

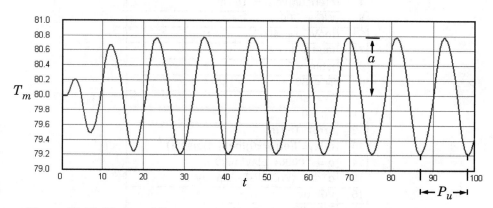

Figure 13–25 Measured Temperature T_m during ATV Tuning (File **P13-10A.POL**)

The POLYMATH problem solution file for part (a) is found in directory
Chapter 13 with file named **P13-10A.POL**.

13.11 RESET WINDUP IN A STIRRED TANK HEATER

13.11.1 Concepts Demonstrated

PI control with limits on the control variable and anti-windup logic to improve control action.

13.11.2 Numerical Methods Utilized

Solution of ordinary differential equations. Generation of step functions. Implementation of limits on variables.

13.11.3 Problem Statement

The stirred tank heater described in Problem 13.6 operates at steady state, with the baseline parameters as given in Table 13–6. The output from the heater is limited to 1.5 times the design value ($q \leq 1.5(10000)$ kJ/min).

(a) At time $t = 0$ the inlet temperature is reduced to half of the steady state value and after 30 minutes it is restored to the steady state value. Simulate the system behavior up to $t = 200$ min. Explain why the controller exhibits poor response dynamics in returning the measured process output temperature to the steady-state value.

(b) Many industrial controllers have anti-windup provisions. One option to prevent reset windup is to hold the error accumulation of the integral action by setting

$$\frac{d}{dt}(\text{errsum}) = 0$$

when the required heat supply exceeds the bounds. Include this feature in your model and repeat the simulation as in part (a). What differences can you notice? Explain your results.

13.11.4 Solution (Suggestion)

Numerical integration of differential equations within simulation problems that have discontinuous changes in variables is sometimes challenging. For this problem, the POLYMATH *Ordinary Differential Equation Solver* may work best with the RKF56 algorithm. This integration algorithm seems to be best suited to deal with the step changes and the anti-windup logic. Alternately, the error tolerance of a particular algorithm can be reduced by increasing its value. Stiff algorithms have difficulty with this type of problem and may not be suitable.

13.12 TEMPERATURE CONTROL AND STARTUP OF A NONISOTHERMAL CSTR

13.12.1 Concepts Demonstrated

Design of a PI control system subject to controlled variable constraints and multiple load variations. Dynamic response of a continuous stirred tank reactor during start-up of a highly exothermic reaction.

13.12.2 Numerical Methods Utilized

Solution of a system of ordinary differential equations representing a second-order system subjected to various constraints and input step functions.

13.12.3 Problem Statement

The jacketed CSTR that is described in Problem 13.3 and Problem 11.23 is to be operated at the high reaction rate steady state indicated in Table 13–3. The temperature in the reactor is measured by a thermocouple that has the same dynamic characteristics as described in Problem 13.6. The feed flow rate is to be adjusted by a PI flow controller with a control objective that is to maintain the upper steady-state temperature of $T \approx 671.28°$R. Safety considerations require that the reactor temperature must not exceed $T = 680°$R. Potential loads include variations of up to $\pm 10°$R in T_{j0} and T_0 (cooling jacket and reactant feed inlet temperatures respectively) and changes of up to $\pm 5\%$ in the inlet concentration of reactant A (C_{A0}). The cooling water flow rate at steady state (F_{j0}) is 49.9 ft^3/hr, and its maximum flow rate is 60 ft^3/hr.

(a) Design a PI controller which will maintain the high-rate steady state subject to the maximum safety temperature and the maximum cooling water flow rate restrictions. Good set point tracking is desired.

(b) Recommend and simulate a procedure that will start up the CSTR by manipulating the feed flow rate and the cooling water flow rate. The objective is to minimize the time necessary to reach a reactor temperature, and jacket temperature that can be maintained at the high-rate reactor steady state by the controller that was designed in part (a). Safety considerations must never be exceeded at any time.

(c) How long would the control system of part (a) be able to recover from a complete loss of cooling water flow rates (F_{j0} and F_j) under steady-state operation? Safety considerations must never be exceeded at any time.

(d) Repeat part (c) in which the reactor flow rates (F_0 and F) are also zero when the cooling water flow rates become zero. (This simulates loss of all electrical pumps.)

13.13 LEVEL CONTROL OF TWO INTERACTIVE TANKS

13.13.1 Concepts Demonstrated

PI control of a process during start up that involves nonlinear interactions.

13.13.2 Numerical Methods Utilized

Solution of ordinary differential equations. Placing limits and constraints on variables.

13.13.3 Problem Statement (Adapted from Corripio[8])

A series of two tanks as shown in Figure 13–26 are in operation where the inlet flow Q_1 is adjusted by a proportional-integral or PI controller. The controller action is given by

$$Q_1 = Q_{max}\left\{0.5 - K_C\left[(H_2 - H_d) + \frac{1}{\tau_I}\int(H_2 - H_d)dt\right]\right\} \qquad \text{(13-52)}$$

where

H_2 = the water level in the second tank (ft)

H_d = the desired water level (set point) in the second tank (ft)

K_C = controller proportional gain (ft^{-1})

τ_I = integral time (min)

Q_{max} = maximum water inlet flow rate Q_1 (ft^3/min)

Q_1 = the adjusted water inlet flow rate (ft^3 / min) where $0 \le Q_1 \le Q_{max}$

The control system should bring the system to steady state where $H_2 = H_d$ in the shortest time. The initial flow starts when both tanks are empty, and the level in either tank cannot exceed the heights of the tanks given by H_{1max} and H_{1max}.

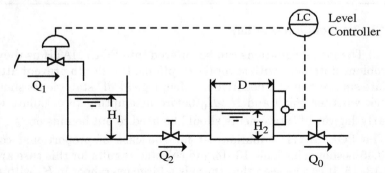

Figure 13–26 Tank Level Control System

Unsteady-state mass balances on each tank lead to the equations

$$\frac{dH_1}{dt} = \frac{4}{\pi d^2}(Q_1 - Q_2) \tag{13-53}$$

and

$$\frac{dH_2}{dt} = \frac{4}{\pi d^2}(Q_2 - Q_0) \tag{13-54}$$

where Q_2 and Q_0 are water flow rates through manual valves (ft³/min). These flow rates can be calculated from the orifice equations of the manual valves

$$Q_2 = C_2\sqrt{H_1 - H_2}$$
$$Q_0 = C_0\sqrt{H_2} \tag{13-55}$$

where C_2 and C_0 are capacity factors for the manual valves (ft³/min·ft$^{1/2}$). The following parameters apply: $H_d = 5.0$ ft, d of each tank = 10 ft, $H_{1max} = H_{2max} = 15$ ft, $Q_{max} = 500$ ft³/min, $C_2 = 63.2$, and $C_0 = 31.6$.

(a)   Use the controller with proportional action only and determine the maximum value of K_c for which the tanks do not overflow during startup to steady state. Verify that $H_2 = H_d$ at steady state.

(b)   Try to eliminate the offset observed in part (a) by including the integral action in the controller's operation. Observe and explain what happens if the integral action is turned on when the flow starts.

(c)   Turn on the integral action of the controller 15 minutes after the flow starts using K_C from part (a) and determine the integral time τ_I that minimizes the time to a stable and non-oscillatory steady state as measured by the integral of the absolute error. This is to be calculated by

$$\int_{15}^{80} |H_2 - H_d| dt$$

13.13.4 Solution (Partial)

(a) The model equations can be entered into POLYMATH as they appear in the problem statement with several exceptions. It order to prevent attempting to calculate square root of a negative number, logical "if" statements should be used to check whether $H_0 > 0$ and $H_1 > H_0$ before attempting to calculate Q_2 and Q_0. Similarly, logical "if" statements should be used to put bounds on Q_1.

The POLYMATH equation set for the case of proportional control with $K_C = 0.45$ is shown in Table 13–14 and selected results for this case are shown in Table 13–15. It can be seen that there is a large overshoot in H_1 with value more than twice higher than the steady-state value. For this value of K_C there is a 15%

offset in the steady-state value of H_2.

The optimal value of K_C can be found by simple trial and error. A small value of K_C causes a small overshoot of H_1 but produces a significant offset in the steady-state value of H_2. The largest value of K_C which does not cause overflow of the first tank is $K_C = 0.48$.

Table 13–14 POLYMATH Program for Proportional Level Control in Two Interactive Tanks

Line	Equation
1	d(H1)/d(t)=4*(Q1-Q2)/(3.1416*d^2)
2	d(H2)/d(t)=4*(Q2-Q0)/(3.1416*d^2)
3	Q2=if (H1>H2) then (63.2*sqrt(H1-H2)) else (0)
4	err=H2-Hd
5	Q0=if (H2>0) then (C0*sqrt(H2)) else (0)
6	Q11=Qmax*(0.5-Kc*(H2-Hd))
7	Q1=if (Q11>Qmax) then (Qmax) else (if (Q11<0) then (0) else (Q11))
8	Kc=0.48
9	Hd=5
10	C0=31.6
11	Qmax=500
12	d=10
13	H1(0)=0
14	H2(0)=0
15	t(0)=0
16	t(f)=50

Table 13–15 Proportional Level Control in Two Interactive Tanks (Selected Variable Values)

Variable	Initial value	Minimal value	Maximal value	Final value
H1	0	0	14.98897	7.158231
H2	0	0	7.706809	5.726585
Q0	0	0	87.7252	75.6197
Q1	500.	0	500.	75.61965
Q11	1450.	-399.6342	1450.	75.61965
Q2	0	0	200.3585	75.61969

(b) The integral of the error, required in Equation (13-52), can be calculated by adding a differential equation to the set shown in Table 13–14:

d(errsum)/d(t)=err
errsum(0)=0

and then modifying line 6 in Table 13–14 to

Q11=Qmax*(0.5-Kc*((H2-Hd)+(1/taul)*errsum))

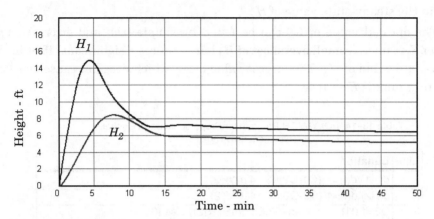

Figure 13–27 PI Control of the Liquid Level in the Two Tanks

Additionally the variable τ_I must be defined.

tauI=20

Adding the integral action eliminates the steady-state offset. This is demonstrated in Figure 13–27 which shows the liquid level in the two tanks and the desired liquid level in the second tank as a function of time when PI controller is used with $K_C = 0.2$ and $\tau_I = 20$. The solution to $t = 80$ indicates that the offset is completely removed, thus $H_2 = H_d$.

Steady state can be reached much faster, without overflowing the first tank, if the integral action of the controller is turned on some time after the flow starts. This option is investigated in part (c).

www

The POLYMATH problem solution file for part (a) is found in directory Chapter 13 with file named **P13-13A.POL**.

13.14 PI CONTROL OF FERMENTER TEMPERATURE

13.14.1 Concepts Demonstrated

Process dynamics and control aspects of temperature control for a batch fermentation process.

13.14.2 Numerical Methods Utilized

Solution of a system of ordinary differential equations.

13.14.3 Problem Statement (adapted from Dunn et al.[9])

Large-scale fermentation requires adequate temperature control due to the exothermic nature of the fermentation process. This is typically accomplished by control of the cooling water temperature to a jacket which partially surrounds the fermenter volume as is shown in Figure 13–28.

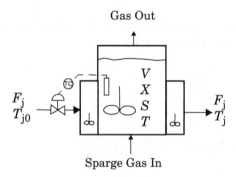

Figure 13–28 Fermenter with Temperature Control

A simple model for this fermenter assumes that the fermentation rate is unaffected by modest changes in temperature. Thus a cell balance on the batch fermenter broth volume yields

$$\frac{dX}{dt} = \mu_{net}X \quad \text{I. C.} \ X = X_0 \text{ at } t = 0 \tag{13-56}$$

Similarly, a balance on the substrate in the fermenter broth volume gives

$$\frac{dS}{dt} = -\frac{\mu_{net}X}{Y_{X/S}} \quad \text{I. C.} \ S = S_0 \text{ at } t = 0 \tag{13-57}$$

where

$$\mu_{net} = \frac{\mu_m S}{K_s + S} \tag{13-58}$$

An energy balance on the fermenter includes the exothermic heat of reaction and the heat transfer to the cooling water. Both the fermenter and the jacket are considered to be completely mixed. The parameters of the fermentation rate expression, Equation (13-58), are considered to be relatively unaffected by temperature changes over the narrow range that the energy balance will be applied. Thus

$$\frac{dT}{dt} = -\frac{\mu_{net}X\Delta H_R}{Y_{X/S}\rho C_p} - \frac{UA(T-T_j)}{V\rho C_p} \quad \text{I. C. } T = T_0 \text{ at } t = 0 \qquad (13\text{-}59)$$

An energy balance on the jacket gives

$$\frac{dT_j}{dt} = \frac{F_j(T_{j0}-T_j)}{V_j} + \frac{UA(T-T_j)}{V_j\rho_j C_{pj}} \quad \text{I. C. } T_j = T_{j0} \text{ at } t = 0 \qquad (13\text{-}60)$$

Implementation of combined proportional and integral control on the fermenter temperature by manipulating the jacket flow rate can be described by

$$F_j = F_{j0} + K_c(T-T_{SET}) + \frac{K_c}{\tau_I}\int_0^t (T-T_{SET})dt \qquad (13\text{-}61)$$

The integral in Equation (13-61) can be expressed by defining a variable ε which represents deviation from the set point such that

$$\frac{d\varepsilon}{dt} = (T-T_{SET}) \quad \text{I. C. } \varepsilon = 0 \text{ at } t = 0 \qquad (13\text{-}62)$$

Thus Equation (13-61) becomes

$$F_j = F_{j0} + K_c(T-T_{SET}) + \frac{K_c}{\tau_I}\varepsilon \qquad (13\text{-}63)$$

Note that F_j cannot be negative.

The fermenter with the P-I control system can be described by Equations (13-56) to (13-60) and (13-62) to (13-63). The parameter definitions and values for the baseline case of fermenter operation are summarized in Table 13–16.

Table 13–16 Baseline Fermenter System and Control Parameters (POLYMATH Variable Name Provided as Used in File **P13-14A.POL**)

C_p = Cp =1 kcal/kg·°C	C_{pj} = Cpj =1 kcal/kg·°C	ΔH_R = deltaHR = −0.466 kcal/ kg substrate
F_{j0} = Fj0 = 2 m³/h	K_c = Kc = 0.8 m³/h·°C	K_s = Ks = 0.1 kg/m³
μ_m = mum = 1 h⁻¹	ρ = rho = 1 kg/m³	ρ_j = rhoj = 1 kg/m³
τ_I = taul = 0.8 h	T_{j0} = Tj0 = 10°C	T_{SET} = Tset = 25°C
UA = UA = 0.4 kcal/kg	V = V = 20 m³	V_j = Vj = 1 m³
Y_{xs} = Yxs = 0.5 kg cells/kg substrate		

(a)   Determine the fermentation time in hours for baseline operation
      when the substrate concentration is reduced to a concentration level
      of 0.1 kg/m^3 under open-loop operation (K_c = 0). Plot the fermenter
      temperature profile during this time period.

(b)   What is the maximum cooling water volumetric flow rate required for
      the baseline operation (with PI control) when the initial fermentation
      broth temperature, T_0, is at 26°C?

(c)   Repeat part (b) when the initial fermentation broth temperature is
      24°C and plot the fermenter temperature profile.

(d)   Determine the optimal initial value of T_0, the initial fermenter broth
      temperature, for the baseline conditions when the maximum cooling
      water flow rate is limited to 40 m^3/h. The objective function, OF, is
      the minimization of the following integral over the reaction time nec-
      essary to reduce the substrate to 0.1 kg/m^3.

$$OF = \int (T - T_{SET})^2$$

13.14.4 Solution (Comments and Suggestions)

(a) The fermentation time is determined in the solution file **P13-14A.POL** by
integration of the reaction time until the substrate concentration is reduced to
below the desired substrate concentration level. At this time, further integration
of the fermentation time is set to zero which results in calculation of the fermentation time as long as the calculations are continued beyond this time. This logic
is found in the problem solution file.

(d) Differentiation of the objective function, OF, with respect to t yields a differential equation that can be solved with the other equations of this problem to
yield the value of the OF at the end of the fermentation time.

The problem solution file for part (a) is found in directory Chapter 13 and
designated **P13-14A.POL**.

13.15 INSULIN DELIVERY TO DIABETICS USING PI CONTROL

13.15.1 Concepts Demonstrated

Model-based control system design of insulin delivery with both P and PI control for various individuals who have Type I diabetes mellitus.

13.15.2 Numerical Methods Utilized

Simulation of a system of ordinary differential equations which model the nonlinear control system and human body response to typical food consumption.

13.15.3 Problem Statement

Blood glucose in the body is regulated by insulin that is produced and introduced into the blood by the pancreas. Consumption and digestion of food causes the glucose level in the blood to increase which in turn causes the body to increase insulin production. Diabetic individuals are not able to produce enough of the needed insulin. For type I diabetes, the needed insulin is provided by several insulin injections during the day in order to provide the needed insulin to the body to break down the glucose. Newly developing technology allows automated introduction of insulin with infusion pumps. An implanted insulin sensor may allow an individual to achieve a normal lifestyle when there is proper control of the insulin delivery into the bloodstream with the infusion pump.

A widely-used model for the effects of insulin infusion and the time-dependent consumption of food (glucose) is given by three differential equations (Bergman et al.[10]).

$$\frac{dG}{dt} = -p_1 G - X(G + G_b) + \frac{G_{\text{meal}}}{V_1} \tag{13-64}$$

$$\frac{dX}{dt} = -p_2 X + p_3 I \tag{13-65}$$

$$\frac{dI}{dt} = -n(I + I_b) + \frac{U}{V_1} \tag{13-66}$$

where G and I represent the deviations in the blood glucose and insulin concentrations in the blood. The variable X is proportional to the insulin concentration in a "remote" compartment for modeling purposes. G_{meal} is the current input of glucose due to food consumption. U is the infusion rate of insulin in mU/min as determined by the control system. G_b and I_b are the steady-state or "baseline" values of glucose and insulin concentrations when G_{meal} is zero. V_1 represents the body blood volume, and the parameters p_1, p_2, p_3, and n depend upon the characteristic metabolic parameters of the diabetic individual. Representative parameter values are summarized in Table 13–17 as given by Bequette,[1] p. 696. Typical glucose intake rates for three meals a day plus an afternoon snack are summarized in Table 13–18 as adapted from Gannt.[11]

Table 13–17 Baseline System and Model Parameters[1]

G_b = 4.5 mmol/L	I_b = 15 mU/L
p_1 = 0 min^{-1}	p_2 = 0.025 min^{-1}
p_3 = 0.000013 mU/L	V_1 = 12 L
n = 5/54 min	

Table 13–18 Typical Glucose Intake for a 24-Hour Day Starting at 6 am, where t = 0 min.

G_i = 0.96 mmol/min for

$120 \leq t < 140$

0.72 for

$360 \leq t < 390$

0.36 for

$540 \leq t < 560$

1.2 for

$720 \leq t < 760$

0.0 for all other t

Application of the PI controller action to the manipulated insulin infusion rate U yields the following equation

$$U = U_b + K_c G + \frac{K_c}{\tau_I} \int_0^t G dt \qquad (13\text{-}67)$$

where K_c is the proportional gain of the controller, G is the glucose concentration deviation variable, and τ_I is the integral time constant or reset time.

As discussed in Problem 13.6, the integral in Equation (13-67) can be calculated by defining a new variable, errsum, as

$$\frac{d}{dt}(\text{errsum}) = G \qquad \text{I. C. errsum} = 0 \text{ at } t = 0 \text{ (steady state)} \qquad (13\text{-}68)$$

Therefore Equation (13-67) becomes

$$U = U_b + K_c G + \frac{K_c}{\tau_I}(\text{errsum}) \qquad (13\text{-}69)$$

and the variable errsum is calculated from Equation (13-68).

The steady-state value U_b is obtained when all the time derivatives are set to zero and the deviation variables G and I are zero. Thus

$$U_b = n I_b V_1 \qquad \text{(13-70)}$$

Typical units of glucose are reported in the literature as mg/dL. Thus the needed conversion factor needed to report the glucose concentration in these units is

1 mmol/L = 18 mg/dL

Glucose levels below 60 mg/dL (hypoglycemia) lead to loss of consciousness and possibly death.

(a)   Develop a model for blood glucose based on Equations (13-68) to (13-70). The system parameters and operating conditions given in Table 13–17 and Table 13–18. Verify the steady-state (baseline) operation with the model, and then plot the dynamic (open loop) glucose concentration response given by $(G+G_b)$ in mg/dL for a 24-hour period (1440 min) starting from the baseline conditions.

(b)   Implement a proportional controller with $K_c = 12$ mU·L/min·mmol to regulate the infusion rate of insulin U. Plot the expected glucose concentration given by $(G+G_b)$ in mg/dL over the 24-hour period starting at 6 am, and determine the minimum, maximum, and average glucose concentrations in mg/dL.

(c)   Repeat part (b) with a PI controller with $K_c = 12$ mU·L/min·mmol and $\tau_I = 1500$ min.

(d)   Summarize the intervals in minutes for the controller settings of part (c) during which the glucose concentration is less than 70 mg/dL, less than 65 mg/dL, and also less than 60 mg/dL.

(e)   Repeat part (d) for the situations where the dinner meal of Table 13–18 is increase by 10% to $G_{meal} = 1.32$ (large meal) and increased by 20% to $G_{meal} = 1.44$ (very large meal). Comment on the ability of the controller to handle these disturbances.

(f)   Determine controller settings and/or strategies that will better reduce the low glucose intervals of part (d) while also controlling the glucose input of the large and very large meals considered in part (e).

REFERENCES

1. Bequette, W. B. *Process Control, Modeling, Design and Simulation*, Prentice Hall: New Jersey, 2003.
2. Stephanopoulos, G. *Chemical Process Control, An Introduction to Theory and Practice*, Prentice Hall: New Jersey, 1984.
3. Cooper, D. J. *Practical Process Control using Loop-Pro Software*, Control Station, Inc., Tolland, CT, 2005.
4. Ziegler, J. G. and Nichols, N. B. "Optimum Settings for Automatic Controllers," *Trans. ASME, 64*, pp.. 759–768 (1942).
5. Luyben, M. L. and Luyben, W. L. *Essentials of Process Control*, McGraw-Hill, New York, 1997.
6. Astrom, K. J. and Hagglund, K. "Automatic Tuning of Simple Regulators for Phase and Amplitude Margins Specifications," *Proceedings of the IFAC Workshop on Adaptive Systems in Control and Signal Processing*, San Francisco, 1983.
7. Brauner, N., Shacham, M. and Cutlip, M. B. "Application of an Interactive ODE Simulation Program in Process Control Education," *Chem. Eng. Educ., 28* (2), pp. 130–135 (1994).
8. Corripio, A. B. "Simulation of Tank Level Control," pp. 101–115 in Westerberg, A. (Ed.) *Computer Programs for Chemical Engineering Education, Volume III. Control*, CACHE, 1972.
9. Dunn, I. J., Heinzle, E., Ingham, J. and Prenosil, J. E. *Biological Reaction Engineering*, 2nd ed., Weinheim: Wiley-VCH, 2003.
10. Bergman, R. N., Philips, L. S. and Cobelli C. "Physiological Evaluation of Factors Controlling Glucose Tolerance in Man," *J. Clin. Invest., 68*, pp. 1456–1467 (1981).
11. Gannt, J. A., Rochelle, K. A., and Gatzke, E. P., "Type I Diabetic Patient Insulin Delivery Using Asymmetric PI Control." *AIChE Annual Fall Meeting*, Austin, November, 2004.

CHAPTER 14

Biochemical Engineering[*]

14.1 ELEMENTARY STEP AND APPROXIMATE MODELS FOR ENZYME KINETICS

14.1.1 Concepts Demonstrated

Development of kinetic models for enzymatic reactions occurring in batch reactors and simulation of elementary step and Michaelis-Menten models for batch enzymatic reactors.

14.1.2 Numerical Methods Utilized

Solution of simultaneous ordinary differential equations.

14.1.3 Problem Statement

Simplified models for enzyme-catalyzed reactions are based on reactions that involve reversible adsorption of the enzyme E with a substrate S yielding an enzyme-substrate complex ES. This complex then reacts to yield the enzyme E and the product P. This overall reaction is presented in Equation (14-1).

$$S + E \underset{k_{-1}}{\overset{k_1}{\rightleftarrows}} ES \overset{k_2}{\rightarrow} E + P \qquad \text{(14-1)}$$

Elementary Step Model
The steps in this overall reaction and their corresponding elementary reaction rates involving concentrations of the reactants are given by

$$r_1 = k_1[E][S] \qquad \text{(14-2)}$$

$$r_{-1} = k_{-1}[ES] \qquad \text{(14-3)}$$

$$r_2 = k_2[ES] \qquad \text{(14-4)}$$

---

[*] The notation and equations of this chapter closely follow those of Shuler, M. L. and Kargi, F., *Bioprocess Engineering*, 2nd ed., Upper Saddle River, NJ: Prentice Hall, 2002.

These enzymatic reactions typically occur in a batch reactor in which the active enzyme is initially added to the reactor to give a known enzyme concentration $[E_0]$ at time zero. Thus the material balances on the various species yield the following differential equations and typical initial conditions.

$$\frac{d[S]}{dt} = -r_1 + r_{-1} = -k_1[E][S] + k_{-1}[ES] \tag{14-5}$$
$$\text{I. C. } [S] = [S_0] \text{ at } t = 0$$

$$\frac{d[E]}{dt} = -r_1 + r_{-1} + r_2 = -k_1[E][S] + k_{-1}[ES] + k_2[ES] \tag{14-6}$$
$$\text{I. C. } [E] = [E_0] \text{ at } t = 0$$

$$\frac{d[ES]}{dt} = r_1 - r_{-1} - r_2 = k_1[E][S] - k_{-1}[ES] - k_2[ES] \tag{14-7}$$
$$\text{I. C. } [ES] = 0 \text{ at } t = 0$$

$$\frac{d[P]}{dt} = r_2 = k_2[ES] \tag{14-8}$$
$$\text{I. C. } [P] = 0 \text{ at } t = 0$$

Thus the elementary step model for this enzymatic reaction is represented by Equations (14-5) through (14-8), and involves four simultaneous ordinary differential equations and their initial conditions.

Michaelis-Menten and Quasi-Steady-State Models

The Michaelis-Menten approximate solution to the elementary step model assumes equilibrium for the first step so that $r_1 = r_{-1}$. The quasi-steady-state model is obtained when pseudo-steady-state hypothesis (see treatment in Chapter 8) is applied to the formation of the enzyme-substrate complex. Both of these treatments result in the same general form of the reaction rate when rate constants are combined into K:

$$v = \frac{V_m[S]}{K + [S]} \tag{14-9}$$

where v is the enzymatic reaction rate, V_m is the maximum reaction rate (when K is negligible relative to $[S]$) given by $k_2[E_0]$ and K is a combination of rate constants. The quasi-steady-state model is preferred to the Michaelis-Menten model as the derivation is more general.[1]

For the Michaelis-Menten model assumptions

$$K = k_{-1}/k_1 \tag{14-10}$$

$$[ES] = \frac{[E_0][S]}{K + [S]} \tag{14-11}$$

For the quasi-steady-state model assumptions

$$K = (k_{-1} + k_2)/k_1 \tag{14-12}$$

$$[ES] = \frac{[E_0][S]}{\dfrac{k_{-1} + k_2}{k_1} + [S]} \tag{14-13}$$

A material balance on the substrate in the batch reactor yields

$$\frac{d[S]}{dt} = -v \tag{14-14}$$

and the corresponding balance on the reaction product gives

$$\frac{d[P]}{dt} = v \tag{14-15}$$

Integration of Equation (14-14) for batch reactor operation yields

$$t = \frac{[S_0] - [S]}{k_2[E_0]} + \frac{K}{k_2[E_0]} \ln\frac{[S_0]}{[S]} \tag{14-16}$$

(a)   Simulate the operation of the batch reactor with the elementary step model using Equations (14-5) through (14-8) where $k_1 = 1{\times}10^9$ min^{-1}, $k_{-1} = 1.5{\times}10^5$ dm^3/g·min, $k_2 = 1$ min^{-1}, $[E_0] = 1{\times}10^{-5}$ g/dm^3, and $[S_0] = 1{\times}10^{-3}$ g-mol/dm^3. The initial conditions at $t = 0$ are $[P_0] = 0$ and $[ES_0] = 0$. Solve and plot [S] and [P] from t = 0 to t = 200 min.

(b)   Repeat part (a) using the quasi-steady-state model given by Equations (14-12) through (14-15) and compare results with part (a). Are the differences between these two models significant for batch operation calculations?

(c)   Calculate the batch reaction time under the conditions of part (a) that is necessary to convert 99% of the substrate S for both models. Carry out both the numerical solution and the analytical solution for the quasi-steady-state model.

(d)   Repeat parts (a) through (c) when $[E_0] = 1{\times}10^{-3}$ g/dm^3 and $[S_0] = 1{\times}10^{-3}$ g-mol/dm^3 for the elementary step model and the quasi-steady-state model. Solve on the interval from t = 0 to t = 20 min. What do you conclude about the quasi-steady-state model application?

14.1.4   Solution (Partial)

(a)–(c) All of the models can be written in a single program by using the same rate constants and specific parameters for each individual model. This is illustrated in Table 14–1 for the POLYMATH *Simultaneous Differential Solver*.

Table 14–1 Modeling of Enzymatic Reactions in POLYMATH (File **P14-01ABC.POL**)

Line	Equation
1	d(S)/d(t) = -k1 * E * S + k1r * ES # Material balance on substrate S
2	S(0) = 1.E-3
3	d(E)/d(t) = -k1 * E * S + k1r * ES + k2 * ES # Material balance on enzyme E
4	E(0) = 1.E-5
5	d(ES)/d(t) = k1 * E * S - k1r * ES - k2 * ES # Material balance on enzyme-substrate complex ES
6	ES(0) = 0
7	d(P)/d(t) = k2 * ES # Material balance on product P
8	P(0) = 0
9	d(t99)/d(t) = If (S >= (.01 * 1E-3)) Then (1) Else (0) # Calculates time t to 99% substrate conversion.
10	t99(0) = 0
11	# Quasi-steady-state solution
12	nu = Vm * Sq / (K + Sq)
13	Vm = k2 * E0
14	E0 = 1.E-5
15	K = (k1r + k2) / k1
16	ESq = E0 * Sq / ((k1r + k2) / k1 + Sq)
17	Eq = E0 - ESq
18	d(Sq)/d(t) = -nu # Material balance on substrate Sq
19	Sq(0) = 1.E-3
20	d(Pq)/d(t) = nu # Material balance on product Pq
21	Pq(0) = 0
22	d(tq99)/d(t) = If (Sq >= (.01 * 1E-3)) Then (1) Else (0) # Calculates time tq to 99% substrate conversion.
23	tq99(0) = 0
24	k1 = 1E9
25	k1r = 5E5 *.3
26	k2 = 1.0E-0 * 1
27	t(0) = 0
28	t(f) = 200
29	# Analytical quasi-steady-state calculation
30	Sq0 = 1.E-3
31	Sqf =.01 * Sq0
32	t99qCALC = (Sq0 - Sqf) / (k2 * E0) + K / (k2 * E0) * ln(Sq0 / Sqf) # This checks the quasi-steady-state solution.

Note that a simple "if ... then ... else ..." statement is used in the earlier differential equations to stop the increase in the batch reaction time for each model when the desired output concentration of substrate has been reached. This is shown in differential equation (14-17) below where variables time, t, and the time for 99% conversion, $t99$, are identical until the substrate reaches 1% of its final value. Then the time derivative of $t99$ is set to zero. Thus the desired value of conversion is retained as the time t continues.

$$\frac{d}{dt}t99 = \text{If } (S>=(.01*1.E-3)) \text{ Then}(1) \text{ Else } (0) \tag{14-17}$$

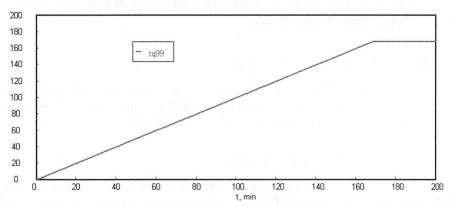

Figure 14–1 Calculation of Batch Reactor Time for 99% Conversion with the Quasi-Steady-State Model (File **P14-01ABC.POL**)

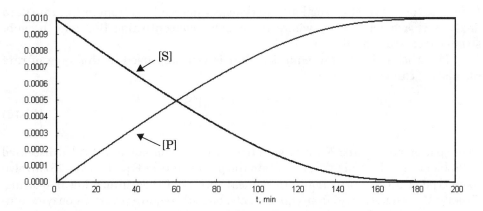

Figure 14–2 Calculation of Substrate and Product Profiles in Batch Reactor for Michaelis Model (File **P14-01ABC.POL**)

The results are presented in Figure 14–1 and Figure 14–2 for $tm99$ as a function of t for part (d) where the calculated batch time for the quasi-steady-model is calculated to be 168.1 min. Note that a "stiff" integration algorithm is needed for this problem. The calculated profiles of substrate and product for the quasi-steady-state model are shown in Figure 14–2.

The problem solution file for parts (a) through (c) is found in directory Chapter 14 and designated **P14-01ABC.POL**.

www

14.2 DETERMINATION AND MODELING INHIBITION FOR ENZYME-CATALYZED REACTIONS

14.2.1   Concepts Demonstrated

Correlation of rate data for an enzyme-catalyzed reaction with Michaelis-Menten kinetics using linearized (Lineweaver-Burk format) and direct nonlinear models that incorporate various modes of enzymatic inhibition.

14.2.2   Numerical Methods Utilized

Multiple linear regression and nonlinear regression of data to determine parameters and confidence intervals for correlation expressions.

14.2.3   Problem Statement

Rate data for enzymatic reactions are typically determined from initial rate data analysis in a batch reactor where the enzyme concentration $[E_0]$ and the substrate concentration $[S]$.

The Michaelis-Menten equation that is typically used in this type of rate correlation can be written as

$$v = \frac{V_m[S]}{K_m + [S]} = \frac{k_2[E_0][S]}{K_m + [S]} \tag{14-18}$$

where parameters k_2 and K_m vary with temperature. This equation is also discussed in Problem 14.1. Typical efforts to obtain the parameters of Equation (14-18) involve some type of a transformation into a linear form and then utilize linear plotting. Usually V_m is determined from experiments, but this requires constant enzyme concentration $[E_0]$. One such transformation to determine V_m and K_m is the Lineweaver-Burk plot that is given by the expression

$$\frac{1}{v} = \frac{1}{V_m} + \frac{K_m}{V_m} \frac{1}{[S]} \tag{14-19}$$

where $1/v$ is plotted versus $1/[S]$ yielding a slope of K_m/V_m and an intercept of $1/V_m$. While this expression is commonly used, it assumes that the error variance of $1/v$ is constant for all data points. This assumption is incorrect and improperly weights the smaller rates in the least-squares objective function for the regression. However, a linear regression of Equation (14-19) is very useful to obtain initial estimates of parameters V_m and K_m for a direct nonlinear regression of Equation (14-18).

Some chemical species can act as inhibitors to the enzymes by binding to the active sites and thus reducing the enzymatic reaction rate. Four models for various types of inhibition and their corresponding rate expressions are given in Table 14–2 where the inhibitor concentration is given by $[I]$. The expressions in

Table 14–2 can be written with $V_m = k_2[E_0]$ in order to deal with various total enzyme concentrations.

Table 14–2 Enzymatic Rate Expressions for Inhibition

Type of Inhibition	Reaction Rate Expression
None	$$v = \dfrac{V_m[S]}{K_m + [S]}$$
Competitive	$$v = \dfrac{V_m[S]}{K_m\left(1 + \dfrac{[I]}{K_I}\right) + [S]}$$
Noncompetitive	$$v = \dfrac{V_m}{\left(1 + \dfrac{[I]}{K_I}\right)\left(1 + \dfrac{K_m}{[S]}\right)}$$
Uncompetitive	$$v = \dfrac{\dfrac{V_m}{\left(1 + \dfrac{[I]}{K_I}\right)}[S]}{\dfrac{K_m}{\left(1 + \dfrac{[I]}{K_I}\right)} + [S]}$$
Substrate Inhibition	$$v = \dfrac{V_m[S]}{K_m + [S] + \dfrac{[S]^2}{K_S}}$$

(a) For the rate data in Table 14–3 with $[I] = 0$, estimate the values of k_2 and K_m for Equation (14-18) at the temperature of 32°C when there is no inhibitor present. A linearized form of the Michaelis-Menten equation (Lineweaver-Burk expression) may be useful to obtain initial estimates for a direct nonlinear regression of the Michaelis-Menten equation.

(b) Repeat part (a) for the data taken at 45°C when there is no inhibitor present.

(c) Utilize all of the data at 32°C to determine the mode of inhibition and to estimate the parameters of this inhibition rate equation.

Table 14–3 Rate Data for Enzyme-Catalyzed Reaction

E_0 g/dm^3	T °C	S g-mol/ dm^3	I g-mol/ dm^3	v g-mol/ dm$^3 \cdot$ min	E_0 g/dm^3	T °C	S g-mol/ dm^3	I g-mol/ dm^3	v g-mol/ dm$^3 \cdot$ min
1.3	32	0.2	0	3.11	1.6	32	0.2	0.4	2.31
1.3	32	0.15	0	2.81	1.6	32	0.15	0.4	2.1
1.3	32	0.1	0	2.4	1.6	32	0.1	0.4	1.75
1.3	32	0.075	0	2.06	1.6	32	0.075	0.4	1.57
1.3	32	0.05	0	1.64	0.8	45	0.2	0	4.36
1.3	32	0.025	0	1.04	0.8	45	0.15	0	3.89
1.3	32	0.01	0	0.467	0.8	45	0.1	0	3.55
1.3	32	0.005	0	0.256	0.8	45	0.075	0	3.07
2.1	32	0.2	0.7	2.33	0.8	45	0.05	0	2.53
2.1	32	0.15	0.7	2.11	0.8	45	0.025	0	1.65
2.1	32	0.1	0.7	1.74	0.8	45	0.01	0	0.777
2.1	32	0.075	0.7	1.54	0.8	45	0.005	0	0.430

The problem data file is found in directory Chapter 14 and designated **P14-02.POL**.

14.2.4   Solution (Partial)

(a) Nonlinear regression of an expression requires good initial estimates for the unknown parameters. The Lineweaver-Burke expression given by Equation (14-19) is a linearized form of the nonlinear Michaelis-Menten equation shown in Equation (14-18). The POLYMATH *Regression and Data Analysis Program* can be used to fit the appropriate data (only the first 8 data points) from the given data file by transforming the data to yield new columns representing the inverses of 1/ v and 1/[S]. These transformed variables can then be utilized in a linear regression as is shown in Figure 14–3. Note that the estimate for the parameter V_m can be calculated from the POLYMATH result for variable a0. Thus $V_m = 1/a0 = 4.2$. Also the parameter K_m is given by a1/a0. Therefore $K_m = a1/a0 = 0.07757$.

Model: INVv = a0 + a1*INVS

$$\frac{1}{v} = \frac{1}{V_m} + \frac{K_m}{V_m}\frac{1}{[S]}$$

Variable	Value	95% confidence
a0	0.2380565	0.0289542
a1	0.0184657	0.0003579

Figure 14–3 Lineweaver-Burke Expression and Linear Regression Result (File **P14-02A.POL**)

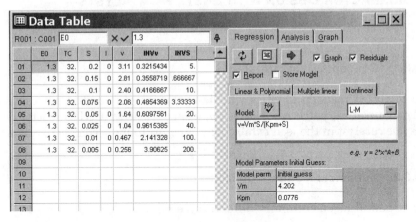

Regression Result   **Model:** v = Vm*S/(Kpm+S)

Variable	Initial guess	Value	95% confidence
Vm	4.202	4.383222	0.1011313
Kpm	0.0776	0.0830883	0.004362

Figure 14–4 Nonlinear Regression of Michaelis-Menten Equation at 32°C
(File **P14-02A.POL**)

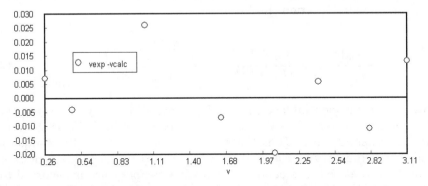

Figure 14–5 Nonlinear Regression of Michaelis-Menten Equation at 32°C
(File **P14-02A.POL**)

The POLYMATH nonlinear regression of the 32°C data with not inhibitor to the Michaelis-Menten equation is accomplished with the problem entry and resulting solution as shown in Figure 14–4. The corresponding residual plot in Figure 14–5 indicates that the errors are randomly distributed and suggests that the regression of the model quite adequate.

The problem solution file for part (a) is found in directory Chapter 14 and designated **P14-02A.POL**.

14.3 BIOREACTOR DESIGN WITH ENZYME CATALYSTS— TEMPERATURE EFFECTS

14.3.1 Concepts Demonstrated

Enzyme-catalyzed reaction kinetics with temperature effects that are caused by thermal denaturation, batch reactor design, ideal chemostat (CSTR) reactor design with enzyme catalyst in the feed exhibiting temperature induced deactivation.

14.3.2 Numerical Methods Utilized

Solution of simultaneous ordinary differential equations.

14.3.3 Problem Statement

Temperature has a very dramatic effect on the kinetics of enzyme-catalyzed reactions as described by the following equations.

The initial portion of the enzyme-catalyzed rate versus temperature curve shown in Figure 14–6 is dependent upon the temperature variation of rate constant k_2 as given in Equation (14-20)

$$v = \frac{V_m[S]}{(K_m + [S])} = \frac{k_2[E_0][S]}{(K_m + [S])} = \frac{A_a e^{-E_a/RT}[E_0(T,t)][S]}{(K_m + [S])} \qquad \textbf{(14-20)}$$

where A_a = pre-exponential factor for k_2, E_a = activation energy for rate constant k_2, R is the gas constant, T is the absolute temperature in K, K_m is the Michaelis-Menten constant, and $[S]$ is the substrate concentration. The temperature dependency of the enzyme concentration (temperature inactivation or thermal denaturation) that becomes dominant at higher temperature is assumed to follow first-order kinetics resulting in the following expression for $E_0(T,t)$.

$$E_0(T,t) = [E_0]e^{-k_d t} \qquad \textbf{(14-21)}$$

with

$$k_d = A_d e^{-E_d/RT} \qquad \textbf{(14-22)}$$

where A_d = frequency factor for the temperature denaturing rate constant and E_d is the activation energy. The Michaelis-Menten constant may also be a function of temperature.

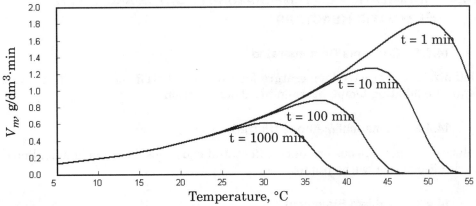

Figure 14–6 Temperature Dependence of an Enzyme-catalyzed Reaction (g/dm^3·min).

(a) Verify a single point on one of the curves of Figure 14–6 where $T \geq 35°C$ and $t > 1$ min. Note that this figure is for $A_a = 6 \times 10^7$ min^{-1}, $E_a = 1.1 \times 10^4$ kcal/g-mol, $A_d = 4 \times 10^{46}$ min^{-1}, $E_d = 7 \times 10^4$ kcal/g-mol, $E_0 = 1$ g/dm^3, and $R = 1.987$ cal/g-mol·K.

(b) Prepare a plot of the maximum enzymatic reaction rate, V_m, as a function of temperature between 5°C and 55°C for the kinetics given in part (a) of this problem when $E_0 = 1.5$ g/dm^3 and the time exposure to temperature is 50 minutes.

(c) Compare the results of (b) to a time exposure of 500 minutes and calculate the percent reduction of V_m at 35°C.

(d) Use the enzymatic reaction rate expression given by Equations (14-20), (14-21), and (14-22) along with the parameters of part (a) to model a batch enzymatic reactor operating at a constant temperature of 35°C. $K_m = 5$ g/dm^3 (constant with T), $E_0 = 1.5$ g/dm^3, and the denaturization begins with the initiation of the batch. The initial substrate concentration is 20 g/dm^3. Determine the time in minutes necessary to achieve 90% conversion of the substrate.

(e) Recommend an optimal constant operating temperature that would maximize the production rate in part (d) for the batch reactor and the corresponding minimum time in minutes to achieve 90% conversion.

14.4 OPTIMIZATION OF TEMPERATURE IN BATCH AND CSTR ENZYMATIC REACTORS

14.4.1 Concepts Demonstrated

Optimization of reactor temperature for batch and CSTR reactors with enzyme catalyst exhibiting temperature-induced deactivation.

14.4.2 Numerical Methods Utilized

Solution of simultaneous ordinary differential equations. Solution of a nonlinear algebraic equation with optimization.

14.4.3 Problem Statement[*]

The enzymatic hydrolysis of lactose in milk can be carried out in an isothermal batch reactor or an isothermal CSTR to provide milk to lactose intolerant individuals. The enzyme lactase for this reaction is competitively inhibited by one of the two products, galactose. Lactose is a disaccharide with one galactose sugar molecule bound to one glucose sugar molecule. The lactase catalyzed reaction is

$$\text{LACTOSE} + H_2O \rightarrow \text{GALACTOSE} + \text{GLUCOSE}$$

The inhibition constant K_I for galactose has the same value as the Michaelis-Menten constant K_m.

Thus the rate expression for this reaction is thus given by

$$v = \frac{V_m[S]}{K_m\left(1 + \frac{[I]}{K_I}\right) + [S]} = \frac{k_2[E_0][S]}{K_m\left(1 + \frac{[I]}{K_m}\right) + [S]} = \frac{k_2[E_0][S]}{K_m + [I] + [S]} \tag{14-23}$$

where the initial condition (batch) or inlet (CSTR) enzyme concentrations are such that

$$k_2[E_0] = 2.71 \times 10^6 e^{-5630/T} \tag{14-24}$$

$$K_m = 2.77 \times 10^2 e^{-3210/T} \tag{14-25}$$

in units of gmol/L·h with T in degrees Kelvin.

The enzyme degradation with time t is described by

$$E_0(T, t) = [E_0]e^{-k_d t} \tag{14-26}$$

in units of g-mol/L where

$$k_d = 6.14 \times 10^{20} e^{-15923/T} \tag{14-27}$$

in units of h^{-1}.

[*] This problem was adapted with permission from one provided by Professor Dhinakar S. Kompala of the University of Colorado from his forthcoming book *Bioprocess Engineering: Fundamentals and Applications.*

(a) Determine the optimal constant temperature of operation for a batch reactor that will hydrolyze 90% of the lactose in milk with an initial concentration of 0.1 gmol/L. What is the needed reaction time for this batch reactor operating at this optimal temperature?

(b) Consider the steady-state operation of a CSTR with continuous feed of milk having a lactose concentration in the feed of 0.1 gmol/L and the same lactase concentration as initially present in the batch reactor of part (a). Eighty percent of the lactose is to be hydrolyzed. The deactivation time for the enzyme in the CSTR feed stream may be considered to be the residence time in the CSTR. One hundred liters of milk are to be treated per hour. What is the optimal temperature in °C and the needed volume of reactor in liters?

14.4.4   Solution (Suggestions)

(a) Material balances for enzymatic reactions in batch reactors are discussed in Problem 14.1.

(b) Material balances for enzymatic reactions in CSTR reactors are discussed in Problem 14.6.

14.5 DIFFUSION WITH REACTION IN SPHERICAL IMMOBILIZED ENZYME PARTICLES

14.5.1 Concepts Demonstrated

Mass transfer of substrate in an enzymatic reaction via an effective diffusivity within a spherical geometry.

14.5.2 Numerical Methods Utilized

Solution of coupled ordinary differential equations that involved split boundary values. The equations also represent the solution of a second-order differential equation via two simultaneous first-order differential equations.

14.5.3 Problem Statement

Enzyme catalysts are often utilized on porous supports in order to contain the enzyme and allow continued catalytic activity. While some of the activity relative to the free enzyme is lost, the remaining catalytic activity can be used in various reactors by retaining the enzyme on the support.

The numerical solutions of these problems in simultaneous diffusion and enzymatic reaction inside porous particles at constant temperature are typically formulated with the isothermal internal effectiveness factor as has been developed in chemical reaction engineering.

$$\eta = \frac{\text{average reaction rate within the particle}}{\text{reaction rate at the concentrations of the particle surface}} \tag{14-28}$$

The effectiveness factor can be calculated by considering the diffusion of the reactant with an effective diffusivity for various geometries of the catalyst support particle. For the case of supported enzyme catalyst, the Michaelis-Menten rate expression can be written as

$$v = \frac{V_m''[\text{S}]}{K_m + [\text{S}]} \tag{14-29}$$

where the units of V_m'' are written as the enzymatic reaction rate per unit volume of the support. V_m'' can also be written as $k_2 E_0''$ where E_0'' is based on a unit volume of the support.

For a spherical support particle that contains enzyme that is uniformly distributed within the particle, a material balance on the differential shell volume shown in Figure 14–7 within the catalyst sphere yields

$$\left. (N_{\text{S}} 4\pi r^2) \right|_r + \frac{V_m''[\text{S}] 4\pi r^2 \Delta r}{K_m + [\text{S}]} = \left. (N_{\text{S}} 4\pi r^2) \right|_{r + \Delta r} \tag{14-30}$$

where N_S is the flux of the substrate and r is the radius of the spherical particle.

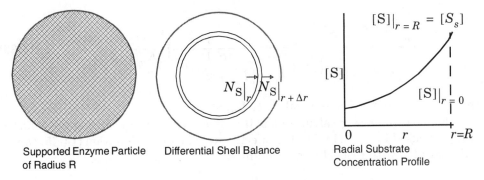

Supported Enzyme Particle of Radius R   Differential Shell Balance   Radial Substrate Concentration Profile

Figure 14–7 Support Particle with Uniformly Distributed Enzyme Catalyst

Rearrangement of this equation and taking the limit as $\Delta r \to 0$ yields

$$\frac{d}{dr}(N_S r^2) = -\frac{V_m''[S]r^2}{K_m + [S]}$$ (14-31)

The initial condition for Equation (14-31) is that there is no flux at the particle center; therefore, N_A or the combined variable $N_A r^2$ is zero at $r = 0$.

The diffusion of substrate S within the particle can be described by Fick's law and can be written as

$$N_S = -D_e \frac{d[S]}{dr}$$ (14-32)

and can be rewritten as

$$\frac{d}{dr}[S] = \frac{N_S}{-D_e}$$ (14-33)

where D_e is the effective diffusivity for the diffusion of substrate S in the porous particle. The boundary condition for this Equation (14-33) is that the [S] at the external particle surface is given by $[S] = [S_s]$ when $r = R$ (the spherical particle radius).

Since the combined variable $(N_S r^2)$ is used in Equation (14-31), an algebraic equation must be calculated during the numerical solution to provide N_S for Equation (14-33). Thus

$$N_S = \frac{(N_S r^2)}{r^2}$$ (14-34)

The effectiveness factor can be calculated from

$$\eta = \frac{\displaystyle\int_0^R V_m'' \frac{[S]}{K_m + [S]}(4\pi r^2)dr}{V_m'' \frac{[S_s]}{K_m + [S_s]}\left(\frac{4}{3}\pi R^3\right)} = \frac{3}{\frac{[S_s]}{K_m + [S_s]}R^3}\int_0^R \frac{[S]}{K_m + [S]}r^2 dr \qquad \text{(14-35)}$$

that is the mathematical equivalent to Equation (14-28).

For convenience in calculation of the effectiveness factor during the solution of Equations (14-31) to (14-34), the effectiveness factor of Equation (14-35) can be differentiated with respect to r to obtain

$$\frac{d\eta}{dr} = \frac{\frac{[S]}{K_m + [S]}3r^2}{\frac{[S_s]}{K_m + [S_s]}R^3} \qquad \text{(14-36)}$$

whose initial condition is $\eta = 0$ and $r = 0$. This differential equation can be solved simultaneously with Equations (14-31) and (14-33). The final value of the effectiveness factor is given by the value of η at $r = R$ provided that the boundary conditions of Equations (14-31) and (14-33) are satisfied.

Similar problem solutions can be obtained for the slab and cylindrical geometries for a variety of reaction rate expressions. The numerical solution of these problems can provide the concentration profiles and the effectiveness factor when the boundary conditions are satisfied. Effectiveness factors for spherical particles containing uniformly dispersed supported enzymes are presented in Figure 14–8 for Michaelis-Menten kinetics.

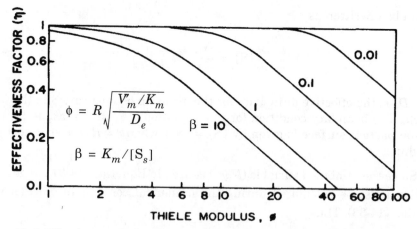

Figure 14–8 Effectiveness Factor for Uniformly Supported Enzymes on Spherical Particles (Reprinted with permission from D. I. C. Wang, et al., *Fermentation and Enzyme Technology*, John Wiley & Sons, New York, p. 329, 1979.)

Consider a uniformly supported enzyme on spherical particles with radius R = 0.001cm. $V_m'' = 0.5$ g–mol/min-cm^3, $D_e = 3 \times 10^{-4}$ cm^2/min, $[S_s] = 1. \times 10^{-4}$ g-mol/cm^3, and $K_m = 0.1[S_s]$.

(a) Calculate the effectiveness factor by simultaneously solving Equations (14-33), (14-34), and (14-36). Plot the $[S_s]$ as a function of radius. Please remember to have a consistent set of units for all variables.

(b) Verify the effectiveness factor calculated from part (a) using Figure 14–8.

(c) Calculate the concentration profile for the substrate as a function of radius and simultaneously determine the effectiveness factor for uniformly supported enzyme catalyst if the enzyme is placed uniformly on only the outer 25% of the particle radius. All other system parameters are the same as given for part (a).

14.5.4   Solution (Partial)

(a) The POLYMATH code for calculating the effectiveness factor η is given in Table 14–4. Note that this solution requires a search to determine the initial condition S(0) in line 12 that yields the value of Ss given in line 9 at the final radius. This split boundary value problem was solved with a simple trial and error procedure.

Table 14–4 Effectiveness Factor Calculation for Supported Enzyme Catalyst Particles (File **P14-05A.POL**)

Line	Equation
1	d(Nsr2) / d(r) = -Vmpp*S*r^2/(Km+S) #Material balance from differential balance within sphere
2	d(S) / d(r) = -Ns/De #Fick's Law for diffusion with constant effective diffusivity
3	d(nu) / d(r) = 3*r^2*(S/(Km+S))/(R^3*Ss/(Km+Ss)) #Calculation of effectiveness factor
4	Ns=Nsr2/r^2 #Calculation of flux
5	Km=0.1*Ss #Given relationship in problem
6	De=3.E-4 #Effective diffusivity in cm2/min
7	R=1.E-3 #Radius in cm
8	Vmpp=0.5 #Vm" in g-mol/min-cm3
9	Ss=1.E-5 #Substrate concentration in g-mol/cm3
10	Thiele=R*(Vmpp/(Km*De))^.5 #Given relationship for Thiele modulus
11	Nsr2(0) = 0 #Initial condition for flux at particle center
12	S(0) =1.4E-20 #Initial condition for substrate concentration at particle center
13	nu(0) = 0 #Initial condition for effectiveness factor calculation
14	r(0) = 1.E-15 #Radius at center of particle
15	r(f) = 1.E-3 #Radius of particle in cm

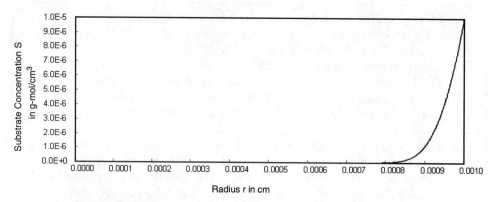

Figure 14–9 Plot of Substrate Concentration versus Particle Radius
(File **P14-05A.POL**)

The plot of the substrate concentration versus the particle radius r is given in Figure 14–9. Note that the substrate concentration is effectively zero for most of the inner particle radius, which indicates poor enzyme utilization and therefore a low effectiveness factor.

The problem solution file for part (a) is found in directory Chapter 14 and designated **P14-05A.POL**.

14.6 MULTIPLE STEADY STATES IN A CHEMOSTAT WITH INHIBITED MICROBIAL GROWTH

14.6.1 Concepts Demonstrated

Calculation of multiple steady-state operating conditions for a biochemical reactor with a substrate-inhibited microbial growth rate expression. Numerical demonstration of the stability of various operating conditions.

14.6.2 Numerical Methods Utilized

Solution of a nonlinear equation with multiple real roots. Solution of an ordinary differential equation.

14.6.3 Problem Statement

A biochemical reaction is occurring in a chemostat according to the following substrate-inhibited rate expression for microbial growth $S + X \rightarrow P + nX$:

$$\mu_g = \frac{\mu_m}{\left(1 + \frac{K_s}{S}\right)\left(1 + \frac{S}{K_1}\right)} \qquad (14\text{-}37)$$

Consider the steady-state and unsteady-state operation of a chemostat or CSTR reactor that utilizes a microorganism that is shown in Figure 14–10. Here F_0 = volumetric flow rate of feed stream, $[S_0]$ designated as S_0 = concentration of substrate in feed stream, $[X_0]$ or X_0 = concentration of cells in the feed stream, F = volumetric

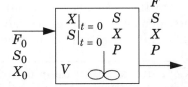

Figure 14–10 CSTR Bioreactor

flow rate of outlet stream, $[X]$ or X = concentration of cells in the outlet stream, $[P]$ or P = concentration of product in the outlet stream and V = volume of liquid in the reactor. Note that for an ideal chemostat or CSTR, complete mixing maintains the concentrations within the reactor and in the outlet stream(s) equal at all times. At initial time, $X|_{t\,=\,0} = X(0)$ = initial concentration of cells in the reactor, and $S|_{t\,=\,0} = S(0)$ = initial concentration of cells in the reactor.

A general unsteady-state material balance on the cells within the chemostat over a time incremental Δt yields

Input + Generation = Output + Accumulation

$$(F_0 X_0)\Delta t + \mu_g V X \Delta t = (FX)\Delta t + (VX)|_{t\,+\,\Delta t} - (VX)|_t \qquad (14\text{-}38)$$

where VX also represents the number of moles of cells in the reactor, N_X. Taking the limit as Δt goes to zero yields the ordinary differential equation

$$\frac{d(VX)}{dt} = \frac{dN_X}{dt} = F_0 X_0 - FX + \mu_g VX \qquad (14\text{-}39)$$

that for constant reactor volume gives

$$\frac{dX}{dt} = \frac{F_0 X_0 - FX}{V} + \mu_g X \tag{14-40}$$

A similar unsteady-state balance on the substrate within the reactor for variable volume yields

$$\frac{d(VS)}{dt} = \frac{dN_S}{dt} = F_0 S_0 - FS - \frac{\mu_g VX}{Y_{X/S}} \tag{14-41}$$

where N_S is the number of moles of substrate in the reactor, and the yield coefficient $Y_{X/S}$ is assumed to be constant and represents the g of cells produced per gram of substrate. For a constant reactor volume, Equation (14-41) becomes

$$\frac{dS}{dt} = \frac{F_0 S_0 - FS}{V} - \frac{\mu_g X}{Y_{X/S}} \tag{14-42}$$

At steady-state operation, the time derivatives of Equations (14-40) and (14-42) are equal to zero; therefore, the following equation can be determined

$$X - X_0 = Y_{X/S}(S_0 - S) \tag{14-43}$$

that also can be determined from an overall material balance at steady state.

For the case of substrate-inhibited rate given by Equation (14-37), the rate expression may become effectively negative order in S. This may lead to multiple steady-state operation for the bioreactor.

Consider a substrate-inhibited microbial growth rate extression in a chemostat where $\mu_m = 0.65 \text{ h}^{-1}$, $K_s = 0.14 \text{ g/dm}^3$, $K_1 = 0.48 \text{ g/dm}^3$, $Y = 0.38$ g cells/g substrate, $V = 2 \text{ dm}^3$, $X_0 = 0$, and $S_0 = 4 \text{ g/dm}^3$. F can be varied between 0 and 0.8 dm^3/h.

(a) What is the minimum value of F in dm^3/h that will give washout of the cells as the only steady state?

(b) What range of F values will result in multiple steady states for the reactor?

(c) Select a value of F that predicts three steady states (including $X = 0$) and simulate the dynamics of the system starting at points just above and below the intermediate "steady-state" value of X to show that this is an unstable steady state. Show typical concentration transients of X and S leading to the low-rate and high-rate steady states from the intermediate "steady-state" operating point. What can be concluded about this intermediate "steady-state" operating point?

(d) Consider the operation of the chemostat at the high-rate steady state with F fixed at 0.4 dm^3/h. What is the maximum time in hours that S_0 in the feed could be temporarily increased to 8 g/dm^3 and then reestablished at 4 g/dm^3 that the reactor will return to the high-rate steady-state operation?

14.6.4 Solution (Suggestions)

(a) At steady state, the time derivative in Equation (14-40) is zero and $F_0 = F$. Algebraic Equation (14-43) relates the variables X and S. Solution of these equations, one nonlinear in X and one rearranged to be explicit in S, can be made for various values of F starting on the high end of the volumetric flow rate. Care should be taken to identify the transition value of F that first leads to multiple solutions.

(b) Further examination of the values of F leading to multiple solutions should identify the region of multiplicity by solving part (a) for different values of F.

(c) The stability of the reactor system can be examined by starting very near the intermediate "steady-state" by solving the differential equations given by Equations (14-40) and (14-42). This might include an initial condition of X just slightly higher than the intermediate "steady-state" solution value and with S at the intermediate "steady-state" solution value. Another initial condition for X might might include a value slightly lower than the intermediate "steady-state" solution value and with S at the intermediate "steady-state" solution value. Note that the integration of the differential equations to large time t will result in a steady state.

(d) Differential Equations (14-40) and (14-42) should be used as the reactor volume is constant. The low-rate steady state needs to be achieved and then the feed stream switched to the higher substrate concentration for a time period that can be optimized to the maximum value of time that the reactor will recover to the high-rate steady state.

14.7 FITTING PARAMETERS IN THE MONOD EQUATION FOR A BATCH CULTURE

14.7.1   Concepts Demonstrated

Fitting of batch reactor data to the semi-empirical Monod equation for microbial growth with an integrated form of the material balance for the reactor.

14.7.2   Numerical Methods Utilized

Nonlinear regression of data to determine kinetic parameters.

14.7.3   Problem Statement

One of the simplest models for the prediction of specific microbial growth rate μ_g can be obtained by applying the Michaelis-Menten kinetics for enzyme reactions to cellular systems with the semi-empirical Monod equation given by

$$\mu_g = \frac{\mu_m S}{K_s + S} \tag{14-44}$$

where μ_m is the maximum of the specific growth rate, K_s is the saturation constant, and S is the concentration of the single growth-limiting substrate.

An unsteady state material balance on a batch reactor on the biomass yields

$$\frac{dX}{dt} = \frac{\mu_m S}{K_s + S} X \tag{14-45}$$

where X is the biomass concentration, typically reported in g/dm^3. A proportional relationship can be assumed between the change in the cell concentration and the substrate concentration within the batch reactor

$$X - X_0 = Y_{X/S}(S_0 - S) \tag{14-46}$$

where the initial values have the subscript "0" and $Y_{X/S}$ is the cell mass yield based on the substrate.

Equation (14-46) can be solved for S and this introduced into Equation (14-45) to yield

$$\frac{dX}{dt} = \frac{\mu_m (Y_{X/S} S_0 + X - X_0) X}{(K_s Y_{X/S} + Y_{X/S} S_0 + X_0 - X)} \tag{14-47}$$

The above differential equation can be integrated for the batch reaction to yield Equation (14-48) given below

$$t = \frac{(K_s Y_{X/S} + Y_{X/S} S_0 + X_0) \ln(X/X_0) - K_s Y_{X/S} \ln[(Y_{X/S} S_0 + X_0 - X)/(Y_{X/S} S_0)]}{\mu_m (Y_{X/S} S_0 + X_0)} \tag{14-48}$$

that can describe the relationship between cell concentration X and batch reaction time t.

Consider a biochemical reaction in which there is no endogenous metabolism. The growth rate is substrate-limited and can be described by the Monod equation. Laboratory data are available for batch reactor experiment as summarized in Table 14–5 where the cell growth is observed. The substrate concentrations are only available at the beginning and end of the batch reactor operation.

(a)   Estimate the cell mass yield based on the substrate.
(b)   Determine the maximum specific growth rate and the saturation constant for this biochemical reaction by using nonlinear regression on the integrated form of the material balance with the Monod rate expression. Report the 95% confidence intervals. Initial parameter estimates can be obtained by solving a set of two nonlinear equations based on two points from the data set.
(c)   Check your regression parameters by numerically integrating the differential equation for the cell concentration and comparing the results with those from the integrated equation at several points.

Table 14–5 Batch Reactor Growth Data

Time t h	Cell Concentration X g/dm^3	Substrate Concentration S g/dm^3
0	5.01	125.0
2.42	5.81	
4.34	6.53	
6.26	7.53	
8.18	8.65	
10.10	9.57	
12.02	11.01	
13.94	11.97	
15.86	13.51	
17.78	14.65	
19.70	15.95	
21.62	17.28	
23.54	17.59	
25.46	18.27	
31.22	20.09	3.71

The problem data file is found in directory Chapter 14 and designated **P14-07.POL**.

14.8 MODELING AND ANALYSIS OF KINETICS IN A CHEMOSTAT

14.8.1   Concepts Demonstrated

Steady-state and unsteady-state material balances on a chemostat for a bioreaction that has an inhibited specific growth rate and a death rate.

14.8.2   Numerical Methods Utilized

Nonlinear regression of steady-state data to determine kinetic parameters. Solution of nonlinear algebraic equations.

14.8.3   Problem Statement

Biomass production is being investigated by variation of the feed substrate concentration, the dilution rate, and the inhibitor concentration in a laboratory chemostat. The specific growth rate is known to be inhibited and also exhibits a death rate. The specific growth rate is given by

$$\mu_{net} = \mu_g - k_d = \frac{\mu_m S}{K_s + S + (IK_s)/K_I} - k_d \tag{14-49}$$

where the growth parameters μ_m, K_s, K_I, and k_d can be determined from data measured in the chemostat.

An unsteady-state material balance on the cells in the culture volume of a chemostat (CSTR) shown in Figure 14–11 over a time interval Δt yields

Input + Generation = Output + Accumulation

$$(F_0 X_0)\Delta t + \mu_{net} VX\Delta t = (FX)\Delta t + (VX)|_{t+\Delta t} - (VX)|_t \tag{14-50}$$

where F_0 is nutrient solution feed flow rate in dm^3/h, X_0 is the cell feed concentration in g/dm^3, X is the cell concentration in the reactor and in the outlet stream in g/dm^3, and V is the liquid phase volume of the culture in dm^3.

The derivative of (VX) can be obtained by rearrangement of Equation (14-50) and taking the limit as $\Delta t \rightarrow 0$. For constant V, nutrient flow rate where $F_0 = F$ and

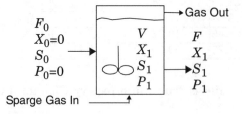

Figure 14–11 Chemostat Flow Diagram

for no cells in the feed ($X_0 = 0$), the resulting differential equation becomes

$$\frac{dX}{dt} = -\frac{FX}{V} + \mu_{net}X = -DX + \left(\frac{\mu_m S}{K_s + S + (IK_s)/K_I} - k_d\right)X \qquad (14\text{-}51)$$

where $D = F/V$ is the dilution rate or inverse of the chemostat residence time. For steady state where $dX/dt = 0$, the dilution rate is given by

$$D = \mu_{net} = \frac{\mu_m S}{K_s + S + (IK_s)/K_I} - k_d \qquad (14\text{-}52)$$

Similarly, an unsteady-state material balance on the substrate in the culture volume of the chemostat (CSTR) and no extra cellular product formation yields

$$F_0 S_0)\Delta t - \frac{\mu_g}{Y_{X/S}^M}VX\Delta t = (FS)\Delta t + (VS)|_{t + \Delta t} - (VS)| \qquad (14\text{-}53)$$

where S_0 is the substrate feed concentration (g/dm^3), S is the substrate concentration in the reactor and in the outlet stream (g/dm^3), and $Y_{X/S}^M$ is the yield coefficient for (g cell/g S). The resulting differential equation becomes

$$\frac{dS}{dt} = \frac{FS_0 - FS}{V} - \left(\frac{\mu_g}{Y_{X/S}^M}\right)X = D(S_0 - S) - \left(\frac{\mu_g}{Y_{X/S}^M}\right)X \qquad (14\text{-}54)$$

At steady state, the time derivative of the above differential equation is zero and thus

$$D(S_0 - S) = \left(\frac{\mu_g}{Y_{X/S}^M}\right)X \qquad (14\text{-}55)$$

Equation (14-52) can be introduced into Equation (14-55) to yield

$$X = Y_{X/S}^M(S_0 - S) \qquad (14\text{-}56)$$

Washout of the chemostat refers to reactor operation in which the dilution rate becomes so large that the cell production is less than the loss of cells in the product stream. This condition is reached when the steady-state concentration of the substrate is equal to the feed substrate concentration as no steady-state production of cell is occurring.

(a) Estimate the value of the yield coefficient $Y_{X/S}^M$ using Equation (14-56) and the data of Table 14–6.

(b) Estimate all of the growth parameters μ_m, K_s, K_I, and k_d of Equation (14-52) by regression of all of the data in Table 14–6. Report the 95% confidence intervals for each parameter.

(c) Determine the smallest dilution rate that will yield 90% of the maximum production rate of the cells where $S_0 = 12$ g/dm^3 and $I = 0.05$ g/dm^3. The production rate is given by the product of D and X.

(d) What is the washout dilution rate for the conditions of part (c)?

The data summarized in Table 14–6 are obtained in a series of steady-state experiments where the dilution rate D, the feed substrate concentration S_0, and the inhibitor concentration I are varied at constant pH and temperature.

Table 14–6 Steady-State Data from a Laboratory Chemostat

Dilution Rate D h^{-1}	Substrate Feed Concentration S_0 g/dm^3	Inhibitor Feed Concentration I g/dm^3	Substrate Concentration S g/dm^3	Cell Concentration X g/dm^3
0.2	15.	0.07	1.05	2.04
0.2	10.	0.04	0.668	1.38
0.2	5.	0	0.156	0.731
0.3	15.	0.07	1.69	2.05
0.3	10.	0.04	1.08	1.35
0.3	5.	0	0.245	0.719
0.4	15.	0.07	2.47	1.93
0.4	10.	0.04	1.59	1.3
0.4	5.	0	0.36	0.69
0.5	15.	0.07	3.54	1.73
0.5	10.	0.04	2.29	1.20
0.5	5.	0	0.514	0.654
0.6	15.	0.07	5.01	1.53
0.6	10.	0.04	3.13	1.04
0.6	5.	0	0.719	0.657

14.8.4   Solution (Suggestions)

(d) The washout dilution rate is determined where the dilution rate D is equal to the net growth rate μ_{net} and $S = S_0$.

The problem data file is found in directory Chapter 14 and designated **P14-08.POL**.

www

14.9 DYNAMIC MODELING OF A CHEMOSTAT

14.9.1 Concepts Demonstrated

Dynamic material balances on a chemostat that has an inhibited specific growth rate and a death rate.

14.9.2 Numerical Methods Utilized

Solution of a system of ordinary differential equations representing dynamic material balances.

14.9.3 Problem Statement

The start-up of the laboratory chemostat involved in the biomass production discussed in Problem 14.8 is being investigated in order to achieve steady-state operation with minimal time. This will be useful in the start-up of a larger sized bioreactor system that can also be modeled as a CSTR.

The specific growth rate for this chemostat is known to be inhibited and also exhibits a death rate that is given by

$$\mu_{\text{net}} = \mu_b - k_d = \frac{\mu_m S}{K_s + S + (IK_s)/K_I} - k_d \tag{14-57}$$

where the growth parameters have been determine to be $\mu_m = 1.6$ h^{-1}, $K_s = 0.90$ g/dm^3, $K_I = 0.02$ g/dm^3, and $k_d = 0.04$h^{-1}. $Y_{X/S}^M = 0.15$ g cell/g S. There is no extra cellular product formation.

When the chemostat is filling, the differential equation for culture volume V is given by

$$\frac{dV}{dt} = F_0 \tag{14-58}$$

where F_0 is the nutrient volumetric flow rate.

The differential equation for a cell balance that is given by Equation (14-51) for the cells with a constant culture volume must be modified to allow the culture volume to change. Thus

$$\frac{dN_X}{dt} = -FX + \mu_{\text{net}}VX = -FX + \left(\frac{\mu_m S}{K_s + S + (IK_s)/K_I} - k_d\right)VX \tag{14-59}$$

where the derivative variable N_X represents the number of cells in the culture within the chemostat and is equal to (VX). X can be calculated from

$$X = \frac{N_X}{V} \tag{14-60}$$

Similarly the differential equation for the substrate balance given in Equation (14-54) must also be altered to allow the culture volume to change. Thus

$$\frac{dN_S}{dt} = F_0 S_0 - FS - \left(\frac{VX}{Y_{X/S}^M} \right) \left(\frac{\mu_m S}{K_s + S + (IK_s)/K_I} \right) \qquad \text{(14-61)}$$

where the derivative variable N_S represents the amount of substrate in the culture within the chemostat and is equal to (VS). Thus S can be calculated from

$$S = \frac{N_S}{V} \qquad \text{(14-62)}$$

A similar unsteady-state balance yields the amount of non-reacting inhibitor in the culture with an inlet concentration of I_0.

$$\frac{dN_I}{dt} = F_0 I_0 - FI \qquad \text{(14-63)}$$

where

$$I = \frac{N_I}{V} \qquad \text{(14-64)}$$

(a) Calculate the expected dynamics of the chemostat when the start-up is from an initial charge of cells to the chemostat during start-up, where the initial charge is 0.1 dm^3 containing 0.1 g cells/dm^3. The substrate stream with $S_0 = 12$ g/dm^3 is then introduced into to a 1 dm^3 chemostat at 0.25 dm^3/h. There is no inhibitor during this inoculation start-up phase. The chemostat remains in the semi-batch mode with no outlet flow-rate until the broth volume reaches 1 dm^3, and then the feed stream is stepped to 1.0 dm^3/h. An output flow controller always keeps the broth volume constant at 1 dm^3. Plot the concentration versus time profiles for X and S for the initial 10 hours of operation. The chemostat may be considered to be a CSTR, and the specific growth rate given by Equation (14-57) is valid during the entire transient period.

(b) What are the steady-state output cell and substrate concentrations resulting from the start-up operation in part (a)?

(c) What is the start-up time for part (a) that is required for the cells to reach 99% of the steady-state output cell concentration?

(d) For the chemostat initially operating at steady state, the substrate feed stream is switched to one that contains $S_0 = 15$ g/dm^3 and an inhibitor at $I_0 = 0.05$ g/dm^3 with a total flow rate of 1.1 dm^3/h. Calculate the new steady-state values of X and S. Determine the transient time in hours for the cell concentration to fall to 101% of this new steady-state value.

14.9.4 Solution (Partial plus Suggestions)

(a) The equations describing the unsteady-state operation of the chemostat (CSTR) are given by differential Equations (14-58), (14-59), (14-61), and (14-63) along with algebraic Equations (14-60), (14-62) and (14-62). These equations and the reactor start-up are integrated into the POLYMATH code found in Table 14–7.

Table 14–7 Dynamic Chemostat Model (File **P14-09AB.POL**)

Line	Equation
1	d(V)/d(t) = If (V) <= (1.0) Then (F0) Else (0) # Chemostat volume in km3l
2	d(Nx)/d(t) = -F * X + (mum * S / (Ks + S + I * Ks / KI) - kd) * V * X # Mass of cells in chemostat in g
3	d(Ns)/d(t) = F0 * S0 - F * S - V * X / Yxs * mum * S / (Ks + S + I * Ks / KI) # Mass of substrate in chemostat in g
4	F = If (V <= 1.0) Then (0) Else (F0) #Flow rate from chemostat in dm3/h
5	F0 =If (V <= 1.0) Then (0.25) Else (1) #Feed rate to chemostat in dm3/hr
6	Yxs = 0.15 #Yield coefficient for cells from substrate in g cell/g S
7	S0 = 12 #Substrate concentration in feed stream in g/dm3
8	kd = 0.04 #Death rate coefficient in h-1
9	KI = 0.02 #Growth parameter constant for inhibitor in g/dm3
10	Ks = 0.9 #Growth parameter constant for substrate in g/dm3
11	mum = 1.6 #Maximum specific growth rate constant in h-1
12	X = Nx / V # Concentration of cells in mg/ml or g/L or g/dm3
13	S = Ns / V # Concentration of substrate in mg/ml or g/L or g/dm3
14	D = F / V #Dilution rate in hr-1
15	I = 0.0 #Inhibitor concentration in mg/ml or g/L or g/dm3
16	t(0) = 0 #Initial condition for time
17	t(f) = 10 #Final condition for time
18	V(0) = .1 #Initial condition for V
19	Nx(0) = .01 #Initial condition for Hx
20	Ns(0) = 0 #Initial condition for Ns

It is always useful to plot the variable(s) that are influenced by the program logic such a the chemostat volume V. The expected dependency upon time t is shown in Figure 14–12.

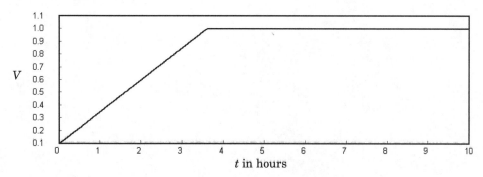

Figure 14–12 Chemostat Volume during Start-up (File **P14-09AB.POL**)

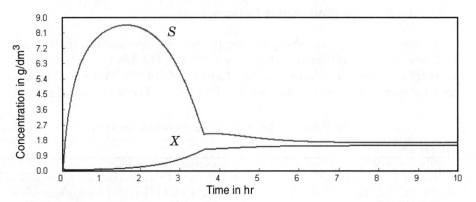

Figure 14–13 Transient Profiles for Substrate S and Cells X during Chemostat Start-up to Steady-State Operation (File **P14-09AB.POL**)

The transients in X and S for the initial 10 hours of operation as shown in Figure 14–13 indicate that steady-state operation has been achieved.

(b) Steady-state operation results when the time derivatives are set to zero. This can be calculated by integrating the differential equation to large time or by solving the system of equations resulting from setting the time derivatives to zero.

(c) and **(d)** Use logical variables to integrate the calculated time as long as the unsteady-state conditions exist.

The problem solution file for part (a) is found in directory Chapter 14 and designated **P14-09AB.POL**.

14.10 PREDATOR-PREY DYNAMICS OF MIXED CULTURES IN A CHEMOSTAT

14.10.1 Concepts Demonstrated

Demonstration of the dynamic behavior of predator-prey in a mixed continuous culture with saturation kinetics based on the Monod model. Equations lead to a variety of dynamic behavior including interesting transients and limit cycles.

14.10.2 Numerical Methods Utilized

Solution of a system of ordinary differential equations.

14.10.3 Problem Statement

The growth of a protozoa (predator) and bacteria (prey) has been studied and modeled by Tusuchiya et al.[2] for a mixed continuous culture. The various material balances on this well-mixed chemostat (CSTR) are given below:

$$\text{Protozoa (Predator):} \quad \frac{dX_p}{dt} = -DX_p + \frac{\mu_{mp} X_b X_p}{K_b + X_b} \tag{14-65}$$

$$\text{Bacteria (Prey):} \quad \frac{dX_b}{dt} = -DX_b + \frac{\mu_{mb} S X_b}{K_S + S} - \frac{1}{Y_{X_p/b}} \frac{\mu_{mp} X_b X_p}{K_b + X_b} \tag{14-66}$$

$$\text{Substrate:} \quad \frac{dS}{dt} = D(S_0 - S) - \frac{1}{Y_{X_b/S}} \frac{\mu_{mb} S X_b}{K_S + S} \tag{14-67}$$

where X_p X_b and S are the concentrations of protozoa, bacteria, and substrate respectively in the chemostat (mg/ml), S_0 is the feed concentration of the glucose substrate (mg/ml), $Y_{X_p/b}$ and $Y_{X_b/S}$ are the yield coefficients for protozoa on bacteria and bacteria on substrate (g/g), μ_{mp} and μ_{mb} are the maximum specific growth rates of protozoa and bacteria respectively (h^{-1}), K_b and K_S are the saturation constants for the bacteria and the substrate, and D is the dilution rate (h^{-1}) that is the inverse of the residence time or holding time. Reported parameters for the growth constants are presented in Table 14–8.

Table 14–8 Values of Growth Parameters at 25°C[2]

Maximum Specific Growth Rate (μ)	Saturation Constant (K)	Yield Coefficients
μ_{mp} = 0.24 h^{-1}	K_b = 4×10^8 bacteria/ml	$Y_{X_p/b}$ = 7.14×10^{-4} protozoa/bacteria
μ_{mb} = 0.25 h^{-1}	K_S = 5×10^{-4} mg of glucose/ml	$Y_{X_b/S}$ = 3.03×10^9 bacteria/mg of glucose

The dynamics of this reacting system are quite interesting, and the transients in concentrations are very dependent on the operating conditions of the chemostat. Experimental conditions can be simulated with a numerical solution where $D = 0.0625$ h^{-1} and $S_0 = 0.5$ with initial conditions as $X_p = 1.4 \times 10^4$, $X_b = 1.6 \times 10^9$, and $S = 0$. The experimental time is quite long, so it is useful to carry a variable for days and to integrate to t = 360 h or 15 days.

The concentration transients of the bacteria, the protozoa, and the substrate are shown in Figures 14–14 to 14–16.

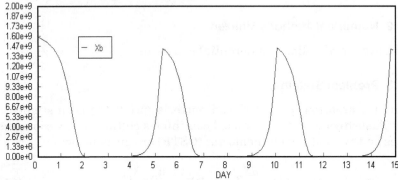

Figure 14–14 Dynamic Response of the Bacteria Concentration, bacteria/ml

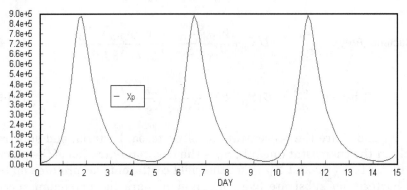

Figure 14–15 Dynamic Response of the Protozoa Concentration, protozoa/ml

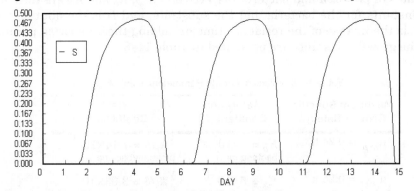

Figure 14–16 Dynamic Response of the Substrate Concentration, mg/ml

Note that the dynamics achieve a repeating cycle that occurs about every five days. Investigations of this dynamic performance closely represent the experimental results for this system.

Stability consideration trends for this dynamic model are indicated in Figure 14–17.

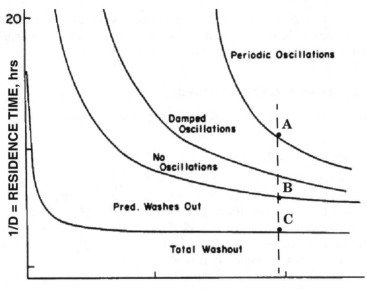

Figure 14–17 Stability Regions Trends for the Predator-Prey Model (Reprinted with permission from *Journal of Bacteriology*, June 1972, p. 1152.)

(a) Reproduce the transients shown in Figures 14–14 to 14–16. What is the cycle time in days that is regularly repeated?

(b) Plot the graphs (limit cycle) of X_p versus X_b and S versus X_b for $S_0 = 0.3$ when the cycle repeats.

(c) For the conditions of part (a), determine the residence time that the system goes from "Periodic Oscillations" to "Damped Oscillations." Note that this boundary is shown schematically as point A in Figure 14–17.

(d) Continue to decrease the residence time to determine point B that is between the "No Oscillations" and "Predator Washes Out."

(e) In a similar manner, determine the residence time for point C that is the boundary between the "Predator Washes Out" and the "Total Washout."

The problem solution file for part (a) is found in directory Chapter 14 and designated **P14-10A.POL**.

14.11 BIOKINETIC MODELING INCORPORATING IMPERFECT MIXING IN A CHEMOSTAT

14.11.1 Concepts Demonstrated

Approximation of imperfect mixing in a chemostat volume by a well-mixed volume and a stagnant volume that sum to the chemostat volume. Determination of overall conversion and washout feed rate to the chemostat.

14.11.2 Numerical Methods Utilized

Solution of a system of simultaneous nonlinear algebraic equations.

14.11.3 Problem Statement

A chemostat is usually considered to be a completely mixed reactor; however, this is not always the case. Consider the situation where the chemostat may be considered to be modeled as a reactor with a completely-mixed volume V_1 that interacts with another completely-mixed but stagnant volume V_2 as shown in Figure 14–18. Volume V_2 may be considered to model the poorly mixed regions within a production fermenter.

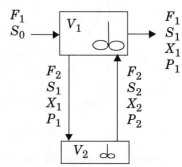

Figure 14–18 Chemostat Reactor Model

The microbial system to be modeled involves substrate S going to product P only under the action of cells X. The following separate balances on the substrate, cells, and product in each reactor utilize Monod kinetics and a cell death rate constant given by k_d.

Steady-State Substrate Balance on Volume V_1

$$\text{Input} + \text{Generation} = \text{Output}$$

$$F_1 S_0 + F_2 S_2 + \frac{1}{Y_{X/S}}\left(\frac{\mu_m S_1}{K_S + S_1}\right) X_1 V_1 = F_1 S_1 + F_2 S_1 \qquad \textbf{(14-68)}$$

Steady-State Substrate Balance on Volume V_2

$$F_2 S_1 + \frac{1}{Y_{X/S}}\left(\frac{\mu_m S_2}{K_S + S_2}\right) X_2 V_2 = F_2 S_2 \qquad \textbf{(14-69)}$$

Steady-State Cell Balance on Volume V_1

$$F_2 X_2 + \left(\frac{\mu_m S_1}{K_S + S_1} - k_d \right) X_1 V_1 = F_1 X_1 + F_2 X_2 \qquad \text{(14-70)}$$

Steady-State Cell Balance on Volume V_2

$$F_2 X_1 + \left(\frac{\mu_m S_2}{K_S + S_2} - k_d \right) X_2 V_2 = F_2 X_2 \qquad \text{(14-71)}$$

Overall Steady-State Material Balance for Product

$$P_1 = Y_{P/S}(S_0 - S_1) \qquad \text{(14-72)}$$

Microbial growth has been studied in a continuous culture and the following parameters were obtained: $\mu_m = 0.2$ h^{-1}, $K_S = 0.2$ g/dm^3, $k_d = 0.002$ h^{-1}, $Y_{X/S} = 0.4$ g cells/g substrate, and $Y_{P/S} = 0.2$ g product/g substrate. Tracer studies have indicated that the incomplete mixing can be described by a well-mixed volume $V_1 = 1.7$ dm^3 and a stagnant volume of $V_2 = 0.3$ dm^3 with a constant flow rate relationship of $F_2 = 0.2\ F_1$ in dm^3/h. The chemostat operation is such that $X_0 = 0$ and $S_0 = 0.6$ g/dm^3 and the endogenous metabolism can be neglected.

(a) Create a single graph of S, X, and P versus the dilution rate defined by $D = F_1/V_1$.

(b) Plot the cell production rate, the product D^*X_1, and the product production rate, the product of D^*P_1, as a function of the dilution rate between 0.05 and 0.130 h^{-1}.

(c) Estimate the dilution rate that will maximize the production rate, DX_1, for the cells and the dilution rate that will maximize the production rate, DP_1, for the product.

14.12 DYNAMIC MODELING OF A CHEMOSTAT SYSTEM WITH TWO STAGES

14.12.1 Concepts Demonstrated

Dynamic modeling of multiple stages for a continuous chemostat system for a microbial reaction with product inhibition.

14.12.2 Numerical Methods Utilized

Solution of a system of simultaneous ordinary differential equations.

14.12.3 Problem Statement

The use of multiple stages in a chemostat system is useful when the concentration of product may inhibit the microbial growth rate. Since the concentration of the product will decrease the growth rate, several reactors in series will produce a higher rate of product because the rate in the upstream reactors are not affected as much by the lower concentration of products. Thus a series of CSTR chemostats should have better productivity than a single CSTR when the same total reactor volume is utilized.

Consider the growth rate of a microbial reaction that is given by the following inhibited rate expression for stage "n"

$$\mu_n = \frac{\mu_m S_n}{K_s + S_n + P_n/K_I} \tag{14-73}$$

where S_n is the substrate concentration, P_n is the product concentration, and K_I is the inhibition constant.

The reaction rate of the product in stage n is dependent upon both growing and non-growing biomass as given by

$$r_{P_n} = (\alpha_n + \beta\mu_n)X_n \tag{14-74}$$

where X is the cell concentration, β is the growth dependent product yield, μ_n is the growth rate from Equation (14-73), and α_n is the inhibited non-growth term given by

$$\alpha_n = \frac{\alpha_0}{1 + P_n} \tag{14-75}$$

with kinetic constant α_0.

The reaction rate for cell growth in the n-th stage is given by

$$r_{X_n} = \mu_n X_n \tag{14-76}$$

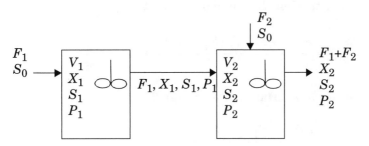

Figure 14–19 Two-Stage Chemostat with Second-Stage Feed

and the substrate rate is represented by

$$r_{S_n} = - \frac{\mu_n X_n}{Y_{X/S}} \qquad (14\text{-}77)$$

A flow diagram of two staged chemostats is shown in Figure 14–19 that includes a separate feed stream input into the second chemostat.

The unsteady-state model development presented in Problem 14.8 can be used to derive the differential equations for the various component material balances for this two-stage chemostat system where volumes V_1 and V_2 are constant with time t.

Dynamic Cell Mass Balance on Stage 1

$$\frac{dX_1}{dt} = - \frac{F_1 X_1}{V_1} + r_{X_1} \qquad (14\text{-}78)$$

Dynamic Substrate Mass Balance on Stage 1

$$\frac{dS_1}{dt} = \frac{F_1(S_0 - S_1)}{V_1} + r_{S_1} - \frac{r_{P_1}}{Y_{P/S}} \qquad (14\text{-}79)$$

Dynamic Product Mass Balance on Stage 1

$$\frac{dP_1}{dt} = \frac{F_1(P_0 - P_1)}{V_1} + r_{P_1} \qquad (14\text{-}80)$$

Dynamic Cell Mass Balance on Stage 2

$$\frac{dX_2}{dt} = \frac{F_1 X_1 - (F_1 + F_2)X_2}{V_2} + r_{X_2} \qquad (14\text{-}81)$$

Dynamic Substrate Mass Balance on Stage 2

$$\frac{dS_2}{dt} = \frac{F_1 S_1 + F_2 S_0 - (F_1 + F_2)S_2}{V_2} + r_{S_2} - \frac{r_{P_2}}{Y_{P/S}} \qquad \textbf{(14-82)}$$

Dynamic Product Mass Balance on Stage 2

$$\frac{dP_2}{dt} = \frac{F_1 P_1 - (F_1 + F_2)P_2}{V_2} + r_{P_2} \qquad \textbf{(14-83)}$$

Overall Production Rate

$$\text{PR} = \frac{P_2(F_1 + F_2)}{(V_1 + V_2)} \qquad \textbf{(14-84)}$$

A chemostat system with two stages and an additional feed stream to the second stage as is shown in Figure 14–19. The following parameters of the microbial growth of this system are as follows: $\mu_m = 0.65$ h^{-1}, $K_s = 0.2$ g/dm^3, $K_I = 1.8$, $\alpha_0 = 0.15$ h^{-1}, $\beta = 0.25$, $Y_{X/S} = 0.3$ g cells/g substrate, and $Y_{P/S} = 0.4$ g product/g substrate. Initial studies of this chemostat system have utilized $V_1 = 2$ dm^3 and $V_2 = 1$ dm^3 with $F_1 = 0.5$ dm^3/h and $F_2 = 0.5$ dm^3/h. The usual substrate stream has $S_0 = 10$ g/dm^3.

(a)   Consider the start-up of the chemostat system with an initial innoculum in chemostat 1 with the following initial conditions at $t = 0$: $X_1 = 1$, $S_1 = 0$, $S_2 = 0$, $P_1 = 0$, $P_2 = 0$. The base flow rates and chemostat volumes are to be used. Plot X_1, X_2, S_1, S_2, P_1, and P_2 during the start-up to steady-state operation. What is the production rate at this steady state?

(b)   What is the value of F_1 for the start-up under the conditions of part (a) that will optimize the steady-state production rate? The ratio of $F_2/F_1 = 1$ is always maintained.

(c)   What is the general effect on the steady-state production rate when the volume of the chemostat 1 is increased and the volume chemostat 2 is decreased by the same amount? The start-up conditions other than reactor volumes are as in part (a); however, The total volume of the two chemostats remains constant: $V_1 + V_2 = 3$ dm^3.

(d)   Suggest operating conditions that will maximize the steady-state production rate of P and indicate any special start-up procedures that may be necessary. Constraints include: $V_1 + V_2 = 3$ dm^3, and initial conditions are such that (X_1 and $X_2 \leq 1$) and (S_1 and $S_2 \leq 1$). There are no constraints on F_1 or F_2 or their ratio.

14.13 SEMICONTINUOUS FED-BATCH AND CYCLIC-FED BATCH OPERATION

14.13.1 Concepts Demonstrated

Use of cyclic fed-batch operation in a semi-batch bioreactor to maximize fermentation products when the bioreaction is inhibited by the substrate.

14.13.2 Numerical Methods Utilized

Solution of simultaneous ordinary differential equations with induced variable cycling that often leads to stiff differential equations.

14.13.3 Problem Statement

Substrate inhibition can be minimized by control of the substrate addition to a batch bioreactor as is shown in Figure 14–20. The first step involves the initiation or start-up of the reactor from an initial innoculum with substrate feed rate of F_I over time interval from $t = 0$ to t_I. The processing is then followed by controlled addition of the substrate at a rate of F_P over time interval from t_I to t_P during which the reactor volume increases from V_I to V_H. The substrate addition is then stopped and the batch bioreactor is harvested with a discharge flow rate of F_H over time interval from t_P to t_H during which the volume decreases from V_H to V_I. In fed-batch operation, the initiation, processing, and harvesting modes are utilized in a single sequence. Cyclic operation involves the repetitive cycle of processing and harvesting modes where n is the cycle number.

Each operational mode of the bioreactor involves unsteady-state balances material balances on the well-mixed culture volume within the reactor. The appropriate differential equations can be derived by consideration of the general balance on this volume over a time interval Δt. This is accomplished below for a balance on the cells in the initiation mode

Input + Generation = Output + Accumulation

$$F_I X_0 \Delta t + \mu_{net} V X \Delta t = F X \Delta t + N_X\big|_{t + \Delta t} - N_X\big|_t \qquad \text{(14-85)}$$

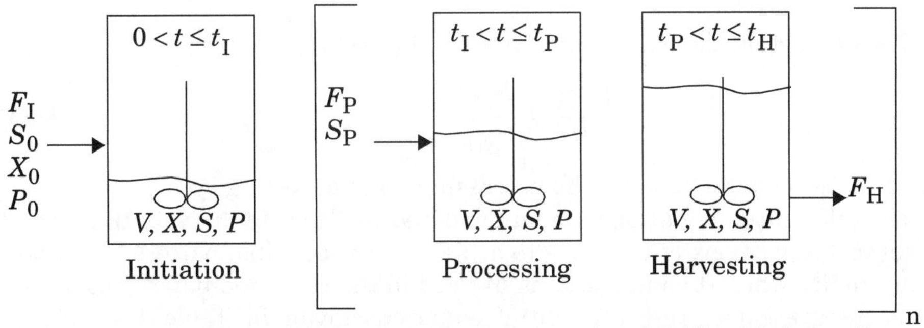

Figure 14–20 Bioreactor in Fed Batch Mode (n = 1) and Cycling Mode (n >1)

where F_I is nutrient solution feed flow rate in L/h, X_0 is the cell feed concentration in g/L, X is the cell concentration in the bioreactor and in the outlet stream in g/L, and V is the volume of the culture in L. Since the culture volume is usually changing with time, it is most convenient to use N_X as the total quantity of cells in Equation (14-85). Rearrangement of this equation and taking the limit as $\Delta t \rightarrow 0$ results in the differential equation

$$\frac{dN_X}{dt} = F_I X_0 + \mu_{net} X V \qquad \text{(14-86)}$$

and the cell concentration can be calculated from the explicit algebraic equation

$$X = \frac{N_X}{V} \qquad \text{(14-87)}$$

Note that the above algebraic equation is necessary for calculation of the cell concentration as the current bioreactor volume V is available from a overall material balance.

Similar material balances can be made on the other reacting components resulting in the following sets of differential and algebraic equations

$$\frac{dN_S}{dt} = F_I S_0 - \frac{\mu_{net} X V}{Y_{X/S}} \qquad \text{(14-88)}$$

$$S = \frac{N_S}{V} \qquad \text{(14-89)}$$

$$\frac{dN_P}{dt} = \frac{\mu_{net} X V}{Y_{X/P}} \qquad \text{(14-90)}$$

$$P = \frac{N_P}{V} \qquad \text{(14-91)}$$

Note that this equation development makes the resulting equation set very general for whatever operation changes the bioreactor culture volume. However, an overall material balance is needed to obtain the differential equation for the culture volume. For example, the initiation mode yields

$$\frac{dV}{dt} = F_I \qquad \text{(14-92)}$$

The production rate can be obtained for the cycling mode from

$$\frac{d\text{PR}}{dt} = \frac{F_H P}{t_{cycle}} \qquad \text{(14-93)}$$

where t_{cycle} is the total time for the processing and harvesting cycle.

The differential equations are summarized in Table 14–9. Note that it will take several repetitions of the cycle to achieve a reproducible pattern of concentration profiles when the bioreactor is utilized in the cyclic fed-batch mode. Continuous integration of the differential equations given in Table 14–9 allows simulation of this cyclic reactor operation.

Table 14–9 Differential Equations for Fed Batch and Cyclic Fed Batch Bioreactors

Differential Equations	Initiation	Processing	Harvesting
$\dfrac{dN_X}{dt} =$	$F_I X_0 + \mu_{net} XV$	$\mu_{net} XV$	$-F_H X + \mu_{net} XV$
$\dfrac{dN_S}{dt} =$	$F_I S_0 - \dfrac{\mu_{net} XV}{Y_{X/S}}$	$F_P S_P - \dfrac{\mu_{net} XV}{Y_{X/S}}$	$-F_H S - \dfrac{\mu_{net} XV}{Y_{X/S}}$
$\dfrac{dN_P}{dt} =$	$\dfrac{\mu_{net} XV}{Y_{X/P}}$	$\dfrac{\mu_{net} XV}{Y_{X/P}}$	$-F_H P + \dfrac{\mu_{net} XV}{Y_{X/P}}$
$\dfrac{dV}{dt} =$	F_I	F_P	$-F_H$

Consider the fed-batch operation of a laboratory pilot plant antibiotic fermentation bioreactor where the glucose is added to the fermentation culture in order to minimize the glucose substrate inhibition. The bioreaction kinetics are: $\mu_m = 0.3$ h^{-1}, $K_S = 1$ g glucose/L, $K_I = 100$ (g glucose/L)^2, $Y_{X/S} = 0.4$ g cells/g glucose, and $Y_{X/P} = 0.15$ g product/g glucose. The cell death rate and endogenous metabolism can be neglected. Thus

$$\mu_{net} = \mu_g = \frac{\mu_m S}{K_S + S + S^2/K_I}$$

The initiation mode is started at $t = 0$ hr with a volume of 0.8 L containing 30 g cells/L and negligible glucose. The glucose concentrations are fixed at $S_0 = S_P = 200$ g glucose/L. Initial operation is such that $F_I = 0.2$ L/hr until $t = 1.0$ hr. Then $F_P = 0.5$ L/hr until $t = 6$ hr and $F_H = 2.5$ L/hr until the culture volume is reduced to $V = 1.0$ L at $t = 7$ hr.

(a) Create a single graph of the concentrations (S, X, and P) within the bioreactor as a function of time for the fed-batch operation to time $t_{H=}\, 7$ hr. This represents the initiation and the first cycle of processing and harvesting.

(b) Repeat part (a) for the case where the initiation is followed by three complete cycles of processing and harvesting. Graph the concentrations (S, X, and P) as a function of time.

(c) Carry out the processing and harvesting cycles of part (b) into the cyclic fed-batch operation where the process repeats itself from cycle to cycle. Calculate the production of P in g/hr under continuous cycling. (Assume the third cycle, n = 3, represents the cyclic operation.)

(d) Optimize the substrate feed concentration to the fed-batch process in part (c) that will maximize the production rate of P. All processing conditions and times remain the same. Calculate the average production rate of P in g/hr under continuous cycling. (Assume the third cycle represents the cyclic operation.)

14.14 OPTIMIZATION OF ETHANOL PRODUCTION IN A BATCH FERMENTER

14.14.1 Concepts Demonstrated

Modeling and optimization of a batch fermenter to maximize the volumetric productivity of ethanol while decreasing the concentration of substrate to a specified level.

14.14.2 Numerical Methods Utilized

Integration of a system of ordinary differential equations.

14.14.3 Problem Statement[*]

The fermentation of glucose, G, to ethanol, E, is an anaerobic biological reaction that can be catalyzed by the yeast cells Saccharomyces cerevisiae, C. The actual reaction stoichiometry on a dry weight basis for glucose is given by

$$G \rightarrow 0.434E + 0.415CO_2 + 0.108C \qquad \text{(14-94)}$$

where any by-products can be considered to be negligible.

The specific rate of ethanol production at 30°C can be written in terms of a modified Monod equation with a term that represents the inhibitory effect of the product ethanol

$$\mu = \mu_m \left[\frac{G}{K_m + G} \right]\left[1 - \frac{E}{E_{max}} \right]^{0.36} \qquad \text{(14-95)}$$

where μ_m = 1.85 g ethanol/g cell·hr, K_m = 0.315 g/L, and E_{max} = 87.5 g/L. G, E, and C represent the concentrations for Equation (14-94) in g/L. Note that the rate of ethanol production goes to zero when the ethanol concentration reaches E_{max}.

A small laboratory fermenter with a volume of 10 L is used, and the initial start-up concentrations in the innoculum are always E = 8.6 g/L and C = 2.1 g/L. The operation of this batch fermenter is to be optimized with respect to the initial glucose concentration that will lead to a maximum in the volumetric productivity of ethanol in g/L·hr. However, the residual glucose concentration at the end of the fermentation must be 2.8 g/L as this is the maximum that can be treated in an existing plant waste treatment facility. There is also a limitation on the cell density the laboratory study of 80 g/L as proper mixing in the industrial-scale fermenters cannot be achieved for higher cell densities.

---

[*] This problem was adapted with permission from the *CACHE Process Design Case Study, Volume 4* authored by Professors Steven E. LeBlanc and Ronald L. Fournier and then student Samer Naser at the University of Toledo.

The average volumetric productivity, AVP, for this laboratory fermenter can be calculated over the fermentation time t_f from

$$\text{AVP} = \int_{0}^{t_f} \mu C \, dt \quad \text{or} \quad \frac{d\text{AVP}}{dt} = \frac{\mu C}{t_f} \qquad \text{(14-96)}$$

The weight percent of glucose in water, WP, is given by the relationship

$$\text{WP} = 100G/(0.378G + 998.215) \qquad \text{(14-97)}$$

14.14.4 Problem Statement

(a) Develop a complete model for the batch fermenter that incorporates the data given and calculates the time to reach a final glucose concentration 2.8 g/L at the end of the fermentation.

(b) Calculations on the batch fermenter of part (a) are summarized in Figure 14–21. Verify the values of the volumetric productivity and the reaction time (residence time) for two different values of the inlet glucose concentrations.

(c) Determine the inlet glucose concentration and reaction time that maximizes the volumetric productivity and plot the time profiles for E, G, and C.

(d) Calculate the inlet glucose concentration in g/L that yields 90% of the maximal volumetric productivity that was found in part (c).

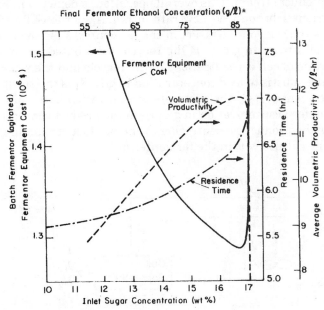

Figure 14–21 Kinetics of Ethanol Production in a Batch Fermenter. (Reprinted with permission from *CACHE Process Design Case Study, Volume 4*, CACHE Corporation, Austin, TX (www.cache.org)).

14.15 Ethanol Production in a Well-Mixed Fermenter with Cell Recycle

14.15.1 Concepts Demonstrated

Modeling and optimization of a well-mixed fermenter (CSTR) with and without cell recycle to maximize the volumetric productivity of ethanol while decreasing the concentration of substrate to a specified level.

14.15.2 Numerical Methods Utilized

Solution of a system of simultaneous algebraic equations with optimization.

14.15.3 Problem Statement

The general overall reaction for ethanol fermentation can be represented as $S \rightarrow P + CO_2 + X$ where S is the substrate, P is the ethanol, CO_2 is product carbon dioxide, and X represents the yeast cells. It is often desirable to recycle the cells from the product stream of CSTR fermenters back into the fermenter in order to increase the reaction rate or to increase the process stability to operational pertubations. This is accomplished by either filtration, settling, or centrifugation of the product stream. This overall operation is depicted in Figure 14–22 where F is nutrient solution feed flow rate in L/h, X_0 is the cell feed concentration in g/L, X_1 is the cell concentration in the fermenter, X_2 is the cell concentration in the outlet stream in g/L, α is the recycle ratio based on volumetric flow rates, C is the ratio of cell concentrations in the recycle and the separator output streams, S_0 is the substrate feed concentration in g/L, S_1 is the substrate concentration in the fermenter, S_2 is the substrate concentration in the product stream, P_0 is the product feed concentration in g/L, P_1 is the product concentration in the fermenter, P_2 is the product concentration in the product stream in g/L, and V is the volume of the culture within the fermenter in L.

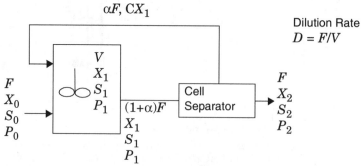

Figure 14–22 CSTR Fermenter with Recycle

Overall Material Balance on Cells

A steady-state material balance on the cells in the overall process that includes the fermenter culture volume V and the recycle stream yields

$$\text{Input + Generation = Output}$$

$$FX_0 + \mu_{net}VX_1 = FX_2 \tag{14-98}$$

where μ_{net} is the net growth rate for the cells in g/dm^3.

Cell Balance on Fermenter

Similarly a cell balance on the fermenter alone gives

$$FX_0 + \alpha FCX_1 + \mu_{net}VX_1 = (1 + \alpha)FX_1 \tag{14-99}$$

Overall Material Balance on Substrate

$$FS_0 - \frac{\mu_{net}VX_1}{Y_{X/S}} = FS_2 \tag{14-100}$$

Substrate Balance on Fermenter

Similarly a substrate balance on the fermenter alone gives

$$FS_0 - \frac{\mu_{net}VX_1}{Y_{X/S}} = (1 + \alpha)FS_1 \tag{14-101}$$

Overall Material Balance on Product

$$FP_0 + \frac{\mu_{net}VX_1}{Y_{X/P}} = FP_2 \tag{14-102}$$

Product Balance on Fermenter

A product balance on the fermenter alone gives

$$FP_0 + \frac{\mu_{net}VX_1}{Y_{X/P}} = (1 + \alpha)FP_1 \tag{14-103}$$

Average Volumetric Productivity

The average volumetric productivity, AVP, for the product P within the fermenter at steady state can be defined by

$$\text{AVP} = \frac{\mu_{net}X_1}{Y_{X/P}} \tag{14-104}$$

Production Rate

The production rate, PR, for product P for the fermenter at steady state in either CSTR or recycle mode is given by

$$PR = F(P_2 - P_0) \qquad (14\text{-}105)$$

Problem Details

Ethanol production as discussed in Problem 14.14 can be reformulated usual biochemical notation (following Shuler and Fikret[1]) for this fermenter where the semi-empirical growth rate for cells in g/L is given by

$$\mu_{net} = 0.4604 \left[\frac{S_1}{K_m + S_1} \right] \left[1 - \frac{P_1}{P_{max}} \right]^{0.36} \qquad (14\text{-}106)$$

where $K_m = 0.315$ g/L, $Y_{X/S} = 0.108$ and $Y_{X/P} = 0.2489$. The product CO_2 does not effect the growth rate and is vented during the fermentation reaction and the dissolved product CO_2 may be neglected.

(a)   Develop a complete steady-state model for a CSTR fermenter with recycle where the biochemical stoichiometry and kinetics are given in Problem 14.14. Glucose concentration in the product stream is to be set at 2.8 g/L as this is the maximum concentration of glucose that can be accepted by an existing waste water treatment facility. The CO_2 can be assumed to be vented to the atmosphere. Cell density within the fermenter cannot exceed 80 g/L. The fermenter feed stream contains no ethanol or cells. The volume of the fermenter is 10 L.

(b)   Plot the ethanol AVP versus the inlet glucose concentration over the interval of 100 to 200 g/L for the CSTR fermenter operating under the conditions of part (a) with no recycle. Determine the optimal inlet glucose concentration that maximizes the AVP and the optimal AVP.

(c)   Repeat part (b) when recycle is employed with $\alpha = 0.15$ and the recycle concentration ratio $C = 5$. What is the optimal inlet glucose concentration that maximizes the AVP and the optimal AVP.

(d)   Calculate the dilution rate D and the recycle ratio α under the conditions of part (a) that will maximize the ethanol AVP. The 10 L fermenter is operated with a feed stream containing only glucose feed at a concentration of 180 g/L, and the recycle concentration ratio is given as $C = 4$. The fermenter operation is subject to the constraints that $X_1 \leq 80$ g/L, $S_1 \leq 2.8$ g/L, and $\alpha \leq 0.25$.

14.16 DYNAMIC MODELING OF AN ANAEROBIC DIGESTER

14.16.1 Concepts Demonstrated

Dynamic model for anaerobic digester incorporating a biochemical gasification reaction and involving unsteady-state material balances in the liquid and gas phases.

14.16.2 Numerical Methods Utilized

Solution of simultaneous ordinary differential equations.

14.16.3 Problem Statement[*]

Anaerobic digestion involves the utilization of bacteria to breakdown organic waste in an oxygen-free process. This biochemical waste treatment process can provide a methane-rich biogas similar to natural gas and liquid/solid products with value as fertilizers and soil conditioners. The digesters experience failures due to wash out of the digester microbes, high organic substrate concentrations leading to high volatile acid concentration that inhibit the methane-forming bacteria, and feed components that are toxic to the bacteria.

The process is assumed to be rate-limited by the conversion of volatile acids within the organic waste by the methane bacteria. The methane bacteria are approximated by $C_5H_7NO_2$ and the volatile acid by acetic acid. The reaction stoichiometry is for the gasification reaction is given by

$$CH_3COOH + 0.032NH_3 \rightarrow 0.032C_5H_7NO_2 + 0.92CO_2 + 0.94CH_4 + 0.096H_2O \text{(14-107)}$$

The nonionized form of the acetic acid (representing all volatile acids) is considered to the substrate and is related to the ionized form by

$$HS \leftrightarrow S^- + H^+ \tag{14-108}$$

The K_a for acetic acid at 38°C is 2.2×10^{-5} and the digester pH is above 6, therefore, most of the acid is in the ionized form such that $[S^-] \approx [S]$. Thus

$$[HS] \approx \frac{[S][H^+]}{K_a} \tag{14-109}$$

The Monod specific growth rate in the digester that includes the substrate inhibition takes the form

$$\mu = \frac{\mu_m}{1 + \dfrac{K_s}{[HS]} + \dfrac{[HS]}{K_I}} \tag{14-110}$$

The bacteria death rate is first order with respect to the toxin concentration $[T]$.

$$r_D = -k_T[T] \tag{14-111}$$

[*] This problem is based on the work of Graef and Andrews,[3] as developed by Bailey and Ollis.[4]

Data from the literature indicate that μ_m = 0.4 day^{-1}, K_s = 0.0333 mmol/dm^3, K_I = 0.667 mmol/dm^3, and k_T = 2.0 day^{-1}.

Carbon dioxide is present in both the liquid phase and the gas phase. It is generated during the biochemical gasification reaction and then is transferred to the gas phase. This process may be described by use of a mass transfer equation

$$T_G = k_La([CO_2]_D - [CO_2]_D^*) \tag{14-112}$$

in which the mass transfer coefficient per unit volume of liquid phase is k_La = 100 (dm^3·day)$^{-1}$, the concentration of dissolved CO_2 in the liquid phase is $[CO_2]_D$, and the concentration of dissolved CO_2 at equilibrium with the gas phase CO_2 is given by $[CO_2]_D^*$.

Henry's law can be utilized to relate the equilibrium concentration of dissolved CO_2 to the partial pressure of CO_2 in the gas phase

$$[CO_2]_D^* = K_H p_{CO_2} \tag{14-113}$$

where K_H = 0.0326 is the Henry's law coefficient in mmol/(dm^3·mmHg) and p_{CO_2} is the partial pressure of CO_2 in the gas phase in mmHg.

In addition to the biochemical reaction, CO_2 can also be generated within the liquid phase by reversible bicarbonate association reaction

$$HCO_3^- + H^+ \leftrightarrow H_2O + CO_2 \tag{14-114}$$

The effective equilibrium constant K_1 for the above equation can be used to form the following expression for $[H^+]$

$$[H^+] = \frac{K_1[CO_2]_D}{[HCO_3^-]} 2 \tag{14-115}$$

where $[CO_2]_D$ is the dissolved CO_2 concentration in mmol/dm^3, $[HCO_3^-]$ is the concentration of bicarbonate ions in mmol/dm^3, and K_1 = 6.5×10^{-7} for an approximately constant water concentration.

Electroneutrality within the liquid phase requires that

$$[C^+] + [H^+] = [A^-] + [HCO_3^-] + [S^-] + [OH^-] + 2[CO_3^{-2}] \tag{14-116}$$

where $[C^+]$ is the total cation concentration that includes calcium, sodium, magnesium, and ammonium. $[A^-]$ is the total anion concentration that includes the concentrations of chlorides, phosphates, sulfide, etc. The normal operation of a digester is such that $[H^+]$, $[OH^-]$, and $[CO_3^{-2}]$ can be neglected and the ionized form of the acids can be approximated by the substrate concentration, $[S^-] \approx [S]$. A new variable that can be defined as the net cationic concentration given by

$$Z = [C^+] - [A^-] \cong [HCO_3^-] + [S] \tag{14-117}$$

will be used to model the digester. In practice, this Z corresponds to the ammonium ion concentration when sulfide concentration is not very large.

Continuous Digester Model at Standard Operating Conditions

Consider the development of a model for a small-scale continuous digester with a liquid volume $V = 10$ dm^3, a gas volume $V_G = 2$ dm^3, and a volumetric feed rate of substrate $F = 1$ dm^3/day. The gas phase has a density given by $\rho_G = 38.91$ mmol/dm^3, a total pressure of $p_T = 760$ mmHg, and a temperature of 38°C. The substrate feed is $[S]_0 = 167$ mmol/dm^3 as acetic acid.

Liquid Volume

The gasification reaction has been studied, and the yield coefficients for the cell mass, the CO_2, and the CH_4 are known to be $Y_{X/S} = 0.032$ mol organism/mol substrate, $Y_{CO_2/X} = 28.8$ mol CH_4/mol substrate, and $Y_{CH_4/X} = 28.8$ mol CO_2/mol substrate. For the liquid phase of volume V, the following equations can be derived:

Cell Material Balance $$\frac{d[X]}{dt} = \frac{F}{V}([X]_0 - [X]) + \mu[X] - k_T[T] \qquad \textbf{(14-118)}$$

Substrate Balance $$\frac{d[S]}{dt} = \frac{F}{V}([S]_0 - [S]) - \frac{\mu}{Y_{X/S}}[X] \qquad \textbf{(14-119)}$$

Net Cation Balance $$\frac{dZ}{dt} = \frac{F}{V}(Z_0 - Z) \qquad \textbf{(14-120)}$$

Toxic Component Balance $$\frac{d[T]}{dt} = \frac{F}{V}([T]_0 - [T]) \qquad \textbf{(14-121)}$$

Rate of CH_4 Formation $$R_{CH_4} = Y_{CH_4/X}\mu X \qquad \textbf{(14-122)}$$

The formation of dissolved CO_2 involves both the biochemical reaction and the bicarbonate association reaction.

Rate of Biochemical CO2 Formation $$R_B = Y_{CO_2/X}\mu X \qquad \textbf{(14-123)}$$

The rate of the bicarbonate association reaction, R_C, is involved in the mass balance on bicarbonate in the constant volume of the liquid phase

$$\frac{d[HCO_3^-]}{dt} = \frac{F}{V}([HCO_3^-]_0 - [HCO_3^-]) - R_C \qquad \textbf{(14-124)}$$

where the subscript "0" indicates feed concentration. Now Equation (14-117) can be rearranged and differentiated with respect to time t to yield

$$\frac{d[HCO_3^-]}{dt} = \frac{dZ}{dt} - \frac{d[S]}{dt} \qquad \textbf{(14-125)}$$

Equations (14-124) and (14-125) can be used to eliminate $\dfrac{d[HCO_3^-]}{dt}$ that results in an equation for R_C.

$$R_C = \frac{F}{V}([HCO_3^-]_0 - [HCO_3^-]) + \frac{d[S]}{dt} - \frac{dZ}{dt} \qquad \textbf{(14-126)}$$

Thus the dissolved CO_2 material balance may be written as

$$\frac{d[CO_2]_D}{dt} = \frac{F}{V}([CO_2]_{D0} - [CO_2]_D) - T_G + R_B + R_C \tag{14-127}$$

Gas Phase

Carbon dioxide is transferred into the gas phase by mass transfer as described by Equation (14-112). CH_4 gas is not soluble in the liquid, so its input to the gas phase is directly via the biochemical gasification reaction. A material balance can be made on the gas phase in terms of CO_2 partial pressure where Q is the total outlet volumetric flow rate in dm^3/day.

$$\text{Gaseous } CO_2 \text{ Material Balance} \qquad \frac{dp_{CO_2}}{dt} = \frac{p_T V T_G}{\rho_G V_G} - \frac{p_{CO_2} Q}{V_G} \tag{14-128}$$

The output volumetric flow rate for CO_2 can be calculated from

$$\text{Output Flow Rate of } CO_2 \qquad Q_{CO_2} = \frac{V T_G}{\rho_G} \tag{14-129}$$

The output volumetric flow rate for CH_4 can be calculated from

$$\text{Output Flow Rate of } CH_4 \qquad Q_{CH_4} = \frac{V Y_{CH_4/X} \mu X}{\rho_G} \tag{14-130}$$

The water partial pressure in the digester gas can be assumed to be the vapor pressure at the digester temperature of 60°C given by p_{H_2O}. Thus the total output flow rate is given by

$$Q = Q_{CO_2} + Q_{CH_4} + \left(\frac{p_{H_2O}}{p_T}\right) Q \tag{14-131}$$

that can be solved for Q yielding

$$\text{Total Output Flow Rate} \qquad Q = \frac{p_T(Q_{CO_2} + Q_{CH_4})}{p_T - p_{H_2O}} \tag{14-132}$$

Steady-State Operation

Some of the steady-state operational variables at the stated standard operating conditions are given in Table 14–10.

Table 14–10 Steady-State Process Variables

Variable	Value	Units
$[X]$	5.280	$mmol/dm^3$
$[S]$	2.014	$mmol/dm^3$
Z	50.	$mmol/dm^3$

Variable	Value	Units
$[T]$	0	mmol/dm^3
$[CO_2]_D$	9.00	mmol/dm^3
$p_{CO_{22}}$	273.14	mmHg

(a) Confirm the steady-state operating values given in Table 14–10 for $[S_0] = 167$ mmol/dm^3 and calculate the daily flow rate of methane, CH_4, in dm^3/day.

(b) Simulate the step change of substrate feed to the reactor from a steady state for the conditions of part (a) to $[S_0] = 300$ mmol/dm^3 at the start of the second day and plot $[S]$ in the digester to a total of 10 days.

(c) Determine the maximum substrate feed concentration for part (b) that will cause washout of the digester and lead to negligible biochemical reaction.

(d) Produce a plot of the CH_4 and CO_2 volumetric flow rates (dm^3/day) in the dry digester outlet gas for the conditions of part (c) where the substrate feed concentration is 90% of the washout feed concentration.

(e) Simulate the response of the digester to a step change to the steady-state operation of part (a) that involves a toxic feed contaminant in the feed at $[T] = 3$ mmol/dm^3 at the start of the second day with a duration of 1 day only. Plot $[S]$ and $[X]$ in the digester to a total of 40 days.

(f) Calculate the maximum number of hours (integer value) that the input involving toxic feed component at the level as given in part (e) can be treated without causing a subsequent washout of the digester.

14.16.4 Solution Suggestions

Numerical solutions should be terminated when then pH is pH ≤ 6 as the model is no longer accurate and the biological reaction will irreversibly stop.

(a) Steady-state operation can be calculated by integration of the system of equations to time that is large enough to allow steady state to be achieved. An alternate approach is to set the time derivatives to zero and solve the resulting system of simultaneous nonlinear equations.

(c) When washout is achieved, the problem becomes difficult to solve even with a stiff integration algorithm. Modest increases in $[S_0]$ can be made until trial and error increases can no longer be integrated.

The problem solution files for part (a) are found in directory Chapter 14 and designated **P14-16A1.POL** and **P14-16A2.POL**. The first file involves differential equations and the second involves nonlinear equations as discussed above in the Solution Suggestions (a).

14.17 START-UP AND CONTROL OF AN ANAEROBIC DIGESTER

14.17.1 Concepts Demonstrated

Dynamic model for anaerobic digester incorporating a biochemical gasification reaction and involving unsteady-state material balances in the liquid and gas phase. Process control algorithms.

14.17.2 Numerical Methods Utilized

Solution of simultaneous ordinary differential equations with logical decision making impacting the dynamics of the numerical solutions.

14.17.3 Problem Statement[*]

Graef and Andrews[3] considered the following four methods of control for anaerobic digesters[4]: (1) gas scrubbing with recycle, (2) addition of base, (3) recycle of microorganisms, and (4) reduction of flow.

Consider the detailed model developed in Problem 14.16 as the basis for investigating control scheme (1) for gas scrubbing with recycle that is shown in Figure 14–23. The effluent gas from the digester is scrubbed to remove some of the CO_2 to control the pH in the digester and affecting the biochemical reaction. The control action consists of removing more CO_2 from the recycle gas causes the carbonic acid concentration to fall and thus raises the pH.

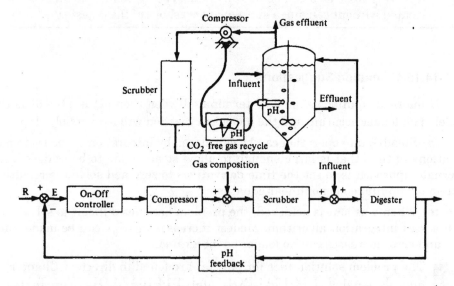

Figure 14–23 Gas Scrubbing with Recycle Flow Diagram from Bailey and Ollis[4]

[*] This problem is based on the work of Graef and Andrews,[3] as developed by Bailey and Ollis.[4]

A control scheme proposed by Graef and Andrews[3] involves two recycle flow rates that are on/off as determined by the measured pH of the digester. The adjustment of recycle gas flow rate Q_R was determined by the logic given below.

if pH >= 7.0 Q_{R1} = 0 L/day

if pH <= 6.75 Q_{R1} = 50 L/day (assumed dry flow rate entering scrubber)

if 6.75 < pH < 7.0 Q_{R1} = current value

if pH >= 7.0 Q_{R2} = 0 L/day

if pH <= 6.65 Q_{R2} = 175 L/day (assumed dry flow rate entering scrubber)

if 6.65 < pH < 7.0 Q_{R2} = current value

$Q_R = Q_{R1} + Q_{R2}$

The results of this control action is shown in Figure 14–24 as reported by Graef and Andrews[3] using the model presented in this problem.

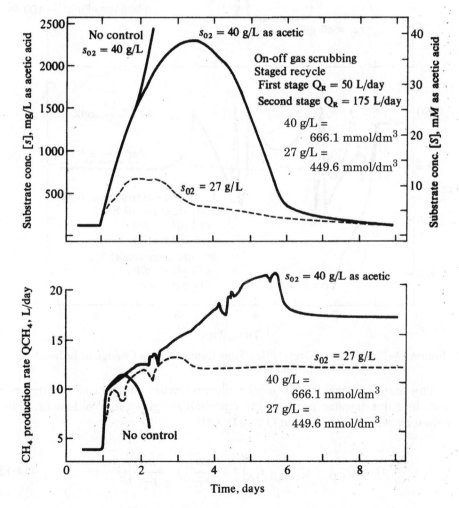

Figure 14–24 Control of Recycle Gas Flow Rate to a Step Change in Influent Substrate at End of First Day[3]

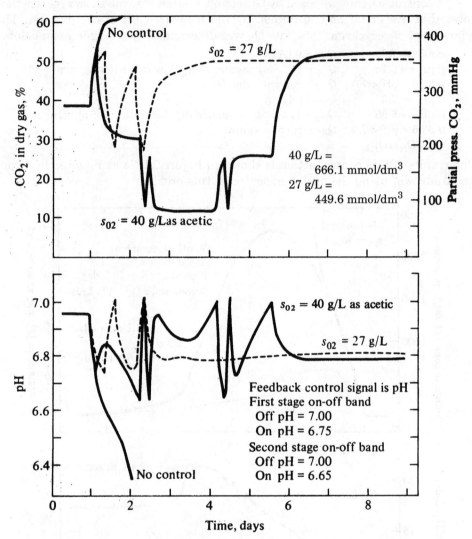

Figure 14-25 Control of Recycle Gas Flow Rate to a Step Change in Influent Substrate at End of First Day[3]

The mathematical model of the digester with addition of the dry recall stream into the digester requires the equation set given in Problem 14.16 to be modified by replacing Equation (14-131) with

$$Q = Q_{CO_2} + Q_{CH_4} + Q_R\left(1 - \frac{p_{CO_2}}{p_T}\right) + \left(\frac{p_{H_2O}}{p_T}\right)Q \qquad \textbf{(14-133)}$$

Note that all of the CO_2 is assumed to be removed in the scrubber and that the recycle flow rate, Q_R, is for the dry flow rate entering scrubber.

Thus Equation (14-132) must be replaced by solving Equation (14-133) for Q.

Total Output Flow Rate

$$Q = \frac{\left[Q_{CO_2} + Q_{CH_4} + Q_R \left(1 - \frac{p_{CO_2}}{p_T} \right) \right]}{\left(1 - \frac{p_{H_2O}}{p_T} \right)} \qquad \textbf{(14-134)}$$

(a) Calculate the final steady-state variables given in Figures 14–24 and 14–25 for $[S_0] = 500$ and 710 mmol/dm^3 and $Q_R = 0$.

(b) Simulate the step change of substrate feed to the reactor from a steady state given in Table 14–10 the conditions of part (a) to $[S_0] = 710$ mmol/dm^3 at the start of the second day and produce the "No control" transients for $[S]$ and Q_{CH_4} similar to those in Figures 14–24 until the pH drops below 6.0.

(c) Implement the control strategy discussed in this problem with $Q_{R1} = 50$ L/day and $Q_{R2} = 175$ L/day for $[S_0] = 500$ mmol/dm^3 at the start of the second day. Plot the transients similar to those given in Figures 14–24 and 14–25 through the eighth day. (Results may not compare completely, but trends should be similar.)

(d) Repeat part (c) for $[S_0] = 710$ mmol/dm^3.

(e) Determine the maximum substrate feed concentration that can be controlled in (d).

(f) Investigate the application of two-mode proportional/integral control on the recycle gas stream where the maximum total Q_R is limited to 250 L/day.

(g) Investigate control based on the recycle of microorganisms with a recycle maximum of 80% of the microorganisms in the digester output stream.

(h) Investigate control based on reducing the flow rate through the digester.

14.17.4 Solution Suggestions

(a) The steady-state variable values are best obtained by solving the system of nonlinear algebraic equations formed by setting the time derivatives equal to zero. Note that the dynamic model may indicate washout during a step change in feed when the digester can tolerate the new feed condition at steady-state operation. See part (a) of Problem 14.16.

The basic problem solution file for parts (a) and (c) are found in directory Chapter 14 and designated **P14-17A1.POL** and **P14-17C.POL** (where the control logic must be entered).

REFERENCES

1. Shuler, M. L. and Kargi, F., *Bioprocess Engineering*, 2nd ed., Upper Saddle River, NJ: Prentice Hall, 2002.
2. Tsuchiya, H. M., Drake, J. F., Jost, J. L., and Fredrickson, A. G., "Predator-Prey Interactions of Dictyostelium discoidem and Escherichia coli in Continuous Culture," *J. Bacterioil.*, *110*, p. 1147 (1972).
3. Graef, S. P. and Andrews, J. F., "Mathematical Modeling and Control of Anaerobic Digestion," *CEP Symp. Ser.*, No. 76, pp. 101-131 (1974).
4. Bailey, J. E., and Ollis, D. F., *Biochemical Engineering Fundamentals*, 2nd ed., New York, McGraw-Hill, pp. 949-954, 1986.

Useful Constants

Ideal Gas

$R = 8314.34$ m$^3 \cdot$ Pa/kg-mol $\cdot$ K

$R = 8314.34$ J/kg-mol $\cdot$ K

$R = 8314.34$ kg $\cdot$ m^2/s$^2 \cdot$ kg-mol $\cdot$ K

$R = 8.31434$ J/g-mol $\cdot$ K

$R = 0.082057$ m$^3 \cdot$ atm/kg-mol $\cdot$ K

$R = 1.9872$ btu/lb-mol $\cdot$ °R

$R = 1545.3$ ft $\cdot$ lb$_f$/lb-mol $\cdot$ °R

$R = 0.7302$ ft$^3 \cdot$ atm/lb-mol $\cdot$ °R

$R = 10.731$ ft$^3 \cdot$ lb$_f$/in$^2 \cdot$ lb-mol $\cdot$ °R

$R = 1.9872$ cal/g-mol $\cdot$ K

$R = 82.057$ cm$^3 \cdot$ atm/g-mol $\cdot$ K

Stefan-Boltzmann

$\sigma = 5.676 \times 10^{-8}$ W/m$^2 \cdot$ K^4

$\sigma = 1.714 \times 10^{-9}$ btu/h $\cdot$ ft$^2 \cdot$ °R^4

Useful Conversion Factors

Temperature

K = °C + 273.15

°R = °F + 459.67

°F = $1.8 \times$ °C + 32

°C = $(5/9) \times$ (°F − 32)

Δ°F = $1.8 \times \Delta$°C

Δ°C = $(5/9) \times \Delta$°F

Mass

1 kg = 1000 g = 2.2046 lb$_m$

1 metric ton = 1000 kg

1 lb$_m$ = 453.59 g = 0.45359 kg

1 short ton = 2000 lb$_m$

1 long ton = 2240 lb$_m$

Length

$1 \text{ m} = 3.2808 \text{ ft} = 100 \text{ cm} = 39.37 \text{ in}$

$1 \text{ } \mu\text{m} = 1 \text{ micron} = 10^{-6} \text{ m} = 10^{-3} \text{ mm} = 10^{-4} \text{ cm}$

$1 \text{ Å} = 10^{-10} \text{ m} = 10^{-4} \text{ } \mu\text{m} = 0.1 \text{ nm}$

$1 \text{ in} = 2.540 \text{ cm}$

$1 \text{ mile} = 5280 \text{ ft}$

Area

$1 \text{ m}^2 = 10^4 \text{ cm} = 10^6 \text{ mm}$

$1 \text{ m}^2 = 10.739 \text{ ft}^2$

$1 \text{ in}^2 = 6.4516 \text{ cm}^2 = 6.4516 \times 10^{-4} \text{ m}^2$

Volume

$1 \text{ m}^3 = 1000 \text{ dm}^3 = 10^6 \text{ cm}^3 = 1000 \text{ liter}$

$1 \text{ m}^3 = 35.313 \text{ ft}^3 = 264.17 \text{ gal (U.S.)}$

$1 \text{ liter} = 1 \text{ dm}^3$

$1 \text{ ft}^3 = 0.028317 \text{ m}^3 = 28.317 \text{ liter}$

$1 \text{ ft}^3 = 7.481 \text{ gal (U.S.)}$

$1 \text{ gal (British)} = 1.2009 \text{ gal (U.S.)}$

$1 \text{ gal (U.S.)} = 3.7854 \text{ liter}$

Force

$1 \text{ N} = 1 \text{ kg} \cdot \text{m/s}^2 = 0.22481 \text{ lb}_f$

$1 \text{ g} \cdot \text{cm/s}^2 = 10^{-5} \text{ N} = 2.2481 \times 10^{-6} \text{ lb}_f$

Pressure

$1 \text{ Pa} = 1 \text{ N/m}^2 = 1 \text{ kg/m} \cdot \text{s}^2$

$1 \text{ bar} = 1 \times 10^5 \text{ Pa}$

$1 \text{ atm} = 1.01325 \times 10^5 \text{ Pa} = 101.325 \text{ kPa}$

$1 \text{ atm} = 760 \text{ mm Hg at } 0°C = 29.921 \text{ in Hg at } 0°C$

$1 \text{ atm} = 33.90 \text{ ft H}_2\text{O at } 4°C$

$1 \text{ atm} = 14.696 \text{ lb}_f/\text{in}^2$

Molar Volume and Density

1 kg-mol of ideal gas at 0°C and 1 atm occupies 22.414 m^3

1 g-mol of ideal gas at 0°C and 1 atm occupies 22.414 liters

1 lb-mol of ideal gas at 32°F and 1 atm occupies 359.05 ft^3

Air: MW = 28.97

Dry air density at 0°C and 1 atm = 1.2929 kg/m^3

Dry air density at 32°F and 1 atm = 0.080711 lb$_m$/ft^3

Water density at 4.0°C = 999.972 kg/m^3 = 62.428 lb$_m$/ft^3

Acceleration of Gravity

$g = 9.80665$ m/s^2

$g = 32.174$ ft/s^2

$g_c = 32.1740$ ft $\cdot$ lb$_m$/lb$_f \cdot$ s^2

Thermodynamics

1 J = 1 N $\cdot$ m= 1 kg $\cdot$ m^2/s^2 = 10^7 g $\cdot$ cm^2/s^2 = 10^7 erg

1 J = 0.73756 ft $\cdot$ lb$_f$

1 kJ = 1000 J = 0.947813 btu = 0.23900 kcal = 239.00 cal

1 btu = 1.05506 kJ = 1055.06 J = 252.16 cal

1 btu = 778.17 ft $\cdot$ lb$_f$

1 kcal = 1000 cal = 4.1840 kJ

1 cal = 4.1840 J

1 J/s = 1 W = 1.3410 $\times$ 10^{-3} hp = 14.340 cal/min

1 hp = 550 ft $\cdot$ lb$_f$/s = 0.7068 btu/s

1 hp $\cdot$ h = 0.7457 kW $\cdot$ h = 2544.5 btu

1 kJ/kg $\cdot$ K = 0.23884 btu/lb$_m \cdot$ °F

1 btu/lb$_m \cdot$ °F = 1 btu/lb$_m \cdot$ °R = 1 cal/g $\cdot$ °C = 1 cal/g $\cdot$ K

1 J/kg = 4.2993 $\times$ 10^{-4} btu/lb$_m$ = 0.33456 ft $\cdot$ lb$_f$/lb$_m$

1 kJ/kg-mol = 2.3901 $\times$ 10^{-4} kcal/g-mol

Fluid Dynamics

1 kg/m $\cdot$ s = 1 N $\cdot$ s/m^2 = 1 Pa $\cdot$ s = 1000 centipoise = 0.67197 lb$_m$/ft $\cdot$ s

1 centipoise = 10^{-3} Pa $\cdot$ s = 10^{-2} g/cm $\cdot$ s = 10^{-2} poise = 2.4191 lb$_m$/ft $\cdot$ h

Heat Transfer

1 W/m $\cdot$ K = 0.57779 btu/h $\cdot$ ft $\cdot$ °F = 2.3901 $\times$ 10^{-3} cal/s $\cdot$ cm $\cdot$ °C

1 btu/h $\cdot$ ft $\cdot$ °F = 4.1365 $\times$ 10^{-3} cal/s $\cdot$ cm $\cdot$ °C = 1.73073 W/m $\cdot$ K

1 W/m$^2 \cdot$ K = 0.17611 btu/h $\cdot$ ft$^2 \cdot$ °F = 0.85991 kcal/h $\cdot$ m$^2 \cdot$ °F

Mass Transfer

$1 \text{ m}^2/\text{s} = 3.875 \times 10^4 \text{ ft}^2/\text{h} = 10^4 \text{ cm}^2/\text{s}$

$1 \text{ m}^2/\text{h} = 10.764 \text{ ft}^2/\text{h} = 10^4 \text{ cm}^2/\text{h}$

$1 \text{ m/s} = 1.1811 \text{ ft/h} = 0.01 \text{ cm/s}$

$1 \text{ kg-mol/s} \cdot \text{m}^2 = 737.35 \text{ lb-mol/h} \cdot \text{ft}^2 = 0.1 \text{ g-mol/s} \cdot \text{cm}^2$

$1 \text{ kg-mol/s} \cdot \text{m}^2 \cdot \text{atm} = 737.35 \text{ lb-mol/h} \cdot \text{ft}^2 \cdot \text{atm} = 0.1 \text{ g-mol/s} \cdot \text{cm}^2 \cdot \text{atm}$

$1 \text{ kg-mol/s} \cdot \text{m}^2 \cdot \text{kPa} = 7.4712 \times 10^4 \text{ lb-mol/h} \cdot \text{ft}^2 \cdot \text{atm} = 10.1325 \text{ g-mol/s} \cdot \text{cm}^2 \cdot \text{atm}$

Useful Finite Difference Approximations

Table A–1 First-Order Finite Difference Approximations

Difference	First-Order Formula	
Forward Difference for First Derivative	$\dfrac{d}{dx}f(x_i) = \dfrac{f(x_{i+1}) - f(x_i)}{\Delta x}$	(A-1)
Backward Difference for First Derivative	$\dfrac{d}{dx}f(x_i) = \dfrac{f(x_i) - f(x_{i-1})}{\Delta x}$	(A-2)
Forward Difference for Second Derivative	$\dfrac{d^2}{dx^2}f(x_i) = \dfrac{f(x_i) - 2f(x_{i+1}) + f(x_{i+2})}{\Delta x^2}$	(A-3)
Backward Difference for Second Derivative	$\dfrac{d^2}{dx^2}f(x_i) = \dfrac{f(x_i) - 2f(x_{i-1}) + f(x_{i-2})}{\Delta x^2}$	(A-4)

Table A–2 Second-Order Finite Difference Approximations

Difference	Second-Order Formula	
Forward Difference for First Derivative	$\dfrac{d}{dx}f(x_i) = \dfrac{-3f(x_i) + 4f(x_{i+1}) - f(x_{i+2})}{2\Delta x}$	(A-5)
Central Difference for First Derivative	$\dfrac{d}{dx}f(x_i) = \dfrac{f(x_{i+1}) - f(x_{i-1})}{2\Delta x}$	(A-6)
Backward Difference for First Derivative	$\dfrac{d}{dx}f(x_i) = \dfrac{3f(x_i) - 4f(x_{i-1}) + f(x_{i-2})}{2\Delta x}$	(A-7)

Difference	Second-Order Formula
Forward Difference for Second Derivative	$\dfrac{d^2}{dx^2}f(x_i) = \dfrac{2f(x_i) - 5f(x_{i+1}) + 4f(x_{i+2}) - f(x_{i+3})}{\Delta x^2}$ **(A-8)**
Central Difference for Second Derivative	$\dfrac{d^2}{dx^2}f(x_i) = \dfrac{f(x_{i+1}) - 2f(x_i) + f(x_{i-1})}{\Delta x^2}$ **(A-9)**
Backward Difference for Second Derivative	$\dfrac{d^2}{dx^2}f(x_i) = \dfrac{2f(x_i) - 5f(x_{i-1}) + 4f(x_{i-2}) - f(x_{i-3})}{\Delta x^2}$ **(A-10)**

Error Functions

Table A–3 Error and Complimentary Error Functions

z	$\text{erf}(z)$	$\text{erfc}(z)$
0	0	1
0.1	0.11246278	0.88753722
0.2	0.22270233	0.77729767
0.3	0.32862638	0.67137362
0.4	0.42839185	0.57160815
0.5	0.52049927	0.47950073
0.6	0.60385538	0.39614462
0.7	0.6778004	0.3221996
0.8	0.7421001	0.2578999
0.9	0.79690728	0.20309272
1	0.84269981	0.15730019
1.1	0.88020404	0.11979596
1.2	0.91031291	0.089687086
1.3	0.93400685	0.065993147
1.4	0.95228401	0.047715994
1.5	0.96610402	0.033895983
1.6	0.97634724	0.023652758
1.7	0.98378931	0.016210692
1.8	0.98908935	0.010910655
1.9	0.99278927	0.0072107315
2.0	0.9953211	0.0046788987

Student's t-Distribution

Table A–4 Student's t-Distribution for 95% Confidence Intervals

Degrees of Freedom	t	Degrees of Freedom	t
1	12.7062	14	2.1448
2	4.3027	15	2.1315
3	3.1824	16	2.1199
4	2.7764	17	2.1098
5	2.5706	18	2.1009
6	2.4469	19	2.0930
7	2.3646	20	2.0860
8	2.3060	21	2.0796
9	2.2622	22	2.0739
10	2.2281	23	2.0687
11	2.2010	24	2.0639
12	2.1788	25	2.0595
13	2.1604		

Data Tables

Table B–1 Vapor Pressure Data for Sulfur Compounds Present in Petroleum[a]
(Files **B-01A.POL** through **B-01E.POL**)[b]

	Temperature, °C				
Pressure mm Hg	Ethane -thiol	1-Propane thiol	2-Propane- thiol	1-Butane -thiol	2-Butane -thiol
149.41	...	24.275	10.697	51.409	38.962
187.57	0.405	29.563	15.770	57.130	44.549
233.72	5.236	34.891	20.899	62.897	50.185
289.13	10.111	40.254	26.071	68.710	55.866
355.22	15.017	45.663	31.282	74.567	61.597
433.56	19.954	51.113	36.536	80.472	67.370
525.86	24.933	56.605	41.833	86.418	73.195
633.99	29.944	62.139	47.175	92.414	79.063
760.00	35.000	67.719	52.558	98.454	84.981
906.06	40.092	73.341	57.985	104.544	90.945
1074.6	45.221	79.004	63.461	110.682	96.963
1268.0	50.390	84.710	68.979	116.863	103.020
1489.1	55.604	90.464	74.540	123.088	109.133
1740.8	60.838	96.255	80.143	129.362	115.287
2026.0	66.115	102.088	85.795	135.679	121.489

[a]Osborn, A. G., and Douglin, D. R., *J. Chem. Eng. Data*, *11*(4), 502–509 (1966).
[b]The files indicated in these appendixes are available from the book's web site, **www.problemsolvingbook.com**, in directory Tables.

Table B–2 Vapor Pressure Data for Sulfur Compounds Present in Petroleum[a]
(Files **B-02A.POL** through **B-02D.POL**)

	Temperature, °C			
Pressure mm Hg	2-Methyl 1-propane -thiol	2-Methyl 2-propane -thiol	1-Pen- tane -thiol	2-Methyl 1-butane -thiol
71.87	...	...	...	51.339
81.64	...	...	...	54.284
92.52	...	...	...	57.243
104.63	...	...	...	60.219
118.06	...	...	...	63.194
132.95	...	...	...	66.193
149.41	42.207	20.496	76.470	69.207
187.57	47.830	25.785	82.569	75.263
233.72	53.498	31.127	88.721	81.361
289.13	59.211	36.519	94.918	87.510
355.22	64.974	41.959	101.167	93.708
433.56	70.780	47.446	107.457	99.955
525.86	76.641	52.983	113.802	106.253
633.99	82.542	58.573	120.193	112.600
760.00	88.493	64.217	126.638	118.999
906.06	94.493	69.908	133.131	125.446
1074.6	100.539	75.654	139.671	131.944
1268.0	106.640	81.449	146.255	138.492
1489.1	112.785	87.294	152.896	145.089
1740.8	118.972	93.188	159.580	151.733
2026.0	125.212	99.138	166.314	158.428

[a]Osborn, A. G., and Douglin, D. R., *J. Chem. Eng. Data*, *11*(4), 502–509 (1966).

Table B–3 Vapor Pressure Data for Sulfur Compounds Present in Petroleum[a]
(Files **B-03A.POL** through **B-03E.POL**)

Pressure mm Hg	Temperature, °C				
	2-Methyl 2-butane thiol	3-Methyl 2-butane thiol	Cyclo pentane thiol	1-Hexane thiol	2-Methyl 2-pentane thiol
71.87	...	42.969	...	80.694	55.855
81.64	...	45.876	...	83.837	58.860
92.52	...	48.791	...	86.991	61.877
104.63	...	51.720	...	90.157	64.907
118.06	...	54.658	...	93.334	67.949
132.95	...	57.613	...	96.530	71.008
149.41	50.888	60.592	80.874	99.733	74.089
187.57	56.725	66.556	87.107	106.168	80.269
233.72	62.625	72.575	93.390	112.658	86.502
289.13	68.578	78.645	99.729	119.198	92.787
355.22	74.579	84.765	106.113	125.789	99.127
433.56	80.638	90.936	112.548	132.429	105.521
525.86	86.749	97.161	119.037	139.121	111.972
633.99	92.914	103.431	125.577	145.866	118.475
760.00	99.132	109.760	132.165	152.659	125.032
906.06	105.401	116.139	138.806	159.507	131.646
1074.6	111.728	122.571	145.501	166.403	138.314
1268.0	118.106	129.051	152.245	173.351	145.037
1489.1	124.537	135.585	159.040	180.349	151.815
1740.8	131.021	142.170	165.887	187.397	158.645
2026.0	137.559	148.805	172.783	194.494	165.531

[a]Osborn, A. G., and Douglin, D. R., *J. Chem. Eng. Data*, *11*(4), 502–509 (1966).

Table B–4 Vapor Pressure Data for Sulfur Compounds Present in Petroleum[a]
(Files **B-04A.POL** through **B-04D.POL**)

	Temperature, °C			
Pressure mm Hg	2,3-Dimethyl 2-butane thiol	Cyclo hexane thiol	Benzene thiol	1-Heptane thiol
71.87	55.814	83.740	...	101.627
81.64	58.867	87.006	...	104.908
92.52	61.931	90.289	...	108.205
104.63	65.011	93.576	...	111.517
118.06	68.099	96.881	...	114.840
132.95	71.208	100.201	...	118.182
149.41	74.334	103.549	114.543	121.546
187.57	80.613	110.259	121.191	128.269
233.72	86.949	117.023	127.897	135.066
289.13	93.338	123.843	134.649	141.911
355.22	99.783	130.719	141.447	148.807
433.56	106.283	137.654	148.294	155.759
525.86	112.843	144.647	155.194	162.758
633.99	119.458	151.695	162.140	169.812
760.00	126.129	158.803	169.137	176.919
906.06	132.858	165.968	176.188	184.082
1074.6	139.644	173.186	183.278	191.292
1268.0	146.492	180.464	190.426	198.551
1489.1	153.391	187.801	197.623	...
1740.8	160.344	195.196	204.867	...
2026.0	167.355	202.645	212.160	...

[a]Osborn, A. G., and Douglin, D. R., *J. Chem. Eng. Data*, *11*(4), 502–509 (1966).

Table B–5 Vapor Pressure of Propane[a] (File **B-05.POL**)

No	Temperature K	Pressure Pa	No	Temperature K	Pressure Pa
1	85.47	1.69E-04	16	231.07	1.01E+05
2	90	9.69E-04	17	240	1.48E+05
3	100	2.51E-02	18	250	2.18E+05
4	110	0.34511	19	260	3.11E+05
5	120	2.948	20	270	4.30E+05
6	130	17.534	21	280	5.82E+05
7	140	78.671	22	290	7.69E+05
8	150	282.35	23	300	9.98E+05
9	160	847	24	310	1.27E+06
10	170	2.20E+03	25	320	1.60E+06
11	180	5.05E+03	26	330	1.98E+06
12	190	1.05E+04	27	340	2.43E+06
13	200	2.01E+04	28	350	2.95E+06
14	210	3.59E+04	29	360	3.55E+06
15	220	6.05E+04			

[a]Haynes, W. M., and Goodwin, R. D., "Thermophysical Properties of Propane from 85 to 700 K at Pressures to 70 MPa," National Bureau of Standards Monograph 170, Boulder, Colorado (1982). (**www.nist.gov**)

Table B-6 Vapor Pressure of Various Substances[a] (Files **B-06A.POL** through **B-06F.POL**)

No.	Pressure	Temperature °C					
	mmHg	Ethane	n-Butane	n-Pentane	n-Hexane	n-Heptane	n-Decane
1	10	−142.82	−77.66	−50.08	−25.09	−2.04	57.70
2	50	−127.31	−55.62	−25.41	1.82	26.81	91.23
3	100	−119.24	−44.15	−12.591	15.782	41.769	108.586
4	150	−114.039	−36.75	−4.330	24.781	51.405	119.757
5	200	−110.104	−31.15	1.920	31.586	56.689	128.196
6	250	−106.898	−26.59	7.008	37.126	64.618	135.062
7	300	−104.173	−22.71	11.333	41.834	69.655	140.893
8	400	−99.665	−16.29	18.486	49.617	77.981	150.527
9	500	−95.981	−11.04	24.329	55.975	84.780	158.389
10	600	−92.840	−6.566	29.311	61.392	90.573	165.085
11	700	−90.087	−2.642	33.676	66.139	95.648	170.948
12	760	−88.580	−0.495	36.065	68.736	98.424	174.155
13	800	−87.626	0.865	37.577	70.380	100.181	176.184
14	900	−85.39	4.05	41.11	74.225	104.291	180.93
15	1000	−83.35	6.96	44.36	77.75	108.06	185.28
16	1200	−79.69	12.18	50.16	84.05	114.79	193.02
17	1500	−74.99	18.87	57.59	92.13	123.40	202.88

[a]Thermodynamics Research Center API44 Hydrocarbon Project, *Selected Values of Properties of Hydrocarbon and Related Compounds*, Texas A&M University, College Station, Texas (1978). (**www.trc.nist.gov**)

Table B–7 Heat Capacity of Gaseous Propane[a] (File **B-07.POL**)

No.	Temperature K	Heat Capacity kJ/kg-mol·K	No.	Temperature K	Heat Capacity kJ/kg-mol·K
1	50	34.06	11	700	142.67
2	100	41.3	12	800	154.77
3	150	48.79	13	900	163.35
4	200	56.07	14	1000	174.6
5	273.16	68.74	15	1100	182.67
6	298.15	73.6	16	1200	189.74
7	300	73.93	17	1300	195.85
8	400	94.01	18	1400	201.21
9	500	112.59	19	1500	205.89
10	600	128.7			

[a]Thermodynamics Research Center API44 Hydrocarbon Project, *Selected Values of Properties of Hydrocarbon and Related Compounds*, Texas A&M University, College Station, Texas (1978). (**www.trc.nist.gov**)

Table B–8 Thermal Conductivity of Gaseous Propane[a] (File **B-08.POL**)

No.	Temperature K	Thermal Conductivity W/m·K	No.	Temperature K	Thermal Conductivity W/m·K
1	231.07	1.14E-02	11	420	3.34E-02
2	240	1.21E-02	12	440	3.63E-02
3	260	1.39E-02	13	460	3.93E-02
4	280	1.59E-02	14	480	4.24E-02
5	300	1.80E-02	15	500	4.55E-02
6	320	2.02E-02	16	520	4.87E-02
7	340	2.26E-02	17	540	5.20E-02
8	360	2.52E-02	18	560	5.53E-02
9	380	2.78E-02	19	580	5.86E-02
10	400	3.06E-02	20	600	6.19E-02

[a]Younglove, B. A., and Ely, J. F., "Thermophysical Properties of Fluids. II. Methane, Ethane, Propane, Isobutane, and Normal Butane," *J. Phys. Chem. Ref. Data*, *16*, 577 (1987).

Table B–9 Viscosity of Liquid Propane at Atmospheric Pressure[a] (File **B-09.POL**)

No.	Temperature K	Viscosity Pa·s	No.	Temperature K	Viscosity Pa·s
1	100	3.82E-03	10	190	3.36E-04
2	110	2.29E-03	11	200	2.87E-04
3	120	1.53E-03	12	220	2.24E-04
4	130	1.08E-03	13	240	1.80E-04
5	140	8.32E-04	14	260	1.44E-04
6	150	6.59E-04	15	270	1.30E-04
7	160	5.46E-04	16	280	1.17E-04
8	170	4.53E-04	17	290	1.05E-04
9	180	3.82E-04	18	300	9.59E-05

[a]Diller, D.E., "Measurements of the Viscosity of Saturated and Compressed Liquid Propane," *J. Chem. Eng. Data*, 27, 240 (1982).

Table B–10 Heat of Vaporization of Propane[a] (File **B-10.POL**)

No.	Temperature K	Heat of Vaporization J/kmol	No.	Temperature K	Heat of Vaporization J/kmol
1	85.47	2.48E+07	16	231.07	1.88E+07
2	90	2.46E+07	17	240	1.83E+07
3	100	2.42E+07	18	250	1.78E+07
4	110	2.37E+07	19	260	1.73E+07
5	120	2.33E+07	20	270	1.67E+07
6	130	2.29E+07	21	280	1.61E+07
7	140	2.25E+07	22	290	1.54E+07
8	150	2.21E+07	23	300	1.46E+07
9	160	2.17E+07	24	310	1.38E+07
10	170	2.13E+07	25	320	1.28E+07
11	180	2.09E+07	26	330	1.18E+07
12	190	2.05E+07	27	340	1.05E+07
13	200	2.01E+07	28	350	8.95E+06
14	210	1.97E+07	29	360	6.81E+06
15	220	1.93E+07			

[a]Haynes, W. M., and Goodwin, R. D., "Thermophysical Properties of Propane from 85 to 700 K at Pressures to 70 MPa," National Bureau of Standards Monograph 170, Boulder, Colorado (1982). (**www.nist.gov**)

Table B–11 Heat Capacity of Various Gases[a] (Files **B-11A.POL** through **B-11D.POL**)

No.	T	Heat Capacity, kJ/kg-mol·K			
	K	Methane	Ethane	Butane	Pentane
1	50	33.26	33.39	38.07	
2	100	33.28	35.65	55.35	
3	150	33.30	38.66	67.32	
4	200	33.51	42.26	76.44	93.55
5	273.16	34.85	49.54	92.30	112.55
6	298.15	35.69	52.47	98.49	120.04
7	300	35.77	52.72	98.95	120.62
8	400	40.63	65.48	124.77	152.55
9	500	46.53	77.99	148.66	182.59
10	600	52.51	89.24	169.28	208.78
11	700	58.20	99.20	187.02	231.38
12	800	63.51	107.99	202.38	250.62
13	900	68.37	115.77	215.73	266.94
14	1000	72.80	122.59	227.36	281.58
15	1100	76.78	128.57	237.48	293.72
16	1200	80.37	133.85	246.27	304.60
17	1300	83.55	138.41	253.93	313.80
18	1400	86.44	142.42	260.58	322.17
19	1500	88.99	145.90	266.40	330.54

[a]Thermodynamics Research Center API44 Hydrocarbon Project, *Selected Values of Properties of Hydrocarbon and Related Compounds*, Texas A&M University, College Station, Texas (1978). (**www.trc.nist.gov**)

Table B–12 Thermal Conductivity of Gases[a] (Files **B-12A.POL** through **B-12G.POL**)

No.	Temp.	Thermal Conductivity x 10^6, cal/s·cm·°C						
	°F	Air	Ammonia	n-Butane	Carbon Dioxide	Ethane	Ethanol	Methane
1	-100					23.97		52.07
2	-40	50.09	43.39		27.90	32.65		61.37
3	-20	52.15	45.87		29.75	35.54		64.55
4	0	54.22	48.35		31.70	38.43		67.86
5	20	56.24	50.83	30.99	33.68	41.33	29.34	71.08
6	40	58.31	53.31	33.06	35.62	44.63	30.99	74.39
7	60	60.34	55.79	35.54	37.61	47.94	32.65	78.11
8	80	62.20	58.68	38.02	39.67	51.24	34.71	81.83
9	100	64.22	61.58	40.91	41.74	54.55	36.78	85.54
10	120	66.04	64.47	43.39	43.81	58.27		89.26
11	200			54.14		74.39		106.62

[a]Weast, R. C. (Ed.), *Handbook of Chemistry and Physics*, 56th ed., CRC Press, Cleveland, Ohio, p. E-2 (1975).

Table B–13 Viscosity of Various Liquids at Atmospheric Pressure[a]
(Files **B-13A.POL** through **B-13F.POL**)

No.	Temp.	Viscosity cp					
	°C	**Methane**	**Ethane**	***n*-Butane**	***n*-Pentane**	***n*-Hexane**	***n*-Heptane**
1	−185	0.225					
2	−180	0.187					
3	−175	0.161	0.982				
4	−170	0.142	0.803				
5	−165	0.127	0.671				
6	−160	0.115	0.572				
7	−155		0.499				
8	−150		0.441				
9	−145		0.395				
10	−140		0.358				
11	−135		0.327				
12	−130		0.300		3.62		
13	−125		0.277		2.88		
14	−120		0.256		2.34		
15	−115		0.237		1.95		
16	−110		0.221		1.66		
17	−105		0.206		1.43		
18	−100		0.194		1.24		
19	−95		0.182		1.09	2.13	
20	−90		0.171	0.63	0.970	1.82	3.76
21	−85		0.162	0.58	0.871	1.58	3.10
22	−80			0.534	0.789	1.38	2.60
23	−75			0.496	0.718	1.22	2.21
24	−70			0.461	0.657	1.09	1.912
25	−65			0.430	0.605	0.975	1.670
26	−60			0.402	0.560	0.885	1.472
27	−55			0.377	0.520	0.807	1.309
28	−50			0.354	0.486	0.739	1.173
29	−45			0.334	0.454	0.681	1.060
30	−40			0.314	0.427	0.630	0.9624
31	−35			0.297	0.402	0.585	0.8793

[a]Thermodynamics Research Center API44 Hydrocarbon Project, *Selected Values of Properties of Hydrocarbon and Related Compounds*, Texas A&M University, College Station, Texas (1978). (**www.trc.nist.gov**)

Table B–14 Heat of Vaporization of Various Compounds[a] (Files **B-14A.POL** through **B-14F.POL**)

No.	Temp. °F	Difluoro-dichloro-methane	Isobutane	Butane	Carbon Disulfide	Carbon Tetra-cloride	Ethyl Ether
1	−40	40.83					
2	−30	40.37					
3	−20	39.89	91.94				
4	−10	39.39	90.56				
5	0	38.87	89.17	94.72	91.94		95.00
6	10	38.32	88.06	93.61	91.39		94.67
7	20	37.74	86.67	92.78	90.67	52.47	94.44
8	30	37.14	85.28	91.94	90.11	52.06	94.11
9	40	36.51	83.89	90.83	89.56	51.78	93.56
10	50	35.84	82.50	89.72	88.89	51.22	93.11
11	60	35.14	81.11	88.61	88.44	50.78	91.89
12	70	34.40	79.72	87.50	87.83	50.04	91.22
13	80	33.62	78.06	86.11	87.17	50.01	90.56
14	90	32.80	76.39	84.44	86.44	49.67	89.72
15	100	31.92	74.72	83.06	85.78	49.28	
16	110	30.99	72.78	81.67	85.11	48.83	
17	120	29.99	70.83	79.72	84.44		
18	130		68.89	78.06			
19	140		66.94	76.39			

[a]Weast, R. C. (Ed.), *Handbook of Chemistry and Physics*, 56th ed., CRC Press, Cleveland, Ohio, p. E-31 (1975).

Table **B–15** Heat Transfer Data External to 3/4-inch OD Tubes[a] (File **B-15.POL**)

Point	Re	Pr	μ/μ_w	Nu	Point	Re	Pr	μ/μ_w	Nu
1	49000	2.3	0.947	277	9	346	273	0.29	49.1
2	68600	2.28	0.954	348	10	122.9	1518	0.294	56
3	84800	2.27	0.959	421	11	54.0	1590	0.279	39.9
4	34200	2.32	0.943	223	12	84.6	1521	0.267	47
5	22900	2.36	0.936	177	13	1249	107.4	0.724	94.2
6	1321	246	0.592	114.8	14	1021	186	0.612	99.9
7	931	247	0.583	95.9	15	465	414	0.512	83.1
8	518	251	0.579	68.3	16	54.8	1302	0.273	35.9

[a]Williams, R. B., and Katz, D. L., *Trans. ASME*, *74*, 1307–1320 (1952).

Table **B–16** Heat Transfer Data for 21° API Oil[a] (File **B-16.POL**)

Point	Re	Pr	μ/μ_w	Nu	Point	Re	Pr	μ/μ_w	Nu
1	368	545	0.109	15.7	14	67.2	520	0.161	12.5
2	381	535	0.112	14.1	15	269	507	0.121	13.5
3	875	345	0.076	16.0	16	655	436	0.114	15.3
4	645	348	0.074	15.0	17	22.6	3350	3.43	25.0
5	90.4	385	0.094	10.5	18	23.8	3160	5.28	21.0
6	90.8	390	0.101	10.7	19	20.9	3620	6.02	18.0
7	545	151	0.052	14.6	20	26.6	2780	6.30	21.2
8	1005	167	0.050	16.1	21	24.7	3100	6.62	18.0
9	978	168	0.050	16.2	22	29.6	2510	6.62	22.6
10	1523	160	0.051	17.4	23	26.5	2860	8.10	22.5
11	1560	158	0.051	17.3	24	29.2	2470	8.80	21.7
12	2100	159	0.053	21.2	25	27.8	2770	9.75	23.0
13	2110	159	0.056	21.4					

[a]Sieder, E.N., and Tate, G. E., *Ind. and Eng. Chem.*, *28*, 1429 (1936).

Table **B–17** Heat Transfer Data of 24.5° API Oil[a] (File **B-17.POL**)

Point	Re	Pr	μ/μ_w	Nu	Point	Re	Pr	μ/μ_w	Nu
1	61.5	567	5.05	19.2	10	531	509	4.80	42.2
2	64.7	540	4.80	22.6	11	763	497	4.85	45.3
3	128	532	4.95	26.8	12	958	493	4.90	42.3
4	121	560	5.25	21.8	13	969	492	4.90	43.6
5	256	625	4.90	24.2	14	1255	372	3.70	38.3
6	256	522	4.90	27.3	15	1260	371	3.75	41.0
7	335	523	4.90	29.2	16	1090	370	3.75	38.2
8	335	522	4.80	31.4	17	1085	367	3.60	32.3
9	531	513	4.55	40.2					

[a]Sieder, E.N., and Tate, G. E., *Ind. and Eng. Chem.*, *28*, 1429 (1936).

Table **B–18** Heat transfer Data of 16° API Oil[a] (File **B-18.POL**)

Point	Re	Pr	μ/μ_w	Nu	Point	Re	Pr	μ/μ_w	Nu
1	678	523	0.0052	14.2	14	135	2350	0.0152	15.3
2	678	520	0.0057	14.6	15	242	2220	0.0166	16.8
3	992	512	0.0044	16.8	16	58.0	5100	0.0387	15.1
4	1030	507	0.0042	16.8	17	111	5170	0.0438	20.0
5	476	1050	0.0080	15.3	18	49.0	5150	0.0390	13.8
6	356	1050	0.0077	14.6	19	49.8	5150	0.0380	15.1
7	238	1040	0.0069	12.3	20	26.6	4900	0.0334	12.6
8	173	1050	0.0063	11.9	21	11.7	5800	0.0366	10.9
9	49.0	1490	0.0077	9.3	22	6.83	13700	0.0600	11.3
10	40.6	1920	0.0114	9.3	23	6.10	14300	0.0666	11.7
11	68.1	2200	0.0131	11.9	24	3.78	16300	0.0672	11.2
12	93.0	2450	0.0140	13.1	25	3.63	16700	0.0694	11.2
13	97.4	2340	0.0143	12.4					

[a]Sieder, E.N., and Tate, G. E., *Ind. and Eng. Chem.*, *28*, 1429 (1936).

Table B–19 Heat Transfer Data to a Fluidized Bed[a] (File **B-19.POL**)

$\dfrac{h_m D_t}{k_g}$	$\dfrac{D_t}{L}$	$\dfrac{D_t}{D_p}$	$\dfrac{1-\varepsilon}{\varepsilon}\dfrac{\rho_s C_s}{\rho_g C_g}$	$\dfrac{D_t G}{\mu_g}$	$\dfrac{h_m D_t}{k_g}$	$\dfrac{D_t}{L}$	$\dfrac{D_t}{D_p}$	$\dfrac{1-\varepsilon}{\varepsilon}\dfrac{\rho_s C_s}{\rho_g C_g}$	$\dfrac{D_t G}{\mu_g}$
469	0.636	309	833	256	672	0.455	1012	867	338
913	0.636	309	868	555	986	0.451	1012	867	565
1120	0.641	309	800	786	1310	0.455	1012	867	811
234	0.285	309	800	255	1190	0.944	1130	1608	343
487	0.285	309	800	555	1890	0.974	1130	1608	573
709	0.283	309	767	850	2460	0.985	1130	1608	814
581	0.518	683	795	254	915	0.602	1130	1673	343
650	0.521	683	795	300	1260	0.602	1130	1673	485
885	0.524	683	795	440	1690	0.617	1130	1673	700

[a]Dow, W. M., and Jacob, M., *Chem. Eng. Progr. 47*, 637 (1951).

Table B–20 Activity Coefficients for the System Methanol (1) + 1, 1, 1-trichloroethane (2) at 96.7 kPa[a] (File **B-20.POL**)

Point	x_1	γ_1	γ_2	Point	x_1	γ_1	γ_2
1	0.9803	0.9885	7.3622	7	0.5351	1.4500	1.8410
2	0.9601	0.9983	6.3254	8	0.4702	1.6217	1.6208
3	0.9013	1.0081	4.2642	9	0.3721	1.9794	1.3672
4	0.8520	1.0303	3.7621	10	0.2751	2.5370	1.1506
5	0.6904	1.1826	2.4921	11	0.1302	3.9803	1.0170
6	0.5852	1.3761	1.9562	12	0.0603	5.6238	1.0078

[a]Srinivas, Ch., and Venkateshwara Rao, M., *Fluid Phase Equilibria, 61*, 285 (1991).

Table B–21 Activity Coefficients for the System Ethanol (1) + 1, 1, 1-trichloroethane (2) at 96.7 kPa[a] (File **B-21.POL**)

Point	x_1	γ_1	γ_2	Point	x_1	γ_1	γ_2
1	0.1502	2.7815	1.0412	6	0.7499	1.0462	2.1097
2	0.2802	1.8966	1.1917	7	0.8201	1.0121	2.2695
3	0.3403	1.6377	1.2700	8	0.8498	1.0079	2.1820
4	0.5202	1.2138	1.5792	9	0.9402	0.9891	1.9823
5	0.6198	1.1192	1.8292				

[a]Lakshman, V., Venkateshwara Rao, M., and Prasad, D. H. L., *Fluid Phase Equilibria, 69*, 271 (1991).

Table B–22 Activity Coefficients for the System Propanol (1) +1, 1, 1-trichloroethane (2) at 96.7 kPa[a]
(File **B-22.POL**)

Point	x_1	γ_1	γ_2	Point	x_1	γ_1	γ_2
1	0.9754	1.0036	5.2313	8	0.5452	1.4699	1.7810
2	0.9504	1.0084	4.9516	9	0.4451	1.7619	1.5490
3	0.9352	1.0153	5.0295	10	0.3603	2.1246	1.3695
4	0.9004	1.0283	4.6721	11	0.1969	3.2094	1.1698
5	0.8502	1.0502	3.9590	12	0.1302	4.6223	1.0745
6	0.7001	1.1865	2.4264	13	0.0522	6.5437	0.9998
7	0.6282	1.2918	2.0725				

[a]Kiran Kumar, R., and Venkateshwara Rao, M., *Fluid Phase Equilibria*, *70*, 19 (1991).

Table B–23 Reaction Rate Data as Function of Temperature for Hydrogenation of Ethylene[a]
(File **B-23.POL**)

Run	$k \times 10^5$ g-mol/s·atm·cm^3	T °C	Run	$k \times 10^5$ g-mol/s·atm·cm^3	T °C
1	2.70	77	18	1.37	64.0
2	2.87	77	19	0.70	54.5
3	1.48	63.5	20	0.146	39.2
4	0.71	53.3	21	0.159	38.3
5	0.66	53.3	22	0.260	49.4
6	2.44	77.6	23	0.284	40.2
7	2.40	77.6	24	0.323	40.2
8	1.26	77.6	25	0.283	40.2
9	0.72	52.9	26	0.284	40.2
10	0.70	52.9	27	0.277	39.7
11	2.40	77.6	28	0.318	40.2
12	1.42	62.7	29	0.323	40.2
13	0.69	53.7	30	0.326	40.2
14	0.68	53.7	31	0.312	39.9
15	3.03	79.5	32	0.314	39.9
16	3.06	79.5	33	0.307	39.8
17	1.31	64.0			

[a]Smith, J. M., *Chemical Engineering Kinetics*, New York: McGraw-Hill, pp. 41–42, 1970.

Vapor-Liquid Equilibrium Data Tables

Table C–1 VLE Data for Water (1)-Methyl Alcohol (2) at 59.4 °C[a] (File **C-01.POL**)

x_1	P (mm Hg)
0.0	145.40
0.2217	317.00
0.2740	342.40
0.3324	368.70
0.3980	393.60
0.5550	450.60
0.6920	497.20
0.7850	530.40
0.8590	557.00
1.0000	609.30

[a]*International Critical Tables*, 1st ed., Vol III, New York: McGraw-Hill, 1928.

Table C–2 VLE Data for Ethyl Ether (1)-Chloroform (2) at 17 °C[a] (File **C-02.POL**)

x_1	P (mm Hg)
0.0	143
0.1	143
0.2	143
0.3	151
0.4	166
0.5	193
0.6	230
0.7	273
0.8	317
0.9	360
1.0	397

[a]*International Critical Tables*, 1st ed., Vol III, New York: McGraw-Hill, 1928.

Table C–3 VLE Data for Toluene (1)-Acetic Acid (2) at 70 °C[a] (File **C-03.POL**)

x_1	P(mm Hg)	x_1	P(mm Hg)
0.000	136.0	0.5912	223.4
0.125	175.3	0.6620	225.0
0.231	195.6	0.7597	225.1
0.3121	204.9	0.8289	222.7
0.4019	213.5	0.9058	216.6
0.4860	218.9	0.9565	210.7
0.5349	221.3	1.0000	202.0

[a]*International Critical Tables*, 1st ed., Vol III, New York: McGraw-Hill, 1928.

Table C–4 VLE Data for Chloroform (1)-Acetone (2) at 35.17 °C[a] (File **C-04.POL**)

x_1	P(mm Hg)	x_1	P(mm Hg)
0.09	344.50	0.4939	255.40
0.0588	332.40	0.5143	252.80
0.1232	319.70	0.5872	248.40
0.1853	307.30	0.6635	249.20
0.2657	291.60	0.7997	261.90
0.2970	285.70	0.9175	280.10
0.3664	272.20	1.0000	293.10
0.4232	263.20		

[a]*International Critical Tables*, 1st ed., Vol III, New York: McGraw-Hill, 1928.

Miscellaneous Data Tables

Table D–1 Transport Properties of Water in English Units[a] (File **D-01.POL**)

T	ρ	C_p	$\mu \times 10^3$	k	Pr
°F	lb_m/ft^3	btu/lbm·°F	$lb_m/ft·s$	btu/h·ft·°F	
32	62.4	1.01	1.20	0.329	13.3
60	62.3	1.00	0.760	0.340	8.07
80	62.2	0.999	0.578	0.353	5.89
100	62.1	0.999	0.458	0.363	4.51
150	61.3	1.00	0.290	0.383	2.72
200	60.1	1.01	0.206	0.393	1.91
250	58.9	1.02	0.160	0.395	1.49
300	57.3	1.03	0.130	0.395	1.22
400	53.6	1.08	0.0930	0.382	0.950
500	49.0	1.19	0.0700	0.349	0.859
600	42.4	1.51	0.0579	0.293	1.07

[a]Geankoplis, C. J., *Transport Processes and Unit Operations*, 3rd ed., Englewood Cliffs, NJ: Prentice Hall, 1993, with permission.

Table D–2 Transport Properties of Water in SI Units[a] (File **D-02.POL**)

T	T	ρ	C_p	$\mu \times 10^3$	k	Pr
°C	K	kg/m^3	kJ/kg·K	kg/m·s	W/m·K	
0	273.2	999.6	4.229	1.786	0.5694	13.3
15.6	288.8	998.0	4.187	1.131	0.5884	8.07
26.7	299.9	996.4	4.183	0.860	0.6109	5.89
37.8	311.0	994.7	4.183	0.682	0.6283	4.51
65.6	338.8	981.9	4.187	0.432	0.6629	2.72
93.3	366.5	962.7	4.229	0.3066	0.6802	1.91
121.1	394.3	943.5	4.271	0.2381	0.6836	1.49
148.9	422.1	917.9	4.312	0.1935	0.6836	1.22
204.4	477.6	858.6	4.522	0.1384	0.6611	0.950
260.0	533.2	784.9	4.982	0.1042	0.6040	0.859
315.6	588.8	679.2	6.322	0.0862	0.5071	1.07

[a]Geankoplis, C. J., *Transport Processes and Unit Operations*, 3rd ed., Englewood Cliffs, NJ: Prentice Hall, 1993, with permission.

Table D–3 Density and Viscosity of Various Liquids in English Units[a] (Files **D-03A.POL** through **D-03H.POL**)

T	ρ	$\mu \times 10^5$	T	ρ	$\mu \times 10^5$
°F	lb_m/ft^3	$lb_m/ft \cdot s$	°F	lb_m/ft^3	$lb_m/ft \cdot s$
Aniline			30	54.0	2220
60	64.0	305	60	53.0	1110
80	63.5	240	80	52.5	695
100	63.0	180	100	52.0	556
150	61.6	100	150	51.0	278
200	60.2	62	200	50.0	250
250	58.9	42	**Glycerin**		
300	57.5	30	30	79.7	7200
Ammonia			60	79.1	1400
-60	43.9	20.6	80	78.7	600
-30	42.7	18.2	100	78.2	100
0	41.3	16.9	**Kerosene**		
30	40.0	16.2	30	48.8	8.0
60	38.5	15.0	60	48.1	6.0
80	37.5	14.2	80	47.6	4.9
100	36.4	13.5	100	47.2	4.2
120	35.3	12.6	150	46.1	3.2
n-Butyl Alcohol			**Mercury**		
60	50.5	225	40	848	1.11
80	50.0	180	60	847	1.05
100	49.6	130	80	845	1.00
150	48.5	68	100	843	0.960
Benzene			150	839	0.893
60	55.2	44.5	200	835	0.850
80	54.6	38	250	831	0.806
100	53.6	33	300	827	0.766
150	51.8	24.5	400	819	0.700
200	49.9	19.4	500	811	0.650
Hydraulic Fluid (MIL-M-5606)			600	804	0.606
0	55.0	5550	800	789	0.550

[a]Welty, J. R., Wicks, C. E., and Wilson, R.E., *Fundamentals of Momentum, Heat and Mass Transfer,* 3rd. ed., New York: John Wiley, 1984, with permission.

Table D–4 Condenser and Heat-Exchanger Tube Sizes[a]

OD	BWG	Wall Thickness		Inside Diameter	
inches	Number	inches	mm	ID, inches	mm
5/8	12	0.109	2.77	0.407	10.33
	14	0.083	2.11	0.459	11.66
	16	0.065	1.65	0.495	12.57
	18	0.049	1.25	0.527	13.39
3/4	12	0.109	2.77	0.532	13.51
	14	0.083	2.11	0.584	14.83
	16	0.065	1.65	0.620	15.75
	18	0.049	1.25	0.652	16.56
7/8	12	0.109	2.77	0.657	16.69
	14	0.083	2.11	0.709	18.01
	16	0.065	1.65	0.745	18.92
	18	0.049	1.25	0.777	19.74
1	10	0.134	3.40	0.732	18.59
	12	0.109	2.77	0.782	19.86
	14	0.083	2.11	0.834	21.18
	16	0.065	1.65	0.870	22.10
1 1/4	10	0.134	3.40	0.982	24.94
	12	0.109	2.77	1.032	26.21
	14	0.083	2.11	1.084	27.53
	16	0.065	1.65	1.120	28.45
1 1/2	10	0.134	3.40	1.232	31.29
	12	0.109	2.77	1.282	32.56
	14	0.083	2.11	1.334	33.88
2	10	0.134	3.40	1.732	43.99
	12	0.109	2.77	1.782	45.26

[a]See, for example, Perry, R. H., Green, D. W., and Maloney, J. O., Eds., *Chemical Engineers Handbook*, 7th ed., Section 10, New York: McGraw-Hill, 1997.

Nominal Size	Outside Diameter	Schedule Number	Wall Thickness	Inside Diameter
inches	inches		inches	inches
1/8	0.405	40	0.068	0.269
		80	0.095	0.215
1/4	0.540	40	0.088	0.364
		80	0.119	0.302
3/8	0.675	40	0.091	0.493
		80	0.126	0.423
1/2	0.840	40	0.109	0.622
		80	0.147	0.546
3/4	1.050	40	0.113	0.824
		80	0.154	0.742
1	1.315	40	0.133	1.049
		80	0.179	0.957
1-1/4	1.660	40	0.191	1.380
		80	0.145	1.278
1-1/2	1.900	40	0.200	1.610
		80	0.179	1.500
2	2.375	40	0.154	2.067
		80	0.218	1.939
2-1/2	2.875	40	0.203	2.469
		80	0.276	2.323
3	3.500	40	0.216	3.068
		80	0.300	2.900
3-1/2	3.500	40	0.318	3.548
		80	0.237	3.364
4	4.500	40	0.237	4.026
		80	0.337	3.826
5	5.563	40	0.258	5.047
		80	0.375	4.813

Nominal Size	Outside Diameter	Schedule Number	Wall Thickness	Inside Diameter
inches	inches		inches	inches
6	6.625	40	0.280	6.065
		80	0.432	5.761
8	8.625	40	0.322	7.981
		80	0.500	7.625

Table D–5 Effective Surface Roughness[a]

Surface	$\varepsilon(ft)$	$\varepsilon(mm)$
Concrete	0.001-0.01	0.3-3.0
Cast Iron	0.00085	0.25
Galvanized Iron	0.0005	0.15
Commercial Steel	0.00015	0.046
Drawn Tubing	0.000005	0.0015

[a]See, for example, G. G. Brown et. al., *Unit Operations*, New York: John Wiley & Sons, p. 140, 1950.

Physical and Transport Properties

Table E–1 Physical and Transport Properties of Air[a] (File **E-01.POL**)

T	C_p	$\mu \times 10^5$	k	ρ	$(g\beta\rho^2)/\mu^2$
°F	btu/lb$_m$·°F	lb$_m$/ft·s	btu/h·ft·°F	lb$_m$/ft^3	1/°F·ft^3
0	0.240	1.09	0.0132	0.0862	4.39×10^6
30	0.240	1.15	0.0139	0.0810	3.28
60	0.240	1.21	0.0146	0.0764	2.48
80	0.240	1.24	0.0152	0.0735	2.09
100	0.240	1.28	0.0156	0.071	1.76
150	0.241	1.36	0.0167	0.0651	1.22
200	0.241	1.45	0.0179	0.0602	0.84

[a]Welty, J.R., Wicks, C.E., and Wilson, R.E., *Fundamentals of Momentum, Heat and Mass Transfer*, 3rd ed., New York: Wiley, 1984, with permission.

Table E–2 Physical and Transport Properties of Carbon Dioxide and Sulfur Dioxide[a] (Files **E-02A.POL** and **E-02B.POL**)

	T	ρ	C_p	$\mu \times 10^5$	k
	°F	lb$_m$/ft^3	btu/lb$_m$·°F	lb$_m$/ft·s	btu/h·ft·°F
Carbon Dioxide (gas)	0	0.132	0.193	0.865	0.00760
	30	0.124	0.198	0.915	0.00830
	60	0.117	0.202	0.965	0.00910
	80	0.112	0.204	1.00	0.00960
	100	0.108	0.207	1.03	0.0102
	150	0.100	0.213	1.12	0.0115
	200	0.092	0.219	1.20	0.0130
	250	0.0850	0.225	1.32	0.0148
	300	0.0800	0.230	1.36	0.0160
Sulfur Dioxide (gas)	0	0.195	0.142	0.700	0.00460
	100	0.161	0.149	0.890	0.00560
	200	0.136	0.157	1.05	0.00670
	300	0.118	0.164	1.20	0.00790

[a]Welty, J.R., Wicks, C.E., and Wilson, R.E., *Fundamentals of Momentum, Heat and Mass Transfer*, 3rd ed., New York: Wiley, 1984.

Table E–3 Physical and Transport Properties of Kerosene, Benzene and Ammonia[a]
(Files **E-03A.POL** through **E-03C.POL**)

	T	ρ	C_p	$\mu \times 10^5$	k
	°F	lb_m/ft^3	$btu/lb_m \cdot °F$	$lb_m/ft \cdot s$	$btu/h \cdot ft \cdot °F$
Kerosene (liquid)	30	48.8	0.456	800	0.0809
	60	48.1	0.474	600	0.0805
	80	47.6	0.491	490	0.0800
	100	47.2	0.505	420	0.0797
	150	46.1	0.540	320	0.0788
Benzene (liquid)	60	55.2	0.395	44.5	0.0856
	80	54.6	0.410	38	0.0836
	100	53.6	0.420	33	0.0814
	150	51.8	0.450	24.5	0.0762
	200	49.9	0.480	19.4	0.0711
Ammonia (liquid)	0	41.3	1.08	16.9	0.315
	30	40.0	1.11	16.2	0.312
	60	38.5	1.14	15.0	0.304
	80	37.5	1.16	14.2	0.296
	100	36.4	1.19	13.5	0.287
	120	35.3	1.22	12.6	0.275

[a]Welty, J.R., Wicks, C.E., and Wilson, R.E., *Fundamentals of Momentum, Heat and Mass Transfer*, 3rd ed., New York: Wiley, 1984, with permission.

Physical Property Data for Ethane

Table F–1 Ethane Heat Capacity[a] (Small Data Set, File **F-01.POL**)

Temperature K	Ideal Gas Heat Capacity J/kmol·K
50	3.3390E+04
100	3.5650E+04
150	3.8660E+04
200	4.2260E+04
273.16	4.9540E+04
298.15	5.2470E+04
300	5.2720E+04
400	6.5480E+04
500	7.7990E+04
600	8.9240E+04
700	9.9200E+04
800	1.0799E+05
900	1.1577E+05
1000	1.2259E+05
1100	1.2857E+05
1200	1.3385E+05
1300	1.3841E+05
1400	1.4242E+05
1500	1.4590E+05

[a]TRC Thermodynamic Tables-
Hydrocarbons. Ed. M. Frenkel,
NIST, Boulder, CO, Series NSRDS-
NIST- 74.

Table F–2 Ethane Heat Capacity[a] (Large Data Set, File **F-02.POL**)

Temperature K	Ideal Gas Heat Capacity J/kmol·K	Temperature K	Ideal Gas Heat Capacity J/kmol·K
100	3.5698E+04	310	5.3926E+04
110	3.6249E+04	320	5.5178E+04
120	3.6817E+04	330	5.6446E+04
130	3.7401E+04	340	5.7727E+04
140	3.8003E+04	350	5.9017E+04
150	3.8628E+04	360	6.0313E+04
160	3.9279E+04	370	6.1612E+04
170	3.9961E+04	380	6.2913E+04
180	4.0680E+04	390	6.4212E+04
190	4.1439E+04	400	6.5507E+04
200	4.2243E+04	410	6.6798E+04
210	4.3092E+04	420	6.8082E+04
220	4.3989E+04	430	6.9357E+04
230	4.4934E+04	440	7.0624E+04
240	4.5924E+04	450	7.1880E+04
250	4.6959E+04	460	7.3126E+04
260	4.8036E+04	470	7.4360E+04
270	4.9151E+04	480	7.5582E+04
280	5.0302E+04	490	7.6791E+04
290	5.1484E+04	500	7.7987E+04
300	5.2692E+04		

[a]Ingham, H., Friend, D. G., and Ely, J. F., "Thermophysical Properties of Ethane," *J. Phys. Ref. Data, 20* (2), 275-347 (1991).

Table F–3 Ethane Vapor Pressure[a] (File **F-03.POL**)

Temperature K	Vapor Pressure Pa	Temperature K	Vapor Pressure Pa	Temperature K	Vapor Pressure Pa
92	1.7	164	29000	234	800000
94	2.8	166	33000	236	853000
96	4.6	168	38000	238	909000
98	7.2	170	43000	240	967000
100	11	172	49000	242	1028000
102	17	174	55000	244	1092000
104	25	176	62000	246	1159000
106	37	178	70000	248	1229000
108	53	180	79000	250	1301000
110	75	182	88000	252	1377000
112	100	184	98000	254	1456000
114	140	186	109000	256	1538000
116	200	188	122000	258	1623000
118	270	190	135000	260	1712000
120	350	192	149000	262	1804000
122	470	194	164000	264	1900000
124	610	196	181000	266	1999000
126	790	198	198000	268	2103000
128	1000	200	217000	270	2210000
130	1300	202	238000	272	2321000
132	1600	204	260000	274	2436000
134	2000	206	283000	276	2555000
136	2500	208	308000	278	2678000
138	3100	210	334000	280	2806000
140	3800	212	362000	282	2938000
142	4700	214	392000	284	3075000
144	5600	216	423000	286	3216000
146	6800	218	457000	288	3363000
148	8100	220	492000	290	3514000
150	9700	222	530000	292	3671000
152	11000	224	569000	294	3834000
154	13000	226	611000	296	4002000
156	16000	228	654000	298	4176000
158	18000	230	700000	300	4356000
160	21000	232	749000	302	4543000
162	25000				

[a] Ingham, H., Friend, D. G., and Ely, J. F., "Thermophysical Properties of Ethane," *J. Phys. Ref. Data, 20* (2), 275-347 (1991).

Table F–4   Ethane Viscosity[a] (File **F-04.POL**)

Temperature K	Liquid Viscosity Pa·s
100	8.7868E-04
110	6.3750E-04
120	4.8849E-04
130	3.9002E-04
140	3.2112E-04
150	2.7054E-04
160	2.3187E-04
170	2.0130E-04
180	1.7642E-04
184	1.6757E-04
190	1.5564E-04
200	1.3817E-04
210	1.2310E-04
220	1.0990E-04
230	9.8190E-05
240	8.7650E-05
250	7.8060E-05
260	6.9190E-05
270	6.0850E-05
280	5.2790E-05
290	4.4600E-05
300	3.5010E-05
304	2.90E-05

[a] Ingham, H.; Friend, D. G. and Ely, J. F.; "Thermophysical Properties of Ethane," *J. Phys. Ref. Data*, *20*(2), 275-347 (1991).

Problems Listed by Subject Areas

Table I-4 Problems in Heat Transfer

Table I-5 Problems in Mass Transfer

Table I-6 Problems in Chemical Reaction Engineering

Table I-6 (Continued) Problems in Chemical Reaction Engineering

Table I-8 (Continued) Problems in Process Dynamics and Control

Table I-9 Problems in Biochemical Engineering